Christ Church
T014882

KT-385-561

Introduction to Digital Techniques

New Titles in ELECTRONIC TECHNOLOGY

Introduction to Digital Techniques

SECOND EDITION

Dan I. Porat, Ph.D.
Stanford Linear Accelerator Center
Stanford University, California

Arpad Barna, Ph.D.
University of California
Santa Cruz, California

John Wiley & Sons
New York Chichester Brisbane Toronto Singapore

To Jason,
Liana,
Elza,
and Miklos

Library of Congress Cataloging-in-Publication Data

Porat, Dan, I.
Introduction to digital techniques.

(Electronic technology series)
Includes index.
1. Digital electronics. I. Barna, Arpad.
II. Title. III. Series.

TK7868.D5P67 1986 621.3815 85-29499
ISBN 0-471-09187-1
Printed in the United States of America
10 9 8 7 6 5 4 3 2 1

Preface

During the eight years since the publication of the first edition of this book there has been a large increase in the availability and use of digital integrated circuits. Thus, Schottky-diode clamped TTL as well as CMOS circuits have attained widespread use; gate arrays and very-large-scale integrated (VLSI) circuits have been introduced; and a much broader variety of MSI and LSI circuits have become available. This second edition reflects these new developments. Additions have been made throughout the book and less current material has been removed or revised. Some structural changes have also been made: The chapter on logic circuits has been split into the present Chapters 2 and 7, and self-evaluation questions have been moved to follow the text in each chapter.

The book is intended to introduce the student of electronics or computer science to digital techniques and is oriented toward available integrated circuits and the way they are used. It was developed primarily to fulfill the instructional needs of community colleges and four-year technical institutions and is also suitable for self-study. The prerequisites are minimal: a rudimentary knowledge of transistor circuits and a familiarity with elementary algebra.

Compared to other texts, this book offers a thorough, up-to-date treatment in the following areas: TTL including Schottky TTL, ECL, and CMOS including high-speed CMOS; number systems; codes and code converters; a detailed operation of various types of flip-flops; ripple and synchronous counters; and MSI and LSI circuits. A chapter is devoted to the introduction of the principles and structure of computers and microcomputers, and another chapter explores digital-to-analog and analog-to-digital converters. The book benefits from our many years of industrial and teaching experience and is especially suited for those who practice or intend to practice the design, construction, and troubleshooting of digital circuits and digital systems.

Each chapter includes a list of instructional objectives, self-evaluation questions, and problem questions. New concepts are reinforced with worked examples and, where appropriate, with figures and tables. The text contains 206 self-evaluation questions, 207 worked examples, 301 problems, 513 illustrations, and 191 tables.

Chapter 1 presents a historical background; Chapter 2 introduces basic logic circuits using switches, relays, and diodes, as well as diode-transistor logic (DTL). Binary arithmetic is described in Chapter 3, which also introduces octal and hexadecimal numbers. Chapter 4 describes coding to prepare the reader for dealing with BCD and other codes in the remainder of the text. Classical combinational logic design, including simplification methods, are treated in Chapters 5 and 6. Chapter 7 describes present-day logic circuits such as TTL, ECL, and CMOS. This is followed by a description of flip-flops (Ch. 8), counters (Ch. 9), and shift registers and shift-register counters (Ch. 10). Chapter 11 deals with large-scale and very-large-scale integrated circuits that have had a revolutionary impact on digital techniques.

Chapter 12 discusses binary arithmetic circuits and is based on the binary arithmetic presented in Chapter 3 and on the digital techniques acquired up to that point. Chapter 13 presents code converters and digital displays. Computers and microcomputers, including an introduction to programming techniques, are treated in Chapter 14. The interface between the analog and digital worlds is presented in Chapter 15. Chapter 16 discusses practical aspects of digital systems as well as troubleshooting instrumentation.

Chapters 2 through 11 provide the fundamentals and preferably should be taught in the sequence presented in the text. However, TTL (Sec. 7-2) may be equally well covered immediately following DTL (Sec. 2-5); also, conversions between number bases (Sec. 3-5) and alphanumeric codes (Sec. 4-5) may be postponed until the study of Chapter 13. Chapters 12 through 16 may be considered application material and its selection is at the option of the instructor. All material in Chapter 16 may be taught earlier: tolerances, noise margins, loading rules (Sec. 16-2), power distribution (Sec. 16-3), grounding (Sec. 16-4), and Schmitt trigger circuits (Sec. 16-5) following present-day logic circuits (Ch. 7), and monostable multivibrators (Sec. 16-6) and troubleshooting instrumentation (Sec. 16-7) following shift registers and shift-register counters (Ch. 10).

Optional parts of the text are marked by asterisks and may be omitted without loss of continuity at the discretion of the instructor. Answers to odd-numbered problems and a glossary of terms are presented at the end of the book. A Solutions Manual for even-numbered problems is available to qualified instructors.

We thank Elizabeth Doble, Joseph Keenan, Barbara Mele, Hank Stewart, Claire Thompson, and Janice Weisner of Wiley for their splendid cooperation. Thanks are also due to the reviewers whose valuable comments led to many significant improvements. We acknowledge Stanford University and the University of California for providing environments that made writing of this book possible.

Dan I. Porat

Arpad Barna

Contents

*Optional material.

*Optional material.

*Optional material.

*Optional material.

Introduction to Digital Techniques

CHAPTER 1 Perspective

Instructional Objectives

This introductory chapter outlines analog and digital systems, including digital computers, in preparation for the material of forthcoming chapters. After you read it you should be able to:

1. Distinguish between analog and digital quantities.
2. Identify several digital systems.
3. Outline basic properties of the binary number system.
4. Distinguish between small-scale integrated circuits (SSI), medium-scale integrated circuits (MSI), large-scale integrated circuits (LSI), and very-large-scale integrated circuits (VLSI).
5. Compare power dissipations of relays, vacuum tubes, and integrated circuits.
6. Compare operating speeds of relays, vacuum tubes, and integrated circuits.
7. Identify several uses of computers.

1-1 ANALOG AND DIGITAL QUANTITIES

In technical use, the term *analog** refers to *continuous* quantities such as distance, weight, and temperature. In contrast, the term *digital* refers to *discrete* quantities such as the *number of* people, trees, and vehicles. Many measurements may be represented in either analog or digital form. For example, time has been represented over the centuries in analog form. The apparent motion of the sun was utilized first in *sundials*. The introduction of mechanisms led to clocks and watches that provided *continuously* moving hands. This is illustrated on the clock face of Fig. 1-1*a*—note that the same clock face can be also used for displaying time incrementally. In contrast, the display shown in Fig. 1-1*b* is *digital*, displaying time in *discrete* 1-second increments.

*New terms are introduced in italics.

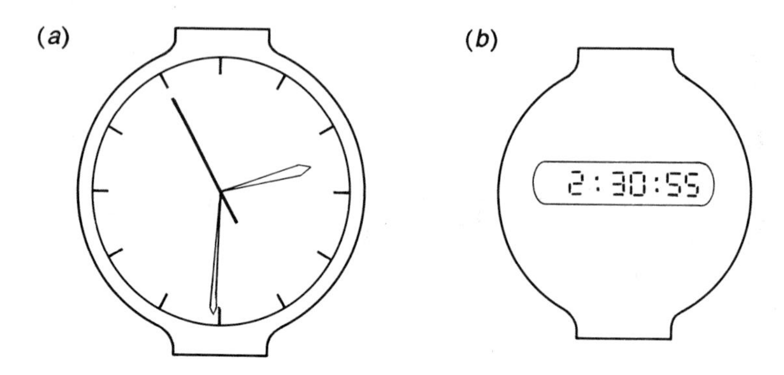

FIGURE 1-1 Two representations of time. (*a*) Analog form. (*b*) Digital form.

1-2 DIGITAL ELECTRONIC SYSTEMS

An electronic system is an orderly collection of interacting building blocks that performs a desired task. A *digital electronic system* performs a task by dealing with *discrete* quantities. There are many digital electronic systems in our everyday life: in addition to digital watches and clocks, there are now numerous digital controllers used in home appliances and automobiles; there are also many calculators and computers.

In a digital electronic system, information is represented by binary digits, *bits*. A bit may assume either one of two values: 0 or 1. A mathematical system that uses only two digits is called a *binary number system*; its foundations were laid by the British mathematician George Boole (1815–1864) in his classic treatise *An Investigation of the Laws of Thought*. This theory, which is based on the concept of TRUE represented by binary 1 and FALSE by binary 0, found a practical application in the 1930s when C. E. Shannon developed a *switching theory* based on Boole's "*Boolean algebra*." At that time switching networks for telephone systems consisted of *relays*, and switching theory aided their design. The successful application of Boolean algebra to switching networks was one of the important breakthroughs in digital information processing.

Boolean algebra was applied to switching networks with the aim of reducing the number of components (relays) without diminishing the functions performed by such a network. Because of the cost, size, power consumption, and limited reliability of relays, it was important to minimize their numbers.

The same criteria of minimization were applied in the design of digital systems using *vacuum tubes* as building blocks. For example, one of the earliest *computers*, the University of Pennsylvania's ENIAC, used as many as 18,000 vacuum tubes. The statistical "mean time between failures" was short as a result of the limited lifetime of the vacuum tube.

The invention of the *transistor* by Bardeen, Brattain, and Shockley in 1948 may be viewed as another major breakthrough in digital information processing. The volume occupied by a transistor is about one hundredth of that occupied by either a relay or a vacuum tube, and a similar ratio holds for the power dissipated by these devices. The transistor also provided the technological base for the development of *integrated circuits*, an ensemble of many transistors that are interconnected to perform a specified function.

1-3 INTEGRATED CIRCUITS

The first integrated circuit (*IC*) was developed in 1959, and it marked the start of a revolutionary change in electronic design. A digital integrated circuit consisted of several transistors, and also of passive components such as resistors, that were located on a semiconductor *substrate* and were interconnected as a *logic gate* (a *switching element* discussed in Chapter 2) to perform a *simple logic function*. Such logic functions implemented simple logic operations and provided building blocks for the realization of switching networks. Further advances in technology enabled fabrication of many transistors on a single substrate. The level of complexity increased, permitting larger amounts and varieties of circuits within a single integrated circuit. The resulting new circuits could count, perform addition, subtraction, or multiplication of two 1-bit numbers, compare them, and also realize more complex logic operations.

The development of technology met the challenge of the next level of complexity. Individual components on the substrate were further decreased in dimensions and the power dissipation of active components was reduced. Circuits with over 100,000 transistors on a single substrate became a reality, and practical limitations were now set primarily by the input and output terminals to the "outside world," rather than by the complexity of the circuit.

Today we distinguish four levels of complexity in digital integrated circuits. The first one is the small-scale integrated circuit, *SSI*, which implements a few simple logic functions. The second one is the medium-scale integrated circuit, *MSI*, which has the capability to perform the equivalent of up to 100 simple logic functions. The third and fourth ones are the large-scale integrated circuits, *LSI*, and the very-large-scale integrated circuits, *VLSI*, which have the capabilities of performing, respectively, the equivalents of up to and over 1000 simple logic functions.

1-4 POWER DISSIPATION AND OPERATING SPEED

Concurrent with the reduction in size, there has been a decrease in the power consumption and an increase in the operating speed of the new semiconductor devices and of the systems that use them. For example, early computers using vacuum tubes occupied large rooms and required tens of kilowatts of electric power as well as sizable air-conditioning equipment to allow this power to be dissipated without raising the room temperature to unbearable levels. Comparable computational capability is obtained today in a desk-top computer that uses integrated circuits and dissipates one thousandth of the former's power.

The operating speed of a switching element can be described by its propagation delay, t: the shorter this is, the more operations the element can carry out in a given time and hence the faster it is. Also, the less power P it uses the more efficient it is in power dissipation. The product of (power dissipation) × (propagation delay), $P \times t$, is thus a good indicator of how efficient a switching element is: the lower this product, the better the efficiency.

For relays operating with a $t = 1/100$ second and with a typical power dissipation of 0.25 watt (W), the product $P \times t = 0.25$ watt (W) × 0.01 second (s) = 0.0025 watt (W) × second (s). In order to compare it with other switching elements, we express this product in W × μs; thus the product becomes 2500 W × μs. For vacuum tubes, t is in the vicinity of 0.1 μs and P is about 2 W, resulting in a product $P \times t = 2$ W × 0.1 μs = 0.2 W × μs; thus a vacuum tube is 12,500 times more efficient than a relay. For an integrated circuit with $t = 10$ ns = 0.01 μs and a power dissipation of 0.25 mW per switching element, the product becomes $P \times t = 0.00025$ W × 0.01 μs = 0.0000025 W × μs; thus, such an integrated circuit is 80,000 times more efficient than a vacuum tube.

1-5 APPLICATIONS

Perhaps the most widespread application of digital integrated circuits today is in computers and in their small siblings such as minicomputers, personal computers, microcomputers, and microprocessors. These computers, in turn, are used in many diverse fields such as information storage, medical instrumentation, traffic control, digital communications, digital sound recording, and tone and voice synthesis.

SUMMARY

The chapter first introduced analog and digital quantities; this was followed by an outline of digital electronic systems and their applications. The binary number system was also briefly mentioned.

Transistors and integrated circuits were referred to, and the terms *small-scale integrated circuit* (SSI), *medium-scale integrated circuit* (MSI), *large-scale integrated circuit* (LSI), and *very-large-scale integrated circuit* (VLSI) were defined. Power dissipation, operating speed, and applications were also briefly discussed.

SELF-EVALUATION QUESTIONS

1. What is an analog quantity?
2. What is a digital quantity?

3. List three techniques for displaying time.
4. Describe a digital electronic system.
5. What is a bit?
6. What is an integrated circuit?
7. What is an SSI?
8. What is an MSI?
9. What is an LSI?
10. What is a VLSI?
11. List three applications of computers.

PROBLEMS

1-1. An integrated circuit has the capability of performing the equivalent of 500 simple logic functions. Based on the data given in the text, is it likely that the integrated circuit is an SSI, an MSI, an LSI, or a VLSI?

1-2. An integrated circuit has the capability of performing the equivalent of 50 simple logic functions. Based on the data given in the text, is it likely that the integrated circuit is an SSI, an MSI, an LSI, or a VLSI?

1-3. Using the data given in the text, estimate the ratio of the $P \times t$ products of relays and integrated circuits.

CHAPTER

2 Basic Logic Circuits

Instructional Objectives

This chapter describes basic logic circuits and their operation. It introduces logic circuits with switches and relays, logic circuits using diodes, and the simplest of logic gates using transistors, the diode-transistor logic (DTL). After you read this chapter you should be able to:

1. Analyze the operation of logic circuits using switches.
***2.** Describe the operation of logic circuits using relays.
3. Prepare truth tables describing the operation of logic circuits.
4. Follow circuit diagrams describing the internal structure of simple logic gates.
5. Describe the operation of diode-transistor logic (DTL) gates.
6. Design wired-OR circuits using DTL gates.

2-1 OVERVIEW

We are all familiar with operations involving switches. For example, in order to have a lamp on, we must have its switch and the breaker turned on. The operation of this simple circuit can be described by a truth table, which is a listing of the four possible combinations of switch and breaker positions and of the resulting states of the lamp. Truth tables are very important in digital logic, and we utilize them throughout the book.

Some of us are also familiar with relays; they are electromechanical devices with capabilities that are broader than those of switches. The limitations of wear and slow speed of switches and relays are overcome by logic gates using semiconductor devices such as diodes and transistors. The simplest logic gates

*Optional material.

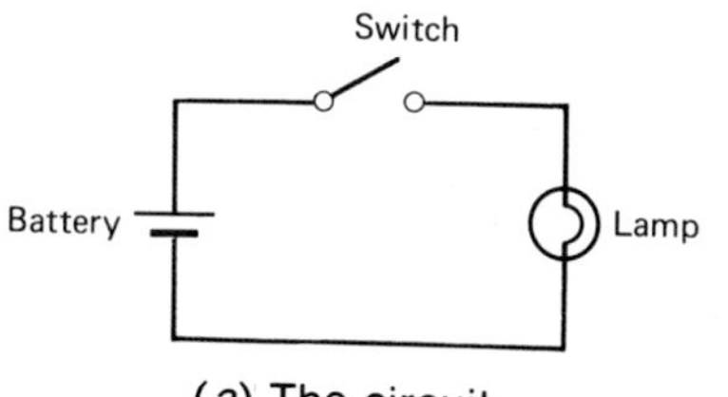

(*a*) The circuit

Switch	Lamp
OPEN	OFF
CLOSED	ON

(*b*) Truth table

FIGURE 2-1 A circuit consisting of a battery, a switch, and a lamp.

using diodes and transistors are introduced in this chapter; additional logic circuits are described in Chapter 7.

2-2 LOGIC CIRCUITS USING SWITCHES

This section illustrates the use of switches in simple logic circuits and describes the operation of these circuits using truth tables.

2-2.1 The Simplest Circuit

Figure 2-1*a* shows a simple *logic circuit* consisting of a battery, a switch, and a lamp. The lamp is OFF when the switch is open and it is ON when the switch is closed. This information is shown in the *truth table* of Fig. 2-1*b*. The left column of the table lists the two possible combinations, or *states*, of the switch and the right column lists the corresponding states of the lamp.

2-2.2 AND Circuits

Figure 2-2*a* shows a logic circuit consisting of a battery, two switches in series, and a lamp. The lamp is ON only when switch *A and* switch *B* are both closed. Hence this circuit is designated as an *AND circuit*. This information is also shown in the truth table of Fig. 2-2*b*. The left two columns of the table list the four possible combinations of switch positions (states) and the rightmost column lists the corresponding states of the lamp.

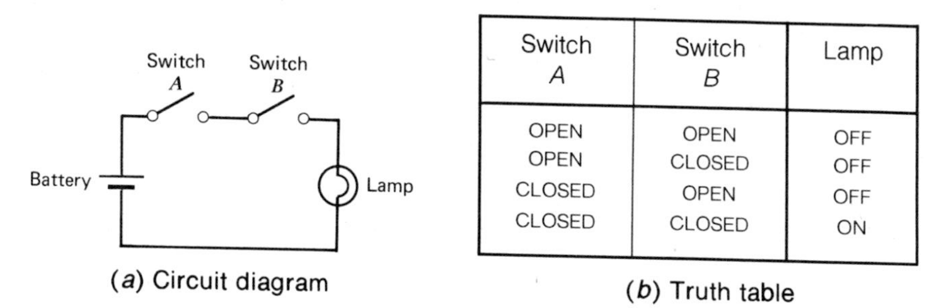

Switch A	Switch B	Lamp
OPEN	OPEN	OFF
OPEN	CLOSED	OFF
CLOSED	OPEN	OFF
CLOSED	CLOSED	ON

(*a*) Circuit diagram (*b*) Truth table

FIGURE 2-2 An AND circuit using switches.

Switch *A*	Switch *B*	Switch *C*	Lamp
OPEN	OPEN	OPEN	OFF
OPEN	OPEN	CLOSED	OFF
OPEN	CLOSED	OPEN	OFF
OPEN	CLOSED	CLOSED	OFF
CLOSED	OPEN	OPEN	OFF
CLOSED	OPEN	CLOSED	OFF
CLOSED	CLOSED	OPEN	OFF
CLOSED	CLOSED	CLOSED	ON

(*a*) Circuit diagram (*b*) Truth table

FIGURE 2-3 An AND circuit using three switches.

Truth tables can also be used to describe the operation of logic circuits having more than two switches in series.

EXAMPLE 2-1 Prepare a truth table for the circuit of Fig. 2-3*a*.

Solution The truth table is shown in Fig. 2-3*b* above. The table consists of four columns and eight rows. We can see that the lamp is ON only when switch *A and* switch *B and* switch *C* are closed. Hence the circuit of Fig. 2-3*a* is an AND circuit.

2-2.3 OR Circuits

Figure 2-4*a* shows a logic circuit that consists of a battery, two switches in parallel, and a lamp. The lamp is ON when either switch *A or* switch *B* is closed or when both switches are closed. Hence this circuit is designated as an *OR circuit*. This information is also shown in the truth table of Fig. 2-4*b*.

As was true for circuits having series switches, truth tables also can be used to describe the operation of circuits with more than two switches in parallel (see Problem 2-2).

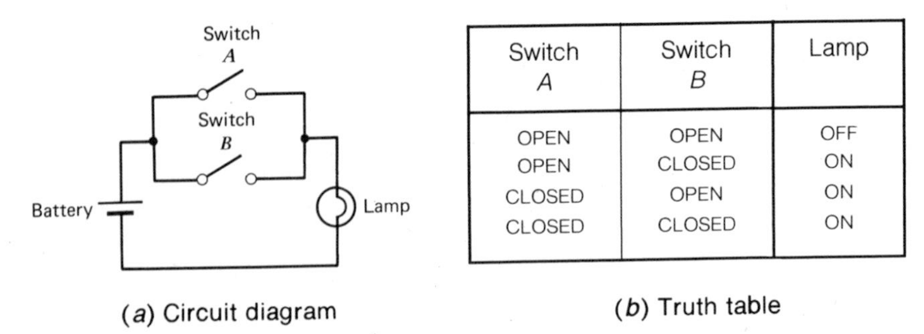

Switch *A*	Switch *B*	Lamp
OPEN	OPEN	OFF
OPEN	CLOSED	ON
CLOSED	OPEN	ON
CLOSED	CLOSED	ON

(*a*) Circuit diagram (*b*) Truth table

FIGURE 2-4 An OR circuit using switches.

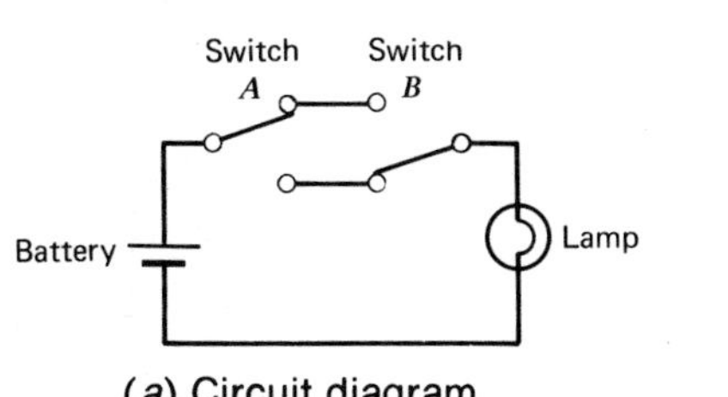

(*a*) Circuit diagram

Switch *A*	Switch *B*	Lamp
DOWN	DOWN	ON
DOWN	UP	OFF
UP	DOWN	OFF
UP	UP	ON

(*b*) Truth table

FIGURE 2-5 A circuit consisting of a battery, two 2-position switches, and a lamp.

2-2.4 Other Circuits

AND circuits and OR circuits are not the only logic circuits that can be built using switches. For example, Fig. 2-5*a* shows a logic circuit that consists of a battery, two 2-position switches, and a lamp. Such circuits are common in buildings where a light is controlled by two switches in different locations: changing the state of either switch changes the state of the lamp. The operation of the circuit also can be described by the truth table of Fig. 2-5*b*.

*2-3 LOGIC CIRCUITS USING RELAYS

This section describes the operation of relays and illustrates their use in logic circuits.

*2-3.1 Basic Operation

The operation of a relay is shown schematically in Fig. 2-6*a*. When current flows through the coil, its magnetic field attracts the iron core, which closes the

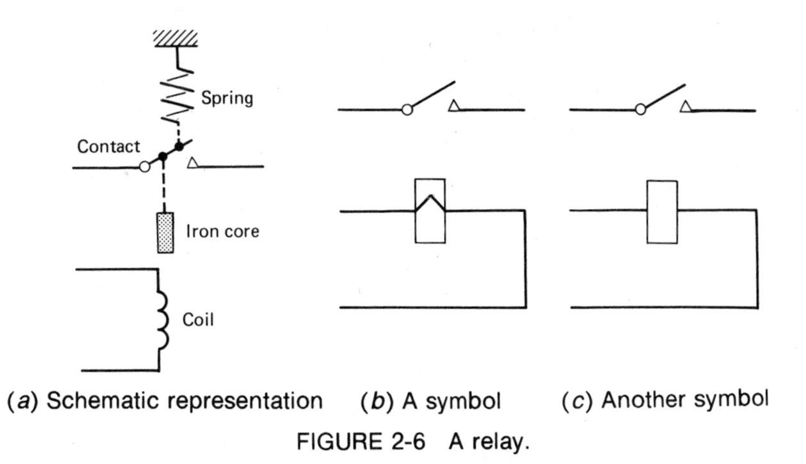

(*a*) Schematic representation (*b*) A symbol (*c*) Another symbol

FIGURE 2-6 A relay.

*Optional material.

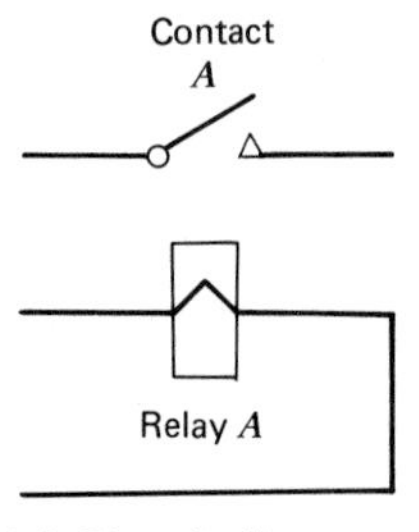

Relay A	Contact A
OFF	OPEN
ON	CLOSED

(*a*) Circuit diagram (*b*) Truth table

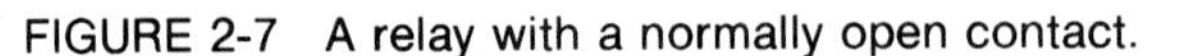
FIGURE 2-7 A relay with a normally open contact.

normally open (*n.o.*) *contact* linked to it. When the coil carries no current, it releases the iron core, which is pulled back by the spring and the contact resumes its open state. Figures 2-6*b* and 2-6*c* show two frequently used relay symbols. In this text we use the symbol of Fig. 2-6*b*.

Figure 2-7 shows the operation of a relay with a normally open (n.o.) contact A: the contact is open when the relay is OFF (not energized) and it is closed when the relay is ON (energized). Figure 2-8 shows the operation of a relay with a *normally closed* (*n.c.*) *contact* $\bar{A}$ (*read*: *A bar*, *A NOT*, or *NOT A*): the contact is closed when the relay is OFF and it is open when the relay is ON.

Normally open and normally closed contacts can be combined on a single relay as shown in the circuit of Fig. 2-9*a*. It consists of relay A, two normally open contacts marked A, and one normally closed contact marked $\bar{A}$. The truth table shown in Fig. 2-9*b* is a combination of the truth tables of Figs. 2-7*b* and 2-8*b*.

A simple use of a relay is shown in Fig. 2-10*a*, where relay A is turned ON by switch S—such use is common when high-power equipment is actuated by a low-power switch. The operation of the circuit is summarized in the truth table of Fig. 2-10*b*. When switch S is open, relay A is OFF, the two contacts A are open, and the contact $\bar{A}$ is closed. When switch S is closed, relay A is ON, the two contacts A are closed, and the contact $\bar{A}$ is open.

Figure 2-11*a* shows a circuit using two relays. When switch S is open, relay A is OFF, contact A is open, and relay B is OFF. When switch S is closed, relay A is ON, contact A is closed, and relay B is ON. The use of contact B is not shown in

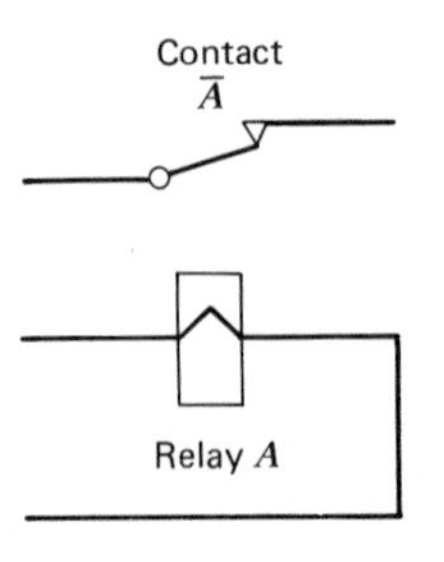

Relay A	Contact $\bar{A}$
OFF	CLOSED
ON	OPEN

(*a*) Circuit diagram (*b*) Truth table

FIGURE 2-8 A relay with a normally closed contact.

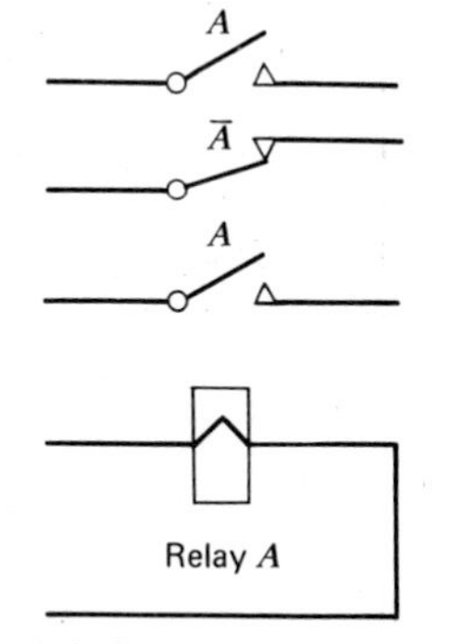

(a) Circuit diagram

Relay A	Contacts A	Contact $\bar{A}$
OFF	OPEN	CLOSED
ON	CLOSED	OPEN

(b) Truth table

FIGURE 2-9 A relay with two normally open contacts and one normally closed contact.

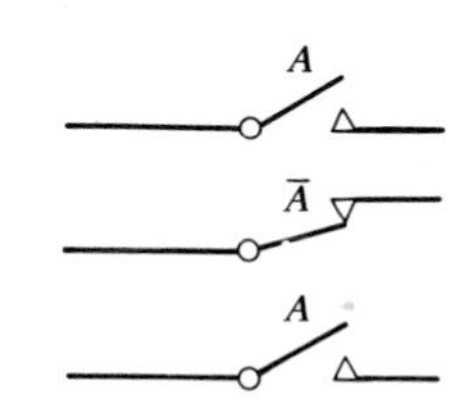

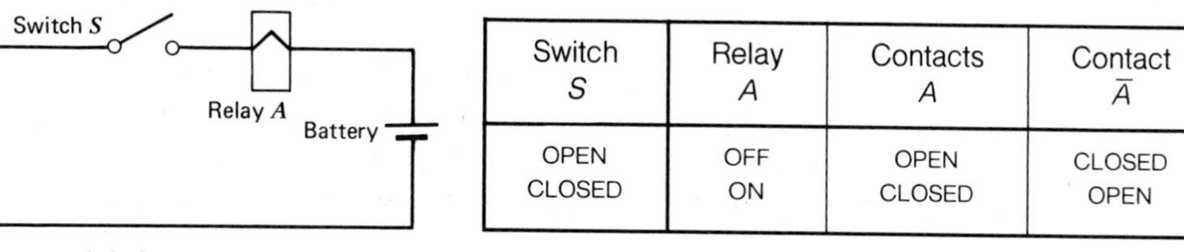

Switch S	Relay A	Contacts A	Contact $\bar{A}$
OPEN	OFF	OPEN	CLOSED
CLOSED	ON	CLOSED	OPEN

(a) Circuit diagram

(b) Truth table

FIGURE 2-10 A circuit consisting of a battery, a switch, and a relay.

Switch S	Relay A	Contact A	Relay B
OPEN	OFF	OPEN	OFF
CLOSED	ON	CLOSED	ON

(b) Truth table

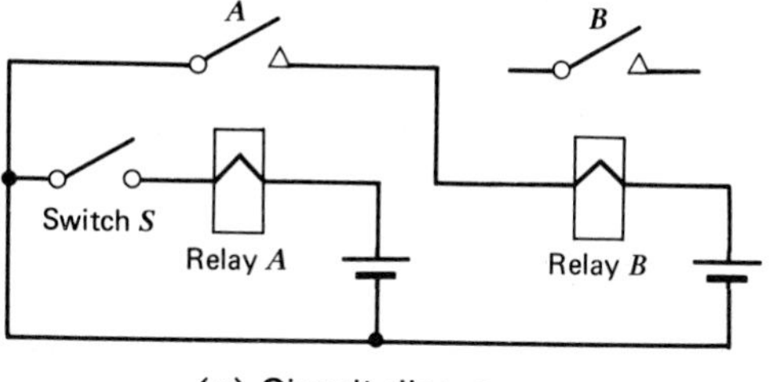

Relay A	Relay B
OFF	OFF
ON	ON

(a) Circuit diagram

(c) Abbreviated truth table

FIGURE 2-11 A circuit with two relays.

Switch S	Relay A	Contact $\overline{A}$	Relay B
OPEN	OFF	CLOSED	ON
CLOSED	ON	OPEN	OFF

(*b*) Truth table

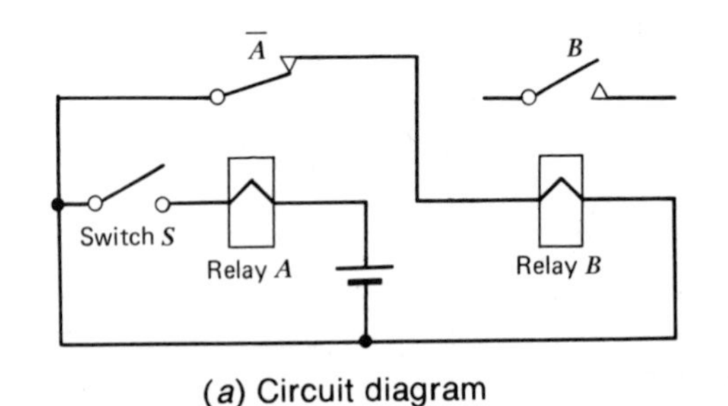

Relay A	Relay B
OFF	ON
ON	OFF

(*a*) Circuit diagram (*c*) Abbreviated truth table

FIGURE 2-12 A circuit with two relays.

the figure. The operation of the circuit is summarized in the truth table of Fig. 2-11*b*. If our interest is limited only to the states of the relays, we may use the abbreviated truth table of Fig. 2-11*c*, which shows only the states of the relays.

Figure 2-12*a* shows another circuit using two relays. In this circuit when switch S is open, relay A is OFF and contact $\overline{A}$ is closed; hence relay B is ON. Also, when switch S is closed, relay A is ON; thus contact A is open and relay B is OFF. The operation of the circuit is summarized in the truth table of Fig. 2-12*b*. If our interest is limited to the states of the relays, we may use the abbreviated truth table of Fig. 2-12*c*, which shows only the states of the relays.

*2-3.2 AND Circuits

Figure 2-13*a* shows a logic circuit using three relays; the connections of relays A and B and the use of relay C are not shown in the figure. Relay C is ON only

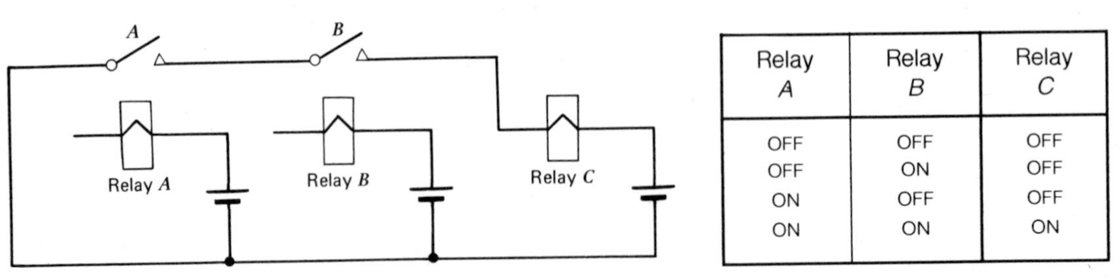

Relay A	Relay B	Relay C
OFF	OFF	OFF
OFF	ON	OFF
ON	OFF	OFF
ON	ON	ON

(*a*) Circuit diagram (*b*) Truth table

FIGURE 2-13 An AND circuit using relays.

*Optional material.

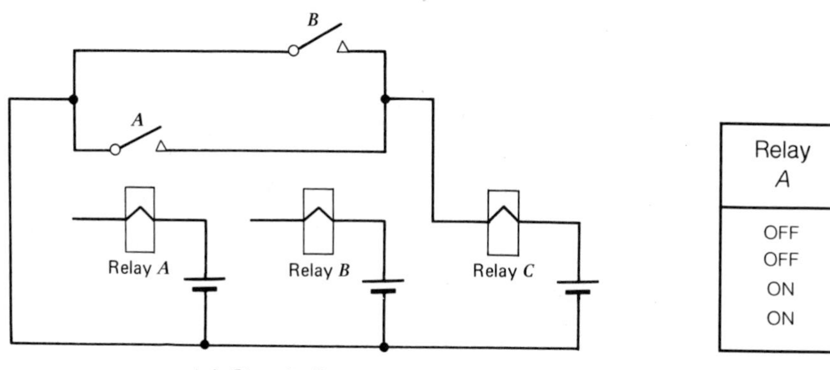

(a) Circuit diagram

Relay A	Relay B	Relay C
OFF	OFF	OFF
OFF	ON	ON
ON	OFF	ON
ON	ON	ON

(b) Truth table

FIGURE 2-14 An OR circuit using relays.

when relay *A and* relay *B* are both ON; hence this circuit is designated an *AND circuit*. This information is also shown in the truth table of Fig. 2-13*b*.

*2-3.3 OR Circuits

Figure 2-14*a* shows a logic circuit using three relays; the connections of relays *A* and *B* and the use of relay *C* are not shown in the figure. Relay *C* is ON when either relay *A or* relay *B* is ON or when both relays *A* and *B* are ON; hence this circuit is designated an *OR circuit*. This information is also shown in the truth table of Fig. 2-14*b*.

*2-3.4 Other Circuits

In addition to the simple logic circuits described above, relays have been used in a wide variety of applications including motor vehicles, automatic elevators, and telephone exchanges. Although they are simple and rugged, the electromechanical nature of relays leads to limitations in their use. The operating life of a relay is limited due to contact wear, metal fatigue, and other mechanical limitations. Another shortcoming for digital logic work is their comparatively slow operating speed, typically in the range of 1 to 100 ms.

These limitations of relays are overcome by *logic gates*, electronic devices with no moving parts. Logic gates are capable of performing the same logic operations as relays, but typically a thousand to a million times faster. Also, because of the absence of moving parts, their reliability is superior to the reliability of relays. The structure and operation of various logic gates are described in the remainder of this chapter and in Chapter 7.

*Optional material.

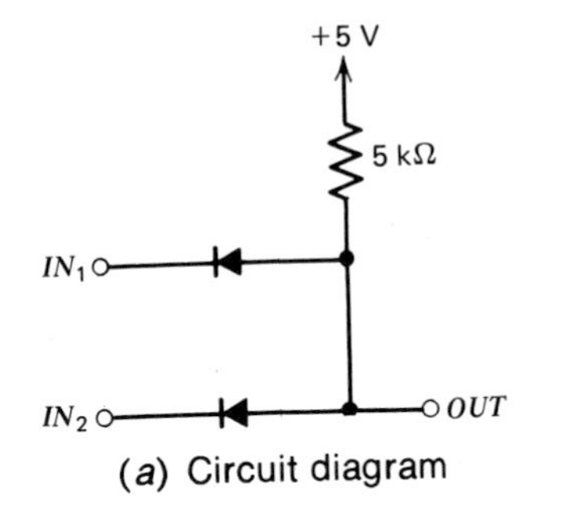

(a) Circuit diagram

IN_1	IN_2	OUT
0 V	0 V	+0.7 V
0 V	+5 V	+0.7 V
+5 V	0 V	+0.7 V
+5 V	+5 V	+5 V

(b) Truth table

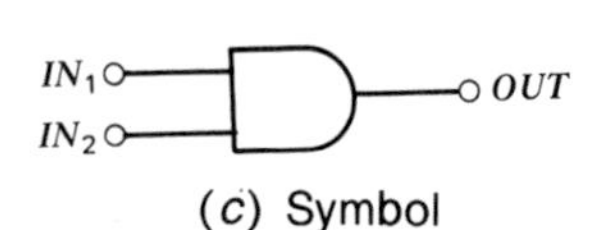

(c) Symbol

FIGURE 2-15 A 2-input AND gate using diodes.

2-4 LOGIC CIRCUITS USING DIODES

This section describes details and operation of logic gates using diodes, as well as resistors.

2-4.1 Diode AND Gates

Figure 2-15*a* shows a *2-input AND gate* using diodes. The operation of the circuit is described in the truth table of Fig. 2-15*b*: the output is at +5 V only when both inputs IN_1 *and* IN_2 are at +5 V; hence the name AND gate.

When both inputs IN_1 and IN_2 are at a voltage of 0 V (first line in the truth table), the current flowing through the 5-kΩ resistor connected to the +5-V supply voltage establishes a voltage at the output equal to the forward voltage drop across the diodes: these are silicon diodes with forward voltage drops of 0.7 V; hence the output is at +0.7 V. Note that the two diodes share the current flowing through the 5-kΩ resistor.

When IN_1 is at a voltage of 0 V and IN_2 is at a voltage of +5 V (second line in the truth table), the upper diode is forward biased (conducting) and the output voltage is at +0.7 V. The lower diode is reverse biased by a reverse voltage of +5 V − 0.7 V = 4.3 V; hence this diode is nonconducting.

The situation is similar in the third line of the truth table but with the roles of the diodes reversed. Note that the conducting diode determines the state of the output.

In the last line of the truth table both inputs IN_1 and IN_2 are at voltages of +5 V. Neither diode is conducting and no current flows through the 5-kΩ resistor. In this case the output is "pulled up" to +5 V by the 5-kΩ resistor.

A logic symbol for the 2-input AND gate is shown in Fig. 2-15*c*. Note that the symbol does not show the connection of the AND gate to +5 V.

Diode gates can also be built with more than 2 inputs, as illustrated in Ex. 2-2 below.

EXAMPLE 2-2

Extend the diode gate circuit, the truth table, and the symbol of Fig. 2-15 to 3 inputs.

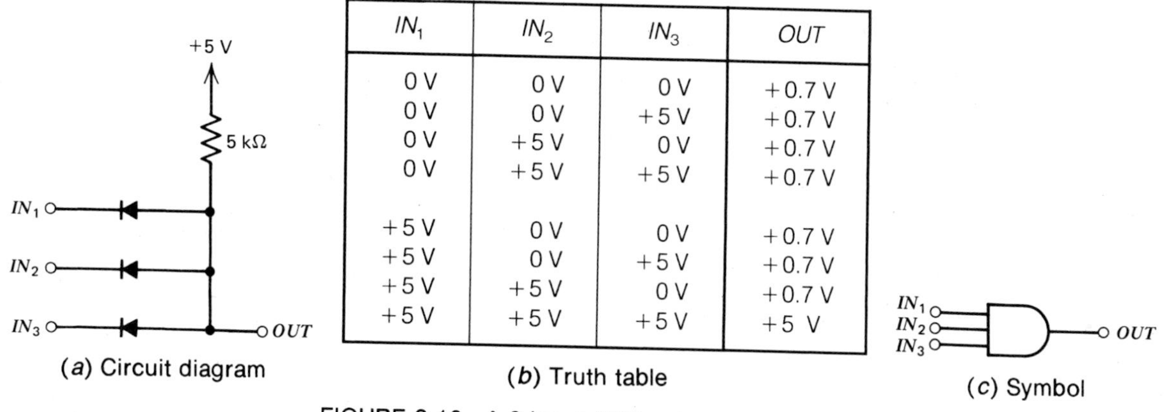

IN_1	IN_2	IN_3	OUT
0 V	0 V	0 V	+0.7 V
0 V	0 V	+5 V	+0.7 V
0 V	+5 V	0 V	+0.7 V
0 V	+5 V	+5 V	+0.7 V
+5 V	0 V	0 V	+0.7 V
+5 V	0 V	+5 V	+0.7 V
+5 V	+5 V	0 V	+0.7 V
+5 V	+5 V	+5 V	+5 V

(*a*) Circuit diagram (*b*) Truth table (*c*) Symbol

FIGURE 2-16 A 3-input AND gate using diodes.

Solution

A 3-input diode AND gate circuit, its truth table, and its symbol are shown in Fig. 2-16.

2-4.2 Diode OR gates

Figure 2-17*a* shows a *2-input OR gate* using diodes. The operation of the circuit is described in the truth table of Fig. 2-17*b*: the output is at +4.3 V when either input IN_1 *or* input IN_2 or both are at +5 V; hence the name OR gate.

When both inputs in Fig. 2-17*a* are at a voltage of 0 V (first line in the truth table), the 5-kΩ resistor connected to ground pulls the output down to 0 V.

When IN_1 is at 0 V and IN_2 is at +5 V (second line in the truth table), the lower diode is forward biased and pulls the output up to a voltage of +5 V − 0.7 V = +4.3 V. The upper diode is thus reverse biased by a reverse voltage of 4.3 V. The situation is similar in the third line of the truth table but with the roles of the diodes reversed. Note that the conducting diode determines the state of the output.

In the last line of the truth table both inputs are at +5 V and they pull the output up via the forward-biased diodes to a voltage of +5 V − 0.7 V = +4.3 V.

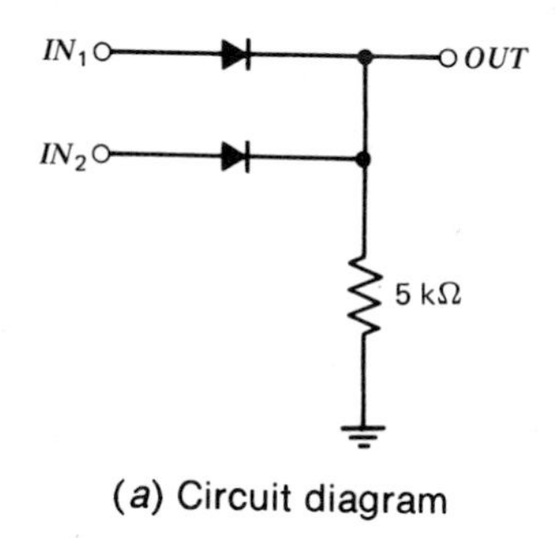

(*a*) Circuit diagram

IN_1	IN_2	OUT
0 V	0 V	0 V
0 V	+5 V	+4.3 V
+5 V	0 V	+4.3 V
+5 V	+5 V	+4.3 V

(*b*) Truth table

IN1
IN2
OUT

(*c*) Symbol

FIGURE 2-17 A 2-input OR gate using diodes.

A symbol for the 2-input OR gate is shown in Fig. 2-17*c*. Note that the symbol does not show the connection of the OR gate to ground.

As was true with AND gates, OR gates can also be built with more than 2 inputs (see Problem 2-6).

2-4.3 Resistive Loading

In both the AND and the OR gate circuits above, the output voltages in the truth tables were shown for no external loads on the outputs of the circuits. A resistive load between the output of the diode AND gate circuit of Fig. 2-15*a* and ground would *lower* the output voltage of the last line in the truth table of Fig. 2-15*b*, as shown in Ex. 2-3 below.

EXAMPLE 2-3

In the diode AND gate circuit of Fig. 2-15*a*, both IN_1 and IN_2 are at a voltage of +5 V. In addition to the circuit elements shown in the figure, a resistor with a value of 15 kΩ is connected between the output of the circuit and ground. Find the voltage V_{OUT} at the output.

Solution

Since both diodes are reverse biased, current flows only through the resistors; hence the output voltage is determined by the voltage divider ratio of the resistors as

$$V_{OUT} = 5\text{ V}\frac{15\text{ k}\Omega}{15\text{ k}\Omega + 5\text{ k}\Omega} = \mathbf{3.75\text{ V}}$$

In the case of the diode OR gate circuit of Fig. 2-17*a*, a resistive load between the output of the circuit and the +5-V power supply would *raise* the output voltage of the first line in the truth table of Fig. 2.17*b*.

2-4.4 Propagation Delays

Thus far we have assumed that the output changes instantaneously to the new value given by the truth table when the voltages change at the inputs of a logic gate. In reality, however, there is always some capacitance in parallel with the output of the logic gate and the change in voltage is never instantaneous. This is illustrated in Ex. 2-4, which follows.

EXAMPLE 2-4

Figure 2-18*a* shows a 2-input diode AND gate circuit with its output loaded by a capacitance of $C = 100$ pF. Find the output voltage V_{OUT} when inputs IN_1 and IN_2 are simultaneously switched from 0 V to +5 V at time $t = 0$.

Solution

Input voltages V_{IN_1} and V_{IN_2} are shown by the heavy solid line in Fig. 2-18*b* above. The capacitive load does not change the steady-state output voltage levels given by the truth table of Fig. 2-15*b*. Hence $V_{OUT} = +0.7$ V as long as the inputs are at 0 V and it ultimately becomes $V_{OUT} = +5$ V.

Now we would have to *derive* the equation describing the transition of output voltage V_{OUT} from its initial value of +0.7 V to its final value of +5 V. Such a derivation is, however, outside the scope of this text; thus we simply

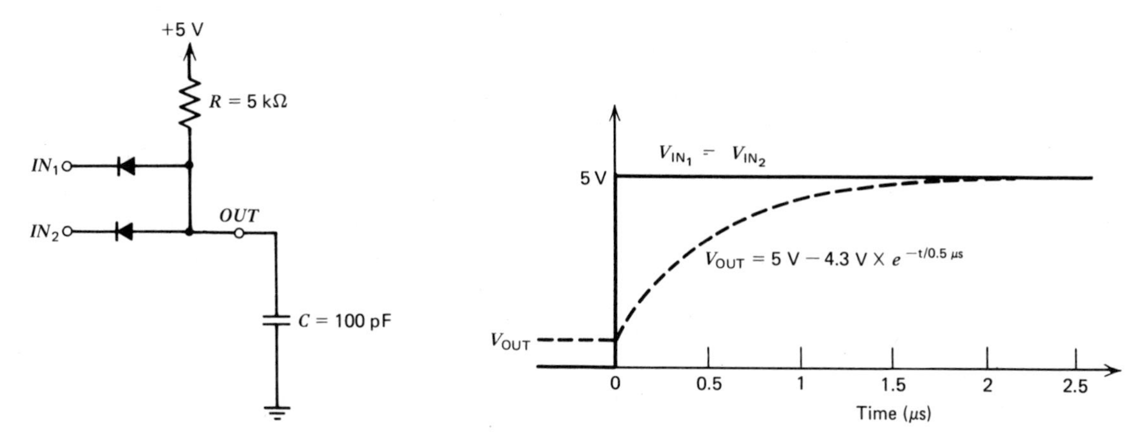

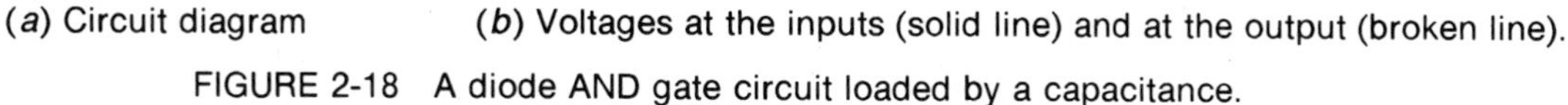
(*a*) Circuit diagram (*b*) Voltages at the inputs (solid line) and at the output (broken line).

FIGURE 2-18 A diode AND gate circuit loaded by a capacitance.

use the result from circuit theory: for times $t \geq 0$, $V_{OUT} = 5\text{ V} - 4.3\text{ V} \times e^{-t/RC}$, where $e = 2.72$, $R = 5\text{ k}\Omega$, and $C = 100\text{ pF}$. Hence we have $V_{OUT} = 5\text{ V} - 4.3\text{ V} \times e^{-t/0.5\,\mu s}$, as shown by the broken line in Fig. 2-18*b*.

A complete description of the output transition for a transition at the inputs is given by the output voltage as a function of time either as a graph or as an equation. However, it is often adequate to describe the transition by a *propagation delay*, which is defined as the time required for the output voltage to reach a specified level (usually near 50% of the logic swing) after a transition at the input.

Propagation delays are, in general, different for positive-going and for negative-going transitions, and they are also influenced by resistive and capacitive loading. We use t_{PLH} for the propagation delay of the transition of the output from its "low" state (+ 0.7 V) to its "high" state (+ 5 V), and t_{PHL} for the propagation delay of the transition of the output from its "high" state (+ 5 V) to its "low" state (+ 0.7 V). (Other abbreviations are also in use.)

EXAMPLE 2-5 Find propagation delay t_{PLH} at the +2.5-V level in the circuit of Fig. 2.18*a*.

Solution The transition of the output from its "low" level of +0.7 V to its "high" level of +5 V is described by the broken line in Fig. 2.18*b*. We can see that this line reaches $V_{OUT} = +2.5$ V at approximately 0.27 μs. Hence $t_{PLH} \cong$ **0.27 μs**.

2-4.5 Availability and Use

Diode gate circuits are rarely used by themselves, but are usually combined with transistor circuitry. There are many such combinations; one of the simplest is described in the next section.

2-5 DIODE-TRANSISTOR LOGIC (DTL)

This section describes *diode-transistor logic* (DTL) *gates*. Although now rarely used in new design, DTL forms the basis of other logic families that are described in Chapter 7.

2-5.1 DTL Inverters

We start the description of DTL gate circuits with the simplest logic gate, the *logic inverter*. The circuit diagram of a logic inverter circuit is shown in Fig. 2.19*a*. The figure also shows operating conditions for an input voltage of 0 V. In this case diode D_1 is conducting; hence the right side of diode D_1 (point x) is at +0.7 V and a current of (5 V − 0.7 V)/4 kΩ = 1.07 mA flows through the 4-kΩ resistor and through diode D_1. The right side of diode D_3 is pulled to ground by the 5-kΩ resistor, leaving only 0.7 V across the *two* diodes D_2 and D_3. The resulting 0.35-V voltages across each of these diodes are insufficient to turn them ON. Hence diodes D_2 and D_3 do *not* conduct, the transistor is turned OFF, and the output is at +5 V.

Operating conditions for an input voltage of +5 V are shown in Fig. 2-19*b*. In this case diode D_1 is reverse biased; hence the current of the 4-kΩ resistor flows through forward-biased diodes D_2 and D_3 into the 5-kΩ resistor and into the base of the transistor. The voltage at the forward-biased base-emitter junction of the transistor is 0.7 V, and diodes D_2 and D_3 drop an additional 1.4 V; hence the voltage on the left side of diode D_2 (point x) is at +2.1 V. Thus the current flowing through the 4-kΩ resistor is (5 V − 2.1 V)/4 kΩ = 0.72 mA. Out of this current, 0.7 V/5 kΩ = 0.14 mA flows through the 5-kΩ resistor and the remaining (0.72 mA − 0.14 mA = 0.58 mA) flows into the base of the transistor. This base current is sufficient to saturate the transistor, bringing the collector voltage to ≈ 0 V; thus the collector current is 5 V/2 kΩ = 2.5 mA.

The operation of the logic inverter is summarized in the truth table of Fig. 2-19*c*. We can see that the output is +5 V for a 0-V input, and the output is 0 V for a +5-V input. It is because of this behavior that the circuit is named a *logic inverter*. Its use is similar to a normally closed contact of a relay, as shown in Fig. 2-12.

In the above we assumed that the input voltage to the circuit is either 0 V or +5 V. The properties of the circuit for input voltages between these two values are described by the *transfer characteristic* of the circuit shown by the heavy solid line in Fig. 2-19*d*. A significant feature is the sharp transition in the vicinity of V_{IN} = 1.5 V. As a result, the output voltage is insensitive to the actual value of the input voltage as long as the latter is either below about +0.8 V or is above 2 V. The sharp transition can be characterized by a *logic threshold voltage* V_{TH}, *defined* as the input voltage that results in an equal output voltage. In Fig. 2-19*d* this is given by the intersection of the transfer characteristic (heavy solid line) with a straight (broken) line drawn at an angle of 45° through the origin. This results in $V_{OUT} = V_{IN} = +1.5$ V, that is, in a $V_{TH} = +1.5$ V.

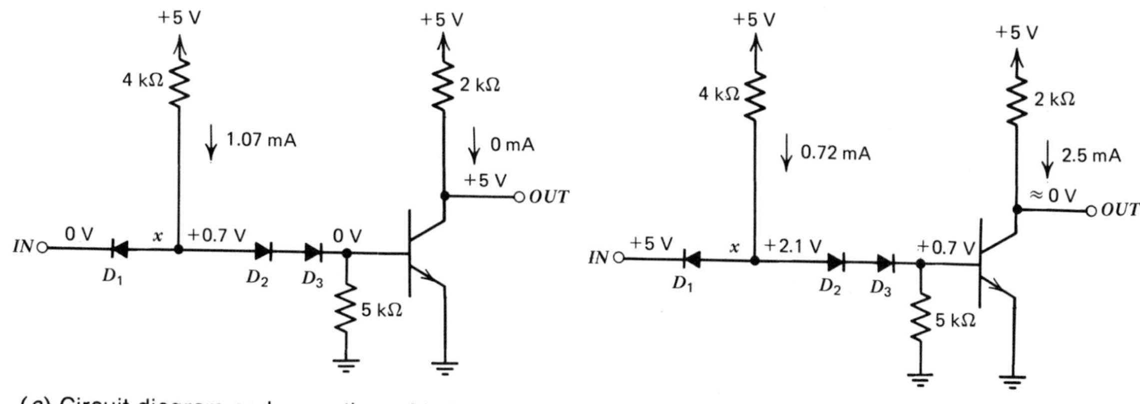

(*a*) Circuit diagram and operation with 0-V input

(*b*) Circuit diagram and operation with +5-V input

IN	*OUT*
0 V	+5 V
+5 V	≈ 0 V

(*c*) Truth table

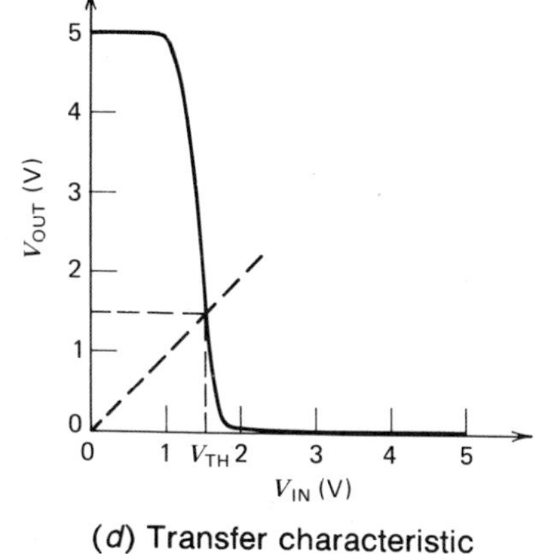

(*d*) Transfer characteristic

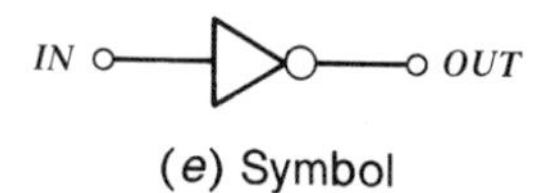

(*e*) Symbol

FIGURE 2-19 Diode-transistor logic (DTL) inverter.

A symbol of the logic inverter is shown in Fig. 2.19*e*. Note that the symbol does not show the connections to ground and to the +5-V power supply.

2-5.2 DTL NAND Gates

The DTL inverter circuit of Fig. 2.19*a* can be extended to a 2-input DTL gate, shown in Fig. 2-20*a*. It consists of a diode AND gate followed by a transistor logic inverter. The voltage and current levels are the same as in Fig. 2-19*a*.

A truth table is shown in Fig. 2-20*b*. The output is 0 V only when IN_1 *and* IN_2 are both at +5 V. Thus the circuit of Fig. 2-20*a* can be thought of as an AND gate followed by a logic inverter (Fig. 2-20*c*). However, because of the widespread use of these logic gates, the symbol of Fig. 2-20*d*, called a *NAND gate* (for NOT AND), is used to replace the two symbols of Fig. 2-20*c*. Hence the logic gate of Fig. 2-20*a* is a *2-input DTL NAND gate* and can be represented by the symbol of Fig. 2-20*d*.

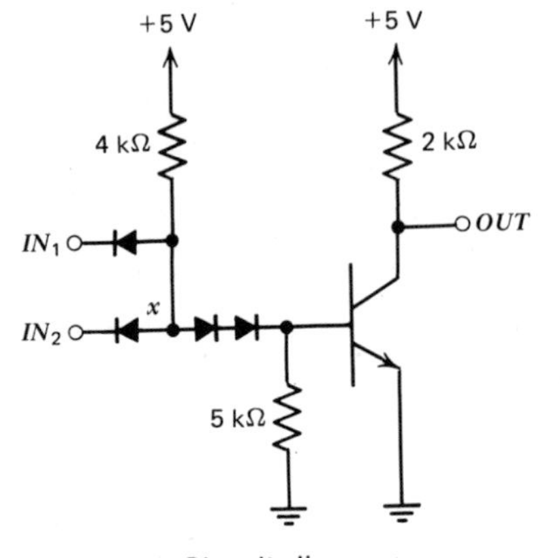

(*a*) Circuit diagram

IN_1	IN_2	OUT
0 V	0 V	+5 V
0 V	+5 V	+5 V
+5 V	0 V	+5 V
+5 V	+5 V	≈ 0 V

(*b*) Truth table

(*c*) Representation as an AND gate followed by a logic inverter

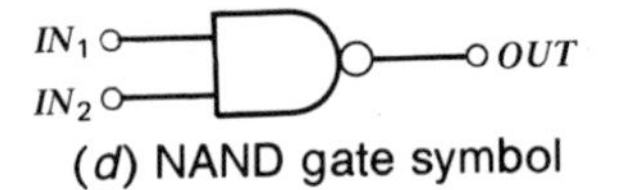

(*d*) NAND gate symbol

FIGURE 2-20 A 2-input DTL NAND gate.

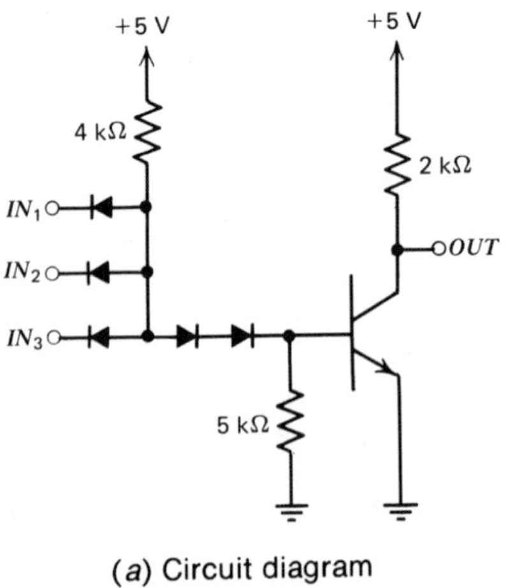

(*a*) Circuit diagram

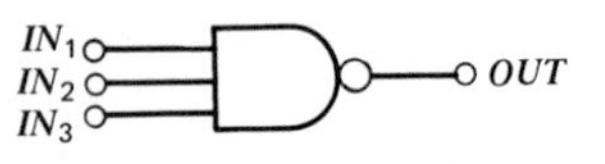

(*b*) Symbol

FIGURE 2-21 A 3-input DTL NAND gate.

The 2-input DTL NAND gate circuit of Fig. 2-20*a* can be extended to more than two inputs by additional diodes at point *x*.

EXAMPLE 2-6

Draw a circuit diagram and a symbol for a 3-input DTL NAND gate.

Solution

Figure 2-21 shows the circuit diagram and the symbol. The preparation of the truth table is left to the reader as an exercise (see Prob. 2-9).

2-5.3 Propagation Delays

The propagation delays of a DTL gate circuit depend on the external capacitive and resistive loads at its output. A propagation delay test circuit is shown in Fig. 2-22, where the output of a DTL gate circuit is loaded by a capacitance *C*, and also by a resistance *R* connected to +4.3 V.

Note that *decreasing* the value of resistance *R increases* the current that flows through it. During a positive-going output transition, this increased current *aids* the pulling up of the output voltage; hence it *reduces* propagation delay t_{PLH}. On the other hand, during a negative-going output transition, the increased current through *R opposes* the pulling down of the output voltage; hence it *increases* propagation delay t_{PHL}. Thus, in order to simulate worst-case operating conditions, t_{PLH} is measured with a high value of *R* and t_{PHL} is measured with a low value of *R*.

EXAMPLE 2-7

The input of a DTL gate circuit can be represented by a resistance of 4 kΩ to a voltage of +4.3 V (see Fig. 2-20*a*). According to the specifications of the logic gates, up to 10 logic gate inputs can be connected to a logic gate output. Find the resistance values to be used in the test circuit of Fig. 2-22 for the measurements of propagation delays t_{PLH} and t_{PHL}.

Solution

For the measurement of t_{PLH} an $R =$ **4 kΩ** is used, since this corresponds to the input of a single logic gate. For the measurement of t_{PHL} an $R =$ **400 Ω** is used, since this represents 10 parallel logic gate inputs.

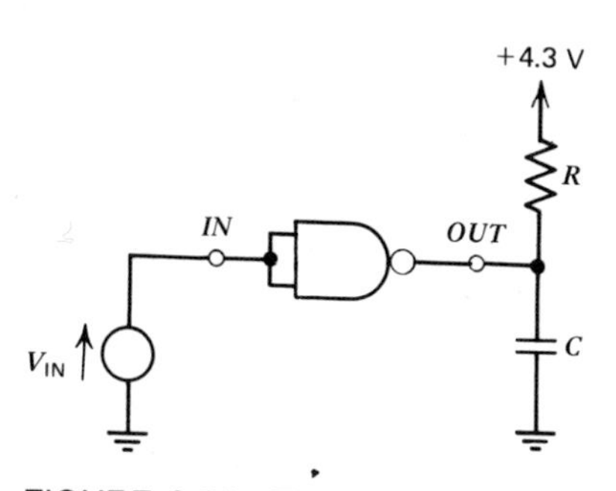

FIGURE 2-22 Test circuit for the measurement of propagation delays in DTL gates.

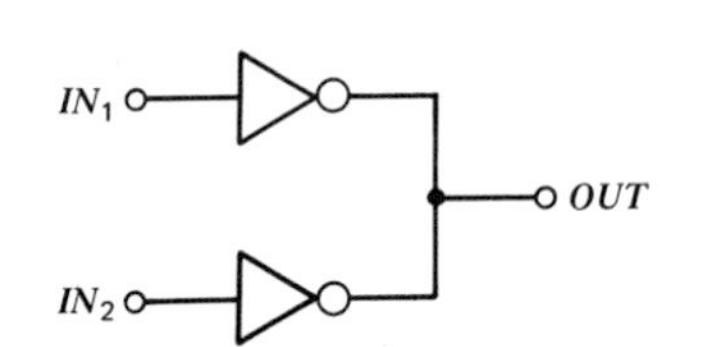

(*a*) Configuration

IN_1	IN_2	OUT
0 V	0 V	+5 V
0 V	+5 V	≈ 0 V
+5 V	0 V	≈ 0 V
+5 V	+5 V	≈ 0 V

(*b*) Truth table

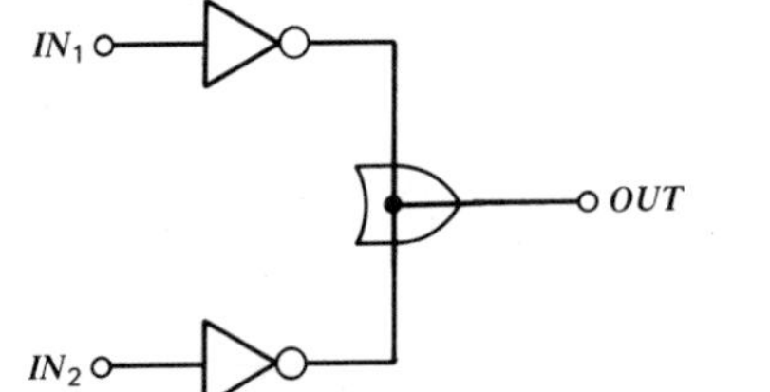

(*c*) Alternative symbol

(*d*) Alternative symbol

FIGURE 2-23 Wired-OR DTL circuits.

2-5.4 Wired-OR DTL Circuits

Consider the circuit of Fig. 2-23*a*, where each logic inverter represents the DTL logic inverter of Fig. 2-19. Even though the circuit consists of two logic inverters, according to the truth table of Fig. 2-23*b*, the output is at 0 V when IN_1 *or* IN_2 or both are at +5 V; we can also see that the output is at +5 V when IN_1 *and* IN_2 are both at 0 V. This circuit configuration, called (somewhat inconsistently) a *wired-OR, wired-AND, intrinsic-OR*, or *collector-dot* circuit, is especially useful when digital data are "fanned in" at various points along a data bus line.

Although the wired-OR configuration of DTL circuits is attained by simply wiring the outputs of the logic gates together (hence the description in Fig. 2-23*a* is adequate), it should be noted that the symbols shown in Figs. 2-23*c* and 2-23*d* are also in use.

2-5.5 Availability and Use

The most widespread integrated circuit (IC) package in use for DTL gate circuits is the 14-pin *dual in-line package* (*DIP*), shown in Fig. 2-24. Two pins are reserved for the connections of the ground and the +5-V power supply voltage, and the remaining 12 pins are available for inputs and outputs. Thus one DIP can accommodate any one of the following combinations: six logic inverters, four 2-input logic gates, three 3-input logic gates, or one 11-input logic gate.

Logic gates that are *expandable* to a larger number of inputs using a separate *expander circuit* are also available. In an expandable logic gate, the common connection of the input diodes (point *x* in Fig. 2-20*a*) is available at a pin for connection to an expander package consisting of additional diodes.

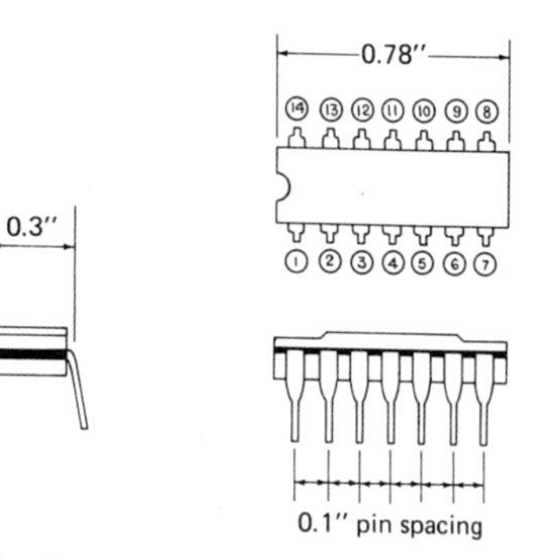

FIGURE 2-24 Outline and pin connections of a 14-pin dual in-line integrated circuit package. Note that pin numbering is counterclockwise and is viewed from the top — not looking from the direction of the pins as is usual for tubes and transistors.

Logic gates that have a 6-kΩ resistor at the output instead of the 2-kΩ resistor, or have no resistor at all, are also available. The use of these circuits reduces the high currents encountered in wired-OR circuits with many outputs connected together.

EXAMPLE 2-8

Figure 2-25 outlines a wired-OR DTL circuit consisting of 20 DTL gates. The maximum current through an output transistor occurs when *that* transistor carries the sum of the currents flowing through *all* 20 output resistors. Find this maximum current if:

a. The 20 logic gates each have a 2-kΩ output resistor as shown in Fig. 2-20*a*.

b. Only one of the 20 logic gates has the 2-kΩ resistor and the remaining 19 logic gates have the resistors replaced by an open circuit.

Solution

When an output transistor is turned ON, the output voltage is ≈ 0 V and the full 5 V appears across each 2-kΩ resistor. Hence *each* resistor carries a current of $5 \text{ V}/2 \text{ k}\Omega = 2.5$ mA.

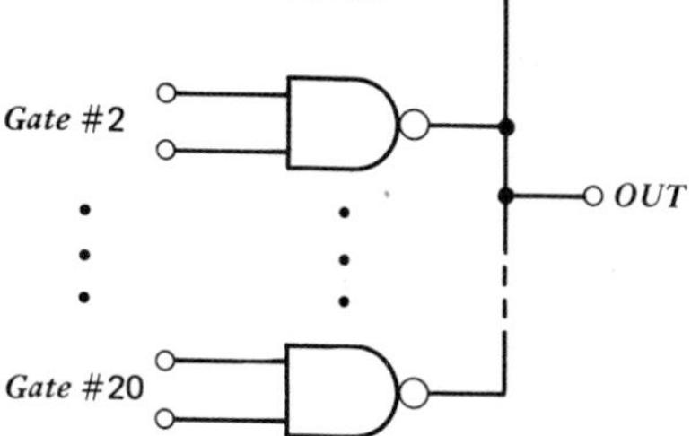

FIGURE 2-25 A wired-OR DTL circuit consisting of 20 DTL gates.

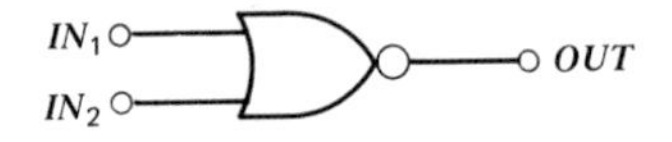

(a) Symbol

IN_1	IN_2	OUT
0 V	0 V	+5 V
0 V	+5 V	≈ 0 V
+5 V	0 V	≈ 0 V
+5 V	+5 V	≈ 0 V

(b) Truth table

FIGURE 2-26 A 2-input NOR gate.

a. The maximum current that would be carried by an output transistor is 2.5 mA × 20 = **50 mA** (this may exceed the current-carrying capability of the transistor).
b. The maximum current is only **2.5 mA**.

Thus far we have discussed only DTL inverters and NAND gates. However, there are also other circuits available, including AND gates, OR gates, as well as *NOR gates*—a NOR gate is equivalent to an OR gate followed by a logic inverter, as illustrated in Fig. 2-26 for a 2-input NOR gate.

Available circuits also include *power logic gates* capable of providing up to 100 mA to an external load, such as a lamp or a relay, connected to +5 V; and also *buffer circuits* capable of providing comparable currents in both directions, for charging and discharging capacitive loads.

Propagation delays of DTL gates are generally in the range of 50 ns to 100 ns, as compared to other gates discussed later that have propagation delays as short as 1 ns to 10 ns. It should be noted, however, that while the slower speed does limit the operating speed of the digital system, the use of DTL gates also leads to some advantages. Since it takes 50 ns to 100 ns to react to a change at its input, a DTL gate is not very sensitive to spurious signals (''spikes'') that are only, say, 10 ns wide. A second advantage is that the demands of DTL circuits on power supply voltage distribution are less stringent than are the demands of faster circuits. (Power distribution is further discussed in Sec. 16-3.)

SUMMARY

The chapter described logic circuits using switches and relays, logic circuits using diodes, and diode-transistor logic (DTL). Truth tables were introduced, and transfer characteristics, power dissipation, operating speed, loading effects, and wired-OR circuits were discussed.

SELF-EVALUATION QUESTIONS

1. What is a truth table?
*__2__. Sketch a relay with one normally open and two normally closed contacts.

*Optional material.

3. Sketch the circuit diagram of a 3-input diode AND gate.
4. Sketch the circuit diagram of a 3-input diode OR gate.
5. Describe the effects of resistive loading in logic circuits using diodes.
6. Define the term propagation delay.
7. What is a transfer characteristic?
8. What is the definition of logic threshold voltage?
9. What is a NAND gate?
10. What is a NOR gate?
11. Sketch a wired-OR DTL circuit and describe its operation.
12. Describe the effects of resistive loading of DTL gates.

PROBLEMS

2-1. Prepare a truth table for the circuit of Fig. P2-1.

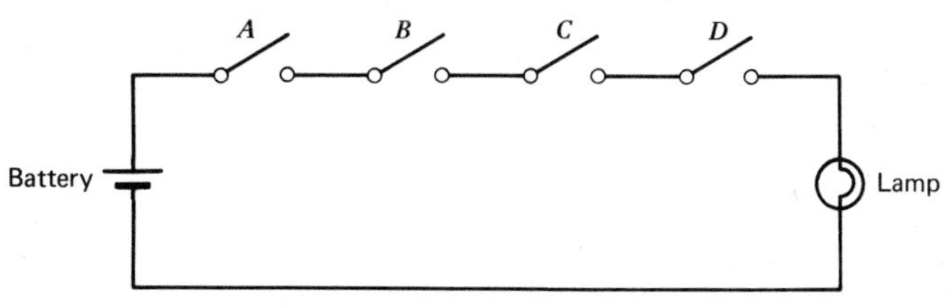

FIGURE P2-1 Circuit in Prob. 2-1.

2-2. Prepare a truth table for the circuit of Fig. P2-2.

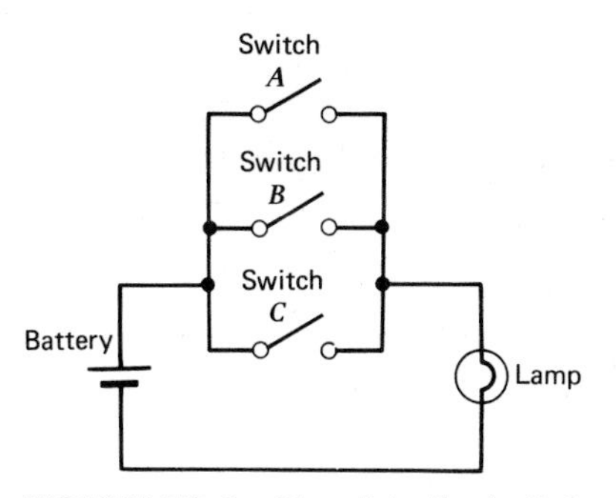

FIGURE P2-2 Circuit in Prob. 2-2.

***2-3.** Prepare a truth table for the circuit of Fig. P2-3. The truth table should show the states of switches S_A and S_B and of relays A, B, and C. Also prepare an abbreviated truth table that shows only the states of relays A, B, and C.

*Optional problem.

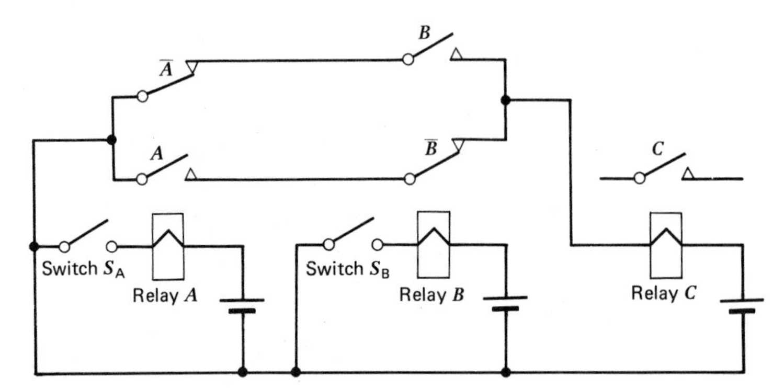

FIGURE P2-3 Circuit in Prob. 2-3.

***2-4.** Replace the contact network of Fig. P2-3 by the network shown in Fig. P2-4. Prepare a truth table that shows the states of switches S_A and S_B and of relays A, B, and C. Also prepare an abbreviated truth table that shows only the states of relays A, B, and C.

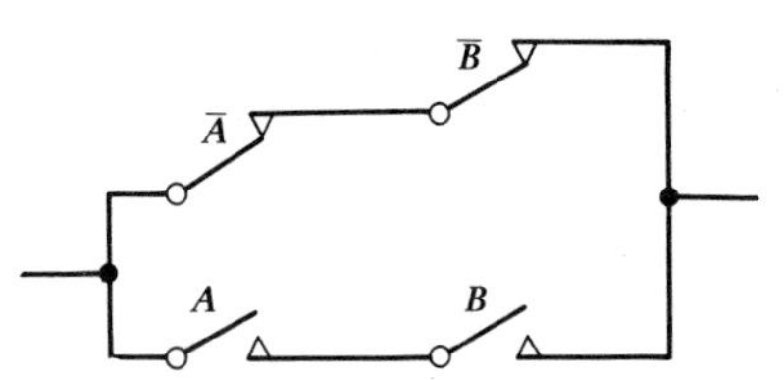

FIGURE P2-4 Contact network in Prob. 2-4.

2-5. Prepare a truth table for a 4-input diode AND gate with inputs IN_1, IN_2, IN_3, and IN_4.

2-6. Figure P2-5 shows the circuit diagram and the symbol of a 3-input diode OR gate. Prepare a truth table for the circuit.

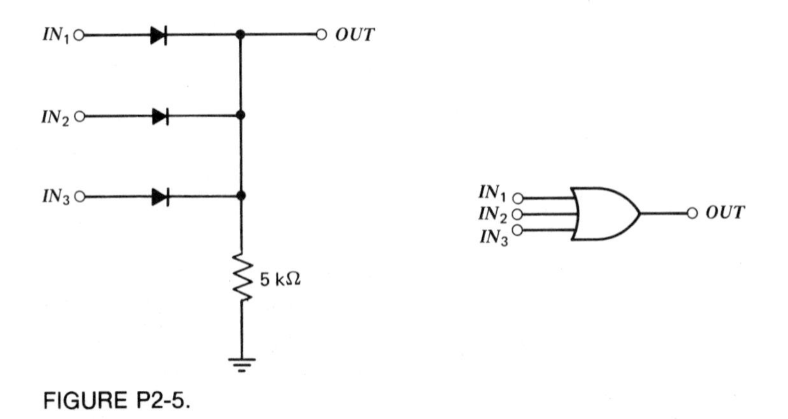

FIGURE P2-5.

2-7. In the diode OR gate circuit of Fig. 2-17*a*, both IN_1 and IN_2 are at a voltage of 0 V. In addition to the circuit elements shown in the figure, a resistor with a value of 15 kΩ is connected between the output of the circuit and the power supply voltage of +5 V. Find the voltage at the output.

2-8. Use the output transition given by the broken line in the diode AND gate circuit of Fig. 2-18*b* and find propagation delay t_{PLH} at the +3-V level in the circuit of Fig. 2-18*a*.

2-9. Prepare a truth table for the 3-input DTL NAND gate of Fig. 2-21*a*.

2-10. Draw a circuit diagram, a truth table, and a symbol for a 4-input DTL NAND gate.

2-11. Replace the lower logic inverter in Fig. 2-23*a* by a NAND gate that has 2 inputs IN_2 and IN_3. What is the voltage at output *OUT* if $IN_1 = +5$ V, $IN_2 = +5$ V, and $IN_3 = 0$ V?

2-12. Replace the lower logic inverter in Fig. 2-23*a* by a NAND gate that has 2 inputs IN_2 and IN_3. What is the voltage at output *OUT* if $IN_1 = 0$ V, $IN_2 = +5$ V, and $IN_3 = 0$ V?

CHAPTER
3 Number Systems

Instructional Objectives

This chapter describes in detail how numerical values may be expressed in various number systems in addition to the decimal number system. After you read it you should be able to:

1. Express any numerical value in the binary number system, which uses only 0s and 1s.
2. Perform binary arithmetic operations.
3. Use the octal and hexadecimal number systems.
4. Convert from one number system to another.
5. Express negative numbers in binary using complement representations.

3-1 OVERVIEW

In everyday representation of numerical values a number is given by a string of decimal digits, possibly preceded by a negative or a positive sign. This is the decimal number system that is discussed first in this chapter.

Chapter 2 introduced logic circuits that were characterized by two states such as OPEN and CLOSED, OFF and ON, or 0 V and +5 V. Such two-state logic circuits are widely used in digital systems for representing data and are described using binary numbers: strings of 0s and 1s—these are discussed next.

In addition to the decimal and binary number systems, the octal and hexadecimal number systems are also in use, and are described here. The chapter is concluded by a discussion of conversion procedures between number systems and by a description of negative numbers and their representations.

3-2 DECIMAL NUMBERS

The *decimal number system* uses 10 different *symbols*: 0, 1, 2, 3, 4, 5, 6, 7, 8, and 9. It also utilizes *positional notation*; this concept requires some clarification. When we look at a decimal number, for example 4328, we interpret it as having a *numerical value* of $(4 \times 1000) + (3 \times 100) + (2 \times 10) + (8 \times 1)$. The multipliers 1000, 100, 10, and 1 derive their values from their positions in the 4-digit number. Therefore we call this number representation positional notation and the numbers 1000, 100, 10, and 1 we call *weights* as shown below:

Decimal Weights:	1000	100	10	1	0.1	0.01	etc.
Powers of 10:	10^3	10^2	10^1	10^0	10^{-1}	10^{-2}	etc.

Note that these weights are all powers of 10. For this reason, the number system using these weights is called the *decimal number system*, as well as the *base-10* or the *radix-10 number system*. (Number systems with bases other than 10 are discussed later in this chapter.)

Each weight is multiplied by a number, the *coefficient*. The coefficients in the above example are, starting with the *most significant* (*leftmost*) *digit*, 4, 3, 2, and 8. Note that the coefficients must be one of the 10 symbols mentioned earlier.

A weight multiplied by a coefficient is called a *weighted coefficient*. The numerical value of a decimal number *is the sum of the weighted coefficients*.

EXAMPLE 3-1

Express the decimal number 670.23 as a sum of weighted coefficients.

Solution

The given number has three digits to the left of the decimal point; thus their weights from left to right (in descending order) are: 10^2, 10^1, and 10^0. To the right of the decimal point the weights have negative exponents: 10^{-1} and

TABLE 3-1
Selected Decimal Numbers (Left Column), Their Weights (Upper Row), and Coefficients (Entries in the Table)

Decimal Number	Thousands 10^3	Hundreds 10^2	Tens 10^1	Units 10^0	Tenths 10^{-1}	Hundredths 10^{-2}	Thousandths 10^{-3}
5670.234	5	6	7	0 .	2	3	4
10			1	0			
0.001				0 .	0	0	1
201.003		2	0	1 .	0	0	3
12.323			1	2 .	3	2	3
5.987				5 .	9	8	7

10^{-2}. The sum is thus

$$(6 \times 10^2) + (7 \times 10^1) + (0 \times 10^0) + (2 \times 10^{-1}) + (3 \times 10^{-2}) = \mathbf{670.23}.$$

As a summary, Table 3-1 shows selected decimal numbers in the left column, the weights in the upper row, and the coefficients as entries in the table.

3-3 BINARY NUMBERS

We have seen that *any* numerical value can be expressed as a decimal number using positional notation, which implies the sum of weighted coefficients (with some extensions, this also holds for negative numbers, as discussed later in this chapter). In the decimal number system these coefficients are the symbols 0 through 9.

It can be shown that any numerical value may also be expressed as a *binary number* using positional notation. The weights in these binary numbers are powers of 2—as compared with decimal numbers where the weights are powers of 10. The coefficients in binary numbers are 0 and 1—as compared with decimal numbers where the coefficients are 0 through 9. The result is the *natural binary number system*, which is also known simply as the *binary number system*, or as the *base-2* or the *radix-2 number system.*

3-3.1 The Binary Number System

The main reason for using binary numbers can be found in the structure of digital systems. Numbers in these systems are often represented as voltages, currents, or magnetization levels. In principle, we could have 10 voltages, currents, or magnetization levels to represent the 10 digits used in the decimal number system. In practice, however, it would require more complex circuitry to distinguish between 10 levels rather than 2. Therefore, we use predominantly the binary number system, which simplifies this problem. This way transistors may be used in their ON or OFF states representing binary 1 or binary 0, or vice versa. Magnetic material, such as recording tape, may represent one binary value when magnetized in one direction and the other binary value when magnetized in the other direction. Other devices, such as light or absence of light, relay contacts closed or open, are also used to represent binary quantities.

As was the case for decimal numbers, a binary number may be represented in *positional notation*: each position has a weight attached to it that is a power of 2. The coefficients that multiply these weights are 0 or 1. In the decimal number system there are 10 symbols and the base is 10. In the *binary number system* there are 2 symbols, 0 and 1, and the base is 2. Also, the word *bit* is commonly used for binary digit (by way of the contraction *BI*nary and digi*T*).

In the binary number system the weights are powers of 2, positive powers for binary integers and negative powers for binary fractions, as shown on facing page:

Binary weights	8	4	2	1	1/2	1/4	1/8	etc.
Powers of 2	2^3	2^2	2^1	2^0	2^{-1}	2^{-2}	2^{-3}	etc.

EXAMPLE 3-2

What is the decimal equivalent of the integer binary number 1011?

Solution

The number has 4 bits. The most significant bit has a weight of 2^3, the next bit a weight of 2^2, and so on. Thus the sum of the weighted coefficients is $(1 \times 2^3) + (0 \times 2^2) + (1 \times 2^1) + (1 \times 2^0) = 8 + 2 + 1 = \mathbf{11}$.

3-3.2 The Range of Expressible Numbers

In general, 2^n different binary numbers can be expressed with n bits. Hence, 2 numbers can be expressed with 1 bit, 4 numbers with 2 bits, 8 numbers with 3 bits, 16 numbers with 4 bits, and so on. As an example, the 8 integer binary numbers that can be expressed with 3 bits are shown below.

Decimal Number	Binary Number
0	000
1	001
2	010
3	011
4	100
5	101
6	110
7	111

Thus, the numbers that can be expressed using 3 bits are limited to the *range* shown in the table above, which consists of the numbers 0 through 7. Since an attempt to express a number *outside* such a range would lead to an erroneous result, in the following we assume that all numbers to be expressed are always *within* such a range.

3-3.3 The Binary Point

Similar to the decimal point, there is a *binary point* in binary numbers to separate the *integer* part of the number from the *fractional* part. Below is shown a *mixed* binary number.

1 0 1 . 0 1 1
↑
binary point

EXAMPLE 3-3

What is the decimal equivalent of the mixed binary number 101.011?

Solution

The integer part has 3 bits, with the most significant bit having a weight 2^2.

Thus the integer part is obtained from the sum of the weighted coefficients as $(1 \times 2^2) + (0 \times 2^1) + (1 \times 2^0) = 5$.

The fractional part has the following weighted coefficients:

$$(0 \times 2^{-1}) + (1 \times 2^{-2}) + (1 \times 2^{-3}) = (1 \times 0.25) + (1 \times 0.125) = 0.375.$$

The decimal equivalent of the binary number 101.011 is thus decimal **5.375**.

3-3.4 Subscripts

Since we are using two number systems, the binary and the decimal, we distinguish between the two by addition of a subscript. Thus 10_{10} is a decimal number that means "10 decimal," while 10_2 is a binary number that is equivalent to "2 decimal." Subscripts are omitted where no confusion can arise.

3-3.5 Binary Arithmetic

Arithmetic operations with binary numbers use a few simple rules. These are described in Sec. 3-3.5.1 through Sec. 3-3.5.4, which follow.

3-3.5.1 Rules for Binary Addition

$$0 + 0 = 0$$
$$0 + 1 = 1$$
$$1 + 0 = 1$$
$$1 + 1 = 0 \text{ carry } 1 = 10_2$$
$$1 + 1 + 1 = 1 \text{ carry } 1 = 11_2$$

EXAMPLE 3-4

Perform the following two binary additions: (**a**) $1010_2 + 1011_2$, (**b**) $11011_2 + 10111_2$.

Solution

a.

```
  1010
 +1011
 -----
 10101
 ↑ ↑
 | └── carry 1 from this bit into the next bit left
 └──── carry 1 from this bit into the next bit left
```

Check: $1010_2 + 1011_2 = 10_{10} + 11_{10} = 21_{10} = 10101_2$.

b.

```
  11011
 +10111
 ------
 110010
 ↑↑↑↑↑
 ||||└── carry 1 from this bit into the next bit left
 |||└─── carry 1 from this bit into the next bit left
 ||└──── carry 1 from this bit into the next bit left
 |└───── carry 1 from this bit into the next bit left
 └────── carry 1 from this bit into the next bit left
```

Check: $11011_2 + 10111_2 = 27_{10} + 23_{10} = 50_{10} = 110010_2$.

3-3.5.2 *Rules for Binary Subtraction*

$$0 - 0 = 0$$
$$0 - 1 = 1 \text{ borrow } 1$$
$$1 - 0 = 1$$
$$1 - 1 = 0$$

EXAMPLE 3-5

Perform the following two binary subtractions: (**a**) $11010_2 - 10101_2$, (**b**) $11100_2 - 10001_2$.

Solution

a.
```
  11010
 -10101
  -----
  00101
   ↑ ↑
   | └─borrow 1 into this bit from the next bit left
   └───borrow 1 into this bit from the next bit left
```

Check: $11010_2 - 10101_2 = 26_{10} - 21_{10} = 5_{10} = 00101_2$.

b.
```
  11100
 -10001
  -----
  01011
     ↑↑
     |└─borrow 1 into this bit from the next bit left
     └──borrow 1 into this bit from the next bit left
```

Check: $11100_2 - 10001_2 = 28_{10} - 17_{10} = 11_{10} = 01011_2$.

3-3.5.3 *Rules for Binary Multiplication*

$$0 \times 0 = 0$$
$$0 \times 1 = 0$$
$$1 \times 0 = 0$$
$$1 \times 1 = 1$$

The mechanics of binary multiplication is similar to that of decimal multiplication. However, in binary multiplication each digit (bit) of the *multiplier* can be only 0 or 1. The *partial product* becomes 0 when the multiplier bit is 0, and it becomes equal to the *multiplicand* when the multiplier bit is 1.

Similar to decimal multiplication, the multiplicand is first multiplied by the *least significant bit*, *lsb*, of the multiplier to obtain a partial product, and the multiplicand is shifted left. Then the shifted multiplicand is multiplied by the next lsb of the multiplier, and the multiplicand is shifted left again. The process is continued until all bits, including the *most significant bit*, *msb*, of the multiplier have participated.

Unlike in the pencil-and-paper method of decimal multiplication, in binary multiplication the partial products are added in at each step because the electronic multiplier circuits are simpler this way.

The above process is illustrated in Ex. 3-6 on the next page.

EXAMPLE 3-6

In a binary multiplication, the multiplicand is 10101_2 and the multiplier is 1001_2. Find the value of the binary product $10101_2 \times 1001_2$.

Solution

```
     10101 × 1001
     ------------
     10101          first partial product
    00000           second partial product
    ------
    010101          sum of the first two partial products
   00000            third partial product
   -------
   0010101          sum of the first three partial products
  10101             fourth partial product
  --------
  10111101          add to obtain result
```

Check: $10101_2 \times 1001_2 = 21_{10} \times 9_{10} = 189_{10} = 10111101_2$.

3-3.5.4 *Rules for Binary Division*

This binary arithmetic operation is carried out step by step as in the pencil-and-paper method of decimal division.

EXAMPLE 3-7

Find the quotient of $1111110_2 \div 110_2$.

Solution

```
        10101
    110/1111110
        110          subtract
        ---
         11          bring down next bit
         111         bring down next bit
         110         subtract
         ---
           11        bring down next bit
           110       bring down next bit
           110       subtract
           ---
           000       0 remainder
```

Thus, $1111110_2 \div 110_2 = \mathbf{10101_2}$.
Check: $1111110_2 \div 110_2 = 126_{10} \div 6_{10} = 21_{10} = 10101_2$.

3-4 OCTAL AND HEXADECIMAL NUMBERS

One inconvenience associated with the use of the binary representation is the large number of digits required to express a given number. Table 3-2 shows representations of selected numbers in various number systems. We can see that, for example, 100_{10} requires 7 digits for representation in the binary number system, which has a base of 2. As we choose a higher base, the number of digits required to express a given number is reduced.

For *octal numbers*, which have a base of 8_{10}, $100_{10} = 144_8$. For *hexadecimal numbers*, which have a base of 16_{10}, $100_{10} = 64_{16}$. Thus there is an incentive to use bases that are higher. As we see later in this chapter, there is

TABLE 3-2
Representation of Selected Numbers in Various Number Systems

Decimal	Binary	Octal	Hexadecimal
0	0	0	0
1	1	1	1
2	10	2	2
3	11	3	3
4	100	4	4
5	101	5	5
6	110	6	6
7	111	7	7
8	1000	10	8
9	1001	11	9
10	1010	12	A
11	1011	13	B
12	1100	14	C
13	1101	15	D
14	1110	16	E
15	1111	17	F
16	10000	20	10
17	10001	21	11
18	10010	22	12
19	10011	23	13
20	10100	24	14
32	100000	40	20
50	110010	62	32
60	111100	74	3C
64	1000000	100	40
100	1100100	144	64
255	11111111	377	FF
1000	1111101000	1750	3E8

also an advantage in using a base that is an integer power of 2, because of the resulting ease of conversion to and from binary numbers.

3-4.1 Octal Numbers

In the *octal* number system, the base is $8_{10} = 2^3$; thus 8 symbols, 0, 1, 2, 3, 4, 5, 6, and 7, are required to express all possible numbers. As before, we make use of positional notation and each position has a weight that is a power of 8.

EXAMPLE 3-8

Find the decimal equivalent of the octal number 157_8.

Solution

$$157_8 = (1 \times 8^2) + (5 \times 8^1) + (7 \times 8^0) = 64 + 40 + 7 = \mathbf{111_{10}}.$$

Selected octal numbers are shown in the third column of Table 3-2.

3-4.1.1 *The Octal Point*

The *octal point* has a function similar to the decimal point: it separates the *integer* part of a number from the *fractional* part. Octal integers have positive powers of 8 associated with them, while octal fractions have negative powers of 8, as shown below:

Octal weights	512	64	8	1	1/8	1/64	1/512	etc.
Powers of 8	8^3	8^2	8^1	8^0	8^{-1}	8^{-2}	8^{-3}	etc.

EXAMPLE 3-9

Find the decimal equivalent of the octal number 17.14_8.

Solution

$$17.14 = (1 \times 8^1) + (7 \times 8^0) + (1 \times 8^{-1}) + (4 \times 8^{-2})$$
$$= 8 + 7 + 0.125 + (4 \times 0.015625) = \mathbf{15.1875}$$

3-4.2 Hexadecimal Numbers

The above considerations for octal numbers hold also for *hexadecimal* numbers, namely that relatively few digits are required to express a numerical value and that hexadecimal numbers are easily converted to and from binary numbers.

In the hexadecimal number system the base is $16_{10} = 2^4$, and 16 different symbols are required to express all possible numbers. These symbols are 0, 1, 2, 3, 4, 5, 6, 7, 8, 9, A, B, C, D, E, and F; their decimal values can be found in Table 3-2, which shows selected hexadecimal numbers in its fourth column.

In a hexadecimal number represented in positional notation, the least significant integer digit has a weight of 16^0 (= 1), the next digit has a weight of 16^1, the next digit 16^2 ($= 256_{10}$), and so on.

EXAMPLE 3-10

Find the decimal equivalent of the hexadecimal integer number $15E_{16}$.

Solution

$$15E_{16} = (1 \times 16^2) + (5 \times 16) + (14 \times 16^0) = 256_{10} + 80_{10} + 14_{10} = \mathbf{350_{10}}.$$

3-4.2.1 *The Hexadecimal Point*

The *hexadecimal point* has a function similar to the decimal point: it separates the *integer* part of a number from the *fractional* part. The first digit to the *right* of the hexadecimal point has a weight of $16^{-1} = 0.0625_{10}$, the next digit to the right has a weight of $16^{-2} = 0.0039_{10}$, and so on.

EXAMPLE 3-11

Find the decimal equivalent of the hexadecimal fractional number $0.A7_{16}$.

Solution

$$0.A7_{16} = (10 \times 16^{-1}) + (7 \times 16^{-2}) = 0.625_{10} + 0.0273_{10} = \mathbf{0.6523_{10}}.$$

3-5 CONVERSIONS BETWEEN NUMBER SYSTEMS

In the preceding sections we have seen how a numerical value can be expressed in different number systems. In the examples we have also carried out conversions to decimal numbers. However, we have not discussed as yet systematic methods for conversion between the various number systems. These conversions are necessary since the human operator deals easily with decimal numbers while most computers operate with binary numbers. Thus at the *input* to such a computer one requires a *decimal-to-binary* conversion, while at the *output* a *binary-to-decimal* conversion is needed. We consider the latter in Sec. 3-5.1 and the former in Sec. 3-5.2.

3-5.1 Binary-to-Decimal Conversion

In what follows here, we describe two methods of binary-to-decimal conversion: one with the sum of weighted coefficients, and one using the "double-dabble" method.

3-5.1.1 Sum of Weighted Coefficients

This method has been demonstrated in Exs. 3-2 and 3-3. It is applicable to integer, fractional, and mixed numbers. It uses a powers-of-2 table (see Appendix) and two basic rules:

1. A 1 in any bit position signifies that the corresponding power of 2 is to be used in the summation.
2. A 0 in any bit position signifies that the corresponding power of 2 is not to be used in the summation.

3-5.1.2 The Double-Dabble Method

A second binary-to-decimal conversion technique applicable to integers is the so-called *double-dabble* method. The steps are as follows:

Step 1. Starting with the *most significant bit*, double that bit.

Step 2. Add this doubled value to the *next* most significant bit and record the sum.

Step 3. Double the sum obtained in step 2.

Step 4 and subsequent steps. Repeat steps 2 and 3 until the *least significant bit* has been added to the previously doubled sum. This terminates the process, which yields the equivalent decimal number.

EXAMPLE 3-12 Convert to decimal the binary number 101101_2.

Solution The process is shown in Fig. 3-1.

Thus, $101101_2 = \mathbf{45_{10}}$.

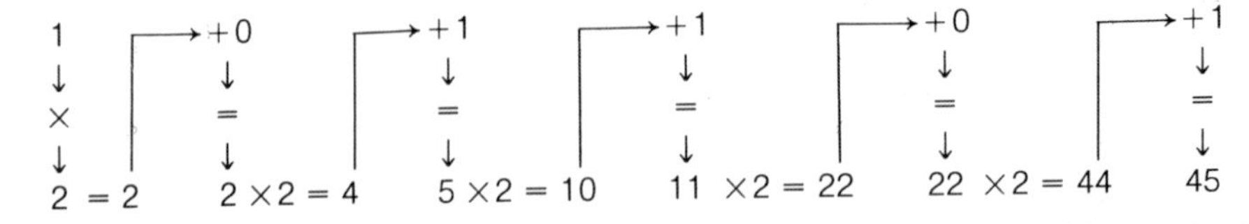

FIGURE 3-1 Binary-to-decimal integer conversion using the double-dabble method.

3-5.2 Decimal-to-Binary Conversion

This section describes four decimal-to-binary conversion methods: two methods applicable to integers, one applicable to fractions, and one applicable to mixed numbers.

3-5.2.1 Conversion of Small Integers

This method of decimal-to-binary conversion is suitable for small integers and is based on the recognition of powers of 2 contained in the given number. The largest power of 2 is recognized first and subtracted from the decimal number. The same process is then applied to the remainder and is repeated on the resulting differences until the least significant bit has been processed.

EXAMPLE 3-13 Convert to binary the decimal number 81_{10}.

Solution

Step 1. The highest power of 2 contained in 81_{10} is $2^6 = 64_{10}$.

Step 2. Subtract $81_{10} - 64_{10} = 17_{10}$.

Step 3. The highest power of 2 contained in 17_{10} is $2^4 = 16_{10}$.

Step 4. Subtract $17_{10} - 16_{10} = 1 = 2^0$.

Step 5. The remainder represents the least significant bit.

Step 6. Account for all the powers of 2 from the largest to the smallest. Fill in 0×2^5, 0×2^3, 0×2^2, and 0×2^1 for all the powers of 2 that were not represented in the five steps above.

Thus, $81_{10} = (1 \times 2^6) + (0 \times 2^5) + (1 \times 2^4) + (0 \times 2^3) + (0 \times 2^2) + (0 \times 2^1) + (1 \times 2^0)$, and in positional notation $81_{10} = \mathbf{1010001_2}$.

3-5.2.2 Conversion of Any Integer

The method described in Sec. 3-5.2.1 above becomes cumbersome for large integers. The method described here is suitable for the conversion of any integer decimal number to binary.

Step 1. Divide the decimal number by 2; the remainder 0 or 1 represents the least significant bit, lsb.

Step 2. Divide the result of the previous operation by 2 again; the remainder represents the next bit (weight $= 2^1$).

Step 3 and subsequent steps. Continue dividing, noting down the remainders until the last result of a division is 0. Write down the remainder that represents the most significant bit, msb.

EXAMPLE 3-14

Convert to binary the decimal integer number 53_{10}.

Solution

$53 \div 2 = 26$, remainder 1 = lsb
$26 \div 2 = 13$, remainder 0
$13 \div 2 = 6$, remainder 1
$6 \div 2 = 3$, remainder 0
$3 \div 2 = 1$, remainder 1
$1 \div 2 = 0$, remainder 1 = msb

Thus, 53_{10} = $\mathbf{110101_2}$ by reading the remainders from *bottom to top.*

3-5.2.3 Conversion of Fractional Numbers

For the conversion of *fractional decimal* numbers to *fractional binary* numbers, use the procedure given below.

Step 1. Multiply the decimal fraction by 2. The result may be either ≥ 1 or < 1. If it is ≥ 1, write down 1 as the most significant fractional bit; if it is < 1, write down 0 as the most significant fractional bit.

Step 2. Multiply only the *fractional* part of the result and again write down 1 or 0 depending on the value of the new result. This represents the next significant bit.

Step 3 and subsequent steps. Repeat step 2; the process is completed when the fractional part is 0 after multiplication by 2, or it is terminated when the desired accuracy has been attained.

EXAMPLE 3-15

Convert to binary the decimal fractional number 0.39_{10}.

Solution

$0.39 \times 2 = 0.78$, the most significant fractional bit is 0
$0.78 \times 2 = 1.56$, the next significant fractional bit is 1
$0.56 \times 2 = 1.12$, the next significant fractional bit is 1
$0.12 \times 2 = 0.24$, the next significant fractional bit is 0
$0.24 \times 2 = 0.48$, the next significant fractional bit is 0
$0.48 \times 2 = 0.96$, the next significant fractional bit is 0
$0.96 \times 2 = 1.92$, the next significant fractional bit is 1
$0.92 \times 2 = 1.84$, the next significant fractional bit is 1
$0.84 \times 2 = 1.68$, the next significant fractional bit is 1
$0.68 \times 2 = 1.36$, the least significant fractional bit is 1

Thus, 0.39_{10} = $\mathbf{0.0110001111_2}$ by reading the bits from *top to bottom.*

Note that we have arbitrarily terminated the process after 10 bits have been accumulated. Thus we incur a round-off error that is smaller than $2^{-10} = 1/1024_{10}$. Rounding off is required when the decimal fraction cannot be exactly represented by a binary fraction containing a practically finite number of bits.

3-5.2.4 Conversion of Mixed Numbers

Conversion of mixed decimal numbers to binary is performed in three steps. First, the integer part is converted using successive division by 2 as in Sec. 3-5.2.2 above. Second, conversion of the fractional part is done by successive multiplication by 2 as in Sec. 3-5.2.3 above. Third, the bits thus obtained from the two steps are put side by side to form the required binary number.

EXAMPLE 3-16 Convert the decimal mixed number 13.625_{10} to a binary number.

Solution

Step 1. Convert the integer part.

$$13 \div 2 = 6 \text{ remainder } 1\ (= \text{lsb})$$
$$6 \div 2 = 3 \text{ remainder } 0$$
$$3 \div 2 = 1 \text{ remainder } 1$$
$$1 \div 2 = 0 \text{ remainder } 1\ (= \text{msb})$$

Integer part: $13_{10} = 1101_2$.

Step 2. Convert the fractional part.

$$0.625 \times 2 = 1.25 \qquad \text{msb} = 1$$
$$0.25 \times 2 = 0.50 \text{ next bit} = 0$$
$$0.50 \times 2 = 1.00 \qquad \text{lsb} = 1$$

The process is terminated because the fractional part of the last multiplication is 0.00. Thus the decimal fraction 0.625 equals 0.101_2.

Step 3. $1101_2 + 0.101_2 = \mathbf{1101.101_2}$.

3-5.3 Binary-to-Octal Conversion

Quite often instructions and data are entered into a computer in the form of *octal* numbers. Compared with using binary numbers, this reduces considerably the tedious task of data entry because of the required lesser number of digits (by a factor of 3) in an octal number that has the same numerical value as the binary number.

Conversion from binary to octal is easy, because base 8 is a binary-derived base, that is, $8 = 2^3$. Therefore, since there are 8 symbols in the octal number system, 0 through 7, all symbols can be represented by 3 bits for the octal numbers $0_8 = 000_2$ through $7_8 = 111_2$.

Conversion from binary to octal is obtained using the following rule: Starting at the binary point, divide the integer part and the fractional part of the binary number into groups of 3 bits. Each group represents one octal digit.

EXAMPLE 3-17

Convert the binary number 111101.1_2 to an octal number.

Solution

Binary:	111	101	.	100
	↓	↓		↓
Octal:	7	5	.	4

Thus, $111101.1_2 = \mathbf{75.4_8}$.

Note that two *trailing zeros* have been added to the *fractional* part of the binary number. These do not change the numerical value of the binary number. Similarly, the addition of *leading zeros* to the *integer* part of the binary number does not change its numerical value either. Also note the drastic reduction in the number of digits required to express a given quantity as the base is increased.

3-5.4 Octal-to-Binary Conversion

This is the inverse process of the conversion shown in Sec. 3-5.3 above. Each octal digit is represented by 3 bits.

EXAMPLE 3-18

Convert the octal number 37.4_8 to a binary number.

Solution

Octal:	3	7	.	4
	↓	↓		↓
Binary:	011	111	.	100

Omitting the leading integer zero and the trailing fractional zeros, we obtain $37.4_8 = \mathbf{11111.1_2}$.

3-5.5 Binary-to-Hexadecimal Conversion

In Sec. 3-5.3 we mentioned that instructions and data are often entered into a computer in the form of *octal* numbers. In many cases, however, *hexadecimal* numbers are used for this purpose. Hence, we have to get acquainted with *binary-to-hexadecimal* conversion methods.

The hexadecimal number system uses $16 = 2^4$ as its base; hence, conversion is easy because exactly 4 bits, 0000_2 through 1111_2, are required to represent each hexadecimal digit. The technique is similar to binary-to-octal conversion shown in Sec. 3-5.3. Thus, conversion from *binary to hexadecimal* is obtained using the following rules: Starting at the binary point divide the *integer* part and the *fractional* part of the binary number into groups of 4 bits. Add leading and trailing zeros where necessary. Each group of 4 bits represents a hexadecimal digit.

EXAMPLE 3-19

Convert the binary number 111100.111_2 to a hexadecimal number.

Solution

	leading zeros			trailing zero
Binary:	0011	1100	.	1110
	↓	↓		↓
Hexadecimal:	3	C	.	E

Thus, 111100.111_2 = $\mathbf{3C.E_{16}}$.

3-5.6 Hexadecimal-to-Binary Conversion

This is the inverse process of the conversion shown in Sec. 3-5.5 above. Each hexadecimal digit is represented by 4 bits.

EXAMPLE 3-20

Convert the hexadecimal number $E8.7_{16}$ to a binary number.

Solution

Hexadecimal:	E	8	.	7
	↓	↓		↓
Binary:	1110	1000	.	0111

Thus, $E8.7_{16}$ = $\mathbf{11101000.0111_2}$.

3-5.7 Decimal-to-Octal Conversion

Conversion from decimal to octal proceeds along similar lines as the decimal-to-binary conversion shown in Sec. 3-5.2.2. The difference is that in the decimal-to-binary conversion of integer numbers we successively divided by 2, whereas for conversion from decimal integers to octal integers we *successively divide by* 8. The remainder of the first division yields the *least significant* octal *digit*, *lsd*, the second division yields the next octal digit, and so on, until the last division, the remainder of which equals the *most significant* octal *digit*, *msd*.

EXAMPLE 3-21

Convert to octal the decimal number 829_{10}.

Solution

$829 \div 8 = 103$ remainder 5 (lsd)
$103 \div 8 = 12$ remainder 7
$12 \div 8 = 1$ remainder 4
$1 \div 8 = 0$ remainder 1 (msd)

Thus, 829_{10} = $\mathbf{1475_8}$.

3-5.8 Octal-to-Decimal Conversion

Conversion of integers from octal to decimal is conducted along similar lines as the *double-dabble* method discussed in Sec. 3-5.1.2.

Step 1. Starting with the *most significant* octal digit, multiply that digit by 8.

Step 2. Add the result of the *next* most significant digit and record the sum.

Step 3. Multiply by 8 the sum obtained in step 2.

Step 4 and subsequent steps. Repeat steps 2 and 3 until the *least significant* octal digit has been added. This terminates the process.

The above procedure is illustrated in Ex. 3-22, which follows.

EXAMPLE 3-22

Convert to decimal the octal number 2053_8.

Solution

$$2 \times 8 = 16$$
$$16 + 0 = 16$$
$$16 \times 8 = 128$$
$$128 + 5 = 133$$
$$133 \times 8 = 1064$$
$$1064 + 3 = 1067$$

Thus, $2053_8 = \mathbf{1067_{10}}$.

Another method for conversion sums the weighted coefficients.

EXAMPLE 3-23

Convert to decimal the octal number 2053_8.

Solution

$$2053_8 = (2 \times 8^3) + (0 \times 8^2) + (5 \times 8^1) + (3 \times 8^0)$$
$$= (2 \times 512) + 0 + 40 + 3 = 1067_{10}.$$

Thus, $2053_8 = \mathbf{1067_{10}}$, the same as in Ex. 3-22 above.

3-6 NEGATIVE NUMBERS

In Sec. 3-3.5 we discussed binary arithmetic, but we avoided the question of negative binary numbers. As in decimal numbers, negative binary numbers do exist and indeed *more than one* (actually *three*) representation is available to express a negative binary number. (The development of three different representations for negative binary numbers originated from the search to simplify the circuitry required for subtraction.)

In all three representations discussed below, positive numbers are expressed in exactly the same way: a leading (leftmost bit) 0 represents the "+" sign, which is followed by the magnitude. Negative numbers, however, have a leading 1 representing the "−" sign, which is followed by (a) the *magnitude*, or (b) the *1's complement*, or (c) the *2's complement* of the binary number to be expressed.

3-6.1 The Range of Expressible Numbers

Table 3-3 shows how numbers are expressed in the three representations using binary numbers with 4 bits. One bit is used as a sign bit, leaving 3 bits available

TABLE 3-3
Positive and Negative Binary Number Representations Using 4 Bits

Decimal	Binary Sign-and-Magnitude	Binary 1's Complement	Binary 2's Complement
−8	—	—	1 000
−7	1 111	1 000	1 001
−6	1 110	1 001	1 010
−5	1 101	1 010	1 011
−4	1 100	1 011	1 100
−3	1 011	1 100	1 101
−2	1 010	1 101	1 110
−1	1 001	1 110	1 111
0	{1 000, 0 000}	{1 111, 0 000}	0 000
1	0 001	0 001	0 001
2	0 010	0 010	0 010
3	0 011	0 011	0 011
4	0 100	0 100	0 100
5	0 101	0 101	0 101
6	0 110	0 110	0 110
7	0 111	0 111	0 111

for expressing the magnitude; this limits the magnitude to the *range* shown in the table. Since an attempt to represent a number *outside* such a range would lead to an erroneous result, in the following we assume that all numbers to be represented are always *within* such a range.

3-6.2 Sign-and-Magnitude Representation

In ordinary decimal arithmetic a negative number such as −345 is represented by the *sign* "−" and the *magnitude* (*absolute value*) 345. Similarly, in binary sign-and-magnitude representation we express a number by a sign followed by the magnitude, as shown in the second column of Table 3-3. The sign-and-magnitude representation is easily readable since the magnitude part is identical for positive and negative numbers. However, implementation of arithmetic circuits is not as easy as with other number representations.

3-6.2.1 Subtraction in the Sign-and-Magnitude Representation

To carry out a subtraction in the sign-and-magnitude representation we have to:

1. Examine which of the two numbers has a greater magnitude.
2. Subtract the number with the *smaller* magnitude *from* that having the *greater* magnitude.
3. Adjust the sign accordingly.

EXAMPLE 3-24

Obtain (**a**) $17_{10} - 25_{10}$ and (**b**) $10001_2 - 11001_2$ ($= 17_{10} - 25_{10}$).

Solution

We notice that the *minuend*, 17_{10}, has a smaller magnitude than the *subtrahend*, 25_{10}. The situation is the same in the subtraction of the binary numbers. We thus carry out the subtractions $25_{10} - 17_{10}$ and $11001_2 - 10001_2$ and apply a negative sign to the end result, in *both* cases, as shown below.

a.

$$\begin{array}{r} 25 \\ -17 \\ \hline \end{array}$$

$= +8_{10}$, apply negative sign $\rightarrow = \mathbf{-8_{10}}$, which is our result.

b. For binary sign-and-magnitude arithmetic we disregard the sign for a moment and carry out the subtraction of the smaller magnitude from the greater.

$$\begin{array}{r} 11001 \\ -10001 \\ \hline 01000 \end{array}$$

We then prefix the minus sign, binary 1, as the most significant bit to obtain the negative result **1 01000** in sign-and-magnitude representation. Note that for clarity we have allowed half a space between the sign bit and the number bits. In actuality, there is no space in the 6-bit number.

Finally, we check: $10001_2 - 11001_2 = 17_{10} - 25_{10} = -8_{10} = 1\,01000$ in sign-and-magnitude representation.

3-6.3 1's Complement Representation

In the *1's complement* representation positive numbers are expressed in the same way as in the sign-and-magntidue representation. The convention of the sign bit is also the same in both representations: namely, 0 for "+" and 1 for "−". However, to express the magnitude part of a negative number in the 1's complement representation we have to take its *1's complement.*

The 1's complement of a binary number is obtained by changing all 1s in the number to 0s and all 0s to 1s.

EXAMPLE 3-25

Find the 1's complements of the binary numbers 0100 and 0010.

Solution

The 1's complement of 0100 is **1011**.
The 1's complement of 0010 is **1101**.

The same process is applicable to fractional and mixed numbers as well.

EXAMPLE 3-26

Find the 1's complement of the binary number 010.01.

Solution

The 1's complement of 010.01 is **101.10**.

1's complements of various numbers are shown in the third column of Table 3-3. We should note that, as compared to other representations, the 1's complement representation receives only limited use in today's digital systems, and for this reason it is not discussed here in more detail.

3-6.4 2's Complement Representation

In the *2's complement* representation positive numbers are expressed in the same way as in the sign-and-magnitude representation. The convention of the sign bit is also the same in both representations: namely, 0 for "+" and 1 for "−". However to express the magnitude part of a negative number in the 2's complement representation we have to take its *2's complement.*

The 2's complement of a number is obtained by first taking the 1's complement of the number and then adding 1 to the result.

EXAMPLE 3-27

Find the 2's complement of the binary numbers 0000, 1010, and 0101.

Solution

Number to be 2's complemented:	0000	1010	0101
1's complement	1111	0101	1010
Add 1	+1	+1	+1
2's complement	**~~1~~0000**	**0110**	**1011**

Note that the 2's complement of 0000 is still 0000 and that we have crossed out the carry that was generated in the course of obtaining the 2's complement.

3-6.4.1 *Subtraction in the 2's Complement Representation*

We can use the 2's complement for binary subtraction by first taking the 2's complement of the subtrahend and then adding it to the minuend.

EXAMPLE 3-28

Using 2's complement addition obtain $10011_2 - 101_2$.

Solution

The subtrahend is 00101; thus its 1's complement is 11010 and its 2's complement is 11010 + 1 = 11011. Prefixing sign bits we get

0 10011	minuend
+1 11011	2's complement of 00101
~~1~~0 01110	result

Note that in 2's complement arithmetic the carry generated from the sign bit is always disregarded. Also note that the result above is *positive*, as can be seen from its most significant sign bit, which is 0.

As a check on the validity of the above we find that

$$\begin{array}{r} 10011_2 = 19_{10} \\ -00101_2 = -5_{10} \\ \hline 01110_2 = 14_{10}. \end{array}$$

No carry is generated when the result of a 2's complement addition is negative, that is, when the sign bit is 1; and the number obtained is in 2's complement form.

EXAMPLE 3-29

Using 2's complement addition obtain $101_2 - 10011_2$.

Solution

First we take the 2's complement of the subtrahend.

10011	subtrahend
01100	1's complement of subtrahend
+1	add
01101	2's complement of subtrahend

Next we prefix the signs and add the minuend to the 2's complement of the subtrahend.

0 00101	minuend
+1 01101	2's complement of subtrahend
1 10010	result (in 2's complement form)

The result is *negative* (the sign bit is 1) and it is therefore in 2's complement form. The decimal equivalent of the result is −14, which may be obtained by taking the 2's complement of the result and prefixing a minus sign.

The fourth column of Table 3-3, shows 2's complements of selected numbers. Note that, unlike in the other two representations, there is only one way of representing zero, namely 0 000.

SUMMARY

Positional notation was first introduced by discussing the decimal number system and was then extended to include the binary, octal, and hexadecimal number systems. The practical advantages of using the binary number system in digital systems were outlined, followed by an introduction of binary arithmetic operations. The use of a variety of number systems necessitated the discussion of conversions, and procedures were given for converting between binary and octal, binary and decimal, binary and hexadecimal, and octal and decimal number systems. Three representations of negative binary numbers were discussed: the sign-and-magnitude, the 1's complement, and the 2's complement representations.

SELF-EVALUATION QUESTIONS

1. Describe the concept of positional notation.
2. What are the weights of integer hexadecimal, decimal, octal, and binary numbers?

3. What are the weights of fractional hexadecimal, decimal, octal, and binary numbers?
4. State the advantages of the octal and hexadecimal number systems.
5. Define binary point.
6. Describe the steps required to convert a mixed binary number to a decimal.
7. Describe the steps required to convert a mixed decimal number to a binary.
8. When does a round-off error occur in conversions between number systems and why?
9. Describe the sign-and-magnitude representation.
10. Define the 1's complement of a number.
11. Define the 2's complement of a number.
12. What are the ranges of numbers that can be expressed using 4 bits in the sign-and-magnitude, 1's complement, and 2's complement representations?

PROBLEMS

3-1. Write the following decimal numbers as sums of weighted coefficients: 4026; 0.521; −12.

3-2. Write the following decimal numbers as sums of weighted coefficients: 896.03; −567; 43.29.

3-3. Write the following binary numbers as sums of weighted coefficients: 1001; 0.1011; 101.01.

3-4. Write the following binary numbers as sums of weighted coefficients: 11011; 101.011; −1010.

3-5. Convert the following binary numbers to their decimal equivalents: 101; 1101; 11101.

3-6. What are the decimal equivalents of the following binary numbers: 10011; 11001; 11100?

3-7. Convert the following binary fractions to their decimal equivalents: 0.1; 0.01; 0.11.

3-8. What are the decimal equivalents of the following binary fractions: 0.001; 0.011; 0.101?

3-9. Convert the following mixed binary numbers to their decimal equivalents: 10.1; 1.01; 11.11; 101.001.

3-10. Convert the following decimal integers to their binary equivalents: 5; 9; 13; 19.

3-11. What are the binary equivalents of the following decimal integers: 21; 24; 31; 126?

3-12. Convert the following decimal fractions to their binary equivalents: 0.5; 0.25; 0.75; 0.125.

3-13. Convert the following mixed decimal numbers to their binary equivalents: 5.5; 9.25; 13.75; 19.125.

3-14. Evaluate (**a**) $1101_2 + 1011_2$; (**b**) $1101_2 - 1011_2$; (**c**) $110_2 \times 101_2$; (**d**) $10010_2 \div 11_2$.

3-15. Evaluate (**a**) $11011_2 + 10110_2$; (**b**) $1011_2 - 1001_2$; (**c**) $111_2 \times 11_2$; (**d**) $11100_2 \div 111_2$.

3-16. Count in octal numbers from 0 up to the equivalent of 20_{10}.

3-17. Find the decimal equivalent of 237_8.

3-18. $143_8 = X_{10}$. $127_{10} = Y_8$. Find X_{10} and Y_8.

3-19. Count in hexadecimal up to the equivalent of 31_{10}.

3-20. What is the decimal equivalent of (**a**) 36_{16}? (**b**) 153_{16}?

3-21. Find the decimal equivalent of (**a**) 54.1_{16}; (**b**) 289_{16}.

3-22. Find the 1's complement of 101.11; 1101; 0.101; 11.01.

3-23. Find the 2's complement of the binary numbers of Prob. 3-22.

3-24. Using 2's complements calculate the following: 10011 − 11001; 11001 − 10011. Check your results by decimal calculation.

3-25. Show how to multiply binary numbers by 2; 4; 8; 16.

3-26. Show how to divide positive binary numbers by 2; 4; 8; 16.

3-27. Show how to multiply octal numbers by 8_{10}; 64_{10}; 512_{10}.

3-28. Show how to divide positive octal numbers by 8_{10}; 64_{10}; 512_{10}.

3-29. The *offset binary code* can be obtained by complementing the msb of the 2's complement code. Convert the decimal number 0_{10} (zero) and -10_{10} to the offset binary code using 5 bits.

3-30. The offset binary code can be obtained by complementing the msb of the 2's complement code. Convert the decimal numbers 5_{10} and -13_{10} to the offset binary code using 5 bits.

CHAPTER 4 Coding

Instructional Objectives

This chapter introduces codes fulfilling specific requirements. After you read it you should be able to:

1. Code decimal numbers with binary-coded decimal (BCD) codes.
2. Perform addition of BCD numbers.
3. Code numbers in the Gray code.
4. Convert from Gray code to binary and vice versa.
5. Establish parity error checks.
6. Describe alphanumeric codes such as the Baudot, Hollerith, ASCII, and EBCDIC codes.

4-1 OVERVIEW

The preceding chapter introduced binary, octal, and hexadecimal numbers. While these provide for efficient representations of data in digital systems, in everyday life we prefer the use of decimal numbers. The resulting gap is bridged by binary-coded decimal (BCD) numbers, which facilitate digit-by-digit representation of decimal numbers in binary systems.

Digital systems often process data originating from measurements of physical quantities, for example, length. Thus we are interested in finding codes that are suited for processing such data. It is for this reason that we introduce and describe unit-distance codes, which facilitate direct conversion of mechanical displacement to digital data.

Notwithstanding the good noise-immunity of present-day digital systems, the processed data are occasionally corrupted by errors. In many applications we would like to detect the presence of errors; this is made possible by the use of error-detecting codes, which are described next.

In addition to numerical data, digital systems also have to process letters of the alphabet and punctuation marks, as well as various control characters required by peripheral devices such as a printer or a magnetic tape unit. Many alphanumeric codes have been developed for representing such information in digital systems; a few of these are described at the end of the chapter.

4-2 BINARY-CODED DECIMAL (BCD) NUMBERS

Chapter 3 introduced binary numbers using positional notation and the binary base. Such numbers required many binary digits to express a numerical value; thus, in the interest of reducing the number of digits, we have introduced *octal* and *hexadecimal* numbers.

Three binary digits were coded to represent one *octal* digit and, similarly, *four* binary digits were coded to represent one *hexadecimal* digit. Though conversions between number systems were made easy by the choice of binary-derived ($2^3, 2^4$) bases, we still have not dealt with the problem that these were binary or binary-derived numbers, while the human operator works best with *decimal* numbers. A good solution to this problem is the use of *binary-coded decimal* (*BCD*) numbers.

4-2.1 Weighted 8-4-2-1 BCD Code

Many binary-coded decimal numbers use 4 bits to express a decimal digit. As was shown in Table 3-2, page 35, 4 bits can represent 16 numbers, 0 through 15. Thus, in a 4-bit BCD code there are 6 *redundant states*, also known as

TABLE 4-1
8-4-2-1 BCD Numbers

Decimal	8-4-2-1 BCD
0	0000
1	0001
2	0010
3	0011
4	0100
5	0101
6	0110
7	0111
8	1000
9	1001
UNALLOWED STATES	1010
	1011
	1100
	1101
	1110
	1111

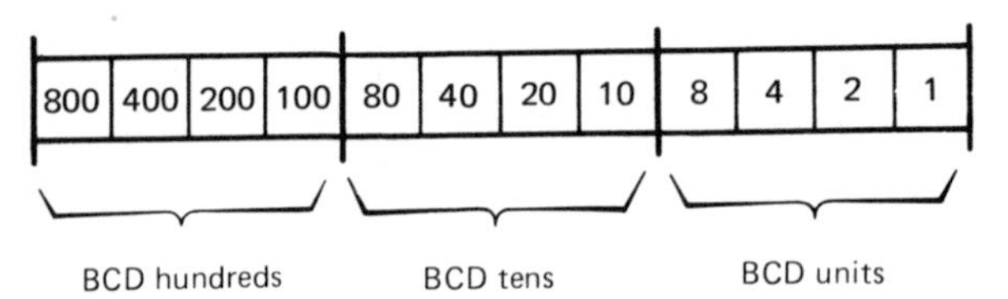

FIGURE 4-1 Weights of the 8-4-2-1 BCD code in a 3-digit decimal number.

unallowed states or *unused states*. This redundancy permits a great variety of BCD codes; however, in this chapter we discuss only a few of these.

The most widely used BCD code is the *8-4-2-1 BCD code*, shown in Table 4-1. The 8-4-2-1 BCD code is a *weighted* code in which the weights (starting with the most significant bit) are 8, 4, 2, and 1. This code is also referred to as the *naturally weighted BCD code*, or *natural BCD code*, because the weights are consecutive powers of 2. Note that each group of 4 bits represents one decimal digit.

To *encode*, or *code*, a multidigit decimal number we require as many groups of 4 bits as there are decimal digits. This is shown in Fig. 4-1 for a 3-digit integer number; however, it is also applicable to any number of digits of fractional and mixed numbers.

EXAMPLE 4-1

Encode the decimal number 863 in the 8-4-2-1 BCD code.

Solution

Decimal:	8	6	3
8-4-2-1 BCD:	**1000**	**0110**	**0011**
Decimal weight:	Hundreds	Tens	Units

BCD encoding involves expressing a decimal number in BCD notation. Conversely, BCD *decoding* involves changing a BCD number into decimal.

EXAMPLE 4-2

Decode the following 8-4-2-1 BCD numbers:

a. 0011 1001 0111, **b.** 0100 1000 0110, **c.** 0101 0111 0100.

Solution

a. 8-4-2-1 BCD: 0011 1001 0111
↓ ↓ ↓
Decimal: 3 9 7

b. 8-4-2-1 BCD: 0100 1000 0110
↓ ↓ ↓
Decimal: 4 8 6

c. 8-4-2-1 BCD: 0101 0111 0100
↓ ↓ ↓
Decimal: 5 7 4

4-2.2 BCD Addition

Addition of 8-4-2-1 BCD numbers involves some complications since the rules of binary arithmetic do not apply. For example, adding $14_{10} + 8_{10}$ in *binary*, we obtain

Decimal		Binary
14_{10}	=	1110
$+\ 8_{10}$	=	+1000
22_{10}	=	**10110**

However, adding these numbers in *8-4-2-1 BCD* yields

Decimal	8-4-2-1	BCD
4_{10}	0001	0100
8_{10}	0000	1000
	0001	**1100**
	Tens	Units

This result is clearly incorrect, since 1100 is not an allowed BCD number (see Table 4-1).

4-2.2.1 Correction in BCD Addition

We obtained above in the result of the units digit the code 1100 that is *not allowed* in 8-4-2-1 BCD. Thus a *correction* is required in this case while other cases may require no correction. The rules for correction are as follows:

1. If the result of an 8-4-2-1 BCD addition equals or is smaller than 1001 ($= 9_{10}$), then no correction is required.
2. If the result of an 8-4-2-1 BCD addition is greater than 1001 ($= 9_{10}$) or if a carry was generated, then add 0110 ($= 6_{10}$).

EXAMPLE 4-3

Add the following 8-4-2-1 BCD numbers:

a. 0101 + 0011, **b.** 0111 + 1000, **c.** 1000 + 1001.

Solution

a.
```
  0101
 +0011
 -----
  1000  result: 5₁₀ + 3₁₀ = 8₁₀
```
(**1000** result: $5_{10} + 3_{10} = 8_{10}$)

No correction is required according to rule 1 above.

b.
```
         0111
        +1000
        -----
         1111  the result is greater than 1001; hence a correction is required
        +0110  add 0110
        -----
  0001   0101  result: 7₁₀ + 8₁₀ = 15₁₀
```
(**0001 0101** result: $7_{10} + 8_{10} = 15_{10}$)

TABLE 4-2
The 2-4-2-1 BCD Code and the Excess-3 BCD Code

Decimal	2-4-2-1 BCD	Excess-3 BCD
0	0000	0011
1	0001	0100
2	0010	0101
3	0011	0110
4	0100	0111
5	1011	1000
6	1100	1001
7	1101	1010
8	1110	1011
9	1111	1100

c.

```
              1000
              1001
        ----------
        0001  0001  carry 1 into the tens digit; hence a correction is required
            + 0110  add 0110 to the BCD digit that generated the carry
        ----------
        0001  0111  result: 8_10 + 9_10 = 17_10
```

(**0001 0111** result: $8_{10} + 9_{10} = 17_{10}$)

4-2.3 Other BCD Codes

Although the 8-4-2-1 natural BCD code is most widely used, other BCD codes have also found applications. Table 4-2 shows two such codes: The *2-4-2-1 BCD code* and the *excess-3* (or *XS-3*) *BCD code*.

Both codes are *self-complementing*: Mirroring around the broken line in Table 4-2 results in complementation. (This property is used in BCD subtractor circuits.)

As is the case for the 8-4-2-1 weighted natural BCD code, the 2-4-2-1 BCD code is also weighted—however, with weights of 2, 4, 2, and 1 instead of 8, 4, 2, and 1.

EXAMPLE 4-4 Express the decimal number 752 in 2-4-2-1 BCD notation.

Solution From Table 4-2, 752_{10} = **1101 1011 0010**.

The excess-3 BCD code is an *unweighted* code that derives its name from the fact that it may be obtained by adding $3_{10} = 11_2$ to each 8-4-2-1 BCD digit.

EXAMPLE 4-5 Obtain 8_{10} in the excess-3 BCD code.

Solution

In the 8-4-2-1 BCD code $8_{10} = 1000$
Add 11

In the excess-3 code 8_{10} = **1011**

Multidigit decimal numbers are encoded into excess-3 BCD digit by digit. Note that there can be no *carry* generated by the addition of $3_{10} = 11_2$ in this encoding process; this is because the maximum possible value of each digit is only $9_{10} = 1001$ BCD and, since $9_{10} + 3_{10} = 12_{10}$, the maximum possible value of each resulting excess-3 BCD digit is only 1100.

EXAMPLE 4-6

Encode 693_{10} in excess-3 BCD.

Solution

Decimal.	6	9	3
	↓	↓	↓
8-4-2-1 BCD:	0110	1001	0011
Add 0011:	0011	0011	0011
Excess-3:	**1001**	**1100**	**0110**.

4-3 UNIT-DISTANCE CODES

Unit-distance codes derive their name from their property that *only one bit* changes between representations of two *consecutive* numbers. This property is *not* present in *all* codes: For example, in natural binary the code for 7_{10} is 0111

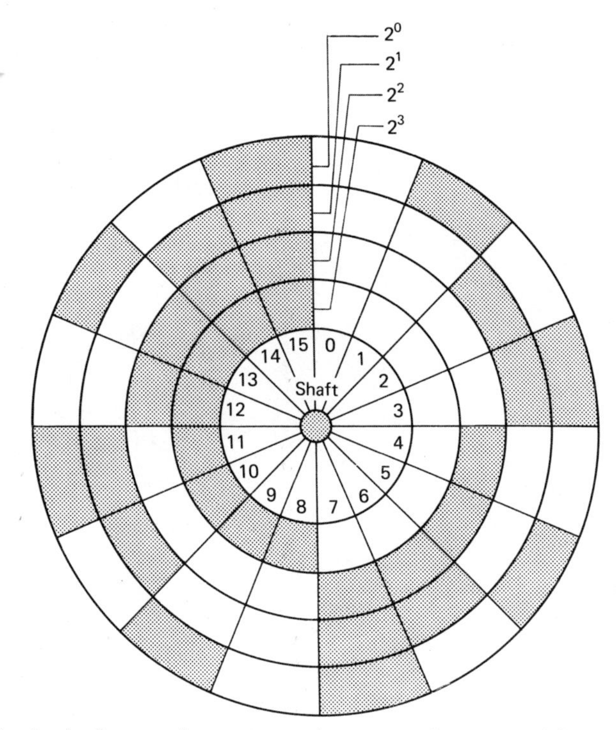

FIGURE 4-2 A shaft encoder pattern that uses the natural binary code and is prone to serious reading errors.

and the code for 8_{10} is 1000; thus four bits change simultaneously when the binary number changes from 7_{10} to 8_{10}. Unit-distance codes are used in instrumentation and, in particular, in *shaft encoders* for measurement of angular displacement and in *linear encoders* for measurement of linear displacement.

An example of a 4-bit *natural binary coded* pattern of a shaft encoder is shown in Fig. 4-2. Each of the four concentric rings, or *zones*, represents a binary weight, and the dark and white areas represent 1s and 0s, respectively. The outmost zone represents the least significant bit (lsb). Four *transducers* (not shown in Fig. 4-2), mechanical or optical, are located along a radius to detect the dark and white areas representing bits of a binary number. (Additional concentric zones and transducers would increase the *resolution* of the encoder.)

Consider, however, the ambiguity arising from the nonzero size and the alignment errors of the transducers when two or more bits are changing simultaneously as, for example, between 0111_2 and 1000_2. If, for instance, the most significant bit (msb) is detected as 1 while the other three bits are still 1, we obtain $1111_2 = 15_{10}$—instead of $0111_2 = 7_{10}$ or $1000_2 = 8_{10}$, either of which would be correct since the encoder is *between* these two positions. Thus the natural binary pattern shown in Fig. 4-2 can lead to serious reading errors.

4-3.1 The Gray Code

The *Gray code*, shown in Table 4-3, is a *unit-distance code* that is commonly used in instrumentation. Note that only one bit changes between *any* two consecutive numbers, including the change between the codes for 15 and 0.

TABLE 4-3
The Gray Code

Decimal	Gray
0	0000
1	0001
2	0011
3	0010
4	0110
5	0111
6	0101
7	0100
8	1100
9	1101
10	1111
11	1110
12	1010
13	1011
14	1001
15	1000

Axis of symmetry (between 7 and 8)

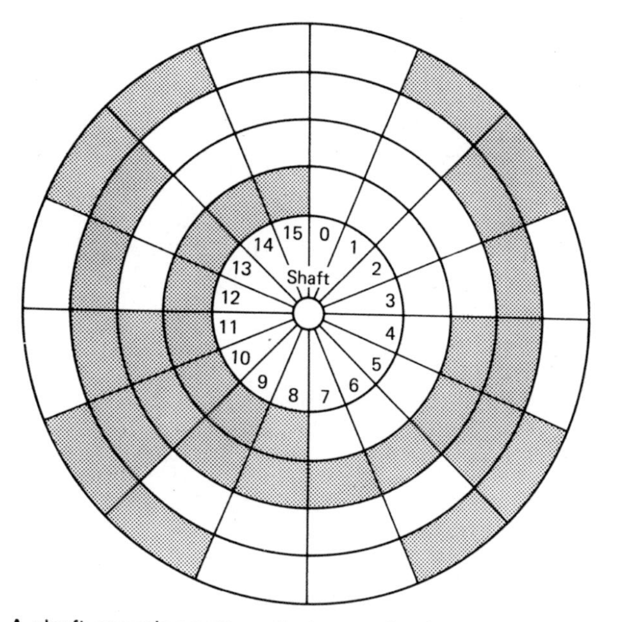

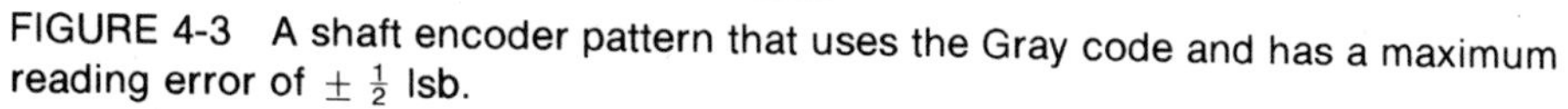
FIGURE 4-3 A shaft encoder pattern that uses the Gray code and has a maximum reading error of $\pm \frac{1}{2}$ lsb.

Also note that an axis of symmetry exists for the last three bits that gives this code the name *reflected binary*.

The resulting shaft-encoder pattern is shown in Fig. 4-3. If we compare Figs. 4-2 and 4-3, we can see that the Gray code is superior for angular measurements.

The Gray code is an *unweighted* code, that is, no relationship exists between the position of a bit and its binary weight. Since the output of Gray-coded encoders is often used for subsequent arithmetic manipulation, it is important to have methods for converting from Gray to binary and from binary to Gray.

4-3.2 Gray-to-Binary and Binary-to-Gray Conversions

Before describing the conversion rules, we first introduce the concept of *modulo-2 addition:* Modulo-2 addition is binary addition disregarding the carry; the symbol for modulo-2 addition is $\oplus$. Thus, $0 \oplus 0 = 0$, $0 \oplus 1 = 1 \oplus 0 = 1$, and $1 \oplus 1 = 0$ (carry disregarded).

4-3.2.1 A Gray-Code to Binary-Code Conversion Procedure

The rules of this Gray-to-binary code conversion procedure are the following:

1. Write down the number in Gray code.
2. The most significant bit, msb, of the binary number equals the msb of the Gray-coded number.

3. Add modulo-2 the msb of the binary number to the next significant bit of the Gray-coded number to obtain the next binary bit.
4. Repeat step 3 until all bits of the Gray-coded number have been added modulo-2. The resulting number is the binary equivalent of the Gray-coded number.

EXAMPLE 4-7

a. Convert the Gray-coded number 1101101 to a binary number.
b. Find its decimal equivalent.

Solution

a.

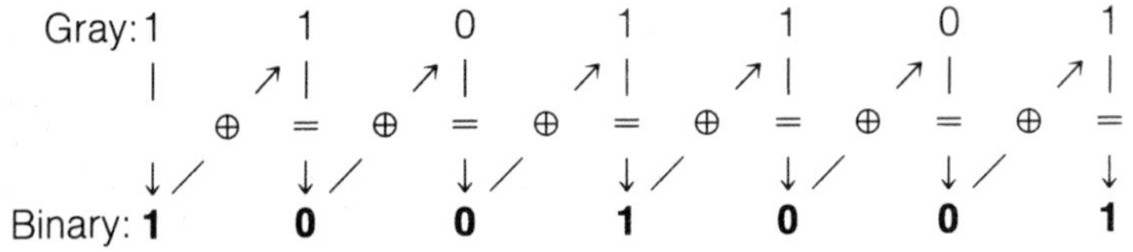

Gray: 1 1 0 1 1 0 1

$\downarrow$ $\oplus$ = $\oplus$ = $\oplus$ = $\oplus$ = $\oplus$ = $\oplus$ = $\downarrow$

Binary: **1** **0** **0** **1** **0** **0** **1**

b. $1001001_2 = 1 \times 2^6 + 1 \times 2^3 + 1 = \mathbf{73_{10}}$.

4-3.2.2 Two Binary-Code to Gray-Code Conversion Procedures

The rules of the first binary-to-Gray code conversion procedure are the following:

1. Write down the number in binary code.
2. The msb of the Gray-coded number equals the msb of the binary number.
3. Starting with the msb of the binary number, add modulo-2 the next significant binary bit. This generates the next significant bit of the Gray-coded number.
4. Repeat step 3 until all binary bits have been added modulo-2. The resulting number is the Gray-coded equivalent of the binary number.

EXAMPLE 4-8

Convert the binary number 1001001 to a Gray-coded number.

Solution

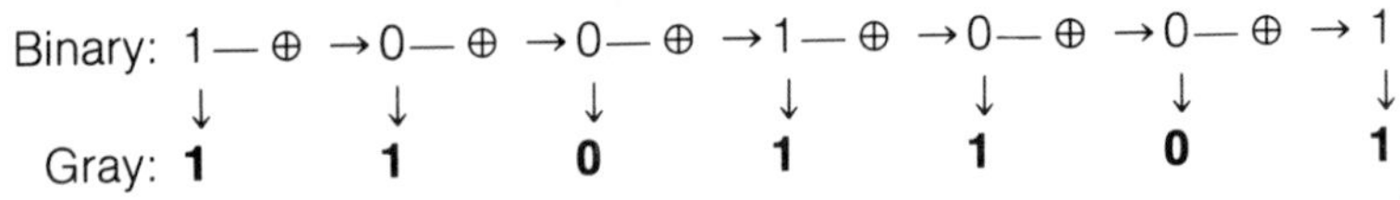

Binary: 1 — $\oplus$ → 0 — $\oplus$ → 0 — $\oplus$ → 1 — $\oplus$ → 0 — $\oplus$ → 0 — $\oplus$ → 1

↓ ↓ ↓ ↓ ↓ ↓ ↓

Gray: **1** **1** **0** **1** **1** **0** **1**

The rules given in 1 through 4 above are also equivalent to the following alternative binary-to-Gray code conversion procedure:

1. Write down the number in binary code.
2. Write down the same number shifted one position to the right.
3. Eliminate the least significant bit (lsb) of the number written down in step 2.
4. Add modulo-2 the numbers obtained in steps 1 and 3.

TABLE 4-4
The Excess-3 Gray Code

Decimal	Excess-3 Gray
0	0010
1	0110
2	0111
3	0101
4	0100
5	1100
6	1101
7	1111
8	1110
9	1010

EXAMPLE 4-9

Convert the binary number 1001001 to a Gray-coded number.

Solution

1001001	binary number
⊕ 100100~~1~~	binary number shifted right and lsb eliminated
1101101	Gray-coded number

4-3.3 The Excess-3 Gray Code

The Gray code was introduced in Sec. 4-3.1 as a unit-distance code in which consecutive coded numbers differ only in one bit. We have also pointed out that the code for 0 is one unit distant from the code for 15. However, in many applications it would be desirable to have a code that is BCD as well as having the property of unit distance. Such a code is shown in Table 4-4 and is referred to as the *excess-3 Gray code*. We can see in Table 4-4 that the codes for 0 and 9 differ only in one bit, and so do all codes for consecutive numbers. Also note the axis of symmetry for the last three bits, indicated by the broken line in Table 4-4.

Outputs of linear or angular encoders may be coded in the excess-3 Gray code to obtain multidigit BCD numbers. This is illustrated in Ex. 4-10 below.

EXAMPLE 4-10

Encode the decimal number 497 in the excess-3 Gray Code.

Solution

From Table 4-4 we derive the following:

Decimal:	4	9	7
	↓	↓	↓
Excess-3 Gray:	**0100**	**1010**	**1111**

4-4 ERROR-DETECTING CODES

Although transmission of information in digital form is more immune to noise than transmission of analog information, the need for error checking does exist. Errors may occur for a variety of reasons such as marginally operating equipment, power-line transients, extraneous noise signals, intermittent faults in the transmitting or receiving equipment, or fading in the transmission path.

Consider first the use of binary numbers, shown in Table 3-2 (page 35). If, for example, an extraneous noise signal changes the second bit from the left in the number 1001 ($= 9_{10}$) from 0 to 1, we obtain 1101 ($= 13_{10}$) and there is no way for us to know that this number is in error. On the other hand, had we used the 8-4-2-1 BCD code, shown in Table 4-1 (page 51), we would have immediately noticed the error because 1101 is not allowed in the 8-4-2-1 BCD code. Thus the 6 redundant states of a 4-bit BCD code enable us to detect *some* errors, but not others.

In general, redundant states, or additional bits, are necessary to detect errors. It can be shown that one additional bit is required to detect one error, two additional bits are required to detect two simultaneous errors, and so on. Also, two additional bits are required to detect the *location* of one error, which enables us to make a correction.

4-4.1 Parity Error Check

In a parity error check, typically 6 to 10 data bits (information bits) are assembled into a *data word* (*information word*) that is followed by a *parity bit*. This results in a *transmitted word* (recorded word) that is one bit longer than the data word. (Circuits for parity generation and detection are discussed in Sec. 6-2.)

Two different parity checks are in common use: the *odd parity* and the *even parity*. For *odd parity*, the additional parity bit is chosen such that the *sum* of all 1s in the data word *and* in the parity bit is *odd*. This is illustrated in Ex. 4-11 below.

EXAMPLE 4-11

Find the odd parity bits for the following data words:
a. 0011, **b**. 0111.

Solution

a. The data word 0011 has an *even* number of 1s. Hence the odd parity bit is 1.

b. The data word 0111 has an *odd* number of 1s. Hence the odd parity bit is **0**.

For *even parity*, the additional parity bit is chosen such that the *sum* of all 1s in the data word *and* in the parity bit is *even*. This is illustrated in Ex. 4-12 below.

EXAMPLE 4-12

Find the even parity bits for the data words of Ex. 4-11.

Solution

a. The even parity bit is **0**.
b. The even parity bit is **1**.

TABLE 4-5
8-4-2-1 BCD Numbers and Their Odd and Even Parity Bits

Decimal	8-4-2-1 BCD	Odd Parity	Even Parity
0	0000	1	0
1	0001	0	1
2	0010	0	1
3	0011	1	0
4	0100	0	1
5	0101	1	0
6	0110	1	0
7	0111	0	1
8	1000	0	1
9	1001	1	0

Table 4-5 shows the odd and even parity bits for the 8-4-2-1 BCD numbers 0 through 9.

Odd parity is more commonly used than even parity. This is because the latter would not recognize a fault condition in which all 0s have been transmitted.

Parity checks are independent of the position of the binary point. Also, they do not detect errors in which 2 bits have been changed *simultaneously*. However, the probability of a 1-bit error is often very small, and in such cases the probability of a 2-bit error is negligible.

In a *digital computer* data are constantly processed and transmitted between parts of the computer, as well as to and from *peripheral devices* such as magnetic tape, magnetic disk, printer, and so on; also, some peripheral devices are more prone to introduce errors than the computer itself. Thus, in such cases 2-bit errors are often *not* negligible and detection of these errors is required.

4-4.2 Block Parity Error Check

The *block parity* error check can detect *2-bit* errors—it is commonly used in peripheral devices. Figure 4-4 shows a *block* of magnetic tape with 9 rows and 7 columns that includes block parity error checking.

The black circles in Fig. 4-4 represent magnetization in one direction for binary 1, and the open circles represent magnetization in the other direction for binary 0. In the block of 9×7 circles, data are contained in the rectangle of 8 rows $\times$ 6 columns, and the remainder of the block contains a *parity row* and a *parity column*. Thus each row has an odd parity bit, the *rightmost* bit, and each column has an odd parity bit, the *bottom* bit.

Each row or column by itself can detect single errors. For example, in the first row of Fig. 4-4 we have the data word 010110 followed by an odd parity bit 0. Any single bit change in the data word 010110 or in the parity bit 0 would indicate an error. A similar situation applies to the columns.

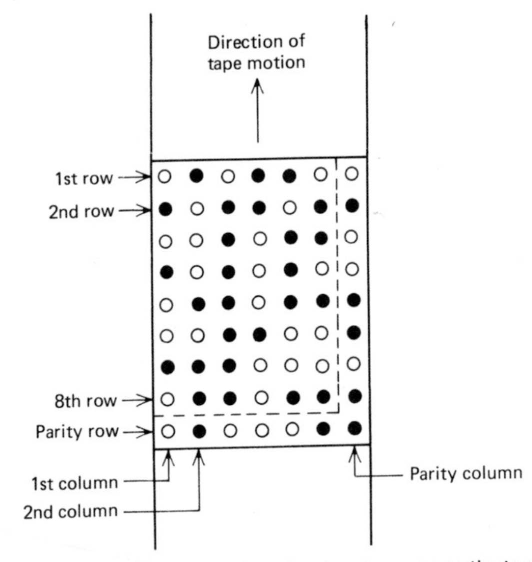

FIGURE 4-4 Block parity check of a magnetic tape.

Consider, however, an error where the third *and* fourth bits changed in the first row of Fig. 4-4, and the data word 010110 became 011010 followed by the unchanged odd parity bit 0. The odd parity bit 0 of the first row would still be correct; however, the odd parity bits of the third and fourth columns would indicate that a 2-bit error had occurred.

4-5 ALPHANUMERIC CODES

Thus far we have discussed codes that can represent only numbers. However, transmission of digital information involves also the characters of the alphabet and punctuation marks, as well as *control characters* that are used for peripheral devices, for example, "carriage return" or "end of tape."

This section describes four *alphanumeric codes*: The early Baudot and Hollerith codes, and also the ASCII and EBCDIC codes that receive widespread use in today's digital computers.

*4-5.1 The 5-Bit Baudot Code

The 5-bit *Baudot code* is named after its French inventor Jean Baudot. It is employed in data communications using a typewriter-like device called a *teletypewriter*. The code is divided into two groups: "*letters*" and "*figures*." The 5 bits allow $2^5 = 32$ different code combinations. Of these 32 combinations, one is used to select "letters" and another one to select "figures."

*Optional material.

TABLE 4-6
Assignment of Bit Combinations for the Baudot Code

			$b_5b_4 = 00$		$b_5b_4 = 01$		$b_5b_4 = 10$		$b_5b_4 = 11$	
b_3	b_2	b_1	Letters	Figures	Letters	Figures	Letters	Figures	Letters	Figures
0	0	0			C.R.	C.R.	T	5	O	9
0	0	1	E	3	D	$	Z	"	B	?
0	1	0	Line Feed		R	4	L	)	G	&
0	1	1	A	—	J	'	W	2	Figures	
1	0	0	Space		N	.	H	=	M	.
1	0	1	S	Bell	F	!	Y	6	X	/
1	1	0	I	8	C	:	P	0	V	;
1	1	1	U	7	K	(	Q	1	Letters	

The assignment of bit combinations is shown in Table 4-6. Thus, if the code for "letters" has been selected, then all codes following this "letters" code are interpreted as *letters*. Conversely, if the code for "figures" has been selected, then all codes following this "figures" code are interpreted as *figures*.

The use of Table 4-6 is illustrated in Ex. 4-13, which follows.

EXAMPLE 4-13

What are the possible interpretations of the Baudot code $b_5b_4b_3b_2b_1 = 10101$?

Solution

a. If the code 10101 appears after the code "letters" has been selected, then it is interpreted as the letter **Y**.

b. If the code 10101 appears after the code "figures" has been selected, then it is interpreted as the numeral **6**.

*4-5.1.1 *Perforated Paper Tape*

Information using the Baudot code is often encoded on *perforated paper tape.* Each character is represented by a serial string of 0s and 1s following a *synchronizing* signal and ending with a *stop* indication. Figure 4-5 shows such a perforated tape, where the black circles indicate holes in the tape (= 1s) and

Figures	—	?	:	$	3	!	&	=	8	'	(	)	.	.	9	Ø	1	4	Bell	5	7	;	2	/	6	"						
Letters	A	B	C	D	E	F	G	H	I	J	K	L	M	N	O	P	Q	R	S	T	U	V	W	X	Y	Z	Blank	L.F.	Space	C.R.	Figures	Letters
b_1	●	●	○	●	●	●	○	○	○	●	●	○	○	○	○	○	●	○	●	○	●	○	●	●	●	●	○	○	○	○	●	●
b_2	●	○	●	○	○	○	●	○	●	●	●	●	○	○	○	●	●	●	○	○	●	●	●	○	○	○	○	●	○	○	●	●
Guide holes	•	•	•	•	•	•	•	•	•	•	•	•	•	•	•	•	•	•	•	•	•	•	•	•	•	•	•	•	•	•	•	•
b_3	○	○	●	○	○	●	○	●	●	○	●	○	●	●	○	●	●	○	●	○	●	●	○	●	●	○	○	○	●	○	○	●
b_4	○	●	●	●	○	●	●	○	○	●	●	○	●	●	●	○	○	●	○	○	○	●	○	●	○	○	○	○	○	●	●	●
b_5	○	●	○	○	○	○	●	●	○	○	○	●	●	○	●	●	●	○	○	●	○	●	●	●	●	●	○	○	○	○	●	●

FIGURE 4-5 Baudot code as represented on perforated paper tape.

*Optional material.

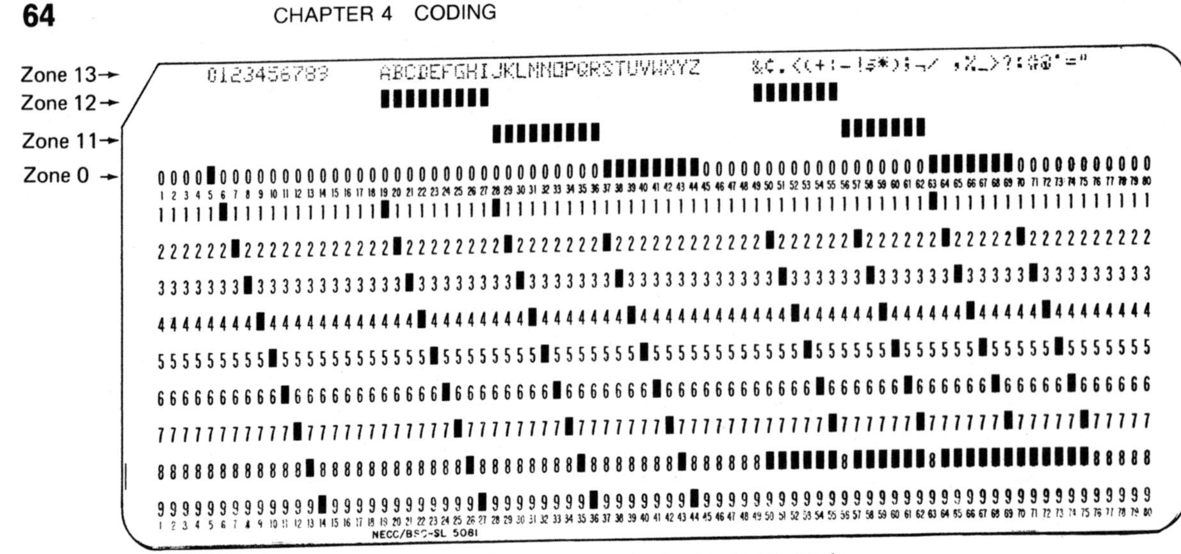

FIGURE 4-6 Punched card with Hollerith code.

the open circles indicate the absence of holes (= 0s). Data can thus be assembled in advance of transmission. A *tape reader* then senses the holes and the absence of holes and generates the appropriate code for transmission, which is usually at a rate of 10 characters per second.

Note that, as can be seen on the perforated paper tape in Fig. 4-5, the 5-bit Baudot code does *not* have a parity check bit.

*4-5.2 The Hollerith Code

This code is used with *punched cards* as input and output medium for digital computers. As shown in Fig. 4-6, a punched card has 80 columns and each column has 12 punching positions: one each for the digits 1 through 9, and one each for *zones* 0, 11, and 12 (zone 13 is reserved for printing).

A digit is recorded by punching a single hole in the corresponding digit position. A letter is a combination of one zone punch and one digit punch in the desired column; for example, the letter *A* is made up of punches in zone 12 and digit 1. Special characters are represented by one, two, or three punches; for example, the symbol + is made up of punches in zone 12 and in digits 6 and 8.

The code assignments are indicated in Fig. 4-6 by black rectangular marks, which represent holes. The code assignments are also described in Table 4-7.

4-5.3 The ASCII Code

The *American Standard Code for Information Interchange* (*ASCII*) uses 7 bits and hence is able to represent $2^7 = 128$ characters. Parity check may be provided by adding an eighth bit.

*Optional material.

TABLE 4-7
Hollerith Code Assignment

No.	Punch in Row	Let-ter	Punch in Rows	Let-ter	Punch in Rows	Let-ter	Punch in Rows	Char-acter	Punch in Rows	Char-acter	Punch in Rows	Char-acter	Punch in Rows	Char-acter	Punch in Rows
0	0							&	12	—	11	/	0,1		
1	1	A	12,1	J	11,1									:	2,8
2	2	B	12,2	K	11,2	S	0,2	¢	12,2,8	!	11,2,8			#	3,8
3	3	C	12,3	L	11,3	T	0,3	.	12,3,8	$	11,3,8	,	0,3,8	@	4,8
4	4	D	12,4	M	11,4	U	0,4	<	12,4,8	*	11,4,8	%	0,4,8	'	5,8
5	5	E	12,5	N	11,5	V	0,5	(	12,5,8	)	11,5,8	—	0,5,8	=	6,8
6	6	F	12,6	O	11,6	W	0,6	+	12,6,8	;	11,6,8	>	0,6,8	"	7,8
7	7	G	12,7	P	11,7	X	0,7	|	12,7,8	⊐	11,7,8	?	0,7,8		
8	8	H	12,8	Q	11,8	Y	0,8								
9	9	I	12,9	R	11,9	Z	0,9								

TABLE 4-8
The 7-Bit ASCII Code

				$b_7b_6 = 00$		$b_7b_6 = 01$		$b_7b_6 = 10$		$b_7b_6 = 11$	
b_4	b_3	b_2	b_1	$b_5 = 0$	$b_5 = 1$	$b_5 = 0$	$b_5 = 1$	$b_5 = 0$	$b_5 = 1$	$b_5 = 0$	$b_5 = 1$
0	0	0	0			SP	0	@	P	‘	p
0	0	0	1			!	1	A	Q	a	q
0	0	1	0			"	2	B	R	b	r
0	0	1	1			#	3	C	S	c	s
0	1	0	0			$	4	D	T	d	t
0	1	0	1			%	5	E	U	e	u
0	1	1	0			&	6	F	V	f	v
0	1	1	1	Control		'	7	G	W	g	w
1	0	0	0	Characters		(	8	H	X	h	x
1	0	0	1			)	9	I	Y	i	y
1	0	1	0			*	:	J	Z	j	z
1	0	1	1			+	;	K	[	k	{
1	1	0	0			,	<	L	\	l	|
1	1	0	1			–	=	M	]	m	}
1	1	1	0			.	>	N	∧	n	~
1	1	1	1			/	?	O	—	o	DEL

The code assignments are described in Table 4-8. A character is defined by 3 *high-order* bits b_7, b_6, and b_5, and by 4 *low-order* bits b_4, b_3, b_2, and b_1. All *control characters* have $b_7 = 0$ and $b_6 = 0$, and *alphanumeric characters* have $b_7 = 0$ and $b_6 = 1$, or $b_7 = 1$ and $b_6 = 0$, or $b_7 = 1$ and $b_6 = 1$. Thus the two most significant bits determine whether a code represents a control character or an alphanumeric character. Note that different peripheral devices require different control characters.

The use of the ASCII code is illustrated in Ex. 4-14, which follows.

EXAMPLE 4-14 What is the ASCII code for the word ASCII?

Solution From Table 4-8 we get the ASCII code:
1000001 1010011 1000011 1001001 1001001

4-5.4 The EBCDIC Code

The *Extended Binary Coded Decimal Interchange Code* (*EBCDIC*) is shown in Table 4-9. It is an 8-bit code; parity check may be provided by adding a ninth bit.

All *control characters* have $b_0 = 0$ and $b_1 = 0$, and *alphanumeric characters* have $b_0 = 0$ and $b_1 = 1$, or $b_0 = 1$ and $b_1 = 0$, or $b_0 = 1$ and $b_1 = 1$. Thus, bits b_0 and b_1 determine whether a code represents a control character or an alphanumeric character. Different peripheral devices require different control characters—these are specified separately.

TABLE 4-9
The 8-Bit EBCDIC Code

<table>
<tr><th></th><th></th><th></th><th></th><th colspan="4">$b_0 b_1 = 00$</th><th colspan="4">$b_0 b_1 = 01$</th><th colspan="4">$b_0 b_1 = 10$</th><th colspan="4">$b_0 b_1 = 11$</th></tr>
<tr><th></th><th></th><th></th><th></th><th colspan="4">$b_2 b_3$</th><th colspan="4">$b_2 b_3$</th><th colspan="4">$b_2 b_3$</th><th colspan="4">$b_2 b_3$</th></tr>
<tr><th>b_4</th><th>b_5</th><th>b_6</th><th>b_7</th><th>00</th><th>01</th><th>10</th><th>11</th><th>00</th><th>01</th><th>10</th><th>11</th><th>00</th><th>01</th><th>10</th><th>11</th><th>00</th><th>01</th><th>10</th><th>11</th></tr>
<tr><td>0</td><td>0</td><td>0</td><td>0</td><td colspan="4" rowspan="16">Control Characters</td><td>SP</td><td>&</td><td>-</td><td></td><td></td><td></td><td></td><td></td><td></td><td></td><td></td><td>0</td></tr>
<tr><td>0</td><td>0</td><td>0</td><td>1</td><td></td><td></td><td>/</td><td></td><td>a</td><td>j</td><td></td><td></td><td>A</td><td>J</td><td></td><td>1</td></tr>
<tr><td>0</td><td>0</td><td>1</td><td>0</td><td></td><td></td><td></td><td></td><td>b</td><td>k</td><td>s</td><td></td><td>B</td><td>K</td><td>S</td><td>2</td></tr>
<tr><td>0</td><td>0</td><td>1</td><td>1</td><td></td><td></td><td></td><td></td><td>c</td><td>l</td><td>t</td><td></td><td>C</td><td>L</td><td>T</td><td>3</td></tr>
<tr><td>0</td><td>1</td><td>0</td><td>0</td><td></td><td></td><td></td><td></td><td>d</td><td>m</td><td>u</td><td></td><td>D</td><td>M</td><td>U</td><td>4</td></tr>
<tr><td>0</td><td>1</td><td>0</td><td>1</td><td></td><td></td><td></td><td></td><td>e</td><td>n</td><td>v</td><td></td><td>E</td><td>N</td><td>V</td><td>5</td></tr>
<tr><td>0</td><td>1</td><td>1</td><td>0</td><td></td><td></td><td></td><td></td><td>f</td><td>o</td><td>w</td><td></td><td>F</td><td>O</td><td>W</td><td>6</td></tr>
<tr><td>0</td><td>1</td><td>1</td><td>1</td><td></td><td></td><td></td><td></td><td>g</td><td>p</td><td>x</td><td></td><td>G</td><td>P</td><td>X</td><td>7</td></tr>
<tr><td>1</td><td>0</td><td>0</td><td>0</td><td></td><td></td><td></td><td></td><td>h</td><td>q</td><td>y</td><td></td><td>H</td><td>Q</td><td>Y</td><td>8</td></tr>
<tr><td>1</td><td>0</td><td>0</td><td>1</td><td></td><td></td><td></td><td></td><td>i</td><td>r</td><td>z</td><td></td><td>I</td><td>R</td><td>Z</td><td>9</td></tr>
<tr><td>1</td><td>0</td><td>1</td><td>0</td><td>¢</td><td>!</td><td></td><td>:</td><td></td><td></td><td></td><td></td><td></td><td></td><td></td><td></td></tr>
<tr><td>1</td><td>0</td><td>1</td><td>1</td><td>.</td><td>$</td><td>,</td><td>#</td><td></td><td></td><td></td><td></td><td></td><td></td><td></td><td></td></tr>
<tr><td>1</td><td>1</td><td>0</td><td>0</td><td><</td><td>*</td><td>%</td><td>@</td><td></td><td></td><td></td><td></td><td></td><td></td><td></td><td></td></tr>
<tr><td>1</td><td>1</td><td>0</td><td>1</td><td>(</td><td>)</td><td>_</td><td>'</td><td></td><td></td><td></td><td></td><td></td><td></td><td></td><td></td></tr>
<tr><td>1</td><td>1</td><td>1</td><td>0</td><td>+</td><td>;</td><td>></td><td>=</td><td></td><td></td><td></td><td></td><td></td><td></td><td></td><td></td></tr>
<tr><td>1</td><td>1</td><td>1</td><td>1</td><td>|</td><td>¬</td><td>?</td><td>"</td><td></td><td></td><td></td><td></td><td></td><td></td><td></td><td></td></tr>
</table>

Note that the numbering conventions for the bit positions within a character are different for the 7-bit ASCII code and for the EBCDIC code. The conventions are:

ASCII 7654321 → lsb
EBCDIC 01234567 → lsb

EXAMPLE 4-15

What is the EBCDIC code for the word EBCDIC?

Solution

From Table 4-9 we get the EBCDIC code:
11000101 11000010 11000011 11000100 11001001 11000011

SUMMARY

After a brief discussion of the need for coding, we introduced two weighted BCD codes followed by the excess-3 BCD code. Unit-distance codes and their applications were treated in a subsequent section that introduced the Gray and excess-3 Gray codes. The parity error check was discussed in connection with error-detecting codes. Finally, four alphanumeric codes used in communications and computer applications were presented.

SELF-EVALUATION QUESTIONS

1. Prepare a table showing the 8-4-2-1 BCD code.
2. Name two weighted BCD codes.
3. Discuss the difference between encoding and decoding.
4. What rules have to be obeyed when adding two 8-4-2-1 BCD numbers?
5. What are the redundant states in the 8-4-2-1 BCD code?
6. What are the redundant states in the 2-4-2-1 BCD code?
7. What is a self-complementing code?
8. How do you obtain an excess-3 coded number?
9. What is a unit-distance code? Where is it used?
10. What is modulo-2 addition? What is the result of $1 \oplus 0 \oplus 1 \oplus 1 \oplus 0 \oplus 1$?
11. What is the purpose of a parity check?
12. Describe the block parity error check.
13. What are alphanumeric codes? Name four of these codes.

*14. How do you distinguish between letters and figures in a Baudot code?

*15. What do the binary words 11011 and 11111 signify in the Baudot code?

16. What is the function of the control characters in the ASCII and EBCDIC codes?

PROBLEMS

4-1. Encode the following decimal numbers into 8-4-2-1 BCD:

a. 957, **b**. 3471, **c**. 892.

4-2. What are the 8-4-2-1 BCD equivalents of the following decimal numbers?

a. 846, **b**. 2360, **c**. 781.

4-3. Decode the following 8-4-2-1 BCD numbers into decimal:

a. 0101 1000, **b**. 1001 0011 0101,
c. 0011 0100 0111 0001.

4-4. What are the decimal equivalents of the following 8-4-2-1 BCD numbers?

a. 0010 0011 1000, **b**. 0001 0011 1001,
c. 0100 0111 0000.

4-5. Add the following numbers using 8-4-2-1 BCD addition:

a. 3 + 4, **b**. 6 + 9, **c**. 46 + 27.

4-6. Encode the following decimal numbers into 2-4-2-1 BCD:

a. 15, **b**. 268, **c**. 973.

*Optional material.

4-7. Decode the following 2-4-2-1 BCD numbers into decimal:

a. 0011 1100, **b.** 1101 0010 1111, **c.** 1100 1011 0011.

4-8. What are the decimal equivalents of the following 2-4-2-1 BCD numbers?

a. 0001 0100 1100, **b.** 1110 0010 1011,
c. 0011 1110 1111.

4-9. What are the decimal equivalents of the following excess-3 BCD numbers?

a. 1100 0100, **b.** 0111 1010 0101, **c.** 1000 0011 1011.

4-10. What are the decimal equivalents of the following excess-3 BCD numbers?

a. 0100 1100, **b.** 0110 0111 1010, **c.** 1011 0011 0101.

4-11. Encode the following decimal numbers into excess-3 BCD:

a. 97, **b.** 852, **c.** 643.

4-12. Encode the decimal number 582 into excess-3 BCD.

4-13. What is the result of $1 \oplus 0 \oplus 1 \oplus 1$?

4-14. What is the result of $1 \oplus 0 \oplus 1 \oplus 1 \oplus 0 \oplus 1 \oplus 0$?

4-15. Convert the Gray-coded number 11001101 to binary.

4-16. Convert the binary number 10001001 to a Gray-coded number.

4-17. What are the odd-parity bits for the excess-3 BCD numbers 0 through 9?

***4-18.** What is the Baudot code for the word CAT?

***4-19.** You find a punched card with the following Hollerith code punched in successive columns: 12,3 11,6 11,4 11,7 0,4 0,3 12,5 11,9. What is the message?

4-20. What is the ASCII code for the words 3 DOGS?

4-21. Decode the following EBCDIC message: 10000101 10010101 10000100.

*Optional material.

CHAPTER 5

Boolean Algebra and Simplification Methods

Instructional Objectives

This chapter lays the foundations for the analysis and synthesis of combinational logic circuits for use in later chapters. After you read it you should be able to:

1. Draw truth tables from logic propositions.
2. Simplify Boolean expressions using theorems of Boolean algebra.
3. Use the three logic operators: AND, OR, and NOT.
4. Design combinational logic circuits.
5. Given a statement or a truth table, find the standard sum-of-products form and the standard product-of-sums form.
6. Draw Karnaugh maps with up to 4 variables.
7. Use Karnaugh maps for the simplification of Boolean functions.
8. Use "don't care conditions" with a Karnaugh map for the simplification of Boolean functions.

5-1 OVERVIEW

Boolean algebra provides a systematic approach for understanding and designing digital systems that use binary logic devices such as switches, relays, and logic gates. It is an algebra that is tailored for binary variables that can be ON or OFF, OPEN or CLOSED, TRUE or FALSE, or 0 or 1. In this chapter these variables are first applied to truth tables and then to basic AND, OR, and NOT operations.

As is also the case in ordinary algebra, Boolean algebra is founded on a set of basic assumptions called axioms (postulates), which are presented next. Following this, the basic rules (theorems) of Boolean algebra are introduced and proven.

With the aid of Boolean algebra we can design logic circuits to suit given specifications. As we see later, simpler Boolean expressions lead to simpler hardware, and it is for this reason that we introduce and discuss simplification methods. Following this, we describe for later use various techniques and forms that are utilized in handling Boolean expressions.

The chapter is concluded by a discussion of the Karnaugh map, which is a graphical tool used for the simplification of Boolean expressions with up to six variables: An introduction of Karnaugh maps and their basic properties is followed by a detailed description of their use for simplification.

5-2 BOOLEAN ALGEBRA

In logic, language *phrases* are combined into sentences and the truth of these sentences is investigated. In order for a sentence to be suitable for logic analysis it must contain an inherent *truth value*, that is, it must be either *TRUE* or *FALSE*. For example, "Bill studies digital logic" is a sentence that contains an inherent truth value since it can be answered by "true" or "false." Such sentences are called *logic propositions* (*logic statements*).

Simple phrases may be combined by using *logic operators* to form more complex phrases, for example, "Bill AND Dick *both* study digital logic." The word AND in the preceding sentence is a logic operator. The logic proposition is TRUE if Bill and Dick *both* study digital logic. It is FALSE if Bill studies digital logic but Dick does not study it, or if Dick studies digital logic and Bill does not study it, or if neither studies digital logic. Thus, at the expense of many words we are able to describe the complete truth about this logic proposition.

Boole, a nineteenth-century mathematician, developed an algebra (*Boolean algebra*) that substituted the phrases by algebraic expressions (*Boolean expressions*). He also developed *theroems* that enable us to handle Boolean expressions, as well as *Boolean functions*, in a way that is somewhat similar to handling expressions and functions in ordinary algebra. In 1938 Shannon recognized the applicability of Boolean algebra to the analysis of *switching functions*. Today Boolean algebra is extensively used in the analysis and synthesis of logic circuits.

5-3 TRUTH TABLES

In Boolean algebra there are two *constants*: 0 and 1. Any number of *variables* may be used and each of these *must assume the values of the constants*. In the following, the variable of an open switch is equal to the constant 0 and the variable of a closed switch to the constant 1.

EXAMPLE 5-1

Two switches represented by variables A and B are connected in *series*. Write down all combinations of the states of the two switches.

Solution

Both switches open: $A = 0, B = 0$
Switch A open, switch B closed: $A = 0, B = 1$
Switch A closed, switch B open: $A = 1, B = 0$
Both switches closed: $A = 1, B = 1$

5-3.1 Truth Tables for One Variable

Example 5-1 above illustrated the statement: "The variables must assume the values of the constants," and it listed all four states of the switches in four sentences. In Boolean algebra such sentences are replaced by a *truth table*. A truth table is a systematic listing of all combinations of the states of the variables, as well as of the corresponding outputs, assigning one row to each combination. A truth table for one variable has two rows, as illustrated in Ex. 5-2 below.

EXAMPLE 5-2

Variable A is used to represent a switch. Draw a truth table that lists all states of A and the resulting outputs; also, draw a circuit diagram.

Solution

The truth table for 1 variable has two rows, as shown in Fig. 5-1*a*. Assuming that the switch is used in the circuit of Fig. 5-1*b*, the light is OFF when switch A is OPEN and the light is ON when switch A is CLOSED. The state of the light, that is the *Output* in the truth table, is shown in the right-hand column of Fig. 5-1*a*. Furthermore, we can substitute the binary constants 0 and 1, respectively, for OPEN and CLOSED, as well as for OFF and ON—this results in the truth table of Fig. 5-1*c*.

5-3.2 Truth Tables for Two Variables

A truth table for 2 variables has $2^2 = 4$ possible combinations. In order to systematically tabulate all four combinations of the variables, it is convenient to

Switch A	*Light*
OPEN	OFF
CLOSED	ON

(*a*) Truth table

(*b*) Circuit described by the truth table of (*a*)

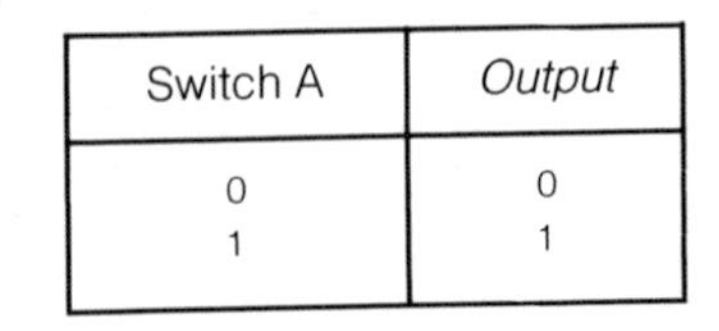

Switch A	*Output*
0	0
1	1

(*c*) Truth table using binary constants

FIGURE 5-1 Illustration of 1-variable truth tables.

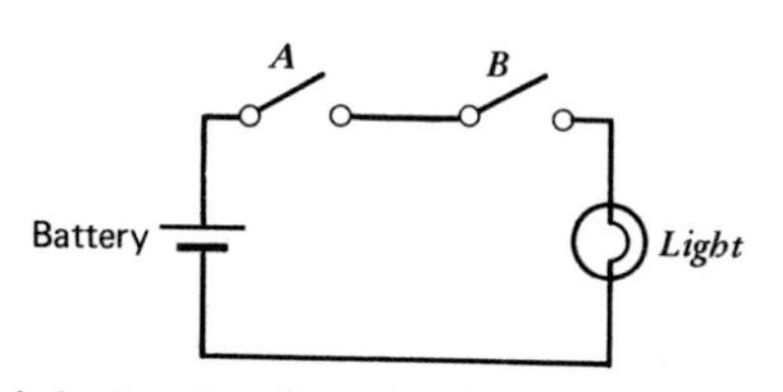

(*a*) A circuit with two switches in series

A	*B*	*Light*
OPEN	OPEN	OFF
OPEN	CLOSED	OFF
CLOSED	OPEN	OFF
CLOSED	CLOSED	ON

(*b*) Truth table

A	*B*	*OUT*
F	F	F
F	T	F
T	F	F
T	T	T

(*c*) Truth table

A	*B*	*OUT*
0	0	0
0	1	0
1	0	0
1	1	1

(*d*) Truth table

FIGURE 5-2 Illustration of two-variable truth tables.

list the rows according to ascending binary numbers, where each number represents a different combination. This is illustrated in Ex. 5-3 below.

EXAMPLE 5-3

Two switches are connected in series as shown in Fig. 5-2*a*. Establish a truth table that shows when the light is ON as a function of the states of the switches.

Solution

We have 2 variables, *A* and *B*; thus the truth table has $2^2 = 4$ rows. Each row describes a different combination of the states of the switches, OPEN or CLOSED. The resulting truth table is shown in Fig. 5-2*b*.

In order to make the truth table more general, we substitute the word "OPEN" by "FALSE" or "F" for short, and the word "CLOSED" by "TRUE" or "T" for short. We also substitute "OFF" by "FALSE" or "F," and "ON" by "TRUE" or "T." The resulting truth table shown in Fig. 5-2*c* provides the same information as Fig. 5-2*b*: The light is TRUE (ON) only when both switches *A* and *B* are TRUE (CLOSED), and the light is FALSE (OFF) for all other combinations of switches.

We can also make a further substitution by changing each F to a binary 0 and each T to a binary 1; this results in the truth table of Fig. 5-2*d*. As we see later, the 0s and 1s allow us to apply Boolean algebra, often resulting in simplified expressions. Note that the combinations in the rows of Fig. 5-2*d* represent ascending binary numbers: 00, 01, 10, and 11.

5-3.3 Truth Tables for Any Number of Variables

We have seen that a truth table for 1 variable has two rows and a truth table for 2 variables has four rows. In general, a truth table for *n* variables has 2^n possible combinations of the states of the variables and hence 2^n rows.

TABLE 5-1
Truth Table for 3 Variables
(The Output Column Is
Not Specified.)

A	*B*	*C*	*Output*
0	0	0	
0	0	1	
0	1	0	
0	1	1	
1	0	0	
1	0	1	
1	1	0	
1	1	1	

EXAMPLE 5-4

Find the number of rows in a truth table

a. With 3 variables,
b. With 4 variables.

Solution

a. A truth table with 3 variables has $2^3 = \mathbf{8}$ rows.
b. A truth table with 4 variables has $2^4 = \mathbf{16}$ rows.

The rows of a truth table for n variables are listed in such a way that the combinations of the variables represent the ascending binary numbers from 0 to $2^n - 1$. As an example, Table 5-1 shows a general form of truth tables for 3 variables; it has $2^3 = 8$ rows and the combinations of the variables represent the binary numbers from 0 to $2^n - 1 = 2^3 - 1 = 7$.

Note that, unlike in the preceding truth tables for 1 and 2 variables, the output column is not specified in Table 5-1. This is because the preceding truth tables described specific circuits and hence specific Boolean functions, while Table 5-1 describes only a general form. The output column of Table 5-1 can be filled in when the Boolean function to be represented by the truth table is specified, either by a Boolean expression or by the description of a circuit.

5-3.4 Listing the States of the Variables in a Truth Table

The procedure outlined below results in a listing of all states of the n variables of a Boolean function as ascending binary numbers.

1. Enter the variables into the headings of the columns in the truth table.
2. Under the rightmost ("least significant") variable, list vertically as many alternate 0s and 1s as required for 2^n rows.
3. Under the next variable list alternately *two* 0s and *two* 1s.
4. Under the next variable list alternately *four* 0s and *four* 1s.

5. Increase in the listing the groups of 0s and 1s by a factor of two for each new variable.
6. The table is complete when all variables have been accounted for.

5-4 LOGIC OPERATORS

Table 5-1 shows a truth table where the output is not specified. This is because the output depends not only on variables A, B, and C, but also on the *logic operator* that defines the specific circuit represented by the truth table. This section introduces three simple logic operators; additional logic operators are introduced later.

5-4.1 The AND Operator

The *AND operation* can be performed by series switches, as illustrated in Fig. 5-3*a*: The voltmeter indicates +5 V if, and only if, *both* switches *A and B* are closed. The AND operation can also be performed by an *AND gate* as illustrated in Fig. 5-3*b*: Output $A \cdot B$ is +5 V if, and only if, *both* inputs *A and B* are at +5 V.

Truth tables for the *AND operator* of a 2-variable function are shown in Figs. 5-3*c* and 5-3*d*. The symbol "·" is used in this book for the AND operation, which is also called *Boolean product.* Note from the truth table that $0 \cdot 0 = 0$, $0 \cdot 1 = 0$, $1 \cdot 0 = 0$, and $1 \cdot 1 = 1$. Also note that the symbol "·" for the Boolean product is often omitted where ambiguities cannot arise.

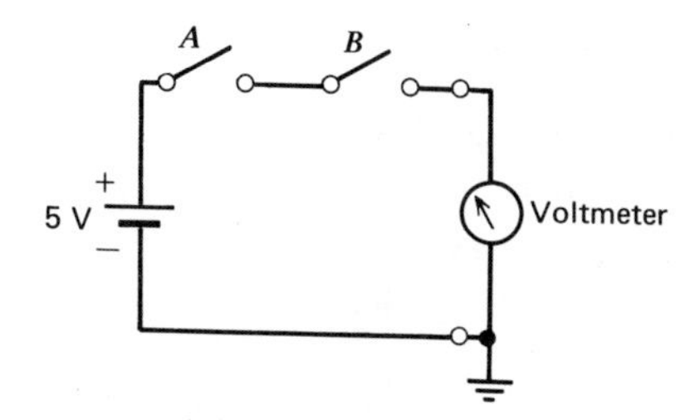

(*a*) Switch circuit

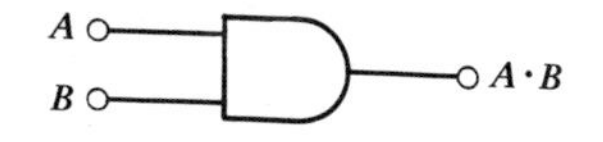

(*b*) AND gate symbol

A	B	$A \cdot B$
0 V	0 V	0 V
0 V	+5 V	0 V
+5 V	0 V	0 V
+5 V	+5 V	+5 V

(*c*) Truth table

A	B	$A \cdot B$
0	0	0
0	1	0
1	0	0
1	1	1

(*d*) Truth table

FIGURE 5-3 The AND operator

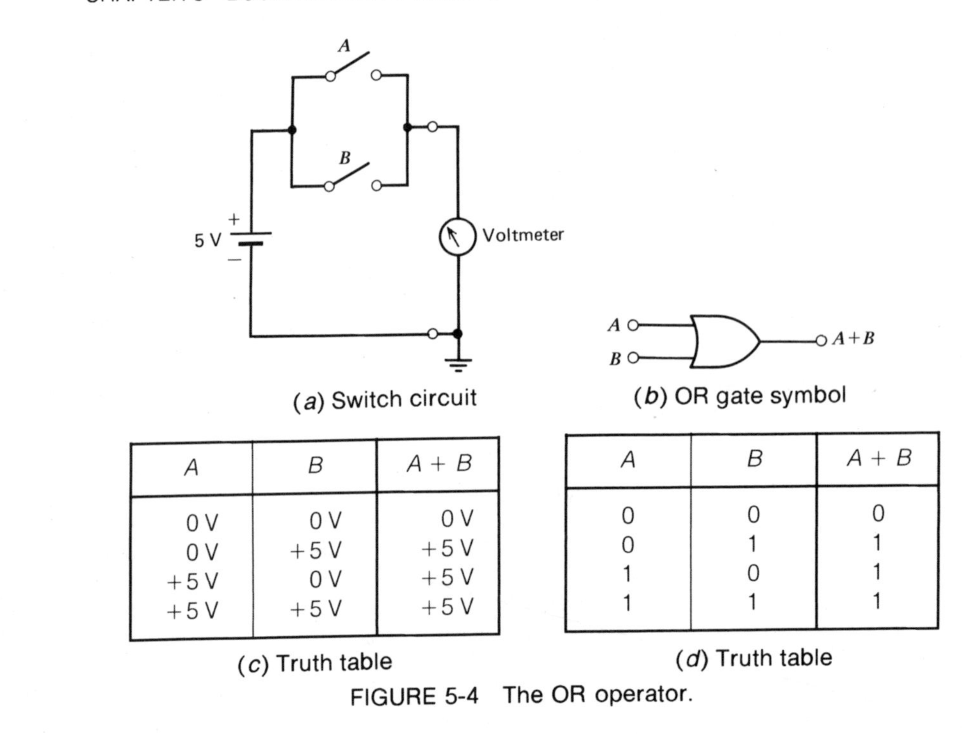

(*a*) Switch circuit

(*b*) OR gate symbol

A	B	$A + B$
0 V	0 V	0 V
0 V	+5 V	+5 V
+5 V	0 V	+5 V
+5 V	+5 V	+5 V

(*c*) Truth table

A	B	$A + B$
0	0	0
0	1	1
1	0	1
1	1	1

(*d*) Truth table

FIGURE 5-4 The OR operator.

5-4.2 The OR Operator

The *OR operation* can be performed by parallel switches, as illustrated in Fig. 5-4*a*: The voltmeter indicates +5 V if either or both switches, *A or B*, are closed. This operation is often referred to as the *INCLUSIVE OR* to distinguish it from the *EXCLUSIVE OR* discussed later. The OR operation can also be performed by an *OR gate*, as illustrated in Fig. 5-4*b*: Output $A + B$ is +5 V if either or both inputs are at +5 V.

Truth tables for the *OR operator* of a 2-variable function are shown in Figs. 5-4*c* and 5-4*d*. The symbol "+" is used in this book for the OR operation, which is also called *Boolean sum*. Note from the truth table that $0 + 0 = 0$, $0 + 1 = 1$, $1 + 0 = 1$, and $1 + 1 = 1$.

5-4.3 The NOT Operator

The *NOT operator* can be realized by use of the *logic inverter* introduced in Sec. 2-5, which is also known as *NOT gate*. Truth tables for the NOT operator are shown in Figs. 5-5*a* and 5-5*b*. Figures 5-5*c* and 5-5*d* show two equivalent symbols of the logic inverter (NOT gate)—these symbols are further discussed in Chapter 6.

An overbar "–" over a variable at the input or at the output of a logic inverter indicates a NOT operation: for example, $\overline{A}$, $\overline{X}$. A NOT operation always adds an overbar over its input variable. Thus, if the variable at the input of a logic inverter has no overbar, as in Fig. 5-5*c*, then its output has an overbar. Also, as is

A	$\bar{A}$
0 V +5 V	+5 V 0 V

(*a*) Truth table

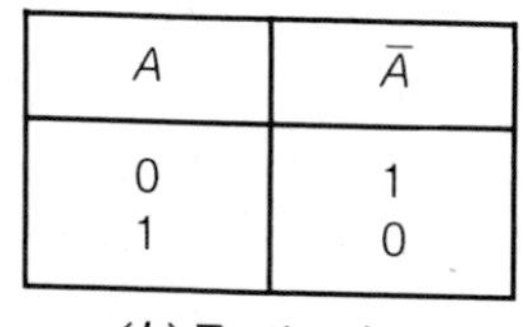

A	$\bar{A}$
0 1	1 0

(*b*) Truth table

(*c*) Logic inverter (NOT gate) symbol

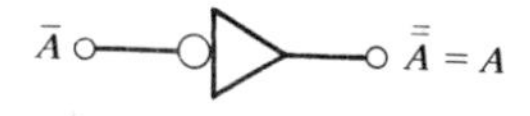

(*d*) Logic inverter (NOT gate) symbol

FIGURE 5-5 The NOT operator.

shown later, a double bar over a variable makes it equivalent to the variable itself: for example, $\bar{\bar{A}} = A$, as in Fig. 5-5*d*.

5-5 POSTULATES OF BOOLEAN ALGEBRA

Postulates are basic assumptions that are also called *axioms*. Boolean algebra is based on 10 postulates, from which all *theorems* (rules) can be derived.

The postulates are listed below in pairs: this is due to the *duality* existing in each pair. For example, changing in Postulate 1a all 0s to 1s and "·" to "+" results in Postulate 1b. Also, Postulate 1a may be obtained from Postulate 1b by similar changes. Such duality also holds for the other postulates, as well as for the theorems that are introduced later.

In what follows here, the postulates are presented and are illustrated by use of switches. Note that an open switch or an open circuit is represented by 0, and a closed switch or a closed circuit by 1.

Postulate 1a: $0 \cdot 0 = 0$
Two open switches in series result in an open circuit.

Postulate 1b: $1 + 1 = 1$
Two closed switches in parallel result in a closed circuit.

Postulate 2a: $0 \cdot 1 = 0$
An open switch in series with a closed switch results in an open circuit.

Postulate 2b: $1 + 0 = 1$
A closed switch in parallel with an open switch results in a closed circuit.

Postulate 3a: $1 \cdot 0 = 0$
A closed switch in series with an open switch results in an open circuit.

Postulate 3b: $0 + 1 = 1$
An open switch in parallel with a closed switch results in a closed circuit.

Postulate 4a: $1 \cdot 1 = 1$
A closed switch in series with a closed switch results in a closed circuit.

Postulate 4b: $0 + 0 = 0$
An open switch in parallel with an open switch results in an open circuit.

Postulate 5a: $\overline{0} = 1$
A switch that is not open is closed.

Postulate 5b: $\overline{1} = 0$
A switch that is not closed is open.

5-6 THEOREMS OF BOOLEAN ALGEBRA

Theorems are basic rules, or laws. This section describes 10 theorems of Boolean algebra and proves them by various means including logic gate diagrams, truth tables, and references to postulates and to previously established theorems.

Theorem 1

The commutative laws

Theorem 1a: $X \cdot Y = Y \cdot X$

Theorem 1a states that the inputs to an AND gate are perfectly interchangeable and that the output is not affected by the order in which the inputs have been written down. See Figs 5-6*a* and 5-6*b* for proof of the theorem.

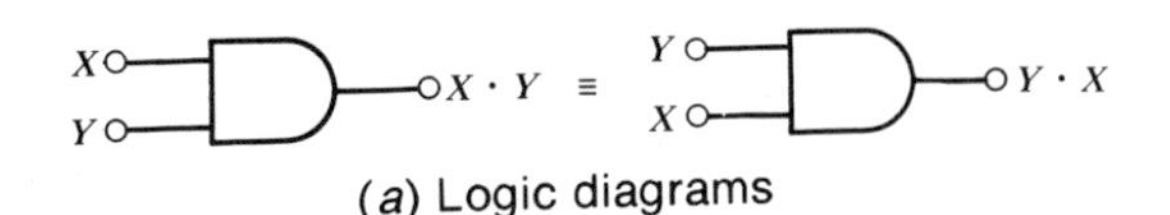

(*a*) Logic diagrams

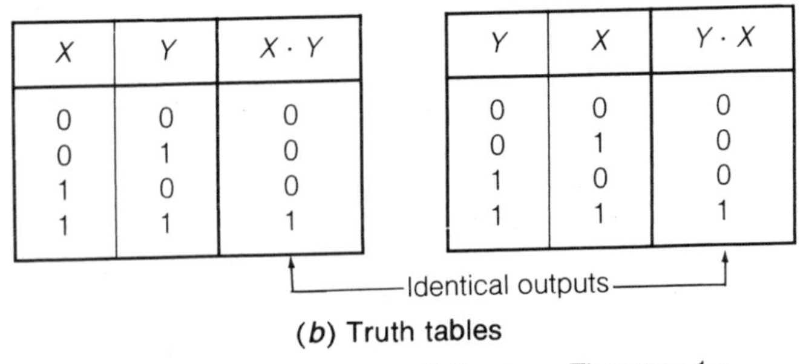

X	Y	$X \cdot Y$
0	0	0
0	1	0
1	0	0
1	1	1

Y	X	$Y \cdot X$
0	0	0
0	1	0
1	0	0
1	1	1

(*b*) Truth tables

FIGURE 5-6 The commutative law, Theorem 1a.

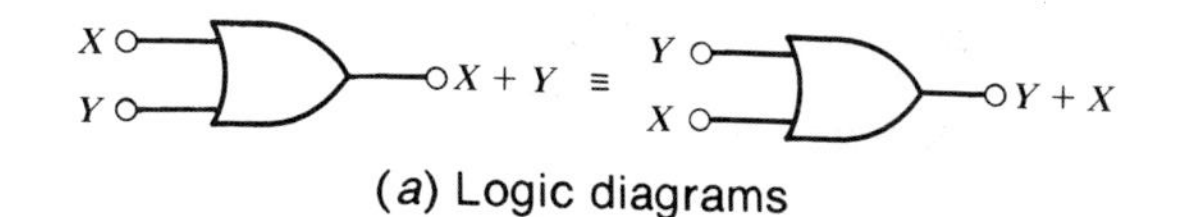

(*a*) Logic diagrams

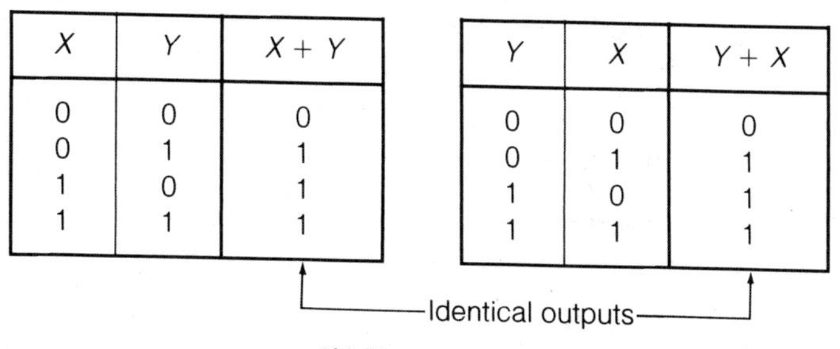

X	Y	X + Y
0	0	0
0	1	1
1	0	1
1	1	1

Y	X	Y + X
0	0	0
0	1	1
1	0	1
1	1	1

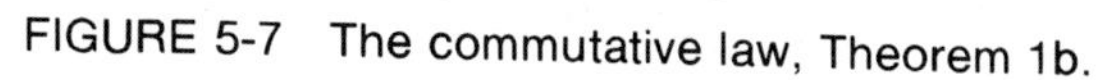

(*b*) Truth tables

FIGURE 5-7 The commutative law, Theorem 1b.

Theorem 1b: $X + Y = Y + X$

This is the dual of Theorem 1a. Its proof is shown in Figs. 5-7*a* and 5-7*b*.

Theorem 2

The associative laws

Theorem 2a: $X \cdot (Y \cdot Z) = (X \cdot Y) \cdot Z$

Theorem 2a states that it makes no difference in what order the variables are applied to an AND expression. Figure 5-8 shows the identity using AND gates.

Theorem 2b: $X + (Y + Z) = (X + Y) + Z$

Theorem 2b states that it makes no difference in what order the variables are applied to an OR expression. Figure 5-9 shows the identity using OR gates.

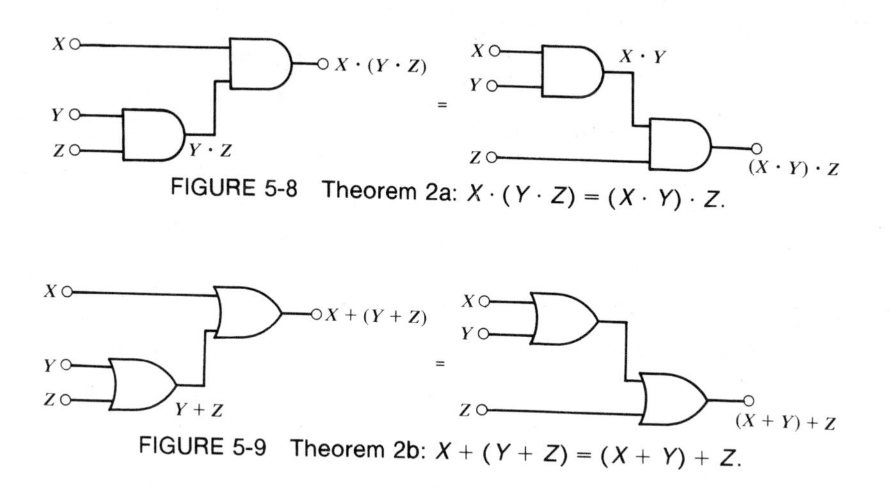

FIGURE 5-8 Theorem 2a: $X \cdot (Y \cdot Z) = (X \cdot Y) \cdot Z$.

FIGURE 5-9 Theorem 2b: $X + (Y + Z) = (X + Y) + Z$.

Theorem 3 **The idempotent laws**

Theorem 3a: $X \cdot X = X$

Theorem 3a states that if a binary value is ANDed with itself, the resulting output has the binary value of the input.

Proof:
For $X = 0$, $X \cdot X = 0 \cdot 0 = 0$
For $X = 1$, $X \cdot X = 1 \cdot 1 = 1$ (see Postulate 4a)

Theorem 3b: $X + X = X$

Theorem 3b states that if a binary value is ORed with itself, the resulting output has the binary value of the input.

Proof:
For $X = 0$, $X + X = 0 + 0 = 0$
For $X = 1$, $X + X = 1 + 1 = 1$ (see Postulate 1b)

Theorem 4 **The laws of identities**

Theorem 4a: $X \cdot 1 = X$

Proof:
For $X = 0$, $X \cdot 1 = 0 \cdot 1 = 0$
For $X = 1$, $X \cdot 1 = 1 \cdot 1 = 1$

Theorem 4b: $X + 0 = X$

Proof:
For $X = 0$, $X + 0 = 0 + 0 = 0$
For $X = 1$, $X + 0 = 1 + 0 = 1$

Theorem 5 **The laws of null elements**

Theorem 5a: $X \cdot 0 = 0$

Proof:
For $X = 0$, $X \cdot 0 = 0 \cdot 0 = 0$
For $X = 1$, $X \cdot 0 = 1 \cdot 0 = 0$

Theorem 5b: $X + 1 = 1$

Proof:
For $X = 0$, $X + 1 = 0 + 1 = 1$
For $X = 1$, $X + 1 = 1 + 1 = 1$

Theorem 6 **The laws of complements**

Theorem 6a: $X \cdot \overline{X} = 0$

Proof:
For $X = 0$, $\overline{X} = 1$; thus $X \cdot \overline{X} = 0 \cdot 1 = 0$
For $X = 1$, $\overline{X} = 0$; thus $X \cdot \overline{X} = 1 \cdot 0 = 0$

Theorem 6b: $X + \overline{X} = 1$

Proof:
For $X = 0$, $\overline{X} = 1$; thus $X + \overline{X} = 0 + 1 = 1$
For $X = 1$, $\overline{X} = 0$; thus $X + \overline{X} = 1 + 0 = 1$

Theorem 7

The laws of absorption

Theorem 7a: $X + X \cdot Y = X$

Proof:

$$\begin{aligned} X + X \cdot Y &= X \cdot (1 + Y) && \text{by factoring} \\ &= X \cdot 1, \quad \text{since } 1 + Y = 1 && \text{by Theorem 5b} \\ &= X, \quad \text{since } X \cdot 1 = X && \text{by Theorem 4a} \end{aligned}$$

Theorem 7b: $X \cdot (X + Y) = X$

Proof:

$$\begin{aligned} X \cdot (X + Y) &= X \cdot X + X \cdot Y && \text{by Boolean multiplication} \\ &= X + X \cdot Y, \quad \text{since } X \cdot X = X && \text{by Theorem 3a} \\ &= X, \quad \text{since } X + X \cdot Y = X && \text{by Theorem 7a} \end{aligned}$$

Theorem 8

The distributive laws

Theorem 8a: $X \cdot (Y + Z) = X \cdot Y + X \cdot Z$

The result on the right-hand side may be obtained by Boolean multiplication of the expression in parentheses by X. This is shown in the truth table of Table 5-2.

Theorem 8b: $(X + Y) \cdot (X + Z) = X + Y \cdot Z$

This theorem can be proven by Boolean algebra, as shown in Ex. 5-5, which follows.

TABLE 5-2
Proof of Theorem 8a: $X \cdot (Y + Z) = X \cdot Y + X \cdot Z$

X	Y	Z	$Y + Z$	$X \cdot (Y + Z)$	$X \cdot Y$	$X \cdot Z$	$X \cdot Y + X \cdot Z$
0	0	0	0	0	0	0	0
0	0	1	1	0	0	0	0
0	1	0	1	0	0	0	0
0	1	1	1	0	0	0	0
1	0	0	0	0	0	0	0
1	0	1	1	1	0	1	1
1	1	0	1	1	1	0	1
1	1	1	1	1	1	1	1

↑ Identical entries in these columns ↑

EXAMPLE 5-5

Prove Theorem 8b using Boolean postulates and previously established theorems.

Solution

$$\begin{aligned}(X+Y)\cdot(X+Z) &= X\cdot X + X\cdot Y + X\cdot Z + Y\cdot Z && \text{by Boolean multiplication}\\ &= X + X\cdot Y + X\cdot Z + Y\cdot Z && \text{since } X\cdot X = X\\ &= X\cdot(1+Y) + X\cdot Z + Y\cdot Z && \text{by factoring}\\ &= X + X\cdot Z + Y\cdot Z && \text{since } X\cdot(1+Y) = X\\ &= X\cdot(1+Z) + Y\cdot Z && \text{by factoring}\\ &= \mathbf{X + Y\cdot Z} && \text{since } X\cdot(1+Z) = X\end{aligned}$$

Theorem 9

The law of double negation

Theorem 9: $\overline{\overline{X}} = X$, where $\overline{\overline{X}} = \overline{(\overline{X})}$

Proof:
For $X = 0$, $\overline{X} = 1$, $\overline{\overline{X}} = \overline{1} = 0$
For $X = 1$, $\overline{X} = 0$, $\overline{\overline{X}} = \overline{0} = 1$

Theorem 10

De Morgan's theorems

De Morgan's theorems are extensively used in Boolean algebra to obtain the complement of an expression or a function, as well as for simplification of Boolean expressions and functions.

Theorem 10a: $\overline{X\cdot Y} = \overline{X} + \overline{Y}$.

This is the first of De Morgan's theorems. It states that the complement of an AND operation equals the OR of the complemented variables. This is shown in Table 5-3.

To implement De Morgan's first theorem change all Boolean products "·" to Boolean sums "+" *and* complement each variable (or constant). If a variable is complemented to begin with, complement it again, obtaining the uncomplemented variable in accordance with Theorem 9 on double negation.

TABLE 5-3
Proof of De Morgan's First Theorem: $\overline{X\cdot Y} = \overline{X} + \overline{Y}$

X	Y	$X\cdot Y$	$\overline{X\cdot Y}$	$\overline{X}$	$\overline{Y}$	$\overline{X}+\overline{Y}$
0	0	0	1	1	1	1
0	1	0	1	1	0	1
1	0	0	1	0	1	1
1	1	1	0	0	0	0

↑ Identical entries in these columns ↑

EXAMPLE 5-6

Apply De Morgan's first theorem to $\overline{X \cdot \overline{Y}}$.

Solution

$$\overline{X \cdot \overline{Y}} = \overline{X} + \overline{\overline{Y}} = \boldsymbol{\overline{X} + Y}$$

De Morgan's first theorem can also be applied to expressions and functions having more than 2 variables.

EXAMPLE 5-7

Apply De Morgan's first theorem to $\overline{W \cdot \overline{X} \cdot Y \cdot \overline{Z}}$.

Solution

$$\overline{W \cdot \overline{X} \cdot Y \cdot \overline{Z}} = \overline{W} + \overline{\overline{X}} + \overline{Y} + \overline{\overline{Z}} = \boldsymbol{\overline{W} + X + \overline{Y} + Z}$$

The theorem may also be applied for the complementation of an expression or a function.

EXAMPLE 5-8

Complement $\overline{X} \cdot Y \cdot \overline{Z}$ by using Theorem 10a.

Solution

The complement of $\overline{X} \cdot Y \cdot \overline{Z}$ is $\overline{\overline{X} \cdot Y \cdot \overline{Z}}$. Using Theorem 10a we get

$$\overline{\overline{X} \cdot Y \cdot \overline{Z}} = \overline{\overline{X}} + \overline{Y} + \overline{\overline{Z}} = \boldsymbol{X + \overline{Y} + Z}.$$

Theorem 10b: $\overline{X + Y} = \overline{X} \cdot \overline{Y}$

This is the second of De Morgan's theorems. It states that the complement of an OR operation equals the AND of the complemented variables. This is shown in Table 5-4.

To implement De Morgan's second theorem, change all Boolean sums "+" to Boolean products "·" *and* complement each variable (or constant). If a variable is complemented to begin with, complement it again, obtaining the uncomplemented variable in accordance with Theorem 9 on double negation.

EXAMPLE 5-9

Apply De Morgan's second theorem to $\overline{X + \overline{Y}}$.

Solution

$$\overline{X + \overline{Y}} = \overline{X} \cdot \overline{\overline{Y}} = \boldsymbol{\overline{X} \cdot Y}.$$

De Morgan's second theorem can also be applied to expressions and functions having more than 2 variables.

TABLE 5-4
Proof of De Morgan's Second Theorem: $\overline{X + Y} = \overline{X} \cdot \overline{Y}$

X	Y	$X + Y$	$\overline{X + Y}$	$\overline{X}$	$\overline{Y}$	$\overline{X} \cdot \overline{Y}$
0	0	0	1	1	1	1
0	1	1	0	1	0	0
1	0	1	0	0	1	0
1	1	1	0	0	0	0

↑ ↑ Identical entries in these columns

EXAMPLE 5-10 Apply De Morgan's second theorem to $\overline{W + \overline{X} + Y + \overline{Z}}$.

Solution

$$\overline{W + \overline{X} + Y + \overline{Z}} = \overline{W} \cdot \overline{\overline{X}} \cdot \overline{Y} \cdot \overline{\overline{Z}} = \boldsymbol{\overline{W} \cdot X \cdot \overline{Y} \cdot Z}$$

The theorem may also be applied for the complementation of an expression or a function.

EXAMPLE 5-11 Complement $\overline{X} + Y + \overline{Z}$ by use of Theorem 10b.

Solution The complement of $\overline{X} + Y + \overline{Z}$ is $\overline{\overline{X} + Y + \overline{Z}}$. Using Theorem 10b we get

$$\overline{\overline{X} + Y + \overline{Z}} = \overline{\overline{X}} \cdot \overline{Y} \cdot \overline{\overline{Z}} = \boldsymbol{X \cdot \overline{Y} \cdot Z}$$

Finally, De Morgan's theorems may be applied to Boolean expressions and functions involving the AND *and* OR operators. In such a case substitute every Boolean product "$\cdot$" by a Boolean sum "+" and every Boolean sum "+" by a Boolean product "$\cdot$"; complement all variables (and constants).

EXAMPLE 5-12 Apply De Morgan's theorems to $\overline{V \cdot \overline{W} + \overline{X} \cdot Y + Z}$.

Solution

$$\overline{V \cdot \overline{W} + \overline{X} \cdot Y + Z} = (\overline{V} + \overline{\overline{W}}) \cdot (\overline{\overline{X}} + \overline{Y}) \cdot \overline{Z} = \boldsymbol{(\overline{V} + W) \cdot (X + \overline{Y}) \cdot \overline{Z}}$$

5-7 SIMPLIFICATION OF BOOLEAN FUNCTIONS

The reason we are interested in the theorems of Boolean algebra introduced above is their use in simplifying Boolean expressions and Boolean functions. (We should also note here that Boolean functions applied to switch networks are often referred to as *switching functions*.)

As was also the case in ordinary algebra, a Boolean function (switching function) can be written in several different forms. Also, it makes good economic sense to reduce a Boolean function to a simple form, since a simpler Boolean function leads to simpler hardware (we find exceptions to this in later chapters).

This section illustrates simplification by algebraic manipulation using theorems of Boolean algebra. Graphic simplification is discussed later in Sec. 5-10 and Sec. 5-11.

EXAMPLE 5-13 Simplify the Boolean function $f = A \cdot B \cdot (C + \overline{A} \cdot \overline{C})$ and show a schematic diagram of the original function and the simplified function.

Solution $A \cdot B \cdot (C + \overline{A} \cdot \overline{C}) = A \cdot B \cdot C + A \cdot B \cdot \overline{A} \cdot \overline{C}$, but $A \cdot \overline{A} = 0$, thus $A \cdot B \cdot (C + \overline{A} \cdot \overline{C}) = A \cdot B \cdot C =$ ***ABC***.

Implementations of the two functions are shown in Fig. 5-10*a* and Fig. 5-10*b*. The two circuits perform identical functions; Fig. 5-10*b*, however, is much simpler.

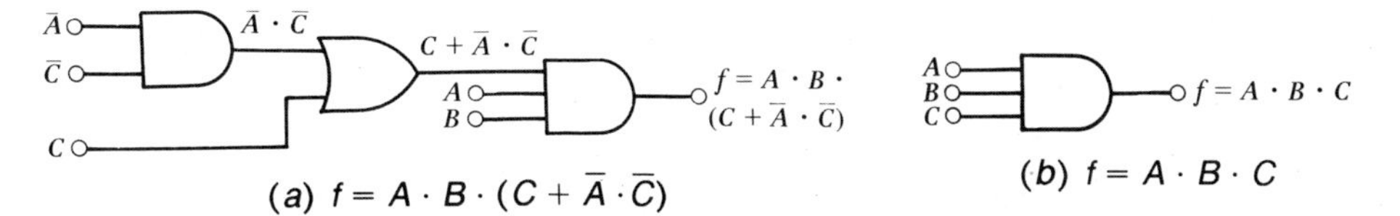

(a) $f = A \cdot B \cdot (C + \bar{A} \cdot \bar{C})$ (b) $f = A \cdot B \cdot C$

FIGURE 5-10 Two combinational logic circuits of Ex. 5-13 yielding identical results.

5-8 ANALYSIS AND SYNTHESIS OF COMBINATIONAL LOGIC CIRCUITS

A *combinational logic circuit* is a network of switches, relays, or logic gates that are interconnected to perform a given function. Also, a combinational logic circuit yields at its output a logic 1 for a defined set of combinations of input variables, and a logic 0 for all other combinations.

The AND, OR, and NOT logic operators were introduced in Sec. 5-4. It can be shown that all possible switching functions can be realized using only the AND, OR, and NOT logic operators, and that all combinational logic circuits can be realized using only AND gates, OR gates, and NOT gates (logic inverters). The two examples in this section include realization of combinational logic circuits using three logic gates; more complex circuits are introduced later.

The result of an *analysis* of a combinational logic circuit is an equation describing the function performed by logic operators, such as AND, OR, and NOT, operating on the input variables of the circuit. This is shown in Ex. 5-14 below; the example also illustrates the equivalence of circuits realized with switches and logic gates.

EXAMPLE 5-14

Derive the Boolean function describing the combinational logic circuits shown in Figs. 5-11*a* and 5-11*b*.

Solution

a. In the switch representation of Fig. 5-11*a*, we note three circuits in series: A in parallel with $\bar{C}$, B, and $\bar{A}$ in parallel with C. The parallel branches are described by the OR operator (see Fig. 5-4), while the series arrangement is described by the AND operator (see Fig. 5-3). Thus

$$f = (\boldsymbol{A} + \bar{\boldsymbol{C}}) \cdot \boldsymbol{B} \cdot (\bar{\boldsymbol{A}} + \boldsymbol{C})$$

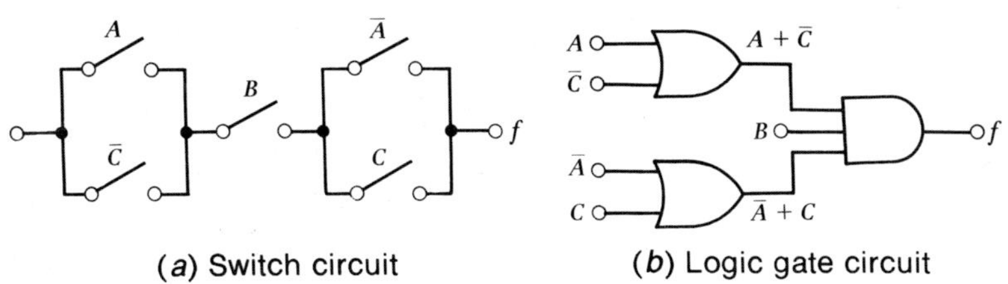

(*a*) Switch circuit (*b*) Logic gate circuit

FIGURE 5-11 Two representations of the Boolean function $f = (A + \bar{C}) \cdot (\bar{A} + C) \cdot B$.

b. To analyze the logic gate circuit we start from the left and write down the result of each operation at the outputs of the respective gates, as shown in Fig. 5-11*b*. The output from the upper OR gate is $A + \overline{C}$, and from the lower OR gate it is $\overline{A} + C$. These, together with the variable B, are ANDed in a 3-input AND gate resulting in

$$f = (\boldsymbol{A} + \overline{\boldsymbol{C}}) \cdot (\overline{\boldsymbol{A}} + \boldsymbol{C}) \cdot \boldsymbol{B}$$

In *synthesis* of a combinational logic circuit we start with a statement (or with a truth table or an equation). We then proceed to interconnect components such as switches, relays, or logic gates to fulfill the specifications of the statement.

EXAMPLE 5-15

Draw the logic circuit that realizes the statement "Electrical engineering students shall take as an elective either Boolean algebra or computer programming."

Solution

Assigning the variable X to students who take Boolean algebra and the variable Y to students who take computer programming, we can reduce the English statement to combinations of Boolean variables. Since we have 2 variables, X and Y, there are four such combinations, as listed below:

a. For $X = 0$, $Y = 0$ the function has an output 0, since it describes those students who take neither Boolean algebra nor computer programming.

b. For $X = 0$, $Y = 1$ the function has an output 1, since it describes those students who take computer programming but do not take Boolean algebra.

c. For $X = 1$, $Y = 0$ the function has an output 1, since it describes those students who take Boolean algebra but do not take computer programming.

d. For $X = 1$, $Y = 1$ the function has an output 0, since it describes those students who take both subjects. This is not according to our specification, which included an either-or statement.

From steps (a) through (d) we obtain the complete truth table. The combination of variables in steps (a) and (d) resulted in the function being "FALSE" = 0, while the combination of variables in steps (b) and (c) resulted in the function being "TRUE" = 1. From this information we draw the truth table, as shown in Fig. 5-12*a*. Furthermore, from the truth table we can derive the Boolean function that is 1 (TRUE) for either $\overline{X} \cdot Y$ or for $X \cdot \overline{Y}$.

X	Y	f
0	0	0
0	1	1
1	0	1
1	1	0

(*a*) Truth table

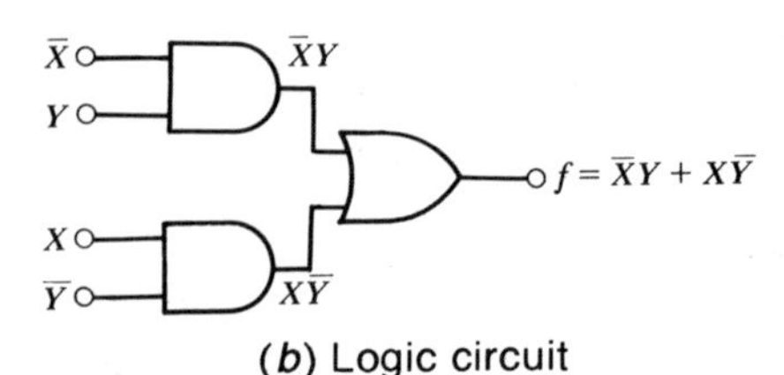

(*b*) Logic circuit

FIGURE 5-12 Synthesis of a logic circuit.

Thus

$$f = \bar{X}Y + X\bar{Y}$$

In Fig. 5-12*b* we show the realization of this function using AND and OR gates.

5-9 STANDARD FORMS

We have seen that a Boolean expression (or function) can be written in several different forms. In what follows here, we describe two forms that are in common use: the sum-of-products form and the product-of-sums form.

5-9.1 Sum-of-Products (SOP) Forms

A *product term* is a product of Boolean variables, for example, $A\bar{B}$, or $WX\bar{Y}Z$. A sum of product terms results in a *sum-of-products* form, also known as *SOP form*; for example, $A\bar{B} + WX\bar{Y}Z$. When a product term includes *all* variables of a Boolean function (some complemented, others uncomplemented), it is called a *standard product term.* A sum of standard product terms results in a *standard SOP form.*

EXAMPLE 5-16

The truth table of a Boolean function is given in Table 5-5. Write the function in a standard SOP form.

Solution

From the truth table, Table 5-5, we derive the following function in standard SOP form:

$$f = \bar{A}BC + A\bar{B}C + ABC.$$

When the standard SOP form of a Boolean function is simplified, we may end up with product terms that do not include *all* Boolean variables, hence are not

TABLE 5-5
Truth Table for Ex. 5-16

A	B	C	f
0	0	0	0
0	0	1	0
0	1	0	0
0	1	1	1
1	0	0	0
1	0	1	1
1	1	0	0
1	1	1	1

standard product terms, but only product terms; also, in such a case their sum is not in a *standard* SOP form, but only in an SOP form.

EXAMPLE 5-17 Simplify the Boolean function of Ex. 5-16.

Solution To simplify the function we add a second term ABC to the right-hand side of the equation:

$$f = \bar{A}BC + A\bar{B}C + ABC + ABC$$

The function has not changed by this addition since from Theorem 3b:

$$ABC + ABC = ABC$$

Thus,

$$\begin{aligned} f &= BC(\bar{A} + A) + AC(\bar{B} + B) && \text{by factoring} \\ &= BC \cdot 1 + AC \cdot 1 && \text{since } \bar{A} + A = 1,\ \bar{B} + B = 1 \\ &= \mathbf{BC + AC} && \text{since } BC \cdot 1 = BC,\ AC \cdot 1 = AC \end{aligned}$$

Neither term in the result has all 3 variables, A, B, and C, hence both terms are only product terms (*not* standard product terms) and the result is only in an SOP form (*not* in a standard SOP form). Also note that the realization of the simplified function results in an AND-OR logic circuit similar to Fig. 5-12*b*.

5-9.2 Product-of-Sums (POS) Forms

A *sum term* is a sum of Boolean variables, for example, $A + \bar{B}$, or $W + X + \bar{Y} + Z$. A product of sum terms results in a *product-of-sums* form, also known as *POS form*; for example, $(A + \bar{B}) \cdot (W + X + \bar{Y} + Z)$. When a sum term includes *all* variables of a Boolean function (some complemented, others uncomplemented), it is called a *standard sum term*. A product of standard sum terms results in a *standard POS form*.

Standard sum terms and standard POS forms are especially useful when the truth table of a Boolean function has fewer 0 outputs than 1 outputs. This is illustrated in Ex. 5-18 below.

EXAMPLE 5-18 The truth table of a Boolean function is given in Table 5-6. Write the function in standard POS form.

Solution The function has 3 variables; hence there are $2^3 = 8$ rows in the table. In the output column we observe five 1s and only three 0s. Remember that 0s represent the combinations where the function is "FALSE" = 0. It might be easier for us to deal with the 0 combinations since these have fewer terms in the truth table. However, dealing with these we obtain the *complement* of the function, $\bar{f}$, rather than f, as

$$\bar{f} = \bar{A}BC + A\bar{B}C + ABC.$$

To obtain f we have to complement $\bar{f}$ since, by Theorem 9, $\bar{\bar{f}} = f$. To complement $\bar{f}$ we are using De Morgan's theorems:

$$\overline{(\bar{f})} = \overline{\bar{A}BC + A\bar{B}C + ABC} = \mathbf{(A + \bar{B} + \bar{C}) \cdot (\bar{A} + B + \bar{C}) \cdot (\bar{A} + \bar{B} + \bar{C})}$$

TABLE 5-6
Truth Table for Ex. 5-18

A	B	C	f
0	0	0	1
0	0	1	1
0	1	0	1
0	1	1	0
1	0	0	1
1	0	1	0
1	1	0	1
1	1	1	0

When the standard POS form of a Boolean function is simplified, we may end up with sum terms that do not include *all* Boolean variables, hence are not *standard* sum terms, but only sum terms; also, in such a case their product is not in a *standard* POS form, but only in a POS form. This is illustrated in Ex. 5-19, which follows.

EXAMPLE 5-19

Simplify the Boolean function of Ex. 5-18.

Solution

The complement $\bar{f}$ of the function of Ex. 5-18 was found to be

$$\bar{f} = \bar{A}BC + A\bar{B}C + ABC.$$

This can be simplified by first adding the term ABC,

$$\begin{aligned}
\bar{f} &= \bar{A}BC + A\bar{B}C + ABC + ABC && \text{since } ABC + ABC = ABC \\
&= BC(\bar{A} + A) + AC(\bar{B} + B) && \text{by factoring} \\
&= BC \cdot 1 + AC \cdot 1 && \text{since } \bar{A} + A = 1 \text{ and } \bar{B} + B = 1 \\
&= BC + AC
\end{aligned}$$

To obtain f we use De Morgan's theorems:

$$f = \bar{\bar{f}} = \overline{BC + AC} = \overline{(BC)} \cdot \overline{(AC)} = (\bar{B} + \bar{C}) \cdot (\bar{A} + \bar{C})$$

Neither term in the result has all 3 variables, A, B, and C, hence both terms are only sum terms (*not* standard sum terms) and the result is only in a POS form (*not* in a standard POS form). Also, the simplified function can be realized by an OR-AND logic circuit as shown in Fig. 5-13.

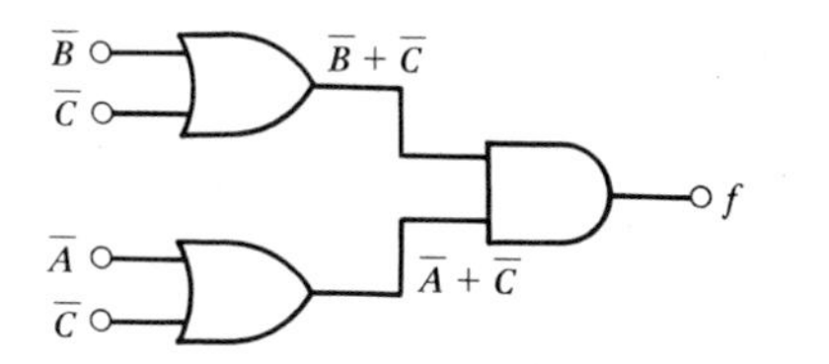

FIGURE 5-13 OR-AND logic circuit.

$\bar{A}\bar{B}$	$\bar{A}B$	AB	$A\bar{B}$

FIGURE 5-14 A Karnaugh map for a 2-variable function and the assignment of cells.

5-10 KARNAUGH MAPS

In Sec. 5-7 above, simplification was attained by algebraic manipulation using theorems of Boolean algebra. Another method used to simplify a Boolean expression or function is a graphic method: the *Karnaugh map*, also known as *K-map*. The Karnaugh map is probably the most extensively used tool for the simplification of Boolean functions with up to six variables. It is a graphic method of presenting information that is easy to use because of the pattern-recognition power of the human mind.

In the following, Karnaugh maps and their use in simplification of Boolean functions are discussed in two parts. This section introduces Karnaugh maps with 2, 3, and 4 variables, while the use of Karnaugh maps in the simplification of Boolean functions is described in Sec. 5-11.

5-10.1 Two-Variable Karnaugh Maps

The Karnaugh map for an n-variable Boolean function consists of 2^n *cells* (squares), and each cell is filled with a standard product term. Thus, a Karnaugh map for 2 variables has 4 cells, as shown in Fig. 5-14. Note that the cells are assigned in such a way that *adjacent* (neighboring) *cells* differ only in the state of *one* variable. Such neighboring terms are called *adjacencies*. Thus $\bar{A}\bar{B}$ and $\bar{A}B$ are adjacencies and so are $\bar{A}B$ and AB, as well as AB and $A\bar{B}$. Interestingly, the two outer terms in the figure are also adjacencies since they represent $\bar{A}\bar{B}$ and $A\bar{B}$.

It is more common to draw a 2-variable Karnaugh map as a *square* array of four terms, as shown in Fig. 5-15*a*. The four terms $\bar{A}\bar{B}$, $\bar{A}B$, $A\bar{B}$, and AB are arranged in such a way that each term has two adjacencies: one horizontally and one vertically. For example, $\bar{A}\bar{B}$ has a horizontal adjacency $A\bar{B}$ and a vertical adjacency $\bar{A}B$.

Instead of writing the standard product terms in the cells, in Fig. 5-15*b* we write the corresponding binary numbers, for example, 00 for $\bar{A}\bar{B}$, 01 for $\bar{A}B$, and so on. Next, we substitute the binary numbers by their decimal equivalents to show more clearly the cell assignments; this is shown in Fig. 5-15*c*. Furthermore, we would like to enter information in the Karnaugh map. To make room for entering 0s and 1s we move the cell designations to the edges of the map, as shown in Fig. 5-15*d*.

The above discussion and Figs. 5-15*a* through 5-15*d* are applicable to the Karnaugh map of *any* 2-variable Boolean function. In order to apply them to a *given* Boolean function, this function has to be first specified. As an example,

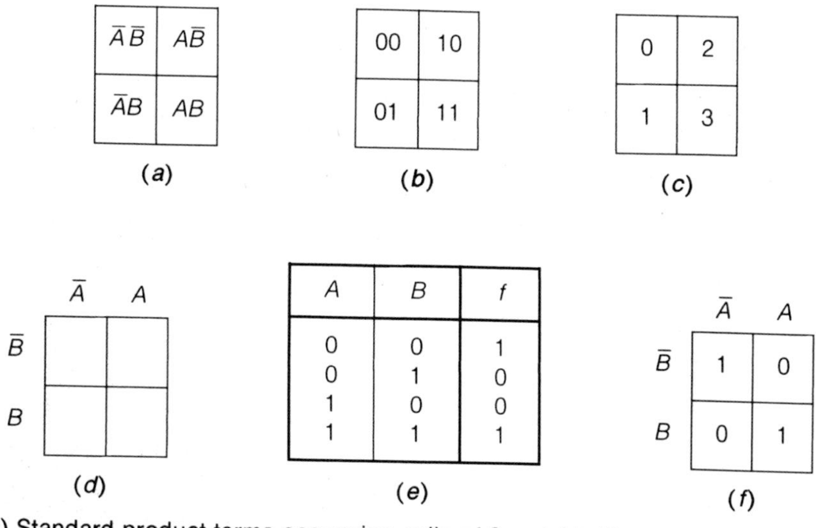

A	B	f
0	0	1
0	1	0
1	0	0
1	1	1

(*a*) Standard product terms occupying cells of 2-variable Karnaugh map

(*b*) Entering into the cells the binary equivalents of the standard product terms

(*c*) Entering into the cells the decimal equivalents of the standard product terms

(*d*) Marking the cell columns and cell rows with labels of the variables

(*e*) Truth table for the Boolean function $f = \bar{A}\bar{B} + AB$

(*f*) Information from the truth table, entered into the Karnaugh map

FIGURE 5-15 Two-variable Karnaugh maps.

below we describe application to the Boolean function specified by the truth table of Fig. 5-15*e*.

The first row in the truth table states that $f = 1$ when $\bar{A} = 1$ and $\bar{B} = 1$, that is, when $\bar{A}\bar{B} = 1$. Hence we enter a 1 in the Karnaugh map cell that represents $\bar{A}\bar{B} = 1$, that is, where the column representing $\bar{A}$ intersects with the row representing $\bar{B}$—this is the cell at the upper left corner in Fig. 5-15*f*.

The next $f = 1$ term in the truth table is in the last row, which states that $f = 1$ when $A = 1$ and $B = 1$, that is, when $AB = 1$. Hence we enter a 1 in the Karnaugh map cell that represents $AB = 1$, that is, where the column representing A intersects with the row representing B—this is the cell at the lower right corner of Fig. 5-15*f*.

We now took care of both $f = 1$ terms in the truth table of Fig. 5-15*e*. There are no other $f = 1$ entries in the truth table, hence all remaining entries in the Karnaugh map of Fig. 5-15*f* are filled with 0s.

5-10.2 Three-Variable Karnaugh Maps

A Karnaugh map for a Boolean function of 3 variables has $2^3 = 8$ cells corresponding to the 8 rows of its truth table. The assignment of the cells is shown in Fig. 5-16*a*.

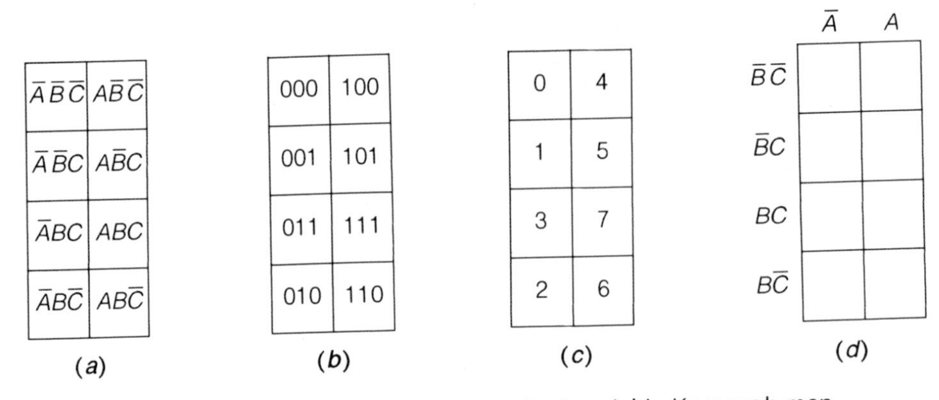

(*a*) Standard product terms occupying cells of a 3-variable Karnaugh map

(*b*) Assignment of binary numbers to replace the standard product terms. Note the Gray code of the assignment

(*c*) Entering into the cells the decimal equivalents of the standard product terms

(*d*) Marking the cell columns and the cell rows with labels of the variables

FIGURE 5-16 Three-variable Karnaugh maps.

Note that each cell is an adjacency to its neighboring cell, *horizontally* and *vertically*. For example, the cell $\bar{A}\bar{B}C$ in Fig. 5-16*a* has three adjacencies. The vertically adjacent cells are $\bar{A}\bar{B}\bar{C}$ (up) and $\bar{A}BC$ (down). Horizontally there is one adjacent cell: $A\bar{B}C$. Notice that each adjacent cell differs from $\bar{A}\bar{B}C$ in only one variable, the change being from a complemented to an uncomplemented variable, or vice versa.

Similar considerations apply to the adjacencies of all cells in the map. Also, the cells at the upper edge of the map are adjacent to the corresponding cells at the lower edge: $\bar{A}\bar{B}\bar{C}$ is adjacent to $\bar{A}B\bar{C}$, and $A\bar{B}\bar{C}$ is adjacent to $AB\bar{C}$.

In Fig. 5-16*b* we assign binary numbers to the standard product terms as follows: $\bar{A}\bar{B}\bar{C} = 000$, $\bar{A}\bar{B}C = 001$, and so on, ending in $ABC = 111$. Note that the cells are not arranged in an ascending order of binary numbers—in order to obtain adjacencies, a Gray code is used (see Sec. 4-3). In Fig. 5-16*c* we enter into the cells the decimal equivalents of the standard product terms. Finally, in Fig. 5-16*d* we move the cell designations to the edges of the map.

5-10.3 Four-Variable Karnaugh Maps

A Karnaugh map for a Boolean function of 4 variables has $2^4 = 16$ cells as shown in Fig. 5-17*a*. Note that each cell has four adjacencies and that the cells at the upper edge of the map are adjacencies of the cells at the lower edge of the map. Also, the cells at the right edge of the map are adjacencies of the corresponding cells at the left edge.

In Fig. 5-17*b* we assign binary numbers to the standard product terms, and in Fig. 5-17*c* we enter the decimal equivalents of the standard product terms.

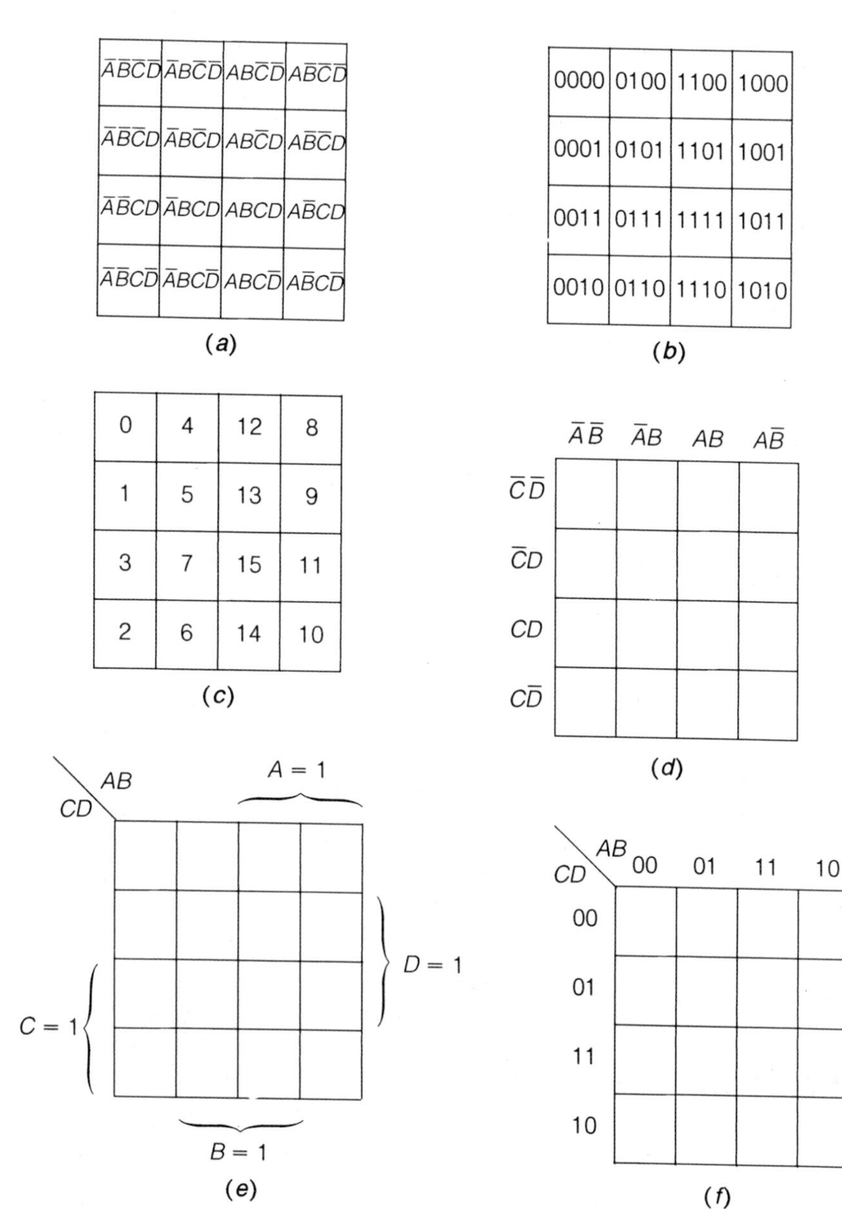

(*a*) Standard product terms occupying cells of a 4-variable Karnaugh map

(*b*) Assignment of binary numbers to replace the standard product terms

(*c*) Assignment of decimal numbers to replace the standard product terms

(*d*) Marking the cell columns and the cell rows with labels of the variables

(*e*) Another method of marking the cell columns and the cell rows

(*f*) Still another method of marking

FIGURE 5-17 Four-variable Karnaugh maps.

Finally in Fig. 5-17*d* we move the cell designations to the edges of the map.

In many places the reader may find different ways of showing cell designations. One common method is shown in Fig. 5-17*e*. At the upper edge of this figure we find two columns bracketed by the term $A = 1$. This signifies that the variable A in the cells under this bracket appears *uncomplemented* in the standard product terms. Conversely, the columns that are not bracketed by $A = 1$ represent standard product terms where A is *complemented*. A similar argument holds for the other variables that are shown by brackets in Fig. 5-17*e*. Still another possibility is shown in Fig. 5-17*f*, where the cells are marked at the edges of the map in Gray-code sequence as 00, 01, 11, 10.

5-11 USE OF KARNAUGH MAPS

In the preceding section we have seen how to assign cells in a Karnaugh map so that neighboring cells are adjacencies. This section describes how adjacencies are put to use in the simplification of Boolean functions with 2, 3, and 4 variables. This is followed by a summary of the steps followed, as well as by a discussion of "don't care" conditions.

5-11.1 Use of 2-Variable Karnaugh Maps

Consider the two standard product terms $A\overline{B}$ and AB that are adjacencies. A Boolean sum of these terms results in the Boolean function $f = A\overline{B} + AB$. This function can be simplified as

$$\begin{aligned} f = A\overline{B} + AB &= A(\overline{B} + B) & \\ &= A \cdot 1 & \text{since } \overline{B} + B = 1 \\ &= A & \text{since } A \cdot 1 = A \end{aligned}$$

Such simplifications are possible whenever there are adjacencies with 1 entries in a Karnaugh map. This is illustrated in Ex. 5-20 below.

EXAMPLE 5-20 Simplify the function $f = \overline{A}\,\overline{B} + A\overline{B}$ using a Karnaugh map.

Solution The truth table for the function is shown in Fig. 5-18*a* and its Karnaugh map in Fig. 5-18*b*: the standard product term $\overline{A}\,\overline{B}$ occupies the cell where $\overline{A}$ inter-

A	B	f
0	0	1
0	1	0
1	0	1
1	1	0

(*a*) Truth table

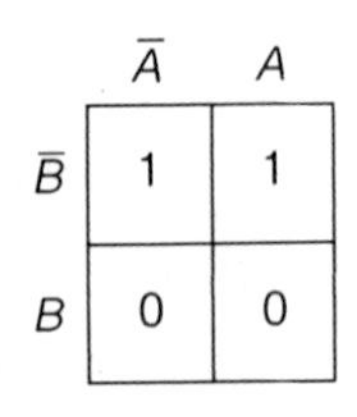

(*b*) Karnaugh map entries

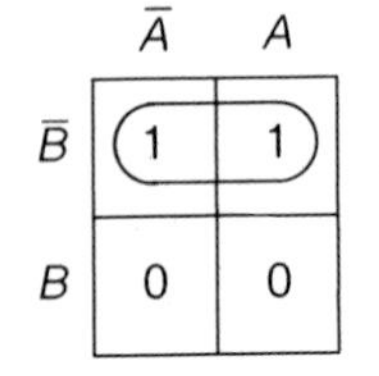

(*c*) Encircling adjacencies

FIGURE 5-18 Two-cell adjacencies.

sects with $\bar{B}$, and the standard product term $A\bar{B}$ occupies the cell where A intersects with $\bar{B}$. The other two standard product terms in the truth table have an output 0 and are thus represented in the map by 0s.

Next we encircle the adjacencies with 1 entries as shown in Fig. 5-18*c*. As we move from one encircled term to the other, we notice that $\bar{A}$ changes to A. The other variable, $\bar{B}$, does not change. This means that the output does not depend on the variable A but only on $\bar{B}$. Hence the solution to Ex. 5-20 is $\boldsymbol{\bar{B}}$.

5-11.2 Use of 3-Variable Karnaugh Maps

We have seen that a variable in a Boolean expression or function can be eliminated when there is a 2-cell adjacency with 1 entries in the Karnaugh map. Similarly, 2 variables in a Boolean expression or function can be eliminated when there is a 4-cell adjacency with 1 entries in the Karnaugh map, as demonstrated in Ex. 5-21 below.

EXAMPLE 5-21

Using a Karnaugh map simplify the Boolean function

$$f = \bar{A}\bar{B}C + \bar{A}BC + A\bar{B}C + ABC$$

Solution

The truth table for the 3-variable function is shown in Fig. 5-19*a*. The output is 1 for the standard product terms $\bar{A}\bar{B}C$, $\bar{A}BC$, $A\bar{B}C$, and ABC. In Fig. 5-19*b* 1s are entered in the corresponding cells of the Karnaugh map and 0s in the remaining cells. We find a 4-cell adjacency that we encircle, as shown in Fig. 5-19*c*. In the encircled group two variables appear complemented *and* uncomplemented and these are A, $\bar{A}$ and B, $\bar{B}$. Therefore the output does not depend on these variables and depends only on C.

Hence the solution to Ex. 5-21 is

$$\bar{A}\bar{B}C + \bar{A}BC + A\bar{B}C + ABC = \mathbf{C}$$

A	B	C	f
0	0	0	0
0	0	1	1
0	1	0	0
0	1	1	1
1	0	0	0
1	0	1	1
1	1	0	0
1	1	1	1

(*a*) Truth table

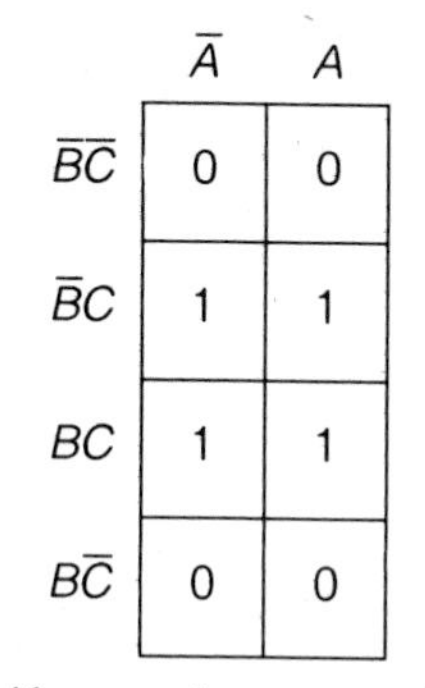

(*b*) Karnaugh map entries

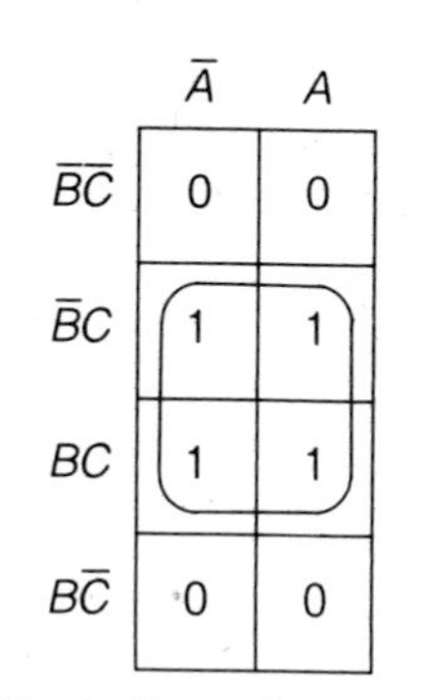

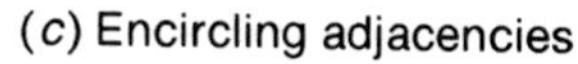

(*c*) Encircling adjacencies

FIGURE 5-19 Four-cell adjacencies.

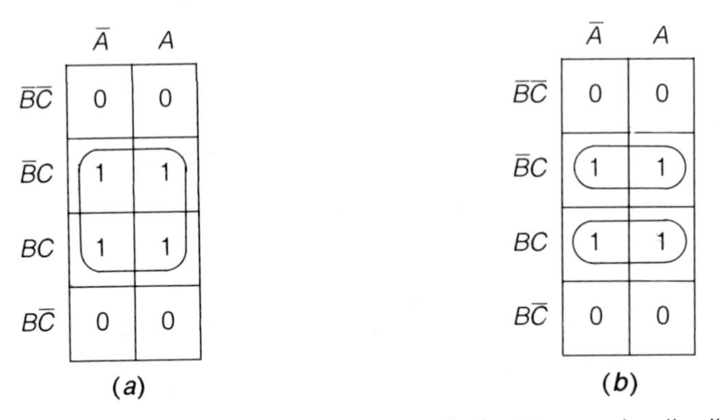

FIGURE 5-20 Two 2-cell adjacencies are equivalent to one 4-cell adjacency.

5-11.2.1 *Proof*

To prove that a 4-cell adjacency with 1 entries eliminates 2 variables, we first copy Fig. 5-19*c* as Fig. 5-20*a*, and then redraw it as shown in Fig. 5-20*b*. The upper 2-cell adjacency with 1 entries in Fig. 5-20*b* represents the term $\bar{B}C$, which is independent of A; the lower 2-cell adjacency with 1 entries represents the term BC, which is also independent of A. When ORed, these two terms result in the function

$$
\begin{aligned}
\bar{B}C + BC &= C(\bar{B} + B) && \text{by factoring} \\
&= C \cdot 1 && \text{since } \bar{B} + B = 1 \\
&= C && \text{since } C \cdot 1 = C
\end{aligned}
$$

Thus we have proven that a 4-cell adjacency with 1 entries eliminates 2 variables.

5-11.3 Use of 4-Variable Karnaugh Maps

When encircling adjacencies, we should always aim to encircle the largest possible number of 1 entries. Consider, for example, the 4-variable Boolean function $f = \bar{A}\bar{B}\bar{C}D + \bar{A}B\bar{C}D + \bar{A}BCD + A\bar{B}\bar{C}D + AB\bar{C}D$. Karnaugh maps for this function are shown in Fig. 5-21*a* and in Fig. 5-21*b*.

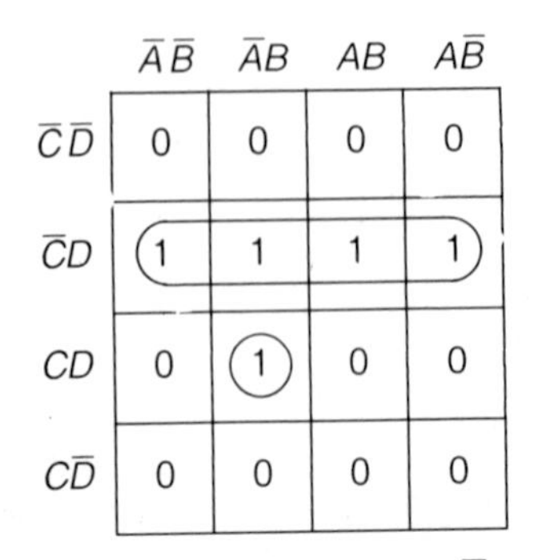

(*a*) Encirclement resulting in $f = \bar{C}D + \bar{A}BCD$

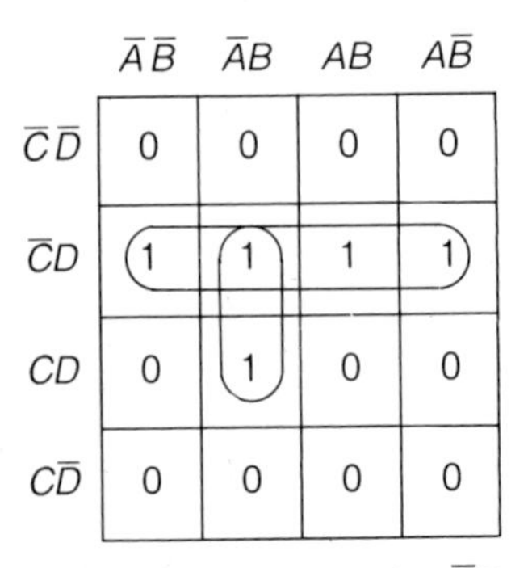

(*b*) Encirclement resulting in $f = \bar{C}D + \bar{A}BD$

FIGURE 5-21 Encirclements in 4-variable Karnaugh maps.

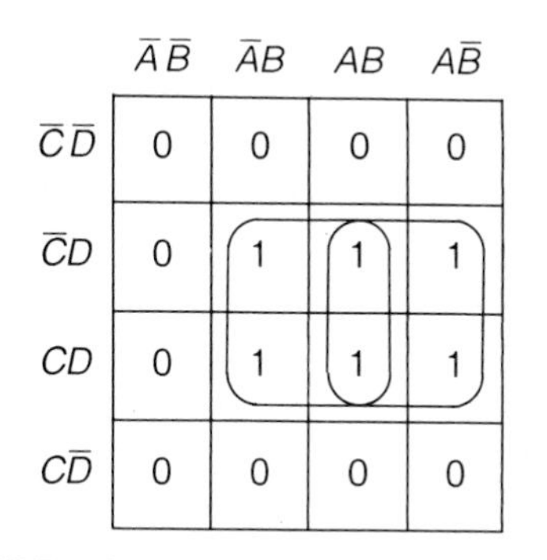

FIGURE 5-22 Karnaugh map for Ex. 5-22.

The encirclements shown in the Karnaugh map of Fig. 5-21*a* result in $f = \bar{C}D + \bar{A}BCD$. However, when a term is encircled twice as shown in Fig. 5-21*b*, the result becomes simpler: $f = \bar{C}D + \bar{A}BD$. Encircling a term twice is permitted because it has the same effect as adding a term to itself as $X + X = X$.

Terms are encircled more than once in Ex. 5-22 and Ex. 5-23, which follow.

EXAMPLE 5-22

Using a Karnaugh map simplify the function

$$f = \bar{A}B\bar{C}D + AB\bar{C}D + A\bar{B}\bar{C}D + \bar{A}BCD + ABCD + A\bar{B}CD.$$

Solution

The Karnaugh map is shown in Fig. 5-22. Two terms are encircled twice. The simplified function is

$$f = \boldsymbol{BD + AD}$$

EXAMPLE 5-23

Simplify the function shown in the map of Fig. 5-23.

Solution

Figure 5-23*a* shows an inefficient solution yielding $f = \bar{C} + BC\bar{D} + \bar{A}\bar{B}C\bar{D}$. By using edge adjacencies and encircling terms more than once as shown in Fig. 5-23*b*, we obtain

$$f = \boldsymbol{\bar{C} + B\bar{D} + \bar{A}\bar{D}}$$

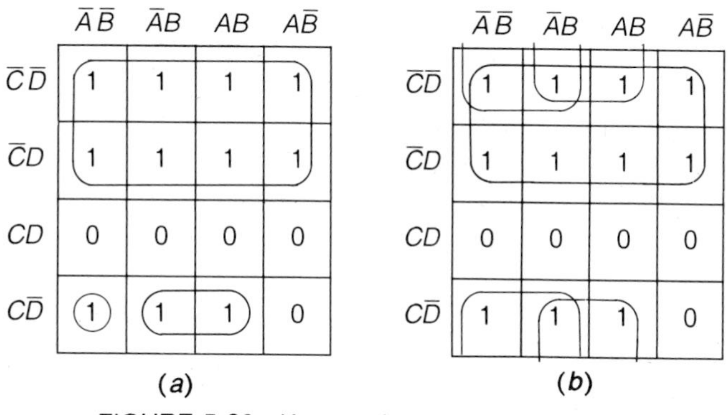

FIGURE 5-23 Karnaugh maps for Ex. 5-23.

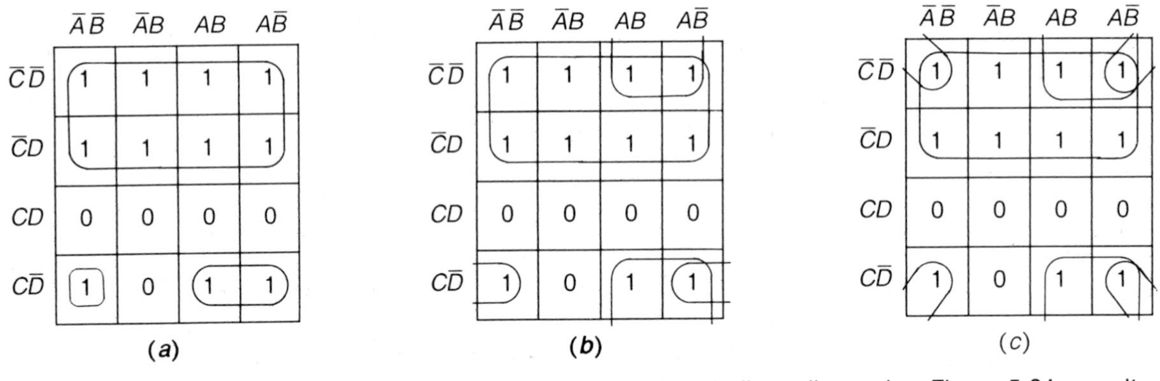

FIGURE 5-24 Three different ways of encircling adjacencies. Figure 5-24*c* results in the most economical implementation.

which is simpler. Note that in both solutions the 8-cell adjacency with 1 entries reduces the first term of the function by 3 variables.

The four corner cells of a map constitute a 4-cell adjacency. Figure 5-24 shows an example with three different ways of encircling the 1 entries. The solution of Fig. 5-24*c* is the most efficient, yielding the simplified function

$$f = \bar{C} + A\bar{D} + \bar{B}\bar{D}$$

where the term $\bar{B}\bar{D}$ is due to the 4-cell corner adjacency.

In Fig. 5-24*c* we encircled some cells twice; also cell $A\bar{B}\bar{C}\bar{D}$ was encircled three times. These *multiple encirclements* were done in order to obtain the largest possible encircled groups of 1 entries. However, there is always *at least one* unnecessary (redundant) term in the result when *all* 1 entries of an encirclement are encircled more than once. This is illustrated in Ex. 5-24 below.

EXAMPLE 5-24

Simplify the function shown in the map of Fig. 5-25*a*.

Solution

We first encircle the 4-cell adjacency with 1 entries, as shown in Fig. 5-25*b*. Next we encircle the four 2-cell adjacencies with 1 entries, as shown in Fig. 5-25*c*. Now all terms are encircled.

However, we can see that *all* terms in the 4-cell encircled adjacency of Fig. 5-25*c* are encircled twice. Thus we can remove the 4-cell encirclement in Fig. 5-25*c*; this results in Fig. 5-25*d*, which is the simplest solution. For comparison note that the less efficient Fig. 5-25*c* results in

$$f = BD + \bar{A}\bar{C}D + AB\bar{C} + ACD + \bar{A}BC$$

Compare this with Fig. 5-25*d*, which yields the simplest solution of

$$f = \boldsymbol{\bar{A}\bar{C}D + AB\bar{C} + ACD + \bar{A}BC}$$

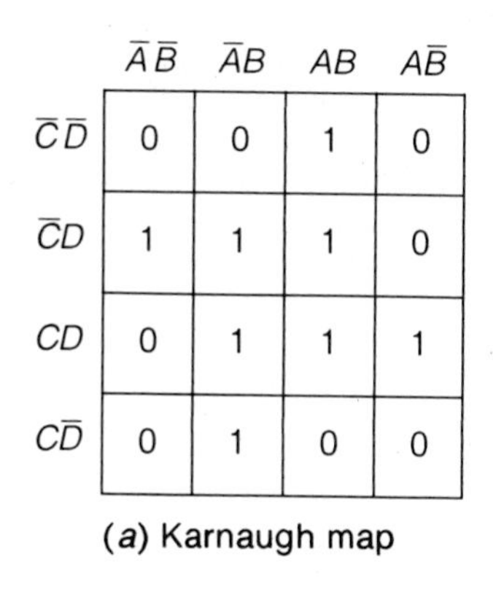

(*a*) Karnaugh map

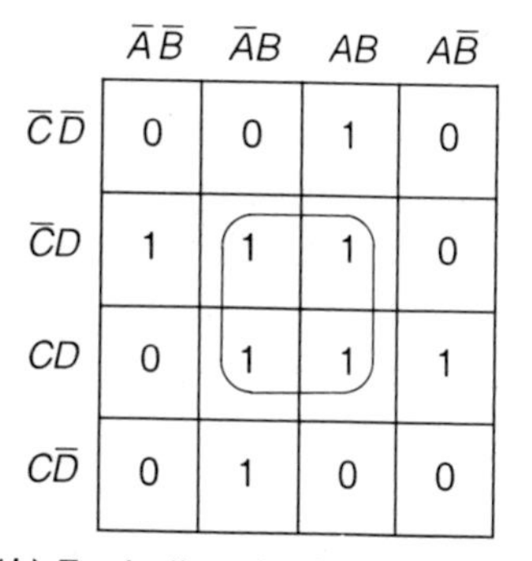

(*b*) Encircling the largest group

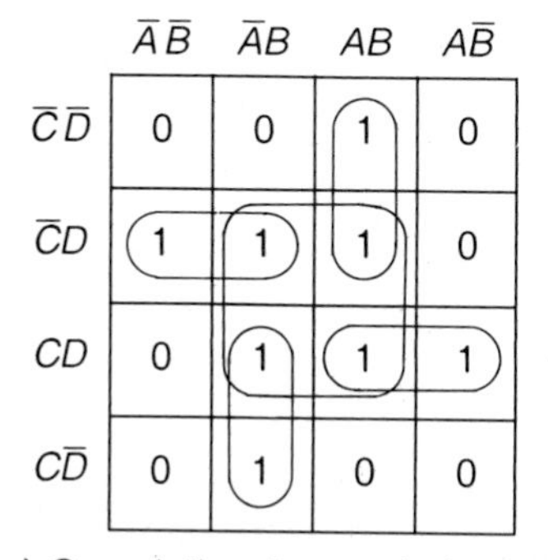

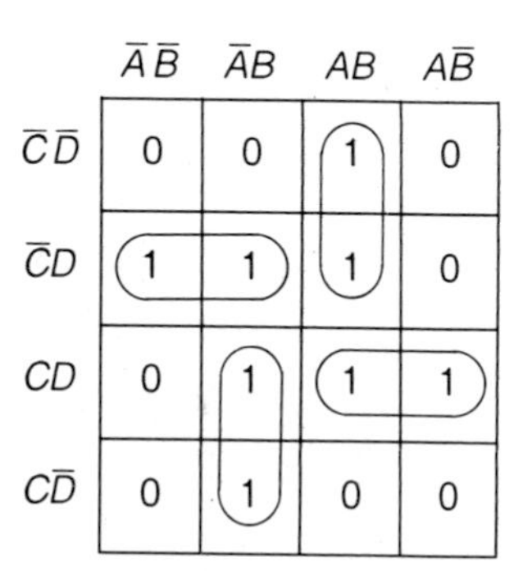

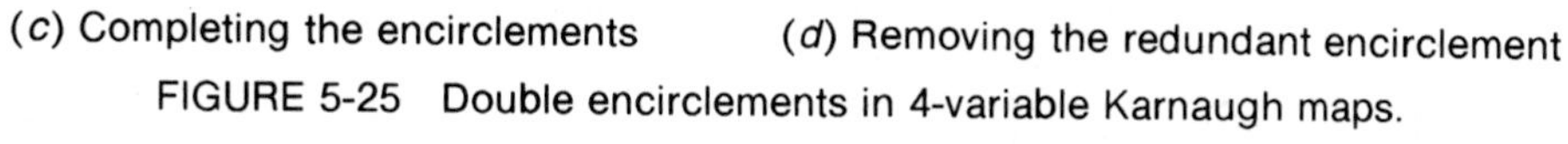

(*c*) Completing the encirclements (*d*) Removing the redundant encirclement

FIGURE 5-25 Double encirclements in 4-variable Karnaugh maps.

5-11.4 Summary of Steps

Here we summarize the steps for simplification of Boolean functions by use of Karnaugh maps.

1. Enter a 1 in the Karnaugh map for each standard product term of the Boolean function or for each 1 output of the truth table. Enter 0s in the remaining cells.
2. Encircle all adjacencies with 1 entries, starting with the largest group and working toward the smallest. Remember to encircle entries more than once where this results in larger groups. Also, use edge and corner adjacencies where possible.
3. Encircle isolated 1 entries, if any should remain.
4. Examine all encircled groups and eliminate those encirclements where *all* 1 entries are encircled more than once.
5. Write down the resulting Boolean function.

TABLE 5-7
Truth Table for the Decoder of the 8-4-2-1 BCD Number 1001

A	B	C	D	Output
0	0	0	0	0
0	0	0	1	0
0	0	1	0	0
0	0	1	1	0
0	1	0	0	0
0	1	0	1	0
0	1	1	0	0
0	1	1	1	0
1	0	0	0	0
1	0	0	1	1

Note that the above procedure does not guarantee a minimized solution. Also, more than one set of solutions is possible, though often these solutions are of comparable simplicity.

5-11.5 Don't Care Conditions

Consider a decoder for the 8-4-2-1 BCD number 1001. The truth table for the decoder, shown in Table 5-7, has 10 rows corresponding to the 10 BCD digits. The six combinations that are not allowed in 8-4-2-1 BCD numbers are 1010, 1011, 1100, 1101, 1110, and 1111. These six combinations cannot appear at the input of the decoder—or if for some reason they do appear, then we do not care whether the output of the decoder is 0 or 1.

The truth table, Table 5-7, has nine rows with 0 outputs and one row with 1 output that represents the number 1001_2. When transferring the information from the truth table to the Karnaugh map, we have to decide how to mark the cells that have unallowed combinations. We could mark each of them with a 0 or with a 1, since these combinations should not occur in any case. It turns out to be best to mark these *don't care conditions* with "x" and choose for each of these x's the binary value 0 or 1 depending on the best grouping we get from such a choice. (Note that, in addition to "x", the symbols "d" and "ϕ" are also in use for *don't care*.)

EXAMPLE 5-25 Draw a Karnaugh map for the function of Table 5-7. Group the 1s with the x's to obtain the largest grouping. What is the simplified Boolean function?

Solution The entries in the Karnaugh map are shown in Fig. 5-26*a*. The don't care conditions are entered in Fig. 5-26*b*, where we also encircled four adjacencies to obtain the simplified Boolean function $f = \boldsymbol{AD}$.

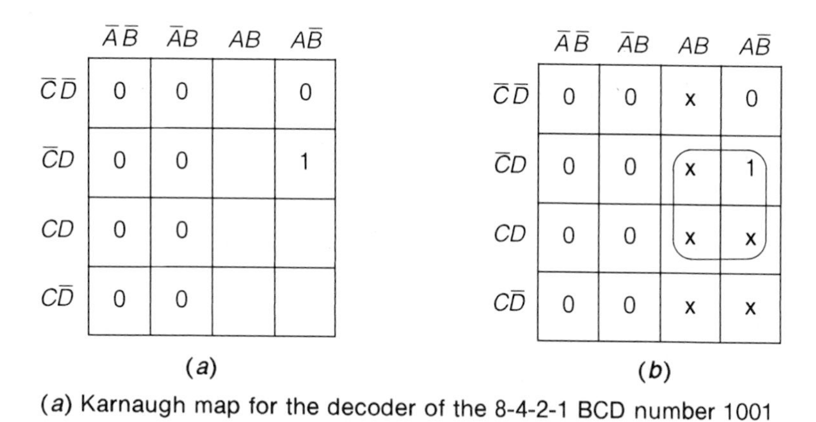

(a) Karnaugh map for the decoder of the 8-4-2-1 BCD number 1001

(b) x's entered as don't care conditions for the unspecified inputs

FIGURE 5-26 Karnaugh maps with don't care conditions.

Below we list the steps for simplification of Boolean functions with don't care conditions by use of Karnaugh maps.

1. Enter a 1 in the Karnaugh map for each standard product term of the Boolean function or for each 1 output of the truth table. Enter x's for all don't care conditions. Enter 0s in the remaining cells.
2. Encircle adjacencies with 1 entries together with as many x's as required to obtain the largest possible encircled groups, starting with the largest group and working toward the smallest. Remember to encircle entries more than once where this results in larger groups. Also, use edge and corner adjacencies where possible. Treat all encircled x's as if they were 1s. Treat all other x's as if they were 0s.
3. Encircle isolated 1 entries, if any should remain.
4. Examine all encircled groups and eliminate those encirclements where *all* 1 entries are encircled more than once (there are no requirements on the encirclement of x entries).
5. Write down the resulting Boolean function.

The above procedure is illustrated in Ex. 5-26, which follows.

EXAMPLE 5-26

Using a Karnaugh map simplify the Boolean function

$$f = \bar{A}\bar{B}\bar{C}D + A\bar{B}\bar{C}D + \bar{A}BCD + ABCD + A\bar{B}C\bar{D}$$

don't cares $\quad \bar{A}B\bar{C}D + AB\bar{C}D + \bar{A}\bar{B}\bar{C}\bar{D} + A\bar{B}\bar{C}\bar{D} + \bar{A}\bar{B}C\bar{D}$

Solution

The Karnaugh map and its encircled terms are shown in Fig. 5-27*a*. Note that don't care terms $\bar{A}B\bar{C}D$ and $AB\bar{C}D$ are used in two 4-cell encirclements. The other three don't care terms are used together with $A\bar{B}C\bar{D}$ in a 4-cell corner

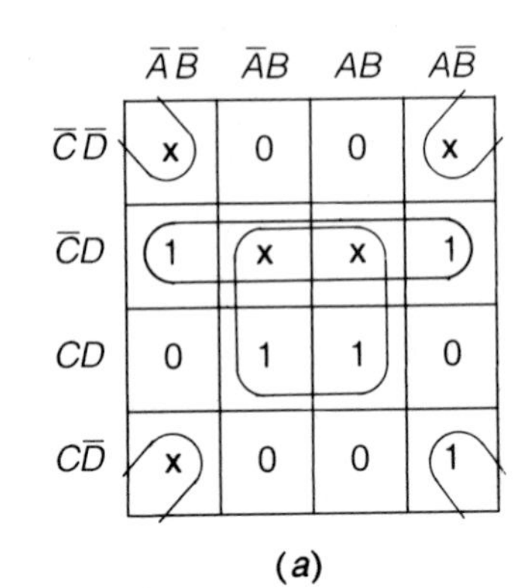

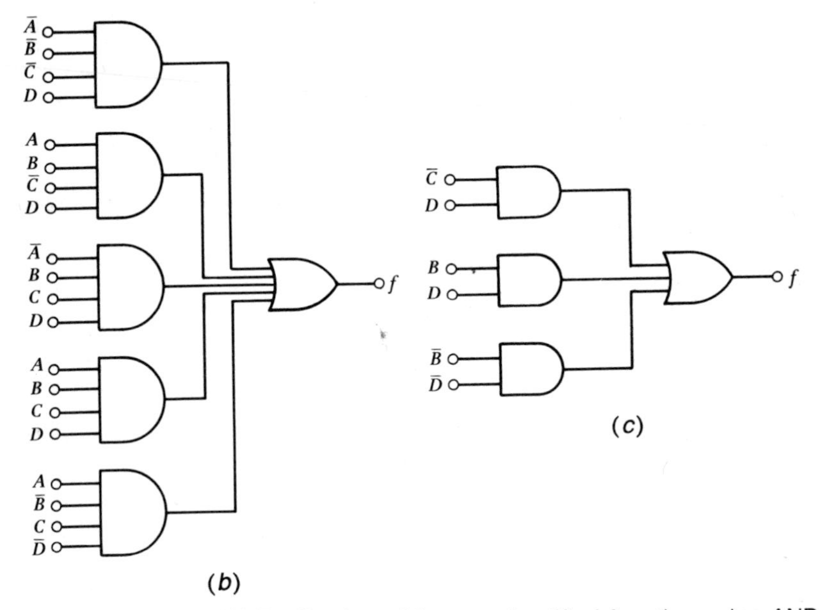

(*a*) Karnaugh map (*b*) Realization of the nonsimplified function using AND and OR gates (*c*) Realization of the simplified function using AND and OR gates

FIGURE 5-27 Simplification using a Karnaugh map with don't care conditions.

adjacency. The result is

$$f = \bar{C}D + BD + \bar{B}\bar{D}$$

The logic circuit of the nonsimplified solution is shown in Fig. 5-27*b*, and of the simplified solution in Fig. 5-27*c*. The latter realization performs an identical function and is simpler and more economical.

SUMMARY

Logic propositions and logic operators were discussed; we have also seen how information containing truth values can be presented in a truth table. The AND,

OR, and NOT logic operators were introduced next. These were followed by 10 postulates (basic assumptions) of Boolean algebra. Theorems of Boolean algebra were developed based on these postulates, and Boolean expressions and functions were simplified based on the theorems.

Analysis and synthesis of combinational logic circuits showed how logic gates may be used to implement logic propositions, truth tables, and Boolean functions. Standard forms were introduced for aiding the handling of Boolean expressions.

The Karnaugh map was introduced and was shown to be a powerful graphic simplification tool. We also saw how unspecified input conditions ("don't cares") could be used in conjunction with Karnaugh maps.

SELF-EVALUATION QUESTIONS

1. What type of sentences contain an inherent truth value?
2. Name three logic operators.
3. What is a postulate?
4. What is a theorem?
5. What useful purpose do the theorems of Boolean algebra serve?
6. Show a convenient way of listing all of the possible combinations in a truth table of any number of variables.
7. What is a Boolean sum? What is a Boolean product?
8. Given a truth table, how do you go about deriving from it a combinational logic circuit?
9. Given a combinational logic circuit in AND-OR form, how do you go about deriving from it a Boolean function?
10. Give an example of a 3-variable Boolean function written in a sum-of-products (SOP) form.
11. Give an example of a 3-variable Boolean function written in a product-of-sums (POS) form.
12. Draw a Karnaugh map for a 4-variable function and enter into each cell the respective standard product term.
13. Can you assign to "don't care conditions" the binary value 0? 1? Either? Neither? Explain.

PROBLEMS

5-1. Draw a truth table for the logic proposition: "Either Apollo 20 or Apollo 21 will be sent to planet Venus."

5-2. Draw a truth table for the switching circuit shown in Fig. P5-1.

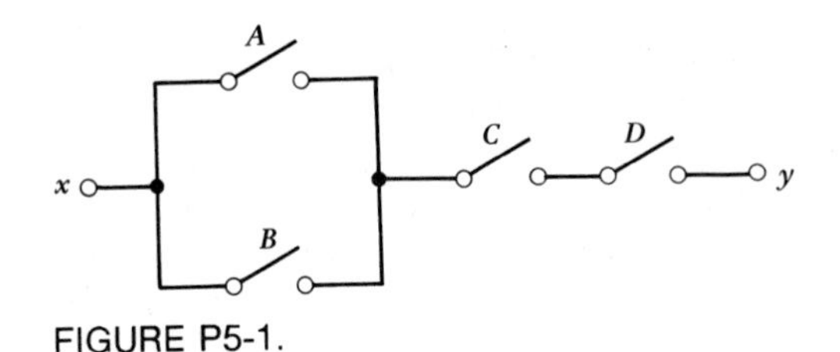

FIGURE P5-1.

5-3. Draw a truth table for the Boolean expression $AB(\overline{B} + C)$.

5-4. Draw a truth table for the Boolean expression $(XY + \overline{Z})(\overline{X} + YZ)$.

5-5. A logic circuit is described by the function $f = A\overline{B} + \overline{A}B$.

a. Draw the truth table. **b**. Draw the logic circuit.

5-6. What is (**a**) the truth table and (**b**) the logic circuit for the Boolean function $f = \overline{A}\,\overline{B} + AB$?

5-7. Establish the Boolean function for the logic circuit given in Fig. P5-2.

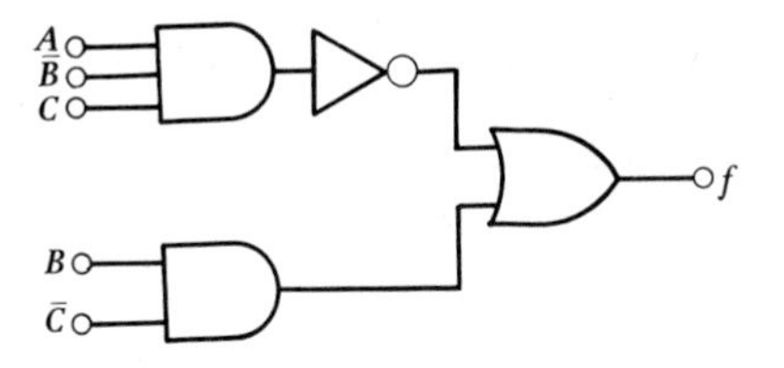

FIGURE P5-2.

5-8. What is the Boolean function for the logic circuit given in Fig. P5-3?

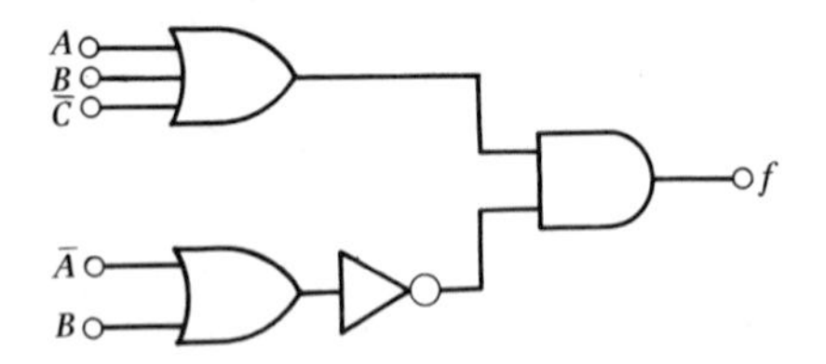

FIGURE P5-3.

5-9. Using truth tables prove De Morgan's first theorem, Theorem 10a.

5-10. Using truth tables prove De Morgan's second theorem, Theorem 10b.

5-11. Obtain the complement of the function $f = (A + \bar{B}) \cdot \overline{(\bar{C} + D)} + E\bar{F}$ using De Morgan's theorems.

5-12. Obtain the complement of the function $f = \overline{(A\bar{B} + \bar{C}D)}(\bar{\bar{E}} + \overline{ABD})$ using De Morgan's theorems.

5-13. Using Boolean theorems simplify the function

$$f = (AB + \bar{C}D)(\bar{A}B + C\bar{D}) + \bar{A}\bar{B}$$

5-14. Using Boolean theorems simplify the function

$$f = W\bar{Z}(\bar{W}\bar{X} + \bar{Y}Z) + \bar{X}Y(\bar{W}X + XZ)$$

5-15. In a function of 3 variables A, B, and C you find a product term AB. Expand it to obtain standard product terms.

5-16. In a function of 3 variables X, Y, and Z you find a sum term $X + \bar{Z}$. Expand it to obtain standard sum terms.

5-17. Using a truth table show that $\overline{X \cdot Y \cdot Z} = \bar{X} + \bar{Y} + \bar{Z}$.

5-18. Using a truth table show that $\overline{A + B + C} = \bar{A} \cdot \bar{B} \cdot \bar{C}$.

5-19. **a**. Using a Karnaugh map simplify the function

$$f = \bar{A}BC + A\bar{B}C + AB\bar{C} + ABC$$

b. Draw the simplified logic circuit.

5-20. **a**. Using a Karnaugh map simplify the function

$$f = A + AB + A\bar{C} + A\bar{B} + \bar{B}\bar{C}$$

b. Draw the simplified logic circuit.

5-21. Using a truth table prove the identity $A + \bar{A}B = A + B$.

5-22. Using a truth table prove the identity $AB(A + B) = AB$.

5-23. Using De Morgan's theorems show that $\overline{AB + \bar{A}\bar{B}} = \bar{A}B + A\bar{B}$.

5-24. Using De Morgan's theorems show that $\overline{A\bar{B} + \bar{A}B} = \bar{A}\bar{B} + AB$.

5-25. Simplify the function $f = AC + AB + B\bar{C}$ using a Karnaugh map.

5-26. Simplify the function

$$f = ABCD + \bar{A}BCD + A\bar{B}CD + AB\bar{C}D + AB\bar{C}\bar{D} + ABC\bar{D}$$

using a Karnaugh map.

5-27. Using a Karnaugh map simplify the function

$$f = \bar{A}B\bar{C}\bar{D} + AB\bar{C}\bar{D} + \bar{A}\bar{B}\bar{C}D + A\bar{B}\bar{C}D + \bar{A}BC\bar{D}$$

don't cares $\bar{A}B\bar{C}D + AB\bar{C}D + ABC\bar{D} + ABCD$

5-28. Using a Karnaugh map simplify the function

$$f = \bar{A}\bar{B}C\bar{D} + \bar{A}\bar{B}CD + \bar{A}BCD + A\bar{B}\bar{C}D + \bar{A}\bar{B}\bar{C}\bar{D} + A\bar{B}\bar{C}\bar{D}$$

don't cares $\quad ABCD + A\bar{B}CD + A\bar{B}C\bar{D} + AB\bar{C}D$

5-29. Simplify the function $f = X + \bar{X}Y$.

5-30. Simplify the function $f = \bar{X}(X + \bar{Y})$.

CHAPTER

6 Combinational Logic Circuits

Instructional Objectives

This chapter introduces additional logic operators and shows how to apply the design procedures of Chapter 5 to these operators. It also describes several medium-scale integrated (MSI) circuits and shows design examples. After you read this chapter you should be able to:

1. Design with NAND, NOR, EXCLUSIVE-OR, and EXCLUSIVE-NOR operators.
2. Apply active-high and active-low logic.
3. Design 2-stage NAND circuits.
4. Design 2-stage NOR circuits.
5. Implement Boolean functions by use of digital multiplexers.
6. Use decoders.
7. Use priority encoders.

6-1 OVERVIEW

This chapter describes simple combinational logic circuits and their use. The basic NAND and NOR gates are discussed, as well as EXCLUSIVE-OR gates, EXCLUSIVE-NOR gates, and parity generation and detection applications. Active-high and active-low logic are described next. This is followed by procedures for implementing Boolean functions by use of NAND, NOR, and AND-OR-INVERT (AOI) gates.

The chapter concludes by discussing three medium-scale integrated (MSI) circuits: the digital multiplexer (data selector) is used for the implementation of Boolean functions as well as for data routing, decoders are suitable for address decoding and for implementing Boolean functions, and priority encoders are used for establishing priority among simultaneous requests in computer systems.

A	*B*	*NAND*
0	0	1
0	1	1
1	0	1
1	1	0

(*a*) Truth table

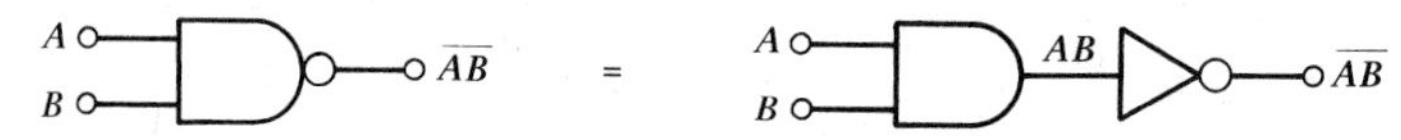

(*b*) Logic symbol (*c*) Equivalent logic circuit using an AND gate followed by a logic inverter

FIGURE 6-1 The NAND operator and the NAND gate.

6-2 ADDITIONAL LOGIC OPERATORS

Section 5-4 introduced the AND, OR, and NOT operators and gates, and characterized them using truth tables. This section introduces the NAND, NOR, EXCLUSIVE-OR, and EXCLUSIVE-NOR operators and gates and shows how to design logic circuits using them.

6-2.1 The NAND Operator and the NAND Gate

The *NAND operator* and the *NAND gate* are described by the truth table of Fig. 6-1*a*, and a logic symbol for the NAND gate is shown in Fig. 6-1*b*. As can be seen in the truth table, the NAND gate performs the function of an AND gate followed by a logic inverter (NOT gate)—this is also shown in Fig. 6-1*c*.

The NAND gate can also be used to perform the NOT operation. This can be done by applying the signal to be inverted to all inputs of the NAND gate simultaneously, as shown for a 2-input NAND gate in Fig. 6-2. Alternatively, the NOT operation can also be attained by applying the signal to be inverted to one input of the NAND gate and connecting all other inputs to logic 1 (see Prob. 6-11).

As we see later, a NAND gate may also be used for an OR operation. Thus, using only NAND gates we can implement the AND, OR, and NOT operations,

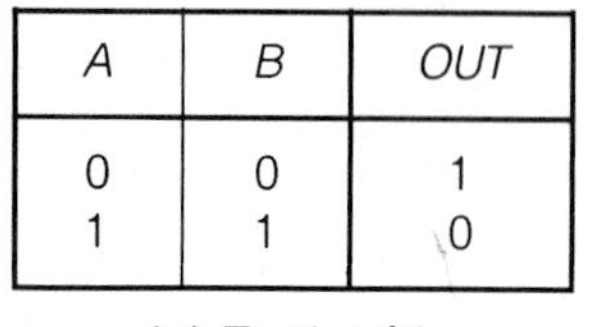

A	*B*	*OUT*
0	0	1
1	1	0

(*a*) Truth table

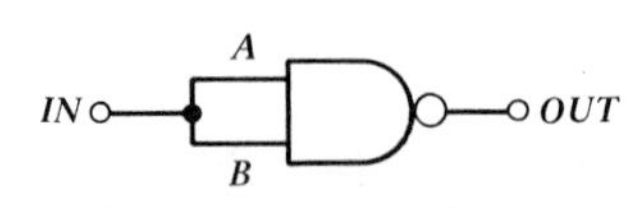

(*b*) Circuit connections

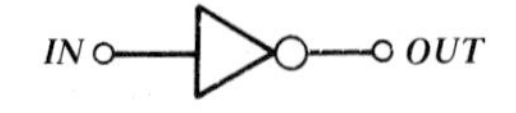

(*c*) Equivalent logic circuit

FIGURE 6-2 The NAND gate used as a logic inverter.

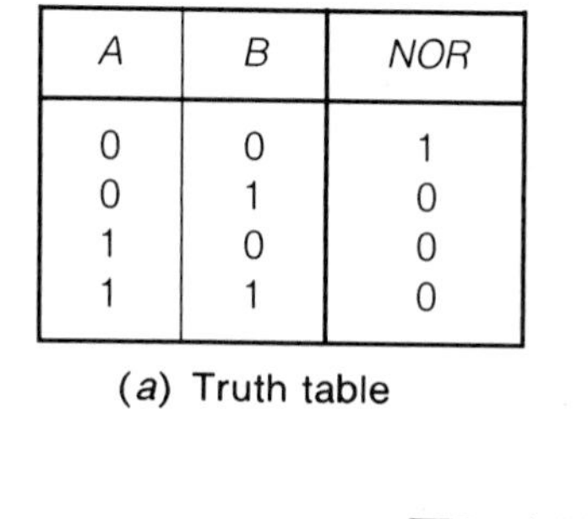

A	B	NOR
0	0	1
0	1	0
1	0	0
1	1	0

(*a*) Truth table

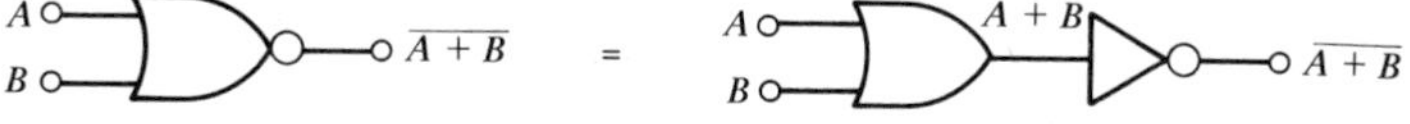

(*b*) Logic symbol (*c*) Equivalent logic circuit using an OR gate followed by a logic inverter

FIGURE 6-3 The NOR operator and the NOR gate.

hence we can also implement all Boolean functions. For this reason NAND gates are widely used in logic circuits, especially in transistor-transistor logic (TTL) described in Chapter 7.

6-2.2 The NOR Operator and the NOR gate

The *NOR operator* and the *NOR gate* are described by the truth table of Fig. 6-3*a*, and a logic symbol for the NOR gate is shown in Fig. 6-3*b*. As can be seen in the truth table, the NOR gate performs the function of an OR gate followed by a logic inverter (NOT gate)—this is also shown in Fig. 6-3*c*.

The NOR gate can also be used to perform the NOT operation. This can be done by applying the signal to be inverted to all inputs of the NOR gate simultaneously, as shown for a 2-input NOR gate in Fig. 6-4. Alternatively, the NOT operation can also be attained by applying the signal to be inverted to one input of the NOR gate and connecting all other inputs to logic 0 (see Prob. 6-12).

As we see later, a NOR gate may also be used for an AND operation. Thus, using only NOR gates we can implement the AND, OR, and NOT operations, hence we can also implement all Boolean functions. For this reason NOR gates are widely used in logic circuits, especially in emitter-coupled logic (ECL) and complementary MOS (CMOS) logic described later in Chapter 7.

A	B	OUT
0	0	1
1	1	0

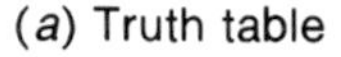

(*a*) Truth table

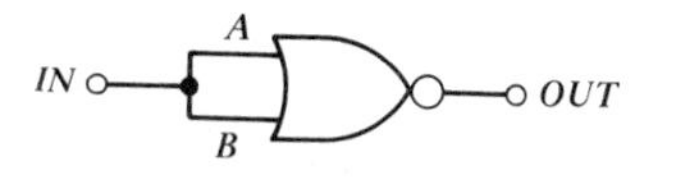

(*b*) Circuit connections

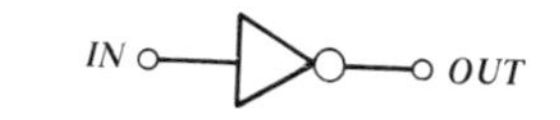

(*c*) Equivalent logic circuit

FIGURE 6-4 The NOR gate used as a logic inverter.

A	*B*	*EXCLUSIVE-OR*
0	0	0
0	1	1
1	0	1
1	1	0

(*a*) Truth table

(*b*) Logic symbol

(*c*) Equivalent logic circuit

FIGURE 6-5 The EXCLUSIVE-OR operator and the EXCLUSIVE-OR gate.

6-2.3 The EXCLUSIVE-OR Operator and the EXCLUSIVE-OR Gate

The *EXCLUSIVE-OR operator* and the *EXCLUSIVE-OR gate* are described by the truth table of Fig. 6-5*a*. A logic symbol for the EXCLUSIVE-OR gate is shown in Fig. 6-5*b*, and an equivalent logic circuit in Fig. 6-5*c*. We can see that the EXCLUSIVE-OR gate performs the function $f = A\overline{B} + \overline{A}B = A \oplus B$, where the expression $A \oplus B$ signifies that an EXCLUSIVE-OR operation is performed on the two variables A and B. The symbol $\oplus$ is the same as was used for modulo-2 addition in Sec. 4-3.

Note the difference between the EXCLUSIVE-OR, which describes the logic proposition of either, or, *but not both*, and the INCLUSIVE-OR (OR for short) which describes the logic proposition of either, or, *or both*. EXCLUSIVE-OR gates are used in digital arithmetic circuits and are available as quadruple 2-input gates in a 14-pin dual in-line package (DIP).

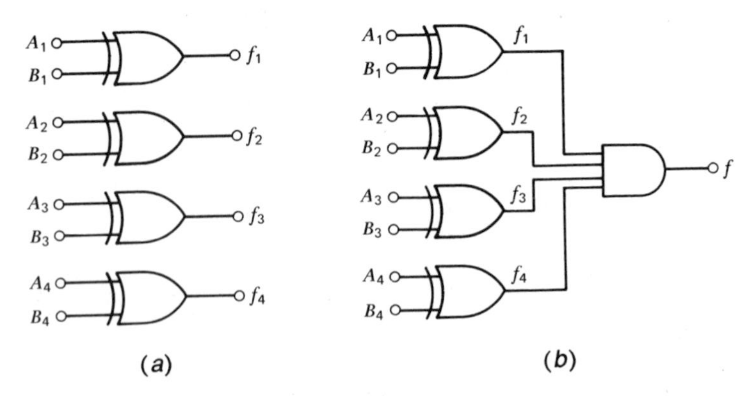

(*a*) Use of the Type 7486 quad EXCLUSIVE-OR gate (*b*) The four outputs f_1, f_2, f_3, and f_4 of the circuit in (*a*) are combined in an AND gate to yield the desired function f

FIGURE 6-6 Solution to Ex. 6-1.

EXAMPLE 6-1

Given two 4-bit words $A_1A_2A_3A_4$ and $B_1B_2B_3B_4$, design a logic circuit that delivers a logic 1 only if the two words are 1's complements of each other.

Solution

We use the Type 7486 quad EXCLUSIVE-OR gate shown in Fig. 6-6*a*. The 4 bits of *A* are applied to one set of the inputs and the 4 bits of *B* to the other set of the inputs. We thus obtain four outputs:

$$f_1 = A_1 \oplus B_1 \qquad f_2 = A_2 \oplus B_2 \qquad f_3 = A_3 \oplus B_3 \qquad f_4 = A_4 \oplus B_4$$

Outputs f_1 through f_4 are at logic 1 only if the input pairs are 1's complements of each other. We apply these outputs to a 4-input AND gate to obtain the solution:

$$\begin{aligned} f &= f_1 \cdot f_2 \cdot f_3 \cdot f_4 \\ &= (\boldsymbol{A}_1 \oplus \boldsymbol{B}_1)(\boldsymbol{A}_2 \oplus \boldsymbol{B}_2)(\boldsymbol{A}_3 \oplus \boldsymbol{B}_3)(\boldsymbol{A}_4 \oplus \boldsymbol{B}_4) \end{aligned}$$

The resulting circuit is shown in Fig. 6-6*b*.

6-2.4 Parity Generation and Detection

Section 4-4 discussed odd and even parity checks. The added check bit was called an *odd parity* bit if the sum of all 1s in the data word *and* in the parity bit was *odd*, and it was called an *even parity* bit if the sum of all 1s in the data word *and* in the parity bit was *even*. The *transmitted word* included the data word and the parity bit, and hence was one bit longer than the data word. In this section we discuss details of parity generation and detection.

In a practical application the transmitted word is sent via a transmission medium to a receiver, where the number of 1s is detected. For example, if *odd* parity is used, then each transmitted word should arrive at the receiver with an *odd* number of 1s—otherwise an error is assumed. (The situation is similar if even parity is used.)

EXCLUSIVE-OR gates may be used for *parity generation* and for *parity detection*. A parity generator for a 4-bit data word is shown in Fig. 6-7*a*, and its truth table in Fig. 6-7*b*. The circuit establishes the parity of the data word and generates the parity bit, which is transmitted together with the data bits.

EXAMPLE 6-2

A parity bit is to be generated for the 4-bit data word 1011. What is the additional parity bit for (**a**) odd parity, (**b**) even parity?

Solution

a. For odd parity the additional parity bit is **0**.
b. For even parity the additional parity bit is **1**.

Figure 6-8 shows a transmitter-receiver channel for 8-bit data words. A parity generator produces the desired parity bit at the transmitting end. The parity bit is sent together with the data bits to the receiving end, where a parity check is carried out. If the parity is *correct*, the *receiver* is activated; if the parity is *incorrect*, an *alarm* is activated. Figure 6-9 shows details of an odd-parity checker for 8-bit data words.

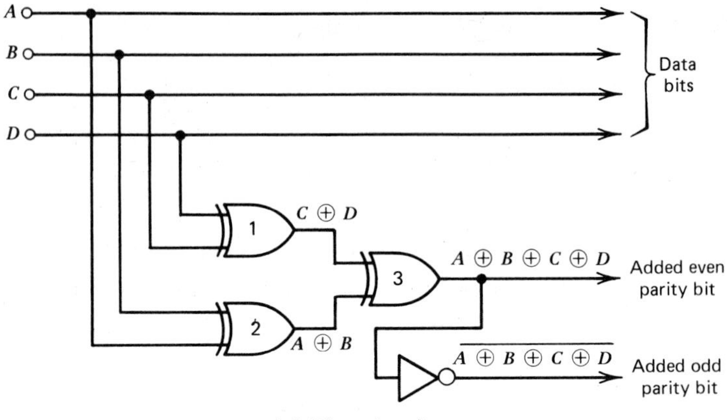

(*a*) The circuit

A	*B*	*C*	*D*	Parity	
				Even	Odd
0	0	0	0	0	1
0	0	0	1	1	0
0	0	1	0	1	0
0	0	1	1	0	1
0	1	0	0	1	0
0	1	0	1	0	1
0	1	1	0	0	1
0	1	1	1	1	0
1	0	0	0	1	0
1	0	0	1	0	1
1	0	1	0	0	1
1	0	1	1	1	0
1	1	0	0	0	1
1	1	0	1	1	0
1	1	1	0	1	0
1	1	1	1	0	1

(*b*) Truth table

FIGURE 6-7 Even/odd parity generator for a 4-bit data word.

6-2.5 The EXCLUSIVE-NOR Operator and the EXCLUSIVE-NOR Gate

The *EXCLUSIVE-NOR operator* and the *EXCLUSIVE-NOR gate* (which is also known as *EQUALITY COMPARATOR* or *COINCIDENCE gate*) are described by the truth table of Fig. 6-10*a*. A logic symbol for the EXCLUSIVE-NOR gate is shown in Fig. 6-10*b*, and an equivalent logic circuit in Fig. 6-10*c*. We can see that the EXCLUSIVE-NOR gate performs the function $f = \overline{A}\,\overline{B} + AB = \overline{A \oplus B}$,

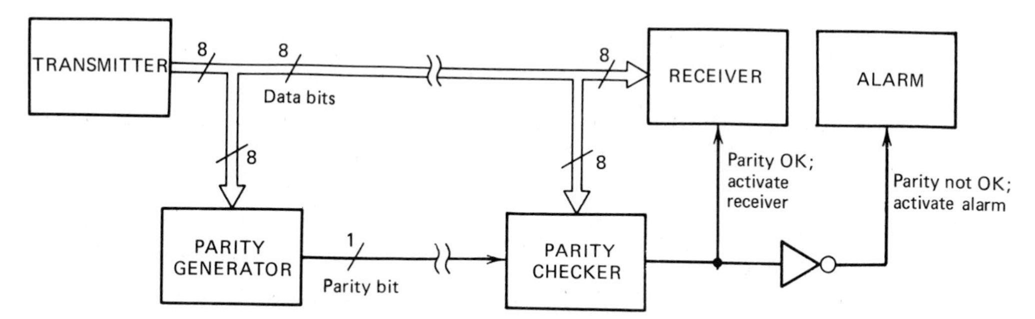

FIGURE 6-8 A transmitter/receiver channel with parity generator and checker for 8-bit data words.

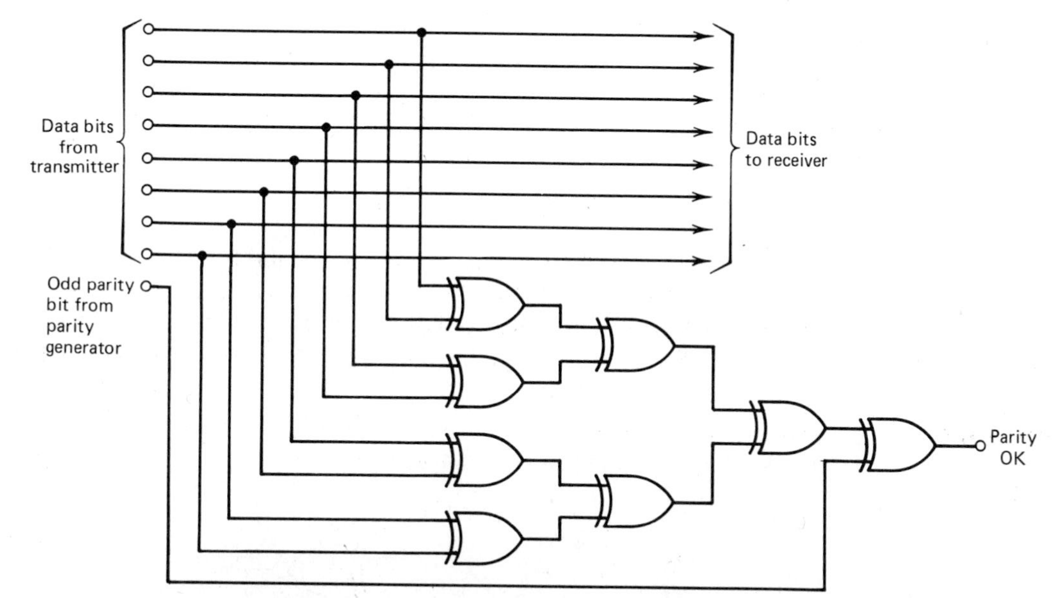

FIGURE 6-9 Odd-parity checker for 8-bit data words.

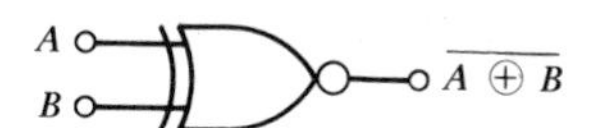

(*c*) Equivalent logic circuit

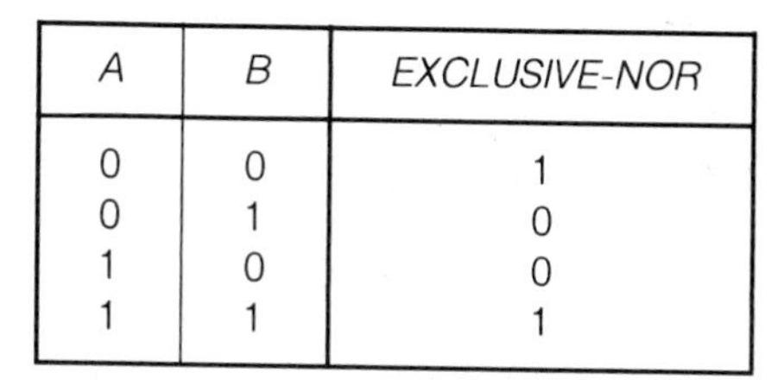

A	*B*	*EXCLUSIVE-NOR*
0	0	1
0	1	0
1	0	0
1	1	1

(*a*) Truth table

(*b*) Logic symbol

FIGURE 6-10 The EXCLUSIVE-NOR operator and the EXCLUSIVE-NOR gate.

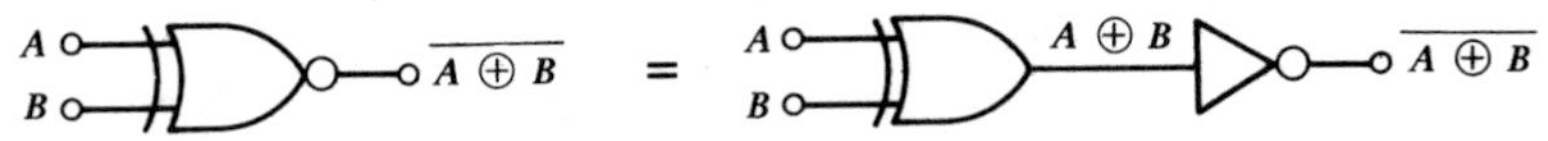

FIGURE 6-11 An EXCLUSIVE-NOR gate is equivalent to an EXCLUSIVE-OR gate followed by a logic inverter.

where the symbol $\overline{A \oplus B}$ signifies that an EXCLUSIVE-NOR operation is performed on the two variables A and B. As indicated by the overbar, the EXCLUSIVE-NOR is the complement of the EXCLUSIVE-OR.

EXAMPLE 6-3

Using De Morgan's theorems show that the EXCLUSIVE-NOR is the complement of the EXCLUSIVE-OR.

Solution

$$
\begin{aligned}
f = \overline{A \oplus B} &= \overline{\bar{A}B + A\bar{B}} \\
&= (A + \bar{B})(\bar{A} + B) \\
&= A\bar{A} + \bar{A}\bar{B} + AB + B\bar{B} \\
&= \bar{A}\bar{B} + AB
\end{aligned}
$$

The logical equivalence is illustrated in Fig. 6-11.

Note that in addition to the symbol that represents the EXCLUSIVE-NOR as the complement of the EXCLUSIVE-OR, the symbol $\odot$ is also in use; that is, $\bar{A}\bar{B} + AB = \overline{A \oplus B} = A \odot B$. EXCLUSIVE-OR and EXCLUSIVE-NOR gates are available as quadruple 2-input gates in a 14-pin dual in-line package (DIP).

EXAMPLE 6-4

Given two 4-bit words $A_1A_2A_3A_4$ and $B_1B_2B_3B_4$, design a logic circuit that delivers a logic 1 if the two words are identical.

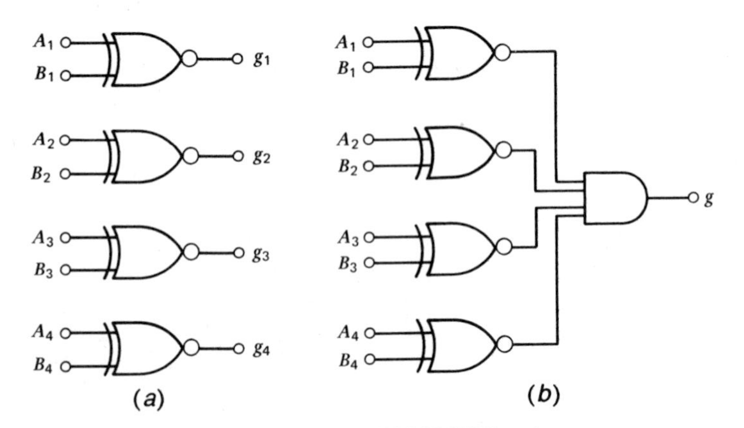

(a) Use of the Type 74LS266 quad EXCLUSIVE-NOR gate

(b) The four outputs g_1, g_2, g_3, and g_4 of the circuit in (a) are combined to yield the desired function g

FIGURE 6-12 Solution to Ex. 6-4.

Solution

We use the Type 74LS266 quad EXCLUSIVE-NOR gate shown in Fig. 6-12*a*. The 4 bits of A are applied to one set of inputs and the 4 bits of B to the other set of inputs. We thus obtain four outputs:

$$g_1 = \overline{A_1 \oplus B_1} \qquad g_2 = \overline{A_2 \oplus B_2}$$
$$g_3 = \overline{A_3 \oplus B_3} \qquad g_4 = \overline{A_4 \oplus B_4}$$

Outputs g_1 through g_4 are at logic 1 only if the input pairs are identical. We apply these outputs to a 4-input AND gate to obtain the solution:

$$\begin{aligned} g &= g_1 \cdot g_2 \cdot g_3 \cdot g_4 \\ &= (\overline{\boldsymbol{A}_1 \oplus \boldsymbol{B}_1})(\overline{\boldsymbol{A}_2 \oplus \boldsymbol{B}_2})(\overline{\boldsymbol{A}_3 \oplus \boldsymbol{B}_3})(\overline{\boldsymbol{A}_4 \oplus \boldsymbol{B}_4}). \end{aligned}$$

The resulting circuit is shown in Fig. 6-12*b*.

6-3 ACTIVE-HIGH AND ACTIVE-LOW LOGIC

In the preceding, truth tables described logic gates by use of logic levels of 0 V and +5 V, as well as logic 0 and logic 1. Here we introduce another pair of logic levels, HIGH and LOW, where HIGH represents the more positive logic level of a logic gate and LOW represents its more negative logic level.

Table 6-1*a* describes a logic circuit by use of HIGH and LOW logic levels. We can see that output *OUT* is HIGH only when both inputs A and B are HIGH. Thus, Table 6-1*a* is applicable to AND gate circuits of diode-transistor logic, DTL (Sec. 2-5), by equating HIGH with +5 V and LOW with 0 V.

Table 6-1*a* is applicable not only to DTL, but to other logic circuits as well. For example, in emitter-coupled logic, discussed later in Chapter 7, the HIGH logic level is −0.8 V and the LOW logic level is −1.6 V. Thus, Table 6-1*a* is applicable to ECL AND gate circuits by equating HIGH with −0.8 V and LOW with −1.6 V.

Table 6-1*b* shows a truth table where the HIGH entries of Table 6-1*a* are represented by logic 1 and the LOW entries by logic 0. We can see that output *OUT* is at logic 1 only when both inputs A and B are at logic 1. This, according to Fig. 5-3*d*, page 75, is consistent with the description of the AND operation we have been using.

The convention of representing HIGH by logic 1 and LOW by logic 0 is called *active-high logic* or *positive logic*. The use of this convention is common in digital systems; however, it is not universal. There is another convention, called *active-low logic* or *negative logic*, where HIGH is represented by logic 0 and LOW is represented by logic 1.

Table 6-1*c* shows a truth table using active-low logic, where the HIGH entries of Table 6-1*a* are represented by logic 0 and the LOW entries by logic 1. We can see that output *OUT* is at logic 1 when either input A or B, or both, are at logic 1. Thus, according to Fig. 5-4*d* (page 76), Table 6-1*c* describes an OR operation—*not* an AND operation.

The logic gate circuit (the actual hardware) described by Table 6-1*a* implements an AND operation in the active-high logic convention (Table 6-1*b*), and it

TABLE 6-1*a*
Truth Table of a Logic Circuit

A	*B*	*OUT*
LOW	LOW	LOW
LOW	HIGH	LOW
HIGH	LOW	LOW
HIGH	HIGH	HIGH

TABLE 6-1*b*
The Truth Table of Table 6-1*a* with Active-High Logic Levels

A	*B*	OUT
0	0	0
0	1	0
1	0	0
1	1	1

TABLE 6-1*c*
The Truth Table of Table 6-1*a* with Active-Low Logic Levels

A	*B*	OUT
1	1	1
1	0	1
0	1	1
0	0	0

implements an OR operation in the active-low logic convention (Table 6-1*c*). These two applications of the same logic circuit are often (but not always) distinguished by use of two different symbols, as shown in Fig. 6-13. Note that both symbols refer to the same logic gate circuit described by Table 6-1*a*.

According to the above, an AND gate in active-high logic is also an OR gate in active-low logic. Similarly, it can be shown that an OR gate in active-high logic is also an AND gate in active-low logic, a NAND gate in active-high logic is also a NOR gate in active-low logic, and a NOR gate in active-high logic is also a NAND gate in active-low logic. However, a logic inverter in active-high logic is also a logic inverter in active-low logic, and the same holds for EXCLUSIVE-OR and EXCLUSIVE-NOR gates.

Active-high and active-low logic symbols for logic inverters were shown, respectively, in Fig. 5-5*c* and Fig. 5-5*d* (page 77). Symbols for other logic gates are shown in Fig. 6-14—note that the two symbols of a row represent the same logic gate circuit (that is, the same actual hardware).

To change from active-high logic to active-low logic, follow the steps given below:

1. Change AND symbols to OR symbols and vice versa. Also change NAND symbols to NOR symbols and vice versa.

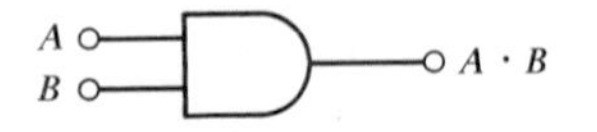

(*a*) AND gate symbol in active-high logic

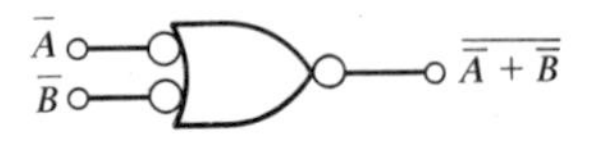

(*b*) OR gate symbol in active-low logic

FIGURE 6-13 Two equivalent symbols for the logic gate circuit described by Table 6-1*a*.

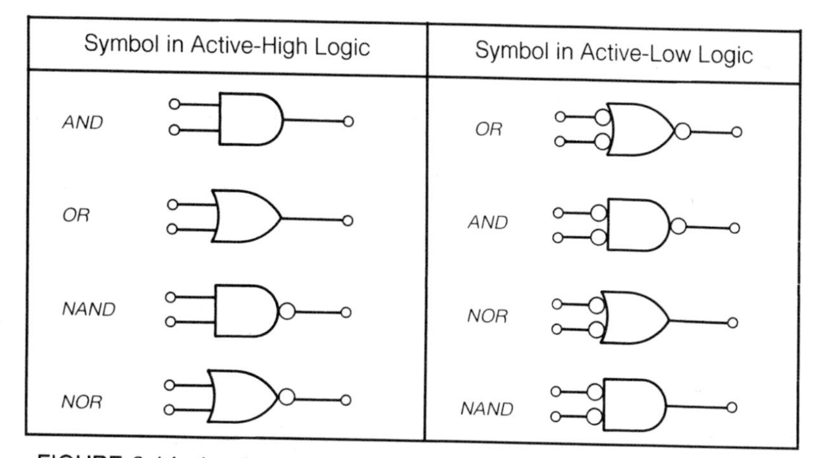

FIGURE 6-14 Logic gate symbols in active-high and active-low logic.

2. Add circles at the input and output terminals wherever there are no circles.
3. Remove circles at the input and output terminals wherever there are circles.

EXAMPLE 6-5

Change a 2-input active-high NAND gate to a 2-input active-low NOR gate.

Solution

Step 1. Change the symbol of the NAND gate to that of a NOR gate.

Step 2. Add circles at the 2 gate inputs.

Step 3. Remove the circle at the gate output.

The resulting symbol is shown at right in the third row of Fig. 6-14.

The active-high logic convention is much more common in digital systems than the active-low logic convention. For this reason, in what follows in this book we use active-high logic unless otherwise indicated.

6-4 TWO-STAGE NAND LOGIC CIRCUITS

Consider the 2-stage NAND logic circuit of Fig. 6-15*a*. The output from gate 1 is $\overline{AB}$ and from gate 2 it is $\overline{CD}$. These applied to gate 3 result at its output in

$$\overline{\overline{AB} \cdot \overline{CD}} = AB + CD$$

Thus, the 2-stage NAND-NAND circuit of Fig. 6-15*a* is equivalent to the 2-stage AND-OR circuit of Fig. 5-12*b* (page 86), and is shown in Fig. 6-15*b*. Also, since gate 3 in Fig. 6-15*a* performs the function of an active-low NOR, we can redraw Fig. 6-15*a* as shown in Fig. 6-15*c*.

From the equivalence of NAND-NAND and AND-OR circuits, we find that Karnaugh map simplification for NAND gates follows the same rules that were given for AND-OR gates in Sec. 5-11. This is illustrated in Ex. 6-6, which follows.

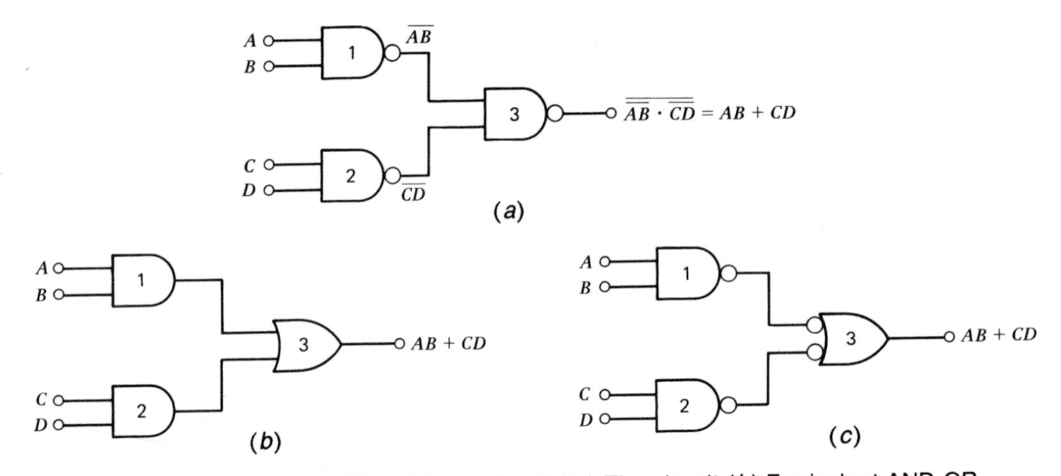

FIGURE 6-15 A 2-stage NAND logic circuit. (*a*) The circuit (*b*) Equivalent AND-OR logic circuit (*c*) Using symbols of active-high NAND gates followed by an active-low NOR gate.

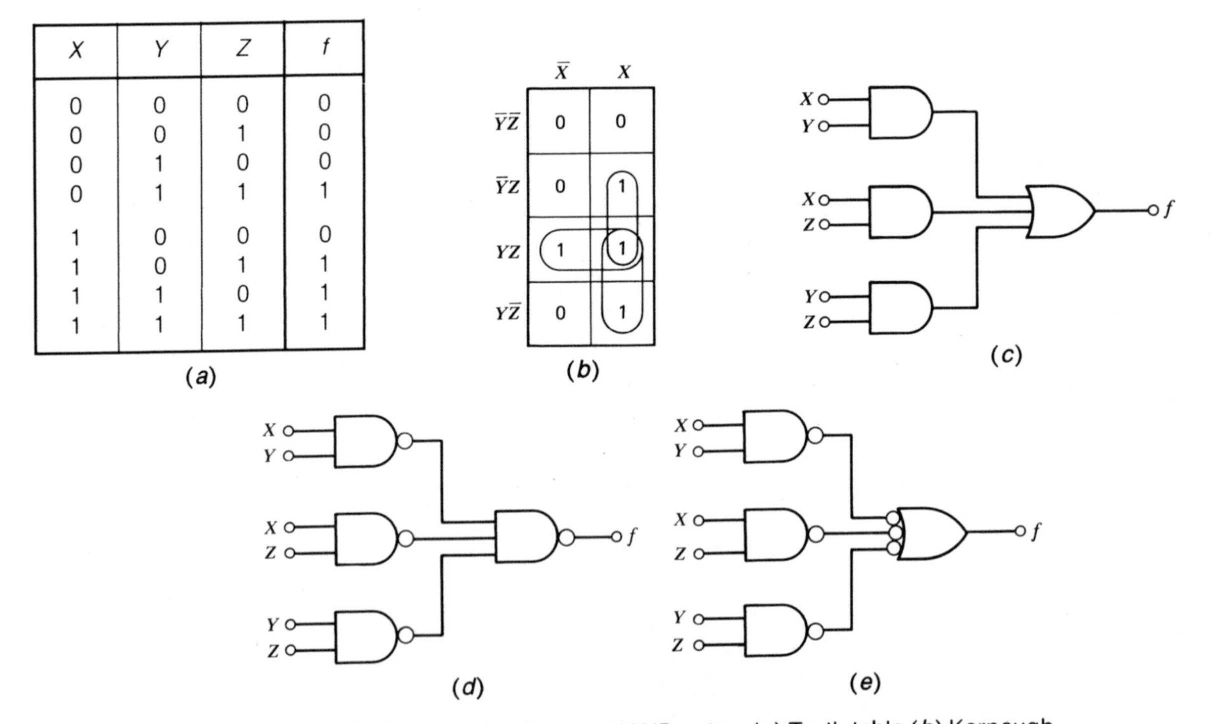

X	Y	Z	f
0	0	0	0
0	0	1	0
0	1	0	0
0	1	1	1
1	0	0	0
1	0	1	1
1	1	0	1
1	1	1	1

FIGURE 6-16 Majority voter circuit using NAND gates. (*a*) Truth table (*b*) Karnaugh map (*c*) Implementation by a 2-stage AND-OR circuit (*d*) Implementation by a 2-stage NAND-NAND circuit (*e*) Using symbols of active-high NAND gates followed by an active-low NOR gate.

EXAMPLE 6-6

Using only NAND gates design a simplified circuit of three variables in which the output always agrees with the majority of the inputs. (This circuit is also called a "majority voter" circuit and is used in high-reliability applications in which all circuits are tripled. The "majority voter" in such a circuit is incorporated to resolve any discrepancies in the outputs in case of failure of one circuit.)

Solution

The truth table of the "majority voter" is shown in Fig. 6-16*a* and its Karnaugh map in Fig. 6-16*b*, from which we derive the function

$$f = XY + XZ + YZ$$

Realization of the function with AND-OR gates is shown in Fig. 6-16*c* and with NAND gates in Fig. 6-16*d*. Finally, in Fig. 6-16*e* we use active-high NAND and active-low NOR logic symbols.

6-5 TWO-STAGE NOR LOGIC CIRCUITS

Consider the 2-stage NOR logic circuit of Fig. 6-17*a*. The output from gate 1 is $\overline{A + B}$ and from gate 2 it is $\overline{C + D}$. These applied to gate 3 result at its output in

$$\overline{\overline{A + B} + \overline{C + D}} = (A + B)(C + D)$$

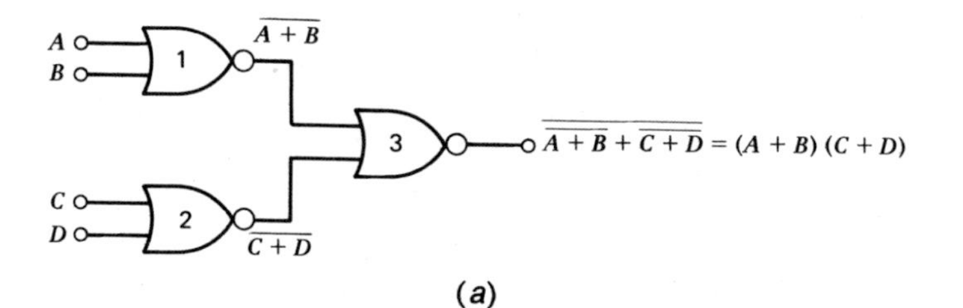

(*a*)

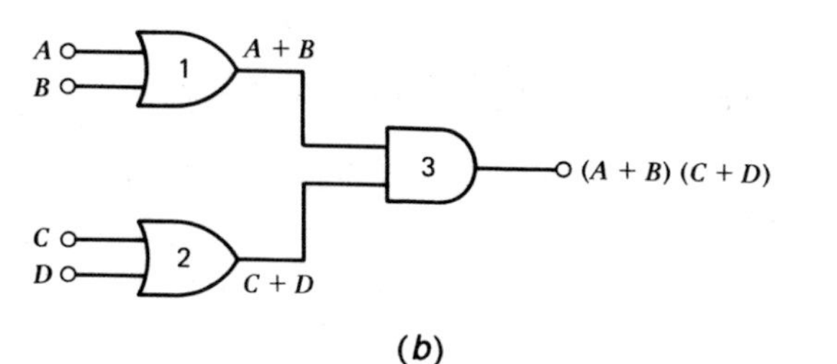

(*b*)

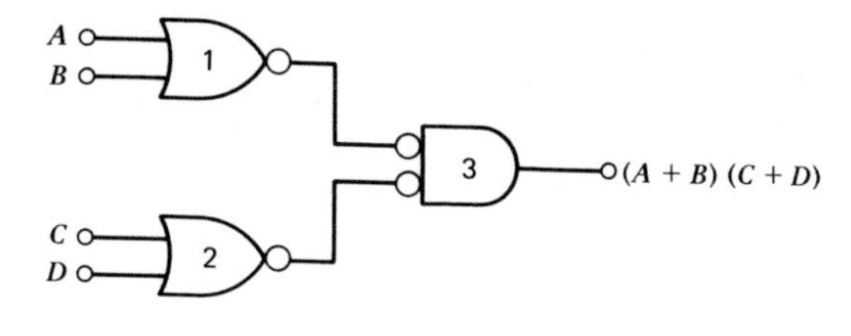

(*c*)

(*a*) The circuit

(*b*) Equivalent AND-OR logic circuit

(*c*) Using symbols of active-high NOR gates followed by an active-low NAND gate

FIGURE 6-17 A 2-stage NOR logic circuit.

X	Y	Z	f
0	0	0	0
0	0	1	0
0	1	0	0
0	1	1	1
1	0	0	0
1	0	1	1
1	1	0	1
1	1	1	1

(a)

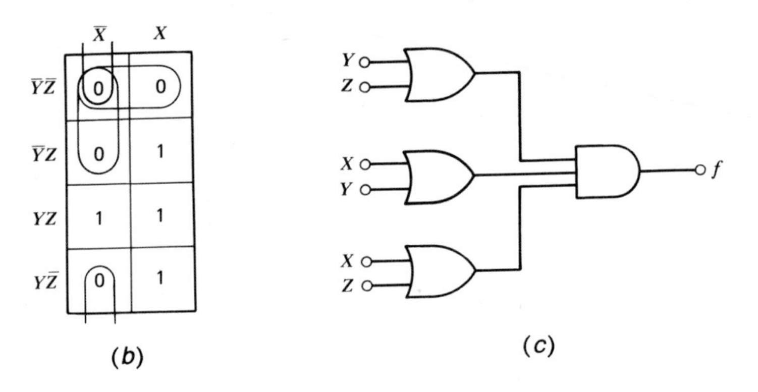

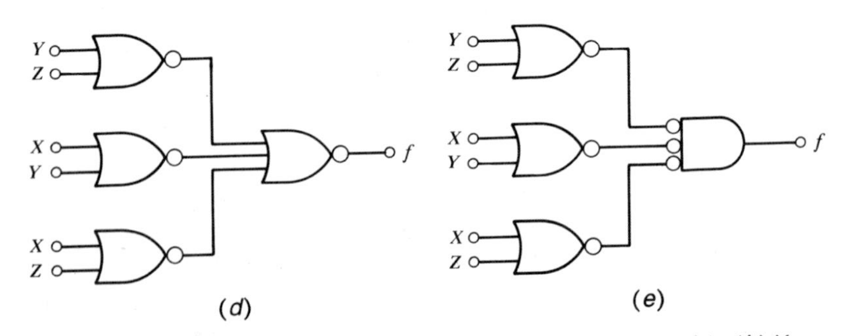

FIGURE 6-18 Majority voter circuit using NOR gates. (a) Truth table (b) Karnaugh map with encircled 0 entries (c) Implementation by a 2-stage OR-AND circuit (d) Implementation by a 2-stage NOR-NOR circuit (e) Using symbols of active-high NOR gates followed by an active-low NAND gate.

Thus, the 2-stage NOR-NOR circuit of Fig. 6-17*a* is equivalent to the 2-stage OR-AND circuit of Fig. 5-13 (page 89) and is shown in Fig. 6-17*b*. Also, since gate 3 in Fig. 6-17*a* performs the function of an active-low NAND, we can redraw Fig. 6-17*a* as shown in Fig. 6-17*c*. Karnaugh map simplification of NOR-NOR circuits is illustrated in Ex. 6-7, which follows.

EXAMPLE 6-7

Using only NOR gates realize a simplified circuit for the "majority voter" of Ex. 6-6.

Solution

We simplify the *complement* of the function, $\bar{f}$, in a Karnaugh map and apply De Morgan's theorems to obtain an OR-AND realization, which is then converted to a NOR-NOR realization. The detailed steps are the following:

Step 1. Draw the truth table as shown in Fig. 6-18*a*.

Step 2. Draw the Karnaugh map as shown in Fig. 6-18*b*.

Step 3. Encircle the *0 entries* to obtain a simplified *complement* of f as $\bar{f} = \bar{Y}\bar{Z} + \bar{X}\bar{Y} + \bar{X}\bar{Z}$.

Step 4. Complement $\bar{f}$ to obtain $\bar{\bar{f}} = f = \overline{\bar{Y}\bar{Z} + \bar{X}\bar{Y} + \bar{X}\bar{Z}}$.

Step 5. Apply De Morgan's theorems to $\bar{\bar{f}} = f$ to obtain a simplified OR-AND realization as $\bar{\bar{f}} = f = (Y + Z)(X + Y)(X + Z)$.

Step 6. Draw the resulting OR-AND realization as shown in Fig. 6-18*c*.

Step 7. From the OR-AND realization of Fig. 6-18*c*, derive the NOR-NOR realization as shown in Fig. 6-18*d*.

Step 8. Redraw the symbol of the output logic gate in Fig. 6-18*d* using an active-low NAND gate symbol as shown in Fig. 6-18*e*.

Note that if we had don't care conditions in the Karnaugh map, in Step 3 we would combine some or all of these with the 0 entries in order to obtain encirclements that are as large as possible.

6-6 AND-OR-INVERT (AOI) GATES

An AND-OR-INVERT (AOI) gate performs the same logic function as two or more AND gates followed by a NOR gate. AOI gates are specified as *m*-wide, *n*-input, where *m* is the number of AND gates and *n* is the number of inputs to each AND gate. Commonly 2 to 4 AND gates are included in an AOI gate, and the number of inputs to each AND gate varies between 2 and 4. A 4-wide 2-input AOI gate is shown in Fig. 6-19; it performs the function

$$f = \overline{AB + CD + EF + GH}$$

The number of AND gates and the number of inputs to each AND gate are limited mainly by the number of pins of the IC package. Thus, for example, the AOI gate of Fig. 6-19 requires 9 pins for the connections of the inputs and the output and 2 pins for power and ground, that is, a total of 11 pins. Similarly, a 4-wide 3-input AOI gate requires a total of 15 pins. Logic functions requiring

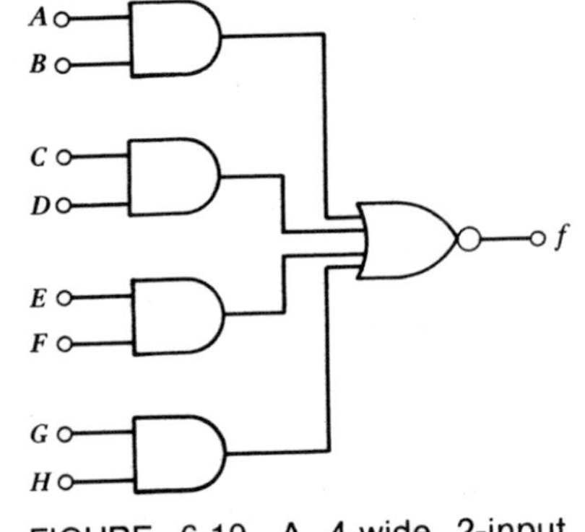

FIGURE 6-19 A 4-wide 2-input AND-OR-INVERT (AOI) gate.

more AND gates than are available in a single IC package may be implemented by the combination of several AOI gates. This can be done using *expandable AOI gates* that can be connected together via their *expander nodes*. Another possibility is to connect the outputs of several AOI gates in parallel in a *wired-OR* configuration (see Sec. 2-5.4).

Karnaugh maps may be used for the simplification of functions when designing with AOI gates. The steps are the same as in an AND-OR implementation. However, since here we have an AND-OR-INVERT instead of an AND-OR, we use the 0 entries of the Karnaugh map for simplification instead of the 1 entries. Also, as before, we can choose for each don't care condition either 0 or 1 depending on the best grouping we would get from such a choice. This is illustrated in Ex. 6-8, which follows.

EXAMPLE 6-8

Implement the 4-variable function

$$f = \bar{A}\bar{B}\bar{C}\bar{D} + \bar{A}\bar{B}C\bar{D} + \bar{A}\bar{B}CD + \bar{A}B\bar{C}\bar{D} + A\bar{B}\bar{C}\bar{D} + A\bar{B}\bar{C}D + A\bar{B}C\bar{D} + ABC\bar{D}$$

don't cares $\bar{A}BCD + AB\bar{C}D$

a. Use a single 4-wide 3-input AOI gate.

b. Use two 2-wide 3-input AOI gates in a wired-OR configuration.

Solution

a. The Karnaugh map of the function is shown in Fig. 6-20*a*, where we also encircled adjacencies by one 4-cell encirclement and four 2-cell encirclements. However, we can see that *all* 0 terms in the 4-cell encirclement are encircled twice; hence, the 4-cell encirclement is not necessary. The resulting simplified function is

$$f = \overline{\bar{A}\bar{C}D + \bar{A}BC + AB\bar{C} + ACD}$$

Figure 6-20*b* shows the resulting implementation using a 4-wide 3-input AOI gate.

b. Using De Morgan's second theorem, the simplified function f can be also written as

$$f = \overline{\bar{A}\bar{C}D + \bar{A}BC} \cdot \overline{AB\bar{C} + ACD}$$

	$\bar{A}\bar{B}$	$\bar{A}B$	AB	$A\bar{B}$
$\bar{C}\bar{D}$	1	1	0	1
$\bar{C}D$	0	0	x	1
CD	1	x	0	0
$C\bar{D}$	1	0	1	1

(*a*) Karnaugh map

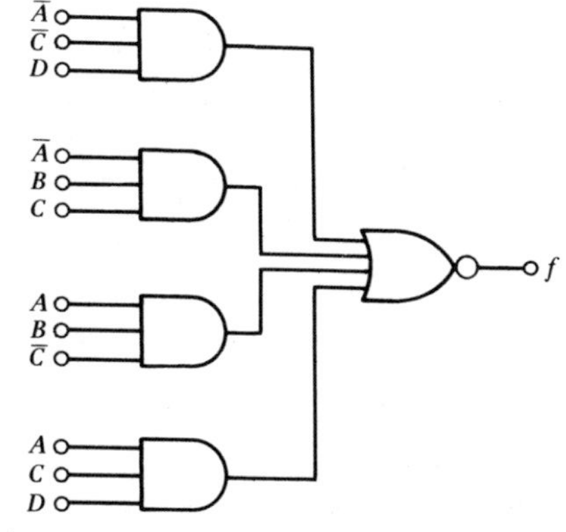

(*b*) Implementation using a 4-wide 3-input AOI gate

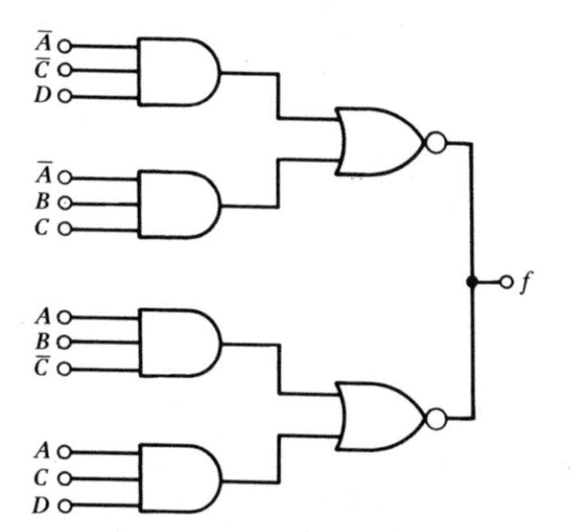

(*c*) Implementation using two 2-wide 3-input AOI gates in a wired-OR configuration

FIGURE 6-20 Solution to Ex. 6-8.

This can be implemented using two 2-wide 3-input AOI gates as shown in Fig. 6-20*c*. Because the outputs of the upper and the lower AOI gates are connected in parallel in a wired-OR configuration, the output is HIGH (logic 1) when $\overline{\bar{A}\bar{C}D + \bar{A}BC}$ *and* $\overline{AB\bar{C} + ACD}$ are *both* HIGH (logic 1)—as required by the function *f*.

6-7 DIGITAL MULTIPLEXERS

A *digital multiplexer* (*MPX* or *MUX*), also known as *data selector*, is a combinational logic circuit that has 1 to 4 *control inputs*, 2 to 16 *data inputs*, and 1 *data output*. (Some units also provide the complemented data output.)

The digital multiplexer performs the function of a single-pole *multiposition switch* in which one of the data inputs is connected to the output at any one time, the particular data input being selected by the state of the control inputs.

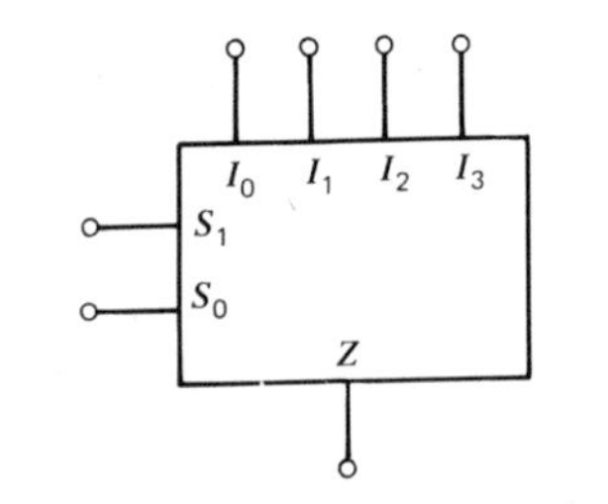

(*a*) Logic symbol

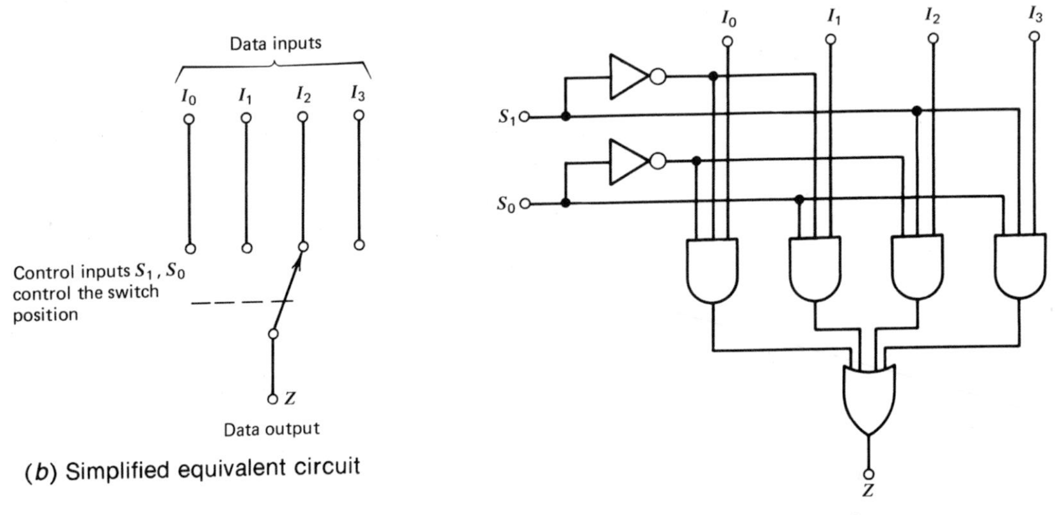

(*c*) Logic circuit

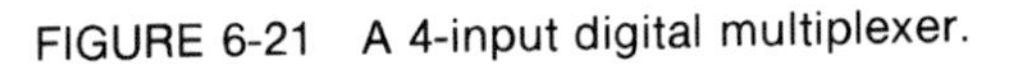
FIGURE 6-21 A 4-input digital multiplexer.

Below we present a detailed discussion of digital multiplexers with 4 data inputs, followed by brief descriptions of those with 2, 8, and 16 data inputs.

6-7.1 Four-Input Digital Multiplexers

This section describes digital multiplexers with 4 data inputs. Structure, data routing applications, use for the implementation of Boolean functions, and availability are discussed.

6-7.1.1 Structure

Figure 6-21*a* shows the logic symbol of a digital multiplexer with 4 data inputs, Fig. 6-21*b* a simplified equivalent circuit, and Fig. 6-21*c* a logic diagram.

Note that, as can be clearly seen in Fig. 6-21*c* but not in Fig. 6-21*b*, a digital multiplexer is *not* suitable for routing data from a single source to several destinations. Thus data inputs I_0, I_1, I_2, and I_3 must be driven and data output Z observed, and not the other way around.

A truth table for a digital multiplexer with 4 data inputs would require 6 input columns: S_0, S_1, I_0, I_1, I_2, and I_3. Such a table would, therefore, have $2^6 = 64$

TABLE 6-2
Function Table of a 4-Input Digital Multiplexer

S_1	S_0	I_0	I_1	I_2	I_3	Z
L	L	L	x	x	x	L
L	L	H	x	x	x	H
L	H	x	L	x	x	L
L	H	x	H	x	x	H
H	L	x	x	L	x	L
H	L	x	x	H	x	H
H	H	x	x	x	L	L
H	H	x	x	x	H	H

rows and would be unwieldy to work with. A little consideration shows that not all the 64 rows are required. For example, when both control inputs S_1 and S_0 are at logic 0, input I_0 is connected to output Z. In this case, the states of inputs I_1, I_2, and I_3 are immaterial and can be considered as don't care conditions. We can thus reduce the 64-row truth table to an 8-row *function table* as shown in Table 6-2, where H stands for HIGH and L for LOW. Note that the function table provides all the information that is contained in a truth table, but it does it in an abbreviated form.

EXAMPLE 6-9

Refer to the logic symbol of Fig. 6-21*a* and to the function table, Table 6-2. What is the state of output Z when control input S_1 is HIGH (H) and control input S_0 is LOW (L)?

Solution

The function table, Table 6-2, shows the above states of the control inputs in its fifth and sixth rows. We can see that in these rows the state of output Z follows the state of data input I_2. Thus the state of output Z is LOW when the state of I_2 is LOW and, conversely, the state of Z is HIGH when the state of I_2 is HIGH. Also, the states of data inputs I_0, I_1, and I_3 do not matter since their columns show x's in the fifth and sixth rows of the function table. Thus $Z = I_2$ for all combinations of all four data inputs I_0, I_1, I_2, and I_3 as long as S_1 is HIGH and S_0 is LOW.

6-7.1.2 *Data Routing*

Digital multiplexers can be used whenever a selection is to be made from several data sources for routing to one destination. Such selection is called *data routing*. In such applications the data sources are connected to the data inputs and the selected source is available at the data output.

6-7.1.3 *Implementation of Boolean Functions*

Digital multiplexers with 4 data inputs may also be used for implementing Boolean functions with 2 or 3 variables. To implement a 2-variable function, we connect control inputs S_1 and S_0 to the 2 variables and connect the data inputs

TABLE 6-3
Truth Table for the Function
$Z = S_1\overline{S}_0 + \overline{S}_1 S_0$

S_1	S_0	Z
L	L	L
L	H	H
H	L	H
H	H	L

to logic LOW (L) or to logic HIGH (H) as required by the specified Boolean function.

EXAMPLE 6-10 Use a 4-input digital multiplexer to implement the function $Z = S_1\overline{S}_0 + \overline{S}_1 S_0$.

Solution Table 6-3 shows a truth table of the function to be implemented. In the first row of this truth table we have $S_1 = \text{L}$ and $S_0 = \text{L}$. According to the first two rows of the function table of a 4-input digital multiplexer, Table 6-2 (page 125), the combination $S_1 = \text{L}$ and $S_0 = \text{L}$ of the control inputs selects data input I_0. The truth table, Table 6-3, shows that we want to have $Z = \text{L}$ for this combination of control inputs. Hence we connect I_0 to L as shown in Fig. 6-22.

In the second row of the truth table we have $S_1 = \text{L}$ and $S_0 = \text{H}$. According to the function table, Table 6-2, this combination selects data input I_1. Hence we connect I_1 to H as required by the second row of the truth table. Also, based on similar reasoning, we connect I_2 to H and I_3 to L. The resulting solution is shown in Fig. 6-22.

To implement a 3-variable Boolean function, we connect 2 of the variables to control inputs S_1 and S_0 and connect the data inputs to LOW, to HIGH, to the third variable, or to the complement of the third variable, as required by the Boolean function to be implemented. To show that any 3-variable Boolean function can be implemented in this way, we reason as follows.

For any given combination of S_1 and S_0, one data input is selected and there are four possibilities that we may want for the output. In the first two of these

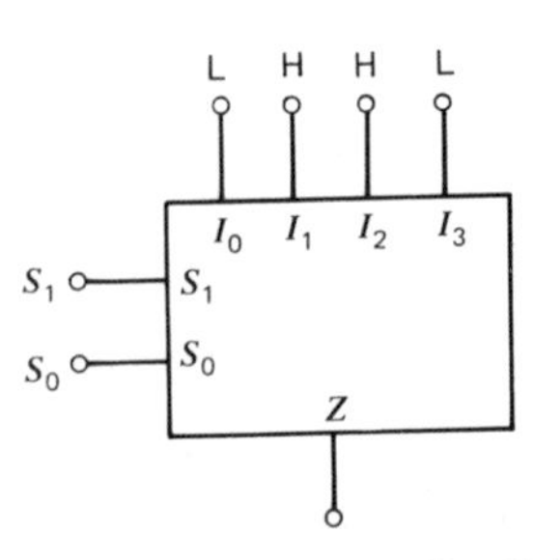

FIGURE 6-22 Solution to Ex. 6-10.

TABLE 6-4
Truth Table for Ex. 6-11

Row	A	B	C	Z
0	0	0	0	0
1	0	0	1	0
2	0	1	0	1
3	0	1	1	1
4	1	0	0	0
5	1	0	1	1
6	1	1	0	1
7	1	1	1	0

possibilities the output is specified to be independent of the third variable: thus the output is to be either LOW, in which case we connect the selected data input to LOW; or the output is to be HIGH, in which case we connect the selected data input to HIGH.

In the second two possibilities the output is *not* independent of the third variable, hence we connect the data input to the third variable or to its complement, whichever is required by the Boolean function to be implemented. The process is illustrated in Ex. 6-11, which follows.

EXAMPLE 6-11 Use a 4-input digital multiplexer to implement the Boolean function specified by the truth table, Table 6-4.

Solution Consider the first two rows of Table 6-4. These are selected by $A = 0$ and $B = 0$, and are shown unshaded in Table 6-5*a*: $A = 0$, $B = 0$; output $Z = 0$ is independent of C. In order to obtain logic 0 at Z, we connect I_0 to logic 0 since I_0 is active when $A = 0$ and $B = 0$ (see Fig. 6-23).

Next, let us consider rows 2 and 3 of Table 6-4. These are selected by $A = 0$ and $B = 1$ and are shown unshaded in Table 6-5*b*: $A = 0$, $B = 1$, and output $Z = 1$ is independent of C. In order to obtain logic 1 at Z we connect I_1 to logic 1 since I_1 is active when $A = 0$ and $B = 1$.

Next, let us consider rows 4 and 5 of Table 6-4. These are selected by $A = 1$ and $B = 0$ and are shown unshaded in Table 6-5*c*: $A = 1$, $B = 0$ while

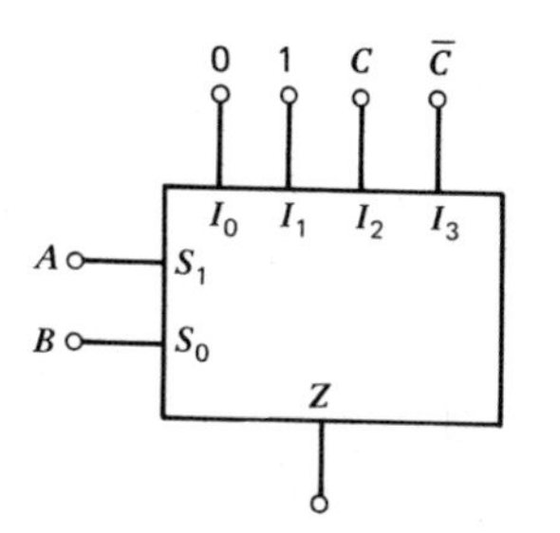

FIGURE 6-23 Solution to Ex. 6-11.

TABLE 6-5

Row	A	B	C	Z
0	0	0	0	0
1	0	0	1	0
2	0	1	0	1
3	0	1	1	1
4	1	0	0	0
5	1	0	1	1
6	1	1	0	1
7	1	1	1	0

(a) In rows 0 and 1 output Z is 0 independent of C.

Row	A	B	C	Z
0	0	0	0	0
1	0	0	1	0
2	0	1	0	1
3	0	1	1	1
4	1	0	0	0
5	1	0	1	1
6	1	1	0	1
7	1	1	1	0

(b) In rows 2 and 3 output Z is 1 independent of C.

Row	A	B	C	Z
0	0	0	0	0
1	0	0	1	0
2	0	1	0	1
3	0	1	1	1
4	1	0	0	0
5	1	0	1	1
6	1	1	0	1
7	1	1	1	0

(c) In rows 4 and 5 output Z has the same binary values as C.

Row	A	B	C	Z
0	0	0	0	0
1	0	0	1	0
2	0	1	0	1
3	0	1	1	1
4	1	0	0	0
5	1	0	1	1
6	1	1	0	1
7	1	1	1	0

(d) In rows 6 and 7 output Z has the same binary values as the complement of C.

output Z in row 4 is 0 and in row 5 it is 1. Thus the output Z in rows 4 and 5 has exactly the same binary values as the variable C has in rows 4 and 5. Hence we connect I_2 to C.

Finally, let us consider the last two rows of Table 6-4. These are selected by $A = 1$ and $B = 1$ and are shown unshaded in Table 6-5d: $A = 1$ and $B = 1$ while output Z in row 6 is logic 1, and in row 7 it is logic 0. Thus the output Z is exactly the *complement* of C in rows 6 and 7. Hence we connect I_3 to $\bar{C}$.

Thus the solution is as shown in Fig. 6-23.

6-7.1.4 Availability

Most commonly, two 4-input digital multiplexers are available in a 16-pin package with the control inputs shared between the two multiplexers. This sharing of control inputs results in some constraints. Thus, in data routing applications the control inputs must be the same for both signals to be routed. Also, in the implementation of Boolean functions we must have at least 2 of the 3 variables identical in the two functions.

TABLE 6-6
Truth Table for the Full-Adder

Row	A	B	C_{IN}	S_{OUT}	C_{OUT}
0	0	0	0	0	0
1	0	0	1	1	0
2	0	1	0	1	0
3	0	1	1	0	1
4	1	0	0	1	0
5	1	0	1	0	1
6	1	1	0	0	1
7	1	1	1	1	1

EXAMPLE 6-12

Table 6-6 shows the truth table of a *full adder* that has three inputs A, B, and C_{IN}, and two outputs S_{OUT} and C_{OUT}. Use a dual 4-input digital multiplexer to implement S_{OUT} and C_{OUT}.

Solution

a. In the S_{OUT} column of Table 6-4 we observe in row 0 logic 0 and in row 1 logic 1; thus $I_{0A} = C_{IN}$. For rows 2 and 3, $I_{1A} = \overline{C}_{IN}$; for rows 4 and 5, $I_{2A} = \overline{C}_{IN}$; and for rows 6 and 7, $I_{3A} = C_{IN}$.

b. For C_{OUT}, $I_{0B} = 0$, $I_{1B} = C_{IN}$, $I_{2B} = C_{IN}$, $I_{3B} = 1$. The implementation of the circuit is shown in Fig. 6-24.

6-7.2 Two-Input, 8-Input, and 16-Input Digital Multiplexers

Four 2-input digital multiplexers are most commonly available in a 16-pin package with the control inputs shared between two or among all four of the multiplexers. This sharing of control inputs results in some constraints. Thus, in data routing applications (discussed in Sec. 6-7.1.2), the control inputs must be the same for the signals that share them. Also, in the implementation of Boolean functions (discussed in Sec. 6-7.1.3), we must have at least one of the 2 variables identical in the functions that share control inputs.

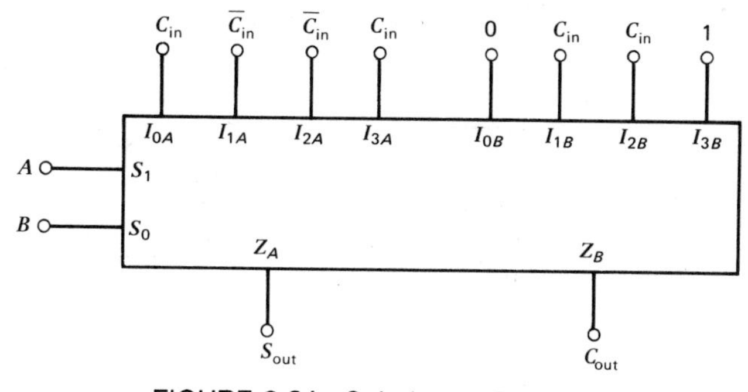

FIGURE 6-24 Solution to Ex. 6-12.

An 8-input digital multiplexer has 3 control inputs. In addition to data routing of 8 inputs, it is suitable for the implementation of Boolean functions with 4 variables.

Sixteen-input digital multiplexers are commonly available in 24-pin packages and have 4 control inputs. In addition to data routing of 16 inputs, they are suitable for the implementation of Boolean functions with 5 variables.

6-8 DECODERS

Decoders are MSI circuits that deliver one unique output for each combination of inputs; they are suitable for address decoding and for the implementation of Boolean functions. In what follows here, we describe 2-line to 4-line decoders and their use, followed by a discussion of other decoders and their application.

6-8.1 Two-Line to Four-Line Decoders

A 2-line to 4-line decoder has 2 *select inputs* and 4 outputs. The states of the 2 select inputs have 4 possible combinations, each selecting one of the 4 outputs.

In an *active-high decoder* the selected output is at logic HIGH and all other outputs are at logic LOW. Active-high decoders find only limited application and for this reason are not further discussed here.

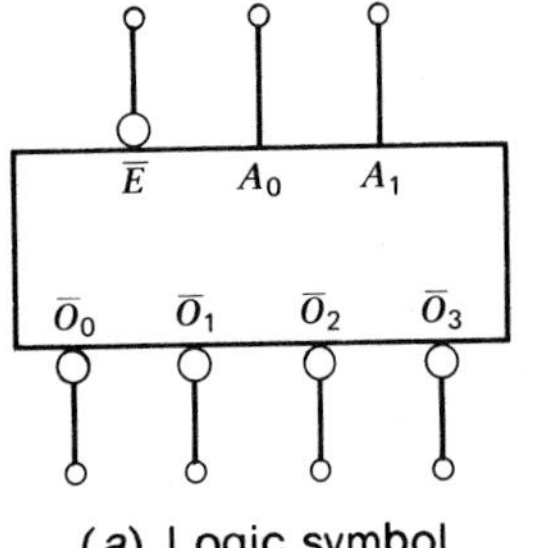

(*a*) Logic symbol

$\overline{E}$	A_1	A_0	$\overline{O}_0$	$\overline{O}_1$	$\overline{O}_2$	$\overline{O}_3$
H	X	X	H	H	H	H
L	L	L	L	H	H	H
L	L	H	H	L	H	H
L	H	L	H	H	L	H
L	H	H	H	H	H	L

(*b*) Function table

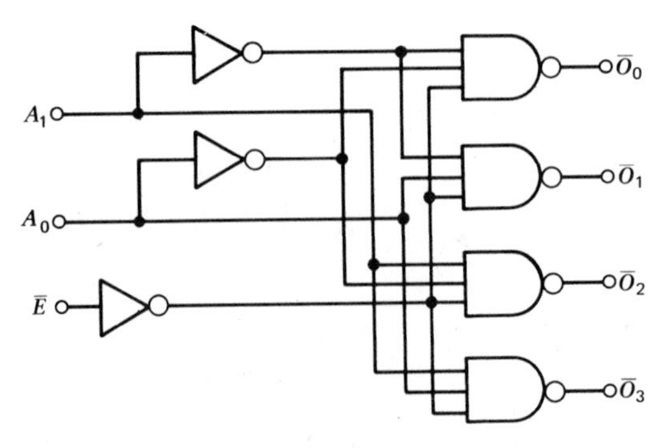

(*c*) Logic diagram

FIGURE 6-25 Two-line to four-line decoder.

In an *active-low decoder* the selected output is at logic LOW (L) and all other outputs are at logic HIGH (H). In addition, often a *strobe* or $\overline{ENABLE}$ ($\bar{E}$) is also provided that can be used to inhibit selection altogether: An $\bar{E}$ = HIGH (H) forces all outputs to HIGH (H) irrespective of the states of the select inputs.

Figure 6-25*a* shows a symbol for an active-low 2-line to 4-line decoder that has 2 select inputs A_1 and A_0, a strobe input $\bar{E}$, and 4 outputs $\bar{O}_0$, $\bar{O}_1$, $\bar{O}_2$, and $\bar{O}_3$. Note the circles ("*bubbles*") at strobe input $\bar{E}$ and at the 4 outputs. Recall that when such a circle is attached to a logic gate, it represents a logic inversion. Thus, for example, attaching a circle to the output of an AND gate makes it equivalent to a NAND gate, that is, to an AND gate followed by a logic inverter. However, the situation is different when a circle is attached to a rectangular symbol of a combinational logic circuit, as in Fig. 6-25*a*. In such cases the circle does *not* represent logic inversion or any other change of function, but is only a reminder of the presence of a complemented (active-low) variable.

Figure 6-25*b* shows a function table. In its first row strobe input $\bar{E}$ = H, hence all 4 outputs are H irrespective of the states of select inputs A_1 and A_0. In the remaining four rows strobe input $\bar{E}$ = L, and one of the 4 outputs is L as governed by the states of select inputs A_1 and A_0.

Figure 6-25*c* shows a logic diagram. The use of the decoder is illustrated in Ex. 6-13 below.

EXAMPLE 6-13

Table 6-7 describes the operation of a remote status display for a copying machine. The left two columns show the four possible states of the machine: A green light is turned on when the machine is available for use, an amber light when it is in use, a red light when it is broken, and a blue light when it is turned off for routine maintenance.

The right two columns of Table 6-7 specify the coding of the four states on the two lines A_1 and A_0. Describe the use of the active-low 2-line to 4-line decoder of Fig. 6-25 for this status monitor.

Solution

We connect the coded status lines A_1 and A_0, respectively, to select inputs A_1 and A_0 of the decoder. We also connect $\bar{E}$ to logic level L.

In the first row of Table 6-7 we have A_1 = L, A_0 = L, and we want the green light to be turned on. According to the second row of the function table of Fig. 6-25*b*, the combination A_1 = L and A_0 = L makes output $\bar{O}_0$ = L; hence we connect the green light between output $\bar{O}_0$ and logic level H.

TABLE 6-7
Status Display for a Copying Machine

STATUS	LIGHT	A_1	A_0
Available	Green	L	L
In use	Amber	L	H
Broken	Red	H	L
Off	Blue	H	H

Similarly, we connect the amber light between $\overline{O}_1$ and H, the red light between $\overline{O}_2$ and H, and the blue light between $\overline{O}_3$ and H.

Note that in a practical situation we are likely to have ground (0 V) for logic L and +5 V for logic H.

6-8.2 Unique Address Decoding

Section 6-8.1 described 2-line to 4-line decoders and their use for generating 4 unique addresses. For 8 unique addresses we require a 3-line to 8-line decoder, and for 16 unique addresses a 4-line to 16-line decoder. Such decoders may be combined when a larger number of unique addresses have to be generated, as illustrated in Ex. 6-14 below.

EXAMPLE 6-14

A 5-bit coded input is to be decoded into 32 unique address lines. Implement the circuit using one 2-line to 4-line decoder and four 3-line to 8-line decoders.

Solution

Refer to Fig. 6-26 in which we utilize a 2-line to 4-line decoder. To the inputs of this decoder we apply the two most significant bits of the number to be decoded. In response to the four possible input combinations, we obtain at outputs $\overline{O}_0$, $\overline{O}_1$, $\overline{O}_2$, and $\overline{O}_3$ four active-low signals. These active-low signals which are exclusive—that is, they can be low only one at a time—are applied to the enable inputs of four 3-line to 8-line decoders. The other three inputs of these decoders are connected to the three least significant bits of the number

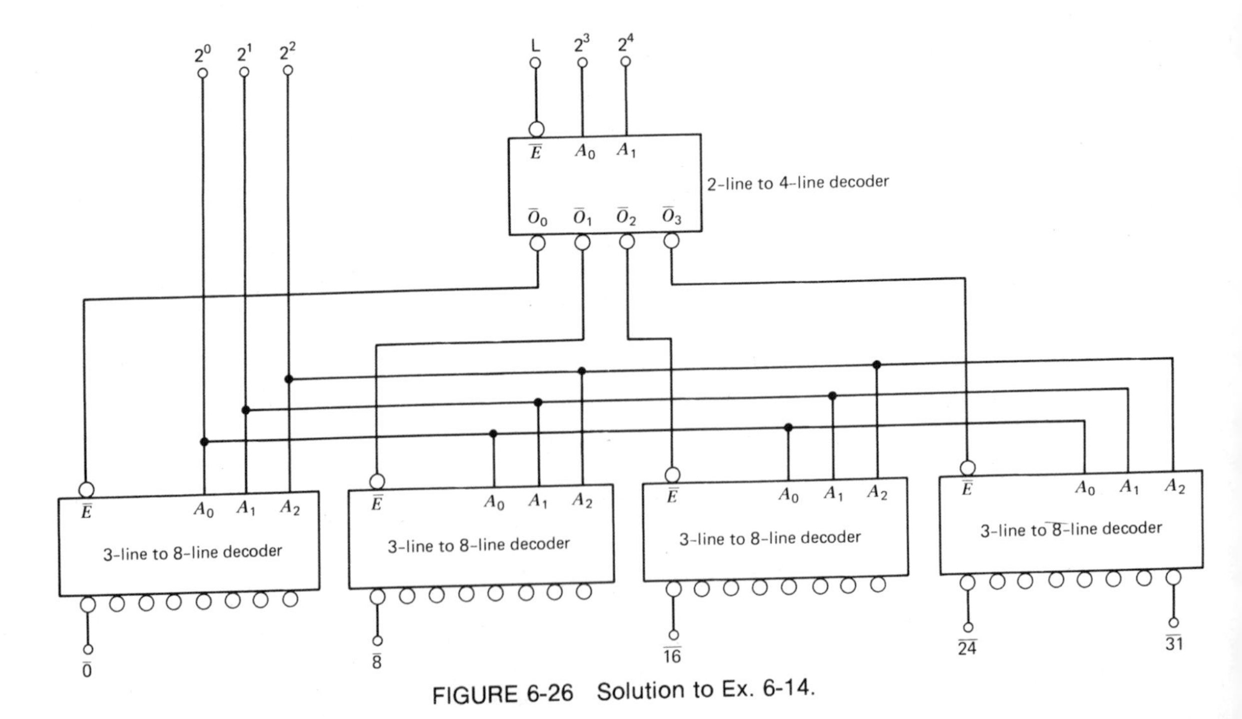

FIGURE 6-26 Solution to Ex. 6-14.

to be decoded. The result is a unique active-low output in response to the 5-bit input combination.

6-8.3 Implementation of Boolean Functions

A 2-line to 4-line decoder can generate all the 4 standard product terms of a function of 2 variables. Similarly, a 3-line to 8-line and a 4-line to 16-line decoder can generate all the 8 and 16 standard product terms of a function of 3 and 4 variables, respectively.

EXAMPLE 6-15

Using a 3-line to 8-line decoder implement the function

$$f = \overline{A}\,\overline{B}\,\overline{C} + \overline{A}B\overline{C} + A\overline{B}\,\overline{C} + ABC$$

Solution

We apply the variables A, B, and C to the three inputs A_0, A_1, and A_2 as shown in Fig. 6-27. The term $\overline{A}\,\overline{B}\,\overline{C}$ generates a low output at $\overline{O}_0$ and, similarly, the other terms of the function generate low outputs at $\overline{O}_2$, $\overline{O}_4$, and $\overline{O}_7$. These 4 outputs are combined in a 4-input negative NOR gate that yields the desired function, as shown in Fig. 6-27. Note that no simplification of the function was attempted, since a simplified expression would *not* have saved any hardware in the implementation.

6-9 PRIORITY ENCODERS

In a typical computer installation we find, in addition to the main computer, peripheral devices such as printers, magnetic tape, magnetic disc, and cathode-ray tube (CRT) displays. These peripheral devices send messages to the

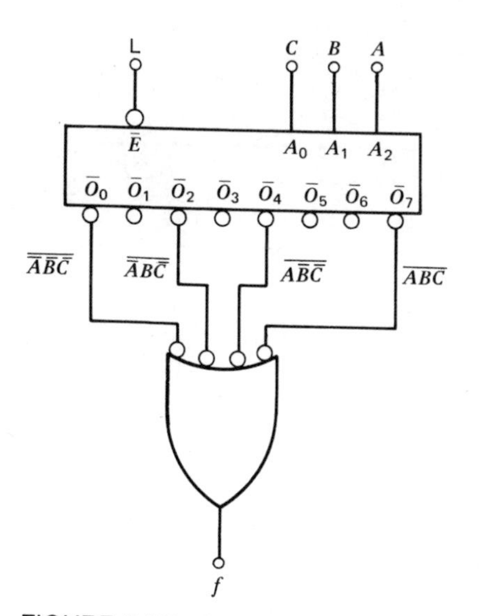

FIGURE 6-27 Solution to Ex. 6-15.

computer at irregular intervals with requests for service. Two types of requests are common:

1. A request to send data *to* the computer.
2. A request to receive data *from* the computer.

It is quite probable for more than one peripheral device to send a request for service at any one time. Therefore, an arbitration circuit is required to determine which peripheral device should have priority, that is, which should be serviced first. A combinational logic circuit that establishes the priority among simultaneous requests is called a *priority encoder*.

Figure 6-28 shows a priority encoder with active-high logic. It has 8 inputs, D_0 through D_7, and 4 outputs, Q_0 through Q_3. A priority is assigned to each input: D_0 has the highest priority and D_7 the lowest. Q_0, Q_1, and Q_2 are the binary-coded outputs representing the location of the highest-priority input that is HIGH (H). When two or more inputs are HIGH (H) simultaneously, outputs Q_0,

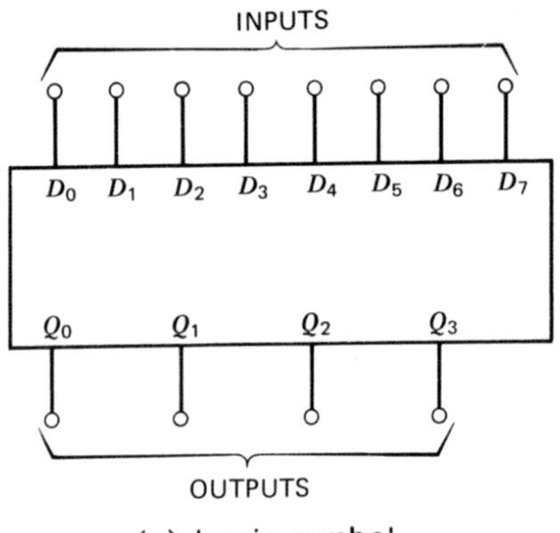

(*a*) Logic symbol

INPUTS								OUTPUTS			
D_0	D_1	D_2	D_3	D_4	D_5	D_6	D_7	Q_3	Q_2	Q_1	Q_0
H	X	X	X	X	X	X	X	H	L	L	L
L	H	X	X	X	X	X	X	H	L	L	H
L	L	H	X	X	X	X	X	H	L	H	L
L	L	L	H	X	X	X	X	H	L	H	H
L	L	L	L	H	X	X	X	H	H	L	L
L	L	L	L	L	H	X	X	H	H	L	H
L	L	L	L	L	L	H	X	H	H	H	L
L	L	L	L	L	L	L	H	H	H	H	H
L	L	L	L	L	L	L	L	L	L	L	L

(*b*) Function table

FIGURE 6-28 Priority encoder.

Q_1, and Q_2 deliver a code corresponding to the highest-priority input that is HIGH (H), while all other inputs are ignored. Output Q_3 is HIGH (H) when *any* input is HIGH (H): This permits direct extension into another priority encoder when more than 8 inputs are required.

EXAMPLE 6-16

The priority encoder of Fig. 6-28 has two HIGH (H) inputs: D_0 and D_6. What is the output code?

Solution

Input D_0 has higher priority than input D_6. Thus, according to the first row in the function table of Fig. 6-28*b*, $Q_2 = Q_1 = Q_0 = \text{L}$; also, $Q_3 = \text{H}$ indicating that at least one HIGH (H) input is being applied to the priority encoder.

SUMMARY

In this chapter we have introduced the following logic operators: NAND, NOR, EXCLUSIVE-OR, EXCLUSIVE-NOR, and AND-OR-INVERT. We have shown that design procedures developed for 2-stage AND-OR circuits are also valid for 2-stage NAND-NAND circuits. Similarly we have shown that design procedures developed for 2-stage OR-AND circuits are also valid for 2-stage NOR-NOR circuits. Active-high and active-low logic were also introduced.

Three classes of medium-scale integrated (MSI) combinational logic circuits (digital multiplexers, decoders, and priority encoders) were introduced and their applications discussed.

SELF-EVALUATION QUESTIONS

1. Draw a truth table, a symbol, and an equivalent circuit for a 2-input NAND gate.
2. Draw a truth table, a symbol, and an equivalent circuit for a 2-input NOR gate.
3. What logic proposition does the EXCLUSIVE-OR gate implement?
4. Why is the EXCLUSIVE-NOR gate also called an EQUALITY COMPARATOR?
5. Discuss parity generation and detection.
6. Describe active-high and active-low logic.
7. Draw the symbols for the following operators: (**a**) active-high AND, OR, NAND, NOR, (**b**) active-low AND, OR, NAND, NOR.
8. Describe the procedure for designing 2-stage NAND circuits using Karnaugh map simplification.
9. Describe the procedure for designing 2-stage NOR circuits using Karnaugh map simplification.
10. Establish a truth table for a 2-wide 2-input AOI gate. How many rows does the table have?
11. Describe the operation of a 4-input digital multiplexer. List two applications of digital multiplexers.

12. Draw the circuit diagram of a 2-line to 4-line decoder with an *ENABLE* input. List two applications of decoders.

13. Describe the operation of a priority encoder. Where is such an encoder commonly used?

PROBLEMS

6-1. Solve the problem of Ex. 6-1 using wired-OR EXCLUSIVE-OR gates. What advantages are derived from wired-OR configuration in this problem?

6-2. Realize a simplified 2-stage NAND circuit for the following function: $f = \bar{A}\bar{B}\bar{C}\bar{D} + \bar{A}B\bar{C}\bar{D} + \bar{A}\bar{B}\bar{C}D + \bar{A}BCD + A\bar{B}CD + \bar{A}BC\bar{D} + ABC\bar{D}$; don't cares: $\bar{A}B\bar{C}D + ABCD + A\bar{B}C\bar{D}$. Draw the circuit diagram.

6-3. Using truth tables show that the active-high EXCLUSIVE-OR gate performs the function of EXCLUSIVE-NOR in active-low logic.

6-4. Using truth tables show that an active-high EXCLUSIVE-NOR gate performs the same function as an active-low EXCLUSIVE-OR gate.

6-5. Using only NAND gates realize a simplified logic circuit for the function $f = A\bar{B}\bar{C} + ABC + \bar{A}B\bar{C} + AB\bar{C}$.

6-6. Using only NOR gates realize a simplified circuit for the function of Prob. 6-5. (*Hint:* Obtain $\bar{f}$ from the 0 entries of a Karnaugh map; then apply De Morgan's theorem.)

***6-7.** Arithmetic circuits require a *conditional complement* when implementing add and subtract functions. Assuming that a variable is applied to one input of an EXCLUSIVE-OR gate, what binary constant has to be applied to the other (control) input to obtain at the output of the gate:

a. The uncomplemented variable?

b. The complemented variable?

***6-8.** Suppose you use an EXCLUSIVE-NOR gate for the solution of Prob. 6-7. What binary constant has to be applied to the control input to obtain at the output of the gate:

a. The uncomplemented variable.

b. The complemented variable.

6-9. Obtain a simplified realization of the 4-variable function $f = \bar{A}\bar{B}\bar{C}D + A\bar{B}\bar{C}D + \bar{A}BCD + A\bar{B}CD$; don't cares: $\bar{A}\bar{B}CD + \bar{A}B\bar{C}D + ABCD$. Use only NAND gates.

6-10. Obtain a simplified realization of the 4-variable function $f = \bar{A}\bar{B}\bar{C}D + \bar{A}\bar{B}CD + \bar{A}\bar{B}C\bar{D} + \bar{A}BCD + A\bar{B}\bar{C}\bar{D} + A\bar{B}CD + A\bar{B}C\bar{D}$; don't cares: $\bar{A}B\bar{C}D + AB\bar{C}D + A\bar{B}\bar{C}D + \bar{A}\bar{B}\bar{C}\bar{D}$. Use only NOR gates. (*Hint:* Ob-

*Optional problem.

tain $\bar{f}$ from the 0 entries of a Karnaugh map; then apply De Morgan's theorem.)

6-11. Show how a 2-input NAND gate may be used as a logic inverter by connecting one gate input to logic 1. Show the truth table of such a gate.

6-12. Show how a 2-input NOR gate may be used as a logic inverter by connecting one gate input to logic 0. Show the truth table of such a gate.

6-13. Figure P6-1 shows a logic diagram of four NAND gates. What function does the circuit perform?

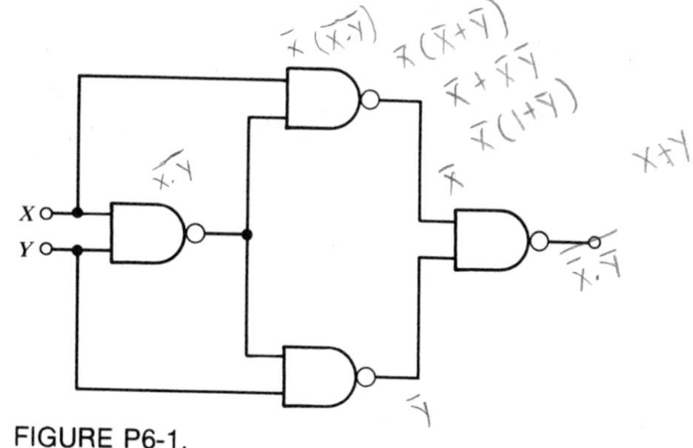

FIGURE P6-1.

6-14. Figure P6-2 shows a logic diagram of four NOR gates. What function does the circuit perform?

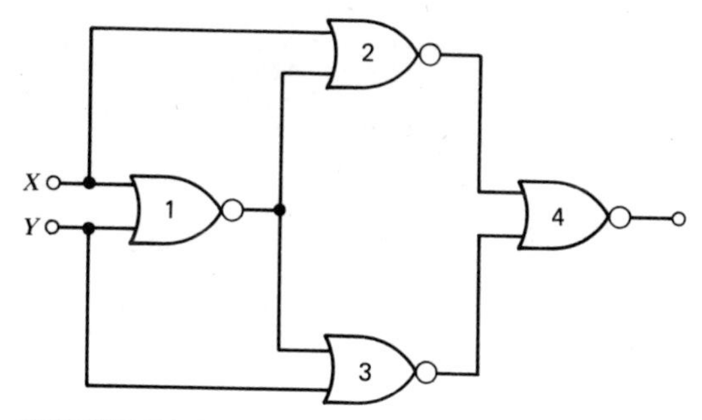

FIGURE P6-2.

6-15. Implement the function $f = \bar{A}B\bar{C}\bar{D} + \bar{A}B\bar{C}D + A\bar{B}\bar{C}\bar{D} + A\bar{B}\bar{C}D$ using an AND gate and an EXCLUSIVE-OR gate.

6-16. Implement the function $f = A\bar{B}\bar{C}\bar{D} + A\bar{B}CD + AB\bar{C}\bar{D} + ABCD$ using an AND gate and an EXCLUSIVE-NOR gate.

6-17. Using a 4-input multiplexer implement the function

$$f = AB\overline{C} + A\overline{B}C + \overline{A}BC + \overline{A}\,\overline{B}C$$

6-18. Using a 4-input multiplexer implement the function

$$f = \overline{A}\,\overline{B}\,\overline{C} + \overline{A}\,\overline{B}C + A\overline{B}\,\overline{C} + ABC$$

6-19. The input combination to the 1-of-32 decoder shown in Fig. 6-26 is H (msb), L, H, L, H (lsb). Which output line is active (L) in response to this input?

6-20. The input combination to the 1-of-32 decoder shown in Fig. 6-26 is H (msb), L, H, H, H (lsb). Which output line is active (L) in response to this input?

***6-21.** What operator does the circuit of Fig. P2-3 (page 26) realize?

***6-22.** What operator does the circuit of Fig. P2-4 (page 26) realize?

*Optional problem.

CHAPTER 7

Present-Day Logic Circuits

Instructional Objectives

This chapter describes logic circuits that are most widely used in current digital systems: transistor-transistor logic (TTL), emitter-coupled logic (ECL), integrated-injection logic (I^2L), metal-oxide-semiconductor (MOS) logic, and complementary MOS (CMOS) logic. After you read this chapter you should be able to:

1. Sketch the circuit diagram of a TTL NAND gate.
2. Describe TTL AND-OR-INVERT gates.
3. Discuss open-collector TTL and 3-state TTL circuits.
4. Compare propagation delays and power dissipations of various TTL circuits.
5. Sketch the circuit diagram of an ECL circuit.
*6. Discuss I^2L circuits.
7. Describe MOS logic circuits.
8. Sketch the circuit diagrams of CMOS NAND and NOR gates.
9. Describe the operation of CMOS transmission gates.
10. Discuss propagation delays of CMOS logic circuits.

7-1 OVERVIEW

This chapter describes the internal circuitry of various logic families used in digital systems. Additional discussions on LSI and VLSI are included in Chapter 11.**

*Optional material.

**The terms SSI, MSI, LSI, and VLSI were introduced in Chapter 1.

Transistor-transistor logic (TTL) is the most widely used logic family in SSI and MSI. However, primarily because of high peak currents drawn from the power supply, its use in LSI and VLSI is limited to interfacing with external circuitry.

Emitter-coupled logic (ECL) and complementary metal-oxide-semiconductor (CMOS) logic are often used in SSI, MSI, and LSI. ECL is noted for short propagation delays, CMOS logic for low power dissipation.

The chapter also briefly describes two other logic families: integrated-injection logic (I^2L) and metal-oxide-semiconductor (MOS) logic. Both of these are used primarily in LSI circuits; also, MOS logic is the principal technology in VLSI circuits.

7-2 TRANSISTOR-TRANSISTOR LOGIC (TTL)

Transistor-transistor logic (*TTL*) is the most widely used logic circuit in SSI and MSI. There are several versions of TTL gates. The "standard TTL" series has maximum *propagation delays* (as defined in Ch. 2) of 22 ns and maximum power dissipations of 20 mW/logic gate. The "low-power TTL" series has maximum propagation delays of 60 ns and maximum power dissipations of 2 mW/logic gate. The "high-speed TTL" series has maximum propagation delays of 10 ns and maximum power dissipations of 40 mW/logic gate.

Propagation delays are further reduced in *Schottky-diode clamped TTL* gates that are also known simply as *Schottky-TTL*. However, this reduction is often accompanied by a slight increase of the lower logic level from a maximum of 0.4 V to a maximum of 0.5 V. This increase of logic level leads to a reduction of *noise margins*—these, as well as power supply and power distribution, are discussed in Ch. 16.

Two versions of Schottky-diode clamped TTL circuits are available. One of these, named "*high-speed Schottky TTL*" or simply "Schottky TTL," has maximum propagation delays of 5 ns and maximum power dissipations of 36 mW/logic gate. The other, named "*low-power Schottky TTL*," has maximum propagation delays of 15 ns and maximum power dissipations of 4 mW/logic gate. In addition to the above, other versions of TTL are also being developed that provide identical or even shorter propagation delays at power dissipations that are further reduced.

Below we first describe various TTL circuits that are available, as well as circuit details of standard TTL circuits. This is followed by a circuit description of Schottky-diode clamped TTL and a comparison of their features with those of standard TTL.

7-2.1 TTL Logic Inverters

The circuit diagram of a "standard" *TTL logic inverter* is shown in Fig. 7-1*a*; the figure also shows operating conditions for an input voltage of 0 V and with an output current of $I_L = 0$. In this case, transistor Q_1 is saturated, Q_2 is cut off, Q_4

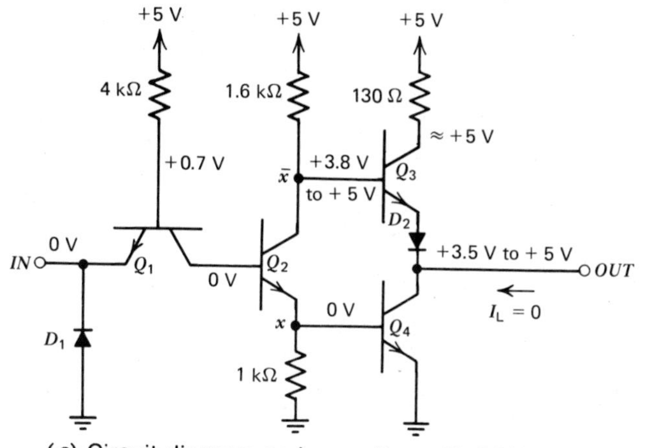

(*a*) Circuit diagram and operation with 0-V input

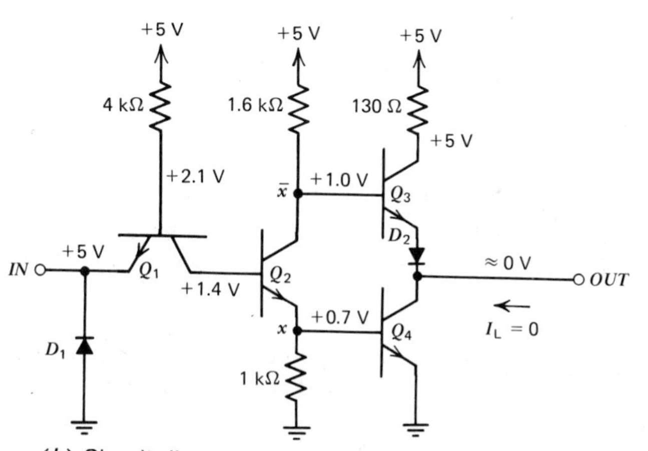

(*b*) Circuit diagram and operation with +5-V input

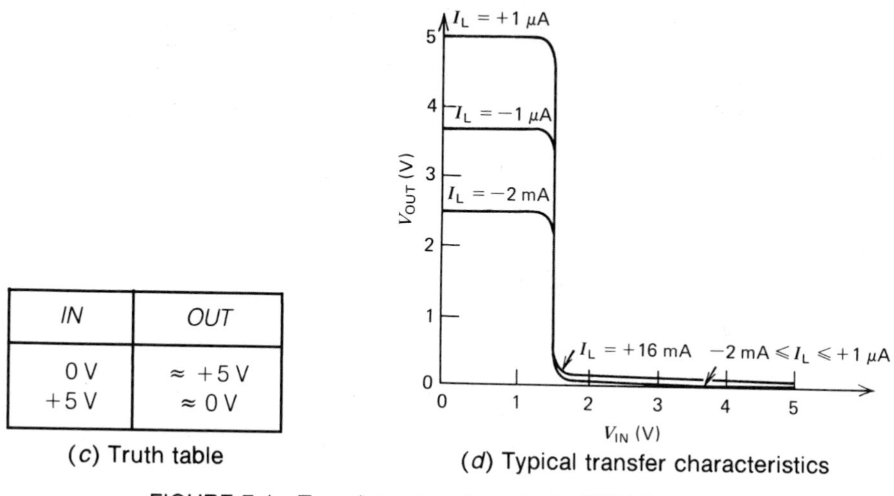

IN	*OUT*
0 V	≈ +5 V
+5 V	≈ 0 V

(*c*) Truth table

(*d*) Typical transfer characteristics

FIGURE 7-1 Transistor-transistor logic (TTL) inverter.

is cut off, and Q_3 pulls the output up to a voltage that is between +3.5 V and +5 V.

Figure 7-1*b* shows the same circuit but with an input voltage of +5 V; again $I_L = 0$. In this case, Q_1 is cut off, Q_2 is saturated, Q_3 is cut off, and the output is pulled down by Q_4 to ≈ 0 V.

The purpose of diode D_1 is the reduction of spurious negative transients ("spikes" or "glitches") at the input; diode D_2 is included to ensure that Q_3 in Fig. 7-1*b* is cut off. A truth table for the circuit is shown in Fig. 7-1*c*.

Note that the circuit shown is but one of the simplest available. With improved manufacturing technology, the inclusion of a larger number of transistors is possible, leading to improved performance—this is further discussed in Sec. 7-2.8 on Schottky-diode clamped TTL.

7-2.2 Loading

Figure 7-1*d* shows the transfer characteristics for the circuit of Figs. 7-1*a* and 7-1*b* with various currents I_L at the output. Note the direction of the output current in these figures: The output current, I_L, is positive when it flows into the circuit ("*sink current*"), and it is negative when it flows out of the circuit ("*load current*").

For input voltages above +2 V, the output voltage of the circuit is near 0 V with currents up to 16 mA flowing into the output. Also, for input voltages less than +0.8 V, the output voltage of the circuit is above +2.4 V with currents up to 2 mA flowing out of the output. These favorable loading properties are the result of having transistor Q_3 pull the output up toward +5 V when the input voltage is below +0.8 V, and having transistor Q_4 pull the output down toward 0 V when the input voltage is above +2 V.

7-2.3 TTL NAND Gates

The TTL inverter circuit shown in Figs. 7-1*a* and 7-1*b* can be extended to a 2-input *TTL NAND gate* using a *2-emitter transistor*, as shown in Fig. 7-2*a*. A 2-emitter transistor symbol and its equivalent circuit are given in Fig. 7-2*b*. A truth table of the circuit is shown in Fig. 7-2*c*.

The circuit of Fig. 7-2*a* can be extended to more than 2 inputs by increasing the number of emitters in transistor Q_1. Logic gates with up to 13 inputs are now available.

7-2.4 Propagation Delays

During transitions the output impedance of a "standard" TTL gate circuit is in the vicinity of $R_{OUT} = 150\ \Omega$. The *intrinsic* propagation delays, that is, the propagation delays for zero load capacitance, are in the vicinity of 15 ns. Propagation delays become longer, however, when the output is loaded by a capacitance. This is illustrated in Ex. 7-1 below.

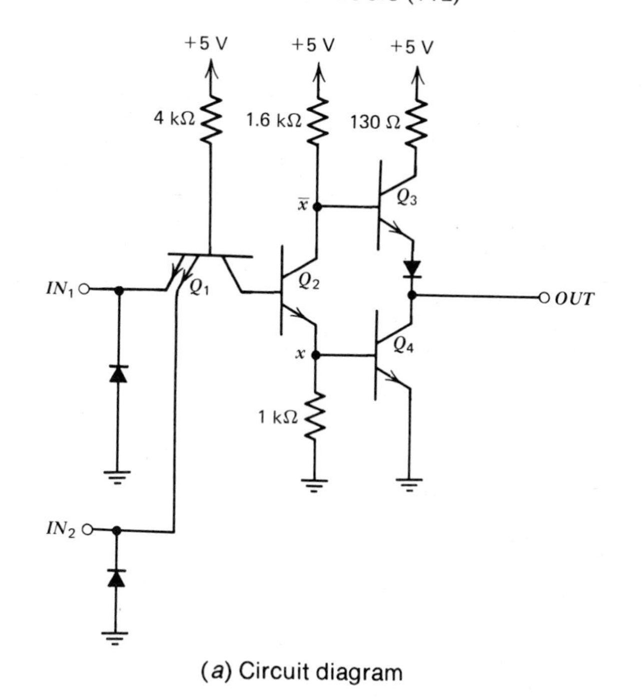

(a) Circuit diagram

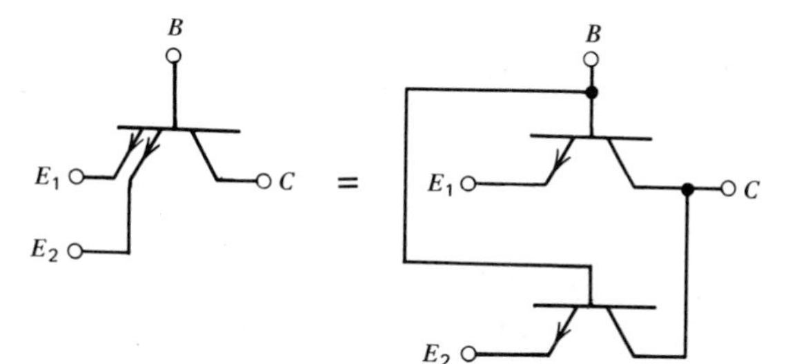

(b) Two-emitter transistor symbol and its equivalent circuit

IN_1	IN_2	OUT
0 V	0 V	≈ +5 V
0 V	+5 V	≈ +5 V
+5 V	0 V	≈ +5 V
+5 V	+5 V	≈ 0 V

(c) Truth table

FIGURE 7-2 A 2-input TTL NAND gate.

EXAMPLE 7-1

The output of the TTL NAND gate shown in Fig. 7-2a is loaded by a capacitance of $C = 100$ pF. The two inputs are simultaneously switched from a voltage of +5 V to a voltage of 0 V. The output impedance during the transition is $R_{\mathrm{OUT}} = 150\ \Omega$. The intrinsic propagation delay $(t_{\mathrm{PLH}})_{\mathrm{intrinsic}} =$ 15 ns. Estimate propagation delay t_{PLH}.

Solution

As was true for the diode gate circuit in Example 2-4, here too we would have to derive the transient of the output voltage from its initial value (≈ 0 V) to its final value (≈ +3.5 V to +5 V). Such a derivation is, however, outside the scope of this book. Instead, we simply state that, as a crude approximation,

$$t_{\mathrm{PLH}} \approx (t_{\mathrm{PLH}})_{\mathrm{intrinsic}} + 0.7\, R_{\mathrm{OUT}}\, C$$

Thus, in our case

$$t_{PLH} \approx 15 \text{ ns} + 0.7 \times 150\ \Omega \times 100 \text{ pF} = \mathbf{25.5\ ns}$$

We have seen that in a DTL gate (Fig. 2-20*a*, page 20) the output is pulled down by a transistor but is pulled up only by a 2-kΩ resistor. As a result, propagation delay t_{PLH} is usually longer than t_{PHL}, especially when the output is loaded by external capacitance. In TTL gate circuits, however, the output is pulled both up (by Q_3) and down (by Q_4), respectively, during positive-going and negative-going output transitions. The circuit is designed such that the resulting impedances are similar in both directions ($R_{OUT} \approx 150\ \Omega$ in standard TTL). Consequently, t_{PLH} and t_{PHL} are comparable even when there is capacitive loading on the output.

7-2.5 AND-OR-INVERT (AOI) Gates

Figure 7-3*a* shows the circuit diagram of an *AND-OR-INVERT* (*AOI*) *gate* that has been obtained by adding the circuitry of Q_1' and Q_2' to the TTL NAND gate circuit of Fig. 7-2*a* at points x and $\bar{x}$. Figure 7-3*b* shows a symbol and Fig. 7-3*c* an equivalent logic circuit.

We can see that the circuit of Fig. 7-3*a* is equivalent to a circuit consisting of two 2-input AND gates followed by a 2-input OR gate and a logic inverter. However, as compared to the circuit of Fig. 7-3*c*, the circuit of Fig. 7-3*a* has the advantage of simplicity, shorter propagation delay, and fewer external connections.

7-2.6 TTL Gates with Open-Collector Outputs

In many cases it would be desirable to use TTL gates in a wired-OR configuration similar to the wired-OR DTL circuit described in Fig. 2-23 (page 22). However, such configurations are not feasible with the TTL output circuitry ("*totem-pole output*") described thus far, because of their low output impedances.

One solution to this problem omits Q_3, D_2, and the 130-Ω resistor in Figs. 7-1*a* and 7-1*b*, resulting in what is known as an *open-collector output* circuit. Such outputs can be connected in parallel in a wired-OR configuration with an external resistor connected between the outputs and +5 V. The resulting propagation delay t_{PLH}, however, becomes long since the output is not pulled up by a transistor during the transition, but only by the external resistor. Nevertheless, TTL gates with open-collector outputs are used in wired-OR applications where propagation delay t_{PLH} is not critical.

7-2.7 TTL Gates with 3-State Outputs

The desire to connect outputs in parallel while retaining the low output impedance of TTL gates led to the development of TTL gates with *3-state outputs*, also known as *3-state TTL gates*. The operation of such a circuit is shown in Fig.

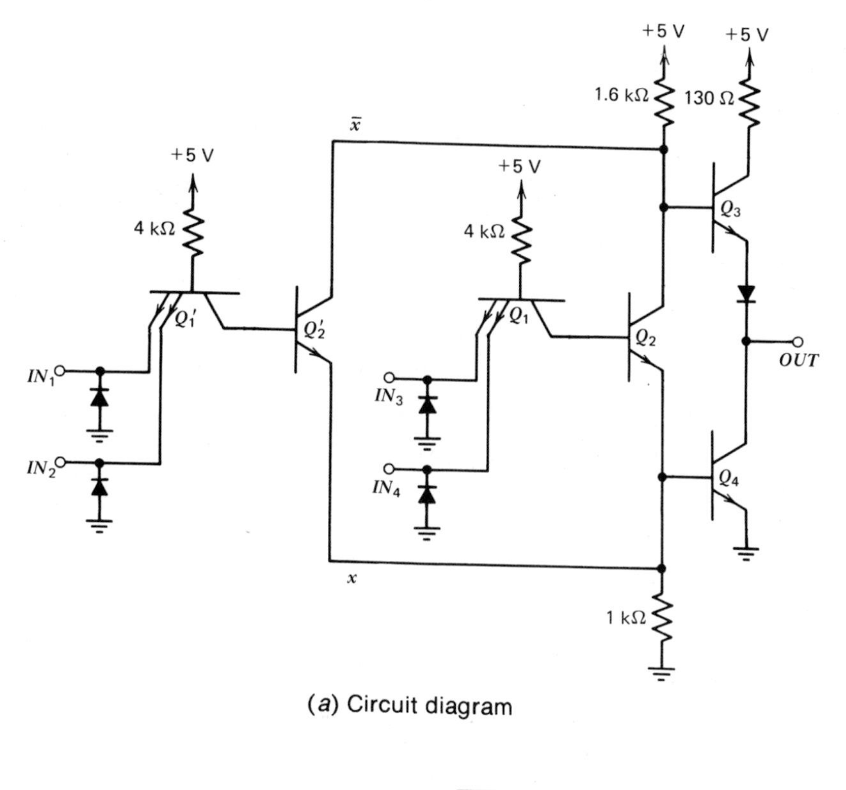

(*a*) Circuit diagram

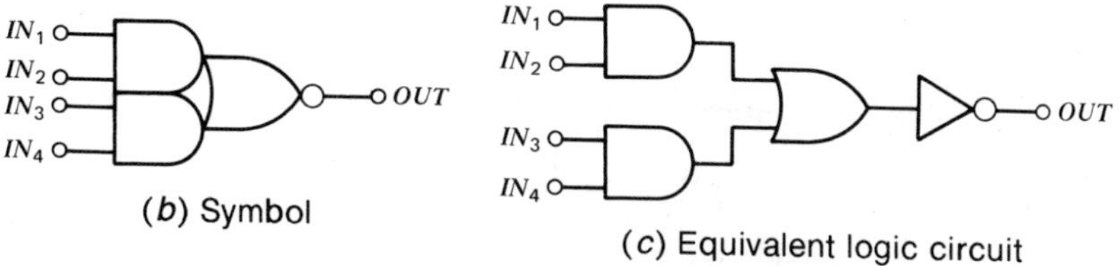

(*b*) Symbol

(*c*) Equivalent logic circuit

FIGURE 7-3 An AND-OR-INVERT (AOI) gate.

7-4*a* for the simplest case of a logic inverter—note that the logic inverter symbol at the *INHIBIT* input of Fig. 7-4*a* stands for the logic inverter circuit of Fig. 7-1*a* or Fig. 7-1*b*. A truth table for the circuit of Fig. 7-4*a* is shown in Fig. 7-4*b*, and a symbol in Fig. 7-4*c*.

We can see that as long as the *INHIBIT* input (also called *DISABLE*, $\overline{ENABLE}$, etc.) is LOW (near 0 V), the circuit operates as a logic inverter. However, when the *INHIBIT* input is HIGH (between +2 V and +5 V), transistors Q_2, Q_3, and Q_4 in Fig. 7-4*a* are cut off, and output *OUT* presents a high impedance irrespective of the state of input *IN*. Thus, the outputs of several circuits of the type shown in Fig. 7-4*a* can be connected in parallel as long as no more than one *INHIBIT* input is LOW at any given time. This is illustrated in Ex. 7-2, which follows.

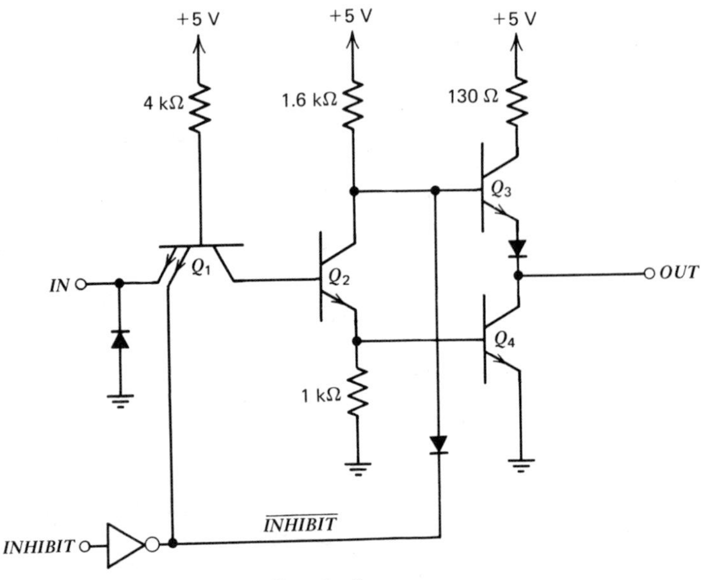

(*a*) Circuit diagram

INHIBIT	*IN*	*OUT*
0 V	0 V	+2.4 V to +5 V
0 V	+5 V	≈ 0 V
+5 V	0 V	Open circuit
+5 V	+5 V	Open circuit

(*b*) Truth table

IN *OUT* *INHIBIT*

(*c*) Symbol

FIGURE 7-4 A 3-state TTL inverter.

EXAMPLE 7-2

Figure 7-5*a* shows two 3-state TTL inverters with their outputs connected in parallel. What is the truth table of the circuit?

Solution

The truth table is shown in Fig. 7-5*b*. Note that the output is not defined in the second and third lines of the truth table. In these lines the two logic inverters are in conflict: one trying to establish an output voltage of ≈ 0 V, the other one an output voltage of +2.4 V to +5 V. These two lines of the truth table would also incur high (≈ 30 mA) currents to be drawn from the +5-V power supply. As a result, the input combinations of the second and third lines of the truth table are *not* permitted.

The 3-state TTL logic inverter circuit shown in Fig. 7-4*a* can be extended to a 3-state TTL NAND gate by use of additional emitters in transistor Q_1. For example, the addition of one more emitter makes Q_1 a three-emitter transistor and results in a 3-state TTL NAND gate with two inputs.

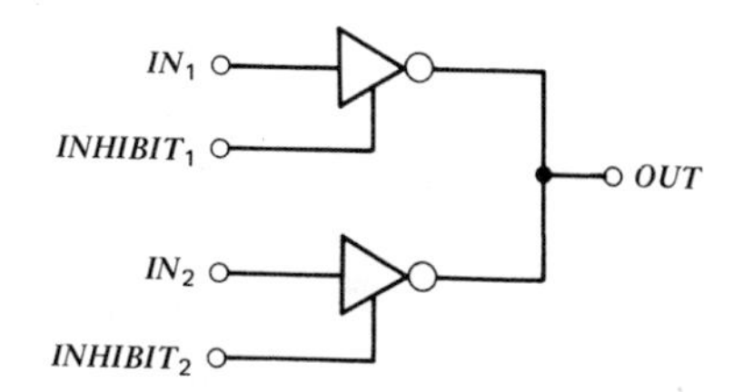

(*a*) The circuit

$INHIBIT_1$	$INHIBIT_2$	IN_1	IN_2	OUT
0 V	0 V	0 V	0 V	+2.4 V to +5 V
0 V	0 V	0 V	+5 V	?
0 V	0 V	+5 V	0 V	?
0 V	0 V	+5 V	+5 V	≈ 0 V
0 V	+5 V	0 V	0 V	+2.4 V to +5 V
0 V	+5 V	0 V	+5 V	+2.4 V to +5 V
0 V	+5 V	+5 V	0 V	≈ 0 V
0 V	+5 V	+5 V	+5 V	≈ 0 V
+5 V	0 V	0 V	0 V	+2.4 V to 5 V
+5 V	0 V	0 V	+5 V	≈ 0 V
+5 V	0 V	+5 V	0 V	+2.4 V to +5 V
+5 V	0 V	+5 V	+5 V	≈ 0 V
+5 V	+5 V	0 V	0 V	Open circuit
+5 V	+5 V	0 V	+5 V	Open circuit
+5 V	+5 V	+5 V	0 V	Open circuit
+5 V	+5 V	+5 V	+5 V	Open circuit

(*b*) Truth table

FIGURE 7-5 Two 3-state TTL inverters with their outputs connected in parallel.

7-2.8 Schottky-Diode Clamped TTL

Compared to the TTL circuits that we have seen thus far, *Schottky-diode clamped TTL* circuits provide about a factor of two reduction in propagation delays at identical power dissipations. This reduction is the result of improved manufacturing technology.

7-2.8.1 Schottky Diodes

One new development has been the introduction of *Schottky diodes*. These diodes have forward voltage drops in the vicinity of 0.3 V, in contrast with other diodes that have forward voltage drops of 0.6 V to 0.7 V.

Figure 7-6 shows a simplified circuit diagram of a *high-speed Schottky TTL* 2-input NAND gate, as well as approximate dc operating conditions with the output at logic LOW. Note the Schottky diodes between collectors and bases of transistors. Because of these, internal collector-base junctions are never strongly

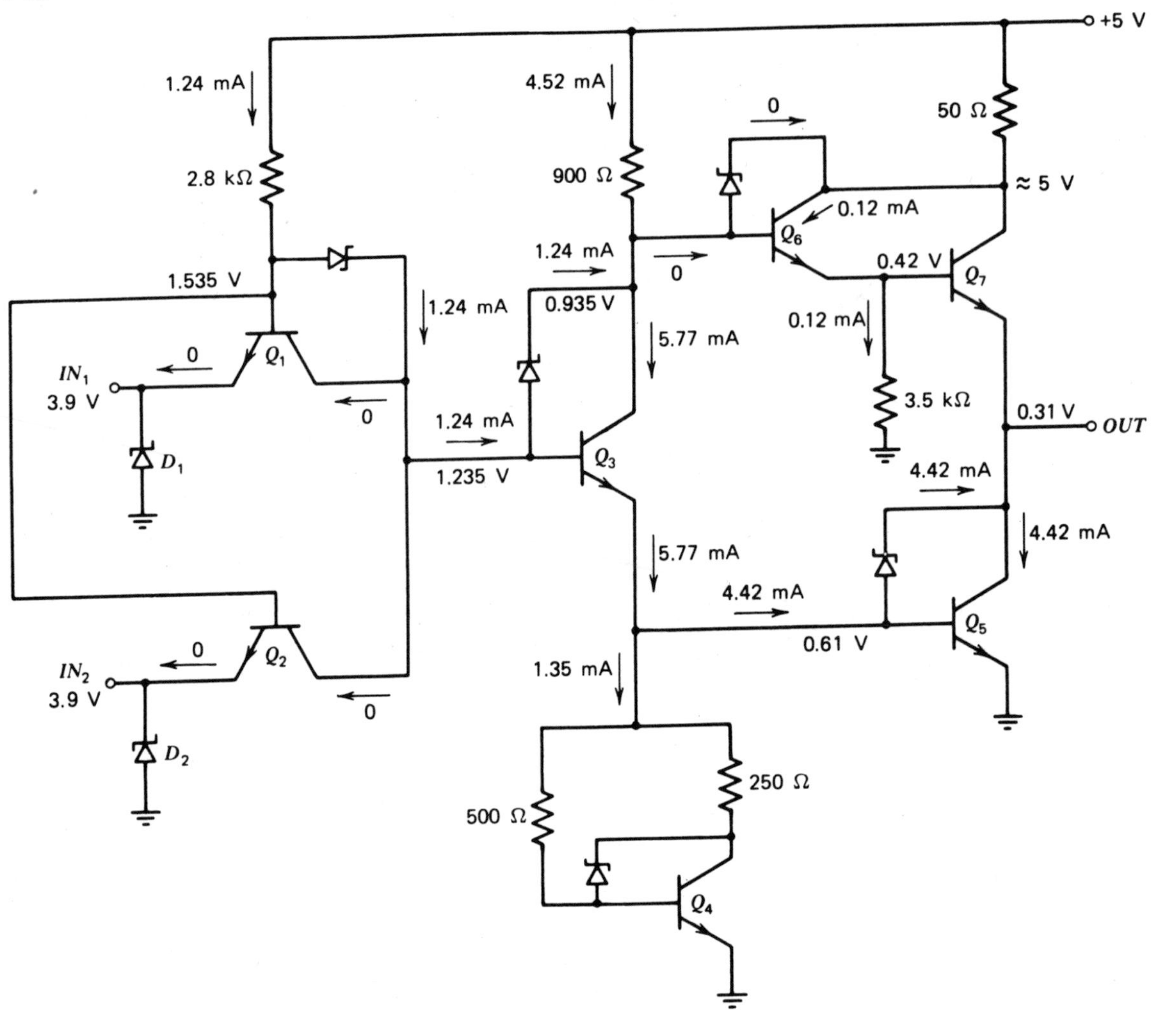

FIGURE 7-6 Simplified schematic diagram of a high-speed Schottky TTL 2-input NAND gate, including approximate dc operating conditions with the output at logic LOW.

forward biased, resulting in reduced stored charges, hence in reduced propagation delays. (Transistor Q_7 has no Schottky diode between its collector and base, since its collector-base junction never gets forward biased.)

*7-2.8.2 Pulldown Circuit

Improved manufacturing technology permits the inclusion of additional circuitry such as the *pulldown circuit* in Fig. 7-6 that consists of transistor Q_4, the Schottky diode between its collector and base, and the 250-Ω and 500-Ω

*Optional material

resistors. This pulldown circuit replaces a single resistor that would have a value of about 450 Ω to carry the 1.35 mA current at 0.61 V.

The advantage of the pulldown circuit can be seen by looking at the state when the output is at logic HIGH as a result of one or both inputs being at logic LOW. In this state transistor Q_3 is cut off and, instead of the 1.35 mA shown in Fig. 7-6, the current flowing into the pulldown circuit is only about 0.01 mA (originating from leakage currents). It can be shown that the resulting voltage across the pulldown circuit is about 0.45 V, while across a 450-Ω resistor it would be 4.5 mV. Thus, instead of a voltage swing of 0.61 V − 4.5 mV = 0.6055 V in the case of the 450-Ω resistor, we have a voltage swing of only 0.61 V − 0.45 V = 0.16 V in the case of the pulldown circuit. This reduced voltage swing results in reduced stored charges, hence in an increase of operating speed.

7-2.8.3 Output Voltage

An unfavorable side effect of the Schottky diode across the lower output transistor Q_5 in Fig. 7-6 is a comparatively high output voltage of 0.31 V that may become as high as 0.5 V when the output is loaded by the inputs of 10 identical logic gates ("fanout of 10"). This is discussed in more detail in Sec. 16-2.

7-2.9 Performance

Table 7-1 summarizes *maximum* propagation delays and power dissipations of various TTL 2-input NAND gates: above the broken line those using a technology without Schottky diodes, below the broken line those using Schottky diodes. The table shows that there is a trade-off between increased operating speed and reduced power dissipation within a given technology.

By comparing the fastest logic gates of the two technologies we can see that the introduction of Schottky diodes led to a reduction of the propagation delay from 10 ns to 5 ns, as well as to a slight reduction of the power dissipation from 40 mW/logic gate to 36 mW/logic gate. As a result, high-speed TTL circuits are now only rarely used in new designs (the same is true for low-power TTL).

TABLE 7-1
Maximum Propagation Delays and Power Dissipations in TTL 2-Input NAND Gates

TTL Version	Propagation Delay	Power Dissipation per Logic Gate
High-speed	10 ns	40 mW
Standard	22 ns	20 mW
Low-power	60 ns	2 mW
High-speed Schottky	5 ns	36 mW
Low-power Schottky	15 ns	4 mW

7-2.10 Availability

In addition to the 14-pin dual in-line package (DIP) of Fig. 2-24 (page 23), similar 16-pin and 20-pin DIPs are often used for TTL, and larger packages are also in use. TTL gates, as well as other TTL circuits described later in this book, are available in such an increasingly large selection that it would be futile to attempt to give even an abbreviated list of them; the reader may look up manufacturers' catalogs for such information.

7-3 EMITTER-COUPLED LOGIC (ECL)

The shortest propagation delays are achieved using *emitter-coupled logic (ECL) circuits.* These circuits are designed such that the collector-to-emitter voltages of the transistors are always above 0.3 V; that is, hard saturation is avoided. This reduces the charge stored in the transistors considerably and hence reduces propagation delays. For example, the fastest available ECL series has maximum propagation delays in the vicinity of 1.5 ns at a maximum power dissipation of 60 mW/logic gate.

Below we first introduce the simplest ECL circuits and some of their applications. This is followed by discussions of circuit improvements and availability.

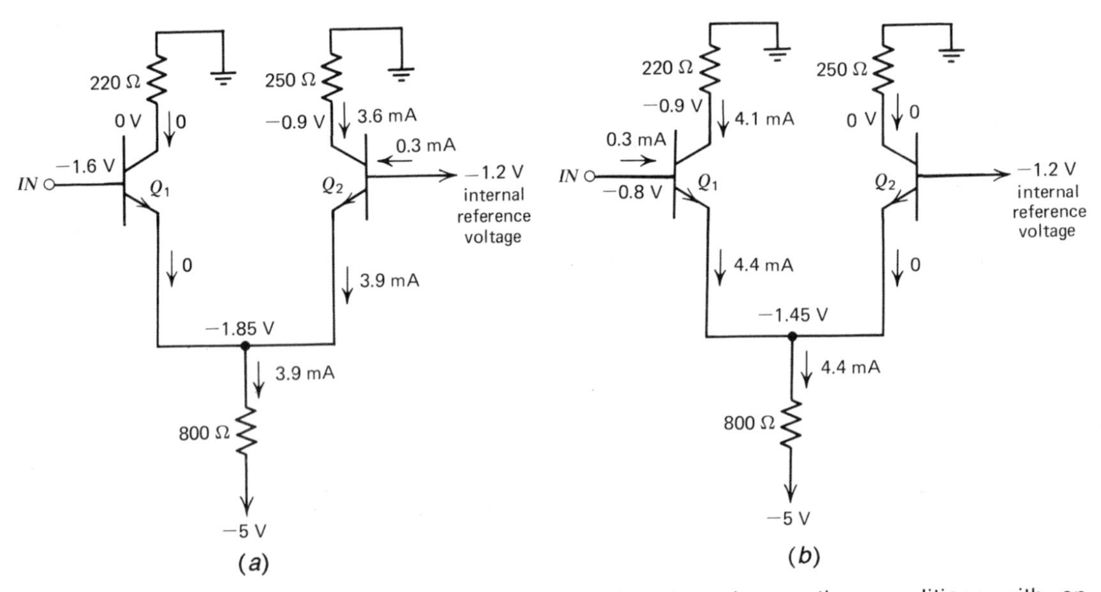

(*a*) Circuit and operating conditions with an input voltage of −1.6 V

(*b*) Circuit and operating conditions with an input voltage of −0.8 V

FIGURE 7-7 Current-switching pair.

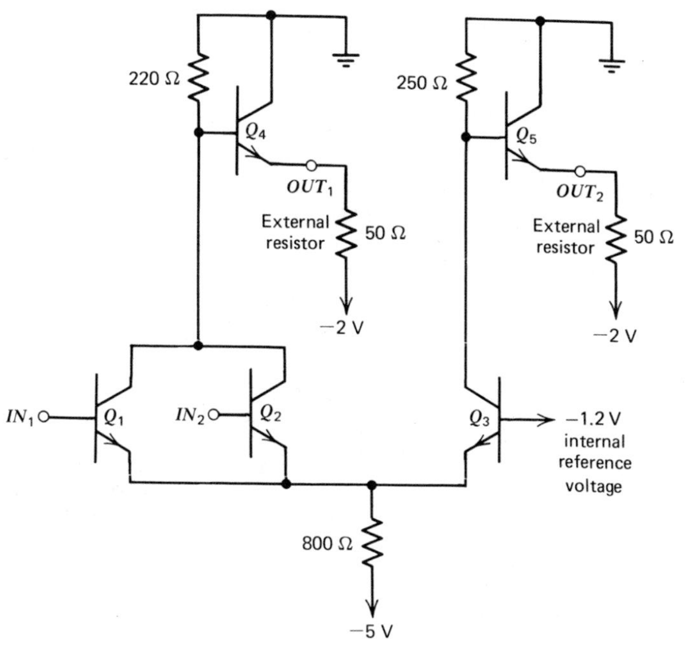

(*a*) Circuit diagram

IN_1	IN_2	OUT_1	OUT_2
−1.6 V	−1.6 V	−0.8 V	−1.6 V
−1.6 V	−0.8 V	−1.6 V	−0.8 V
−0.8 V	−1.6 V	−1.6 V	−0.8 V
−0.8 V	−0.8 V	−1.6 V	−0.8 V

(*b*) Truth table

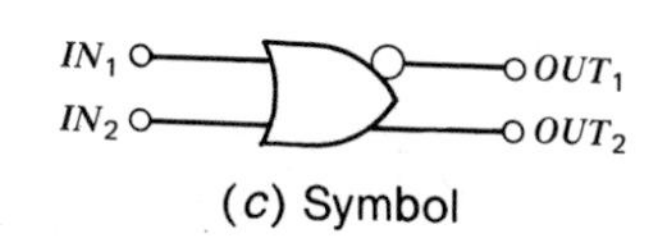

(*c*) Symbol

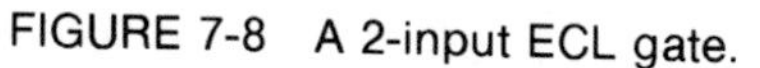

FIGURE 7-8 A 2-input ECL gate.

7-3.1 The Current-Switching Pair

Emitter-coupled logic (ECL) circuits are based on the *current-switching pair*, which consists of two transistors with their emitters connected together. A current-switching pair is shown in Fig. 7-7*a* with an input voltage of −1.6 V and in Fig. 7-7*b* with an input voltage of −0.8 V.

In Fig. 7-7*a* the input voltage of −1.6 V at the base of transistor Q_1 is more negative by 0.4 V than the −1.2-V reference voltage at the base of transistor Q_2. As a result, Q_1 is cut off, the upper end of the 800-Ω resistor is at −1.2 V − 0.65 V = −1.85 V, and the current through the resistor is (5 V − 1.85 V)/800 Ω = 3.9 mA. This current is also the emitter current of Q_2, since Q_1 is cut off. The resulting collector current of Q_2 is approximately 3.6 mA and the voltage across the 250-Ω resistor is ≈ 0.9 V.

In Fig. 7-7*b* the input voltage of -0.8 V at the base of transistor Q_1 is more positive by 0.4 V than the -1.2-V reference voltage at the base of transistor Q_2. As a result, Q_2 is cut off, the upper end of the 800-Ω resistor is at $-0.8\text{ V} - 0.65\text{ V} = -1.45$ V, and the current through the resistor is $(5\text{ V} - 1.45\text{ V})/800\ \Omega = 4.4$ mA. This current is also the emitter current of Q_1, since Q_2 is cut off. The resulting collector current of Q_1 is approximately 4.1 mA and the voltage across the 220-Ω resistor is ≈ 0.9 V.

7-3.2 ECL Gate Circuits

Figure 7-8*a* shows a 2-*input ECL gate circuit.* We can see that transistor Q_1 of Fig. 7-7 is replaced by 2 transistors with their emitters and collectors connected in parallel, making the circuit a 2-input ECL gate. We can also see that the two outputs are fed via *emitter-followers* Q_4 and Q_5, which shift the output voltage levels to -0.8 V and -1.6 V and provide low output impedances. A truth table is shown in Fig. 7-8*b* and a logic symbol in Fig. 7-8*c*.

7-3.3 Wired-OR ECL Circuits

The outputs of ECL gates can be connected in parallel resulting in a wired-OR circuit, shown in Fig. 7-9*a* for 2 logic gates. The inverting outputs of the 2 logic gates are connected in parallel as output OUT_A; an equivalent logic circuit for this output is shown in Fig. 7-9*b*. Also, the noninverting outputs of the logic gates are connected in parallel as output OUT_B; an equivalent logic circuit for this output is shown in Fig. 7-9*c*. Note that the original outputs OUT_1 and OUT_2 as given in the truth table of Fig. 7-8*b* are no longer available.

Figure 7-10*a* shows another wired-OR circuit with 2 ECL gates. Here the inverting output of the upper logic gate and the noninverting output of the lower logic gate are connected in parallel as output OUT_A; an equivalent logic circuit for this output is shown in Fig. 7-10*b*. Also, the noninverting output of the upper

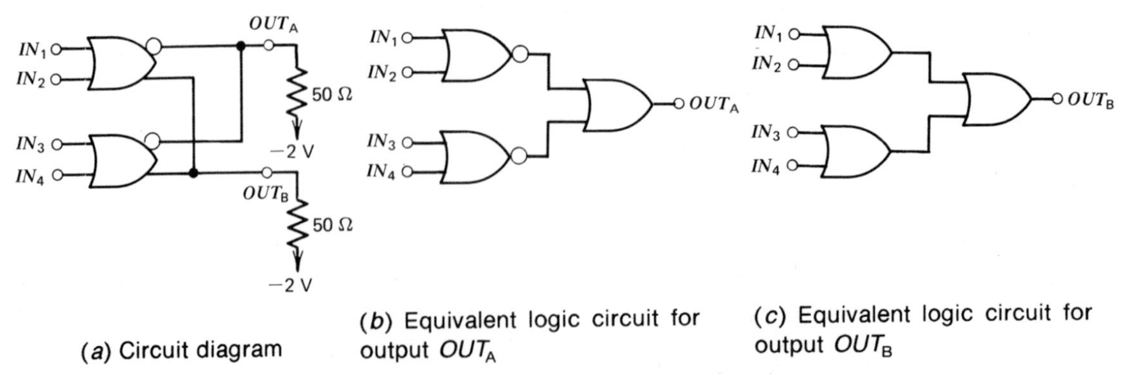

(*a*) Circuit diagram (*b*) Equivalent logic circuit for output OUT_A (*c*) Equivalent logic circuit for output OUT_B

FIGURE 7-9 A wired-OR ECL circuit using 2 logic gates.

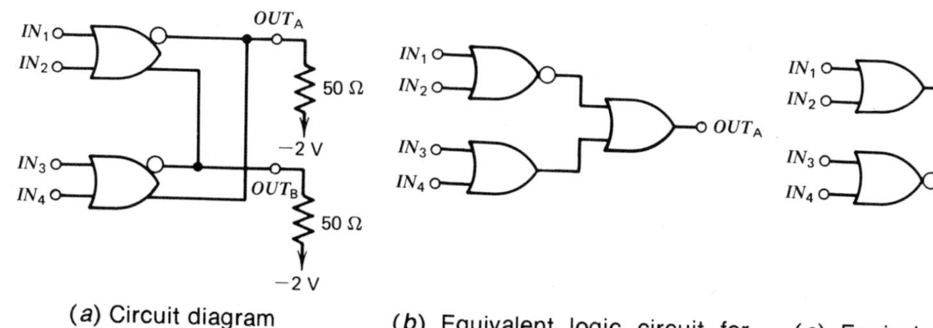

(*a*) Circuit diagram

(*b*) Equivalent logic circuit for output OUT_A

(*c*) Equivalent logic circuit for output OUT_B

FIGURE 7-10 Another wired-OR ECL circuit using 2 logic gates.

logic gate and the inverting output of the lower logic gate are connected in parallel as output OUT_B; an equivalent logic circuit for this output is shown in Fig. 7-10*c*. Note again that the original outputs OUT_1 and OUT_2 as given in the truth table of Fig. 7-8*b* are no longer available.

*7-3.3.1 *Multiple-Output ECL Gates*

In order to facilitate simultaneous wired-OR outputs, ECL gates are also made available with multiple outputs. In these gates each output emitter follower transistor is replaced by a multi-emitter transistor (see Fig. 7-2*b*), resulting in several outputs that can be used independently in wired-OR circuits.

EXAMPLE 7-3

In the 2-input ECL gate shown in Fig. 7-8, each of the two output transistors are replaced by a 3-emitter transistor. A symbol for the resulting ECL gate is shown in Fig. 7-11*a*, and its use is illustrated in Fig. 7-11*b*. Draw equivalent logic circuits for the resulting outputs.

Solution

Equivalent logic circuits for the 8 resulting outputs are shown in Fig. 7-12.

7-3.4 Circuit Improvements

The most widespread ECL circuit improvement is the replacement of the 800-Ω resistor in Fig. 7-7 by a transistor current source (not shown). Additional improvements provide output logic levels that are independent of power supply voltages and of operating temperature.

7-3.5 Availability

The demand for high-speed logic circuits led to the development of a large selection of ECL gates including AND, OR, NAND, NOR, OR-AND, and OR-

*Optional material.

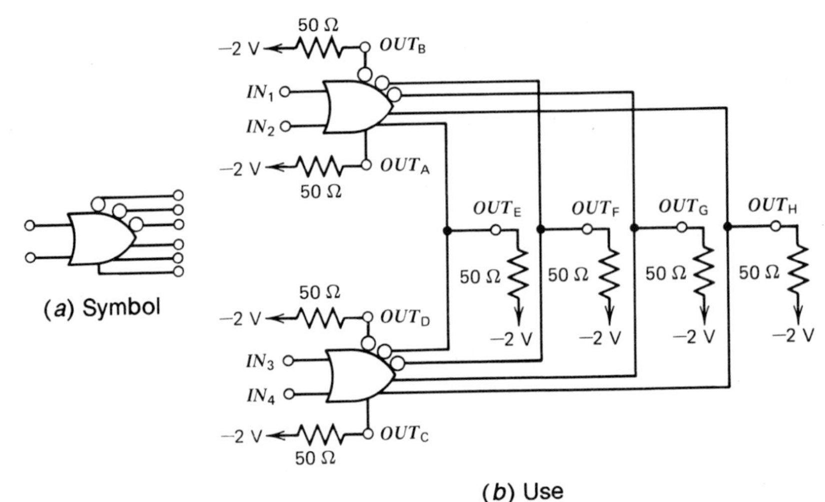

FIGURE 7-11 A multiple-output ECL gate.

AND-INVERT gates, as well as more complex circuits discussed later in the book. As the fastest available logic family, ECL is used whenever the shortest possible propagation delays are desired, although Schottky TTL is often a close choice. However, power requirements of ECL are more steady than of TTL, the latter drawing high peak (transient) currents during switching; this advantage of ECL makes it more suitable for use in LSI gate arrays (see Ch. 11).

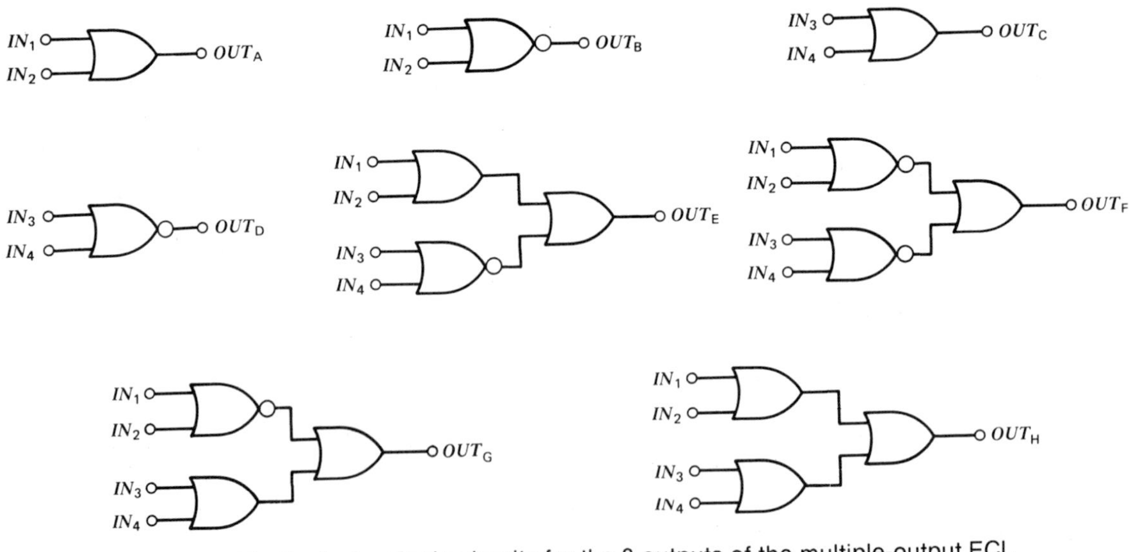

FIGURE 7-12 Equivalent logic circuits for the 8 outputs of the multiple-output ECL circuit of Fig. 7-11*b*.

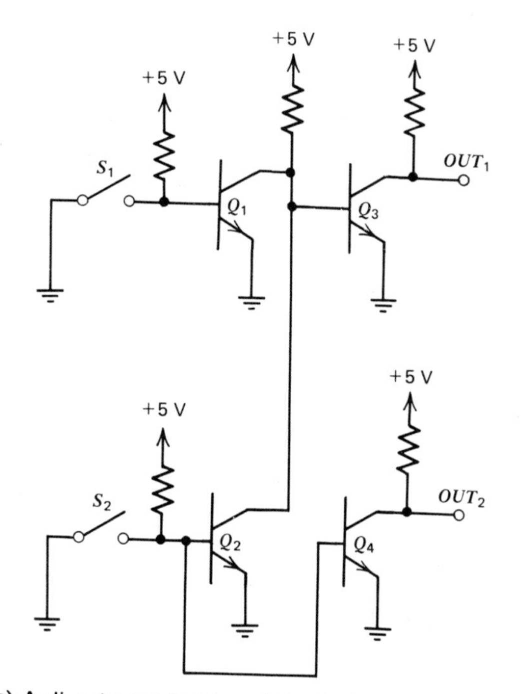

(*a*) A direct-coupled transistor logic (DCTL) circuit

S_1	S_2	Q_1	Q_2	Base of Q_3	OUT_1	OUT_2
OPEN	OPEN	ON	ON	≈ 0 V	+5 V	≈ 0 V
OPEN	CLOSED	ON	OFF	≈ 0 V	+5 V	+5 V
CLOSED	OPEN	OFF	ON	≈ 0 V	+5 V	≈ 0 V
CLOSED	CLOSED	OFF	OFF	≈ +0.7 V	≈ 0 V	+5 V

(*b*) Truth table

FIGURE 7-13 Illustration of the origin of I²L.

*7-4 INTEGRATED INJECTION LOGIC (I²L)

A logic family that finds application primarily in LSI is the *integrated injection logic* (*IIL* or *I²L*). Its origin is shown in Fig. 7-13. Figure 7-13*a* shows a *direct-coupled transistor logic* (*DCTL*) circuit. The inputs of the circuit are connected to switches S_1 and S_2. (In a practical application the inputs would be connected to outputs of preceding circuits instead of to the switches.) A truth table for the circuit is shown in Fig. 7-13*b*.

We can see that in the circuit of Fig. 7-13*a* transistors Q_2 and Q_4 have their emitters and bases connected in parallel; in the I²L integrated circuit process

*Optional material.

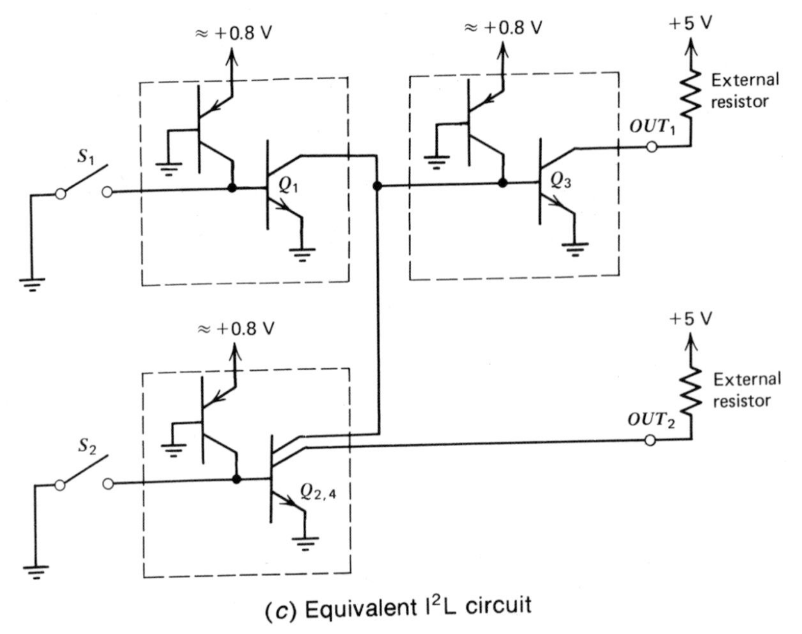

(c) Equivalent I²L circuit

FIGURE 7-13 (*Continued*)

these are made as one transistor with two collectors. Also, the resistors can be substituted by current sources using *pnp* transistors. The resulting circuit, shown in Fig. 7-13*c*, consists of 3 I²L gates, each boxed in by broken lines.

An advantage of the I²L circuit of Fig. 7-13*c* over the circuit of Fig. 7-13*a* is in the integrated circuit (IC) process used for making I²L circuits. The absence of resistors (which are more difficult to make than transistors in an integrated circuit process) and the particular configuration permit a small layout and, hence, a more efficient use of the silicon wafer area.

Another advantage of I²L is that the power dissipation and the propagation delay of a *given circuit* can be altered by varying the +0.8-V voltage that is connected to the emitters of the *pnp* transistors, thus varying their collector currents. The product of the propagation delay and the power dissipation has the dimension of energy (watt × second = joule) and is called the *speed-power product* of the logic gate; ideally, this should be as small as possible. In I²L gates the speed-power product is nearly constant over a wide range of operating currents. This is illustrated in Ex. 7-4, which follows.

EXAMPLE 7-4

When operated at a current of 0.5 mA from a +0.8-V power supply, the propagation delay of an I²L gate is 10 ns. When operated at a current of 0.5 μA from a + 0.6 V power supply, the same logic gate has a propagation delay of 10 μs. Find the speed-power product for both operating currents.

Solution

When operated at a current of 0.5 mA, the power dissipation of the logic gate is 0.8 V × 0.5 mA = 0.4 mW, and the speed-power product is 10 ns × 0.4 mW = 4×10^{-12} W s = **4 pJ** (4 picojoules). When operated at a current of 0.5 μA, the power dissipation of the logic gate is 0.6 V $\times$ 0.5 μA = 0.3 μW, and the speed-power product is 10 μs $\times$ 0.3 μW = 3 $\times$ 10^{-12} W s = **3 pJ** (3 picojoules).

7-5 MOS LOGIC

Another logic family that finds application in LSI is the *metal-oxide-semiconductor (MOS) logic* using either *n*-channel or *p*-channel MOS devices. (Circuits using both *n*-channel and *p*-channel MOS devices are discussed in Sec. 7-6.)

Four MOS device symbols are shown in Fig. 7-14. The complete symbols for *n*-channel and *p*-channel depletion-mode MOS devices are shown in Figs. 7-14*a* and 7-14*b*, respectively. The symbols of Fig. 7-14*c* and 7-14*d* do not distinguish between *n*-channel and *p*-channel devices; also, they do not show the connection of the substrate, which is assumed to be made to the most negative point in the circuit for *n*-channel devices and to the most positive point in the circuit for *p*-channel devices.

7-5.1 MOS Logic Inverters

Figure 7-15*a* shows an MOS logic inverter using *n*-channel devices. Enhancement-mode device Q_1 is an inverter and depletion-mode device Q_2 is its load resistance. Device Q_2 is designed such that its source current is typically one quarter of the current capability of the drain of Q_1 ("*ratio type inverter*"). Thus, when Q_1 is turned ON by an input voltage of $\approx$ +5 V, it easily overpowers Q_2 and pulls the output down to $\approx$ 0 V; when Q_1 is turned OFF by an input voltage of $\approx$ 0 V, the output is pulled up to $\approx$ +5 V by Q_2.

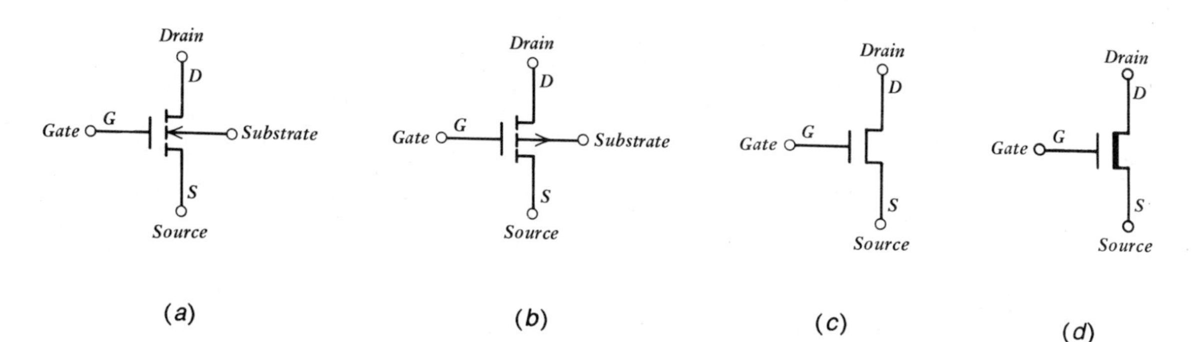

FIGURE 7-14 MOS device symbols. (*a*) *n*-channel enhancement-mode (normally off) device (*b*) *p*-channel enhancement-mode (normally off) device (*c*) Simplified symbol for *n*-channel and *p*-channel enhancement-mode (normally off) devices (*d*) Simplified symbol for *n*-channel and *p*-channel depletion-mode (normally on) devices

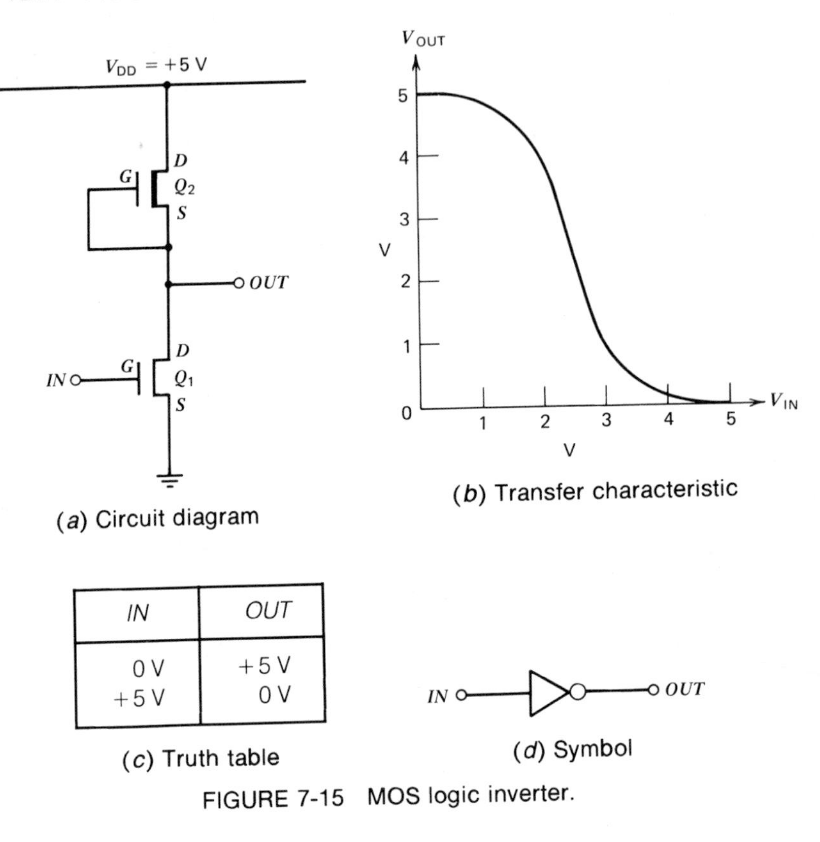

(a) Circuit diagram

(b) Transfer characteristic

IN	OUT
0 V	+5 V
+5 V	0 V

(c) Truth table

(d) Symbol

FIGURE 7-15 MOS logic inverter.

The transfer characteristic of the circuit is shown in Fig. 7-15*b*. We can see that the transition is much softer than either in a DTL gate (Fig. 2-19*d*, page 19) or in a TTL gate (Fig. 7-1*d*). A truth table is shown in Fig. 7-15*c*, and a symbol in Fig. 7-15*d*.

7-5.2 MOS NAND Gates

The MOS logic inverter shown in Fig. 7-15*a* can be extended to a NAND gate as shown in Fig. 7-16*a*. The output is pulled up to $\approx$ +5 V by Q_3 when Q_1 or Q_2 is turned OFF by an input voltage of $\approx$ 0 V. The output is pulled down to $\approx$ 0 V by Q_1 and Q_2 when they are both turned ON by input voltages of $\approx$ +5 V, since the source current of Q_3 is much smaller than the drain current capabilities of Q_1 and Q_2. A truth table is shown in Fig. 7-16*b* and a symbol in Fig. 7-16*c*.

7-5.3 MOS NOR Gates

An MOS NOR gate is shown in Fig. 7-17*a*. The output is pulled up to +5 V by Q_3 when Q_1 and Q_2 are both turned OFF by input voltages of 0 V. The output is pulled down to 0 V by either Q_1 or Q_2 when the input voltage is +5 V; again,

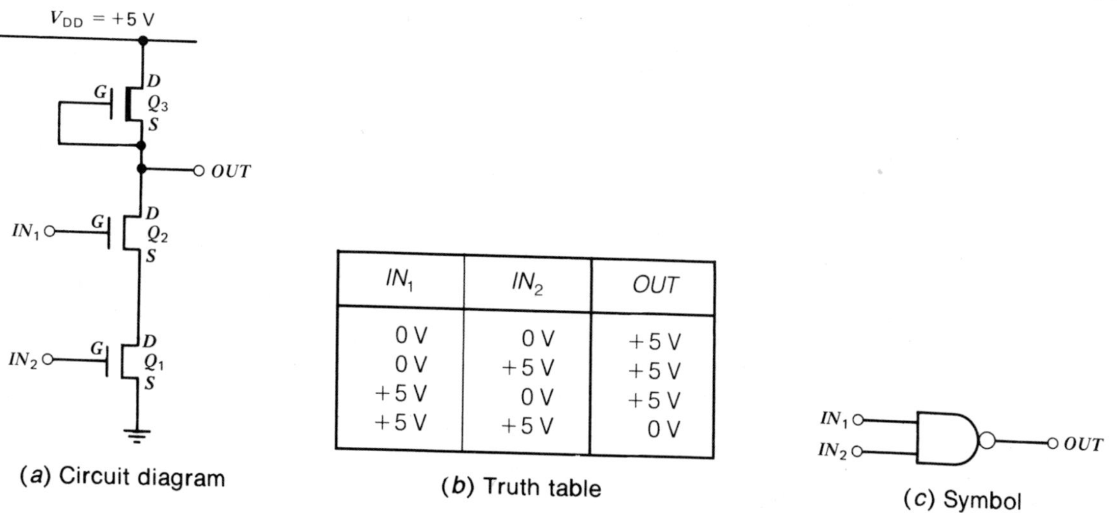

(a) Circuit diagram

IN_1	IN_2	OUT
0 V	0 V	+5 V
0 V	+5 V	+5 V
+5 V	0 V	+5 V
+5 V	+5 V	0 V

(b) Truth table

(c) Symbol

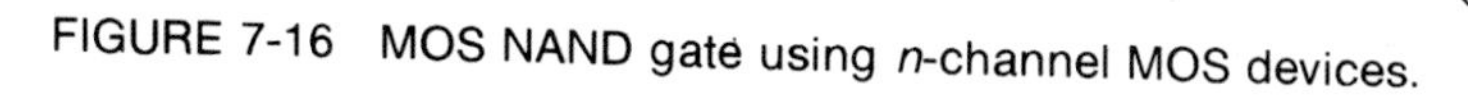

FIGURE 7-16 MOS NAND gate using *n*-channel MOS devices.

the source current of Q_3 is much smaller than the current-carrying capability of Q_1 or Q_2. A truth table is shown in Fig. 7-17*b* and a symbol in Fig. 7-17*c*.

7-5.4 Availability and Use

The principal use of *n*-channel and *p*-channel MOS logic is in LSI and VLSI. These are described in detail in Chapter 11.

7-6 COMPLEMENTARY MOS (CMOS) LOGIC

Low power dissipation is of primary importance in many applications. This led to the development of *complementary metal-oxide-semiconductor (CMOS) logic*, which includes both *n*-channel and *p*-channel devices on the same substrate.

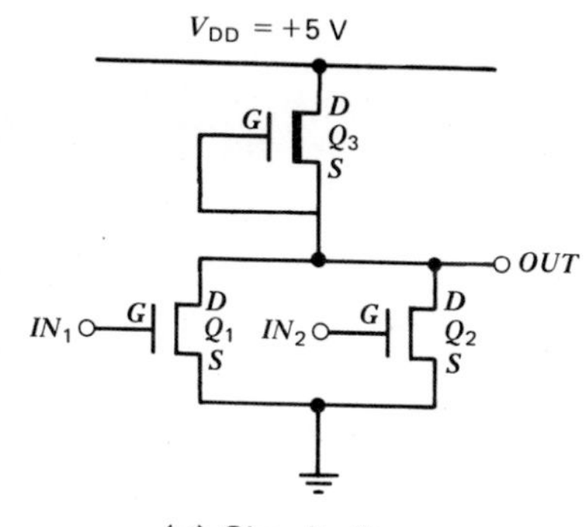

(a) Circuit diagram

IN_1	IN_2	OUT
0 V	0 V	+5 V
0 V	+5 V	0 V
+5 V	0 V	0 V
+5 V	+5 V	0 V

(b) Truth table

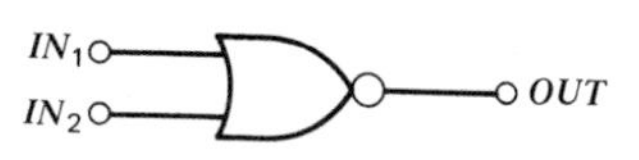

(c) Symbol

FIGURE 7-17 MOS NOR gate using *n*-channel MOS devices.

Although the initial use of the CMOS technology was in low-power applications, it is now also used for high-speed logic circuits with performance comparable to that of TTL. This section introduces the simplest CMOS circuits: logic inverters, NAND and NOR gates, and transmission gates. Additional CMOS circuits are discussed later in the book.

7-6.1 CMOS Logic Inverters

Figure 7-18*a* shows a simplified circuit diagram of an *unbuffered CMOS logic inverter*. It consists of *n*-channel enhancement-mode MOS device Q_1 and *p*-channel enhancement-mode MOS device Q_2.

When the input voltage in Fig. 7-18*a* is at 0 V, Q_1 is turned OFF; also Q_2 is turned ON, pulling the output voltage up to +5 V. When the input voltage is at +5 V, Q_2 is turned OFF; also Q_1 is turned ON, pulling the output voltage down to 0 V. Figure 7-18*b* shows a truth table that includes the states of the MOS

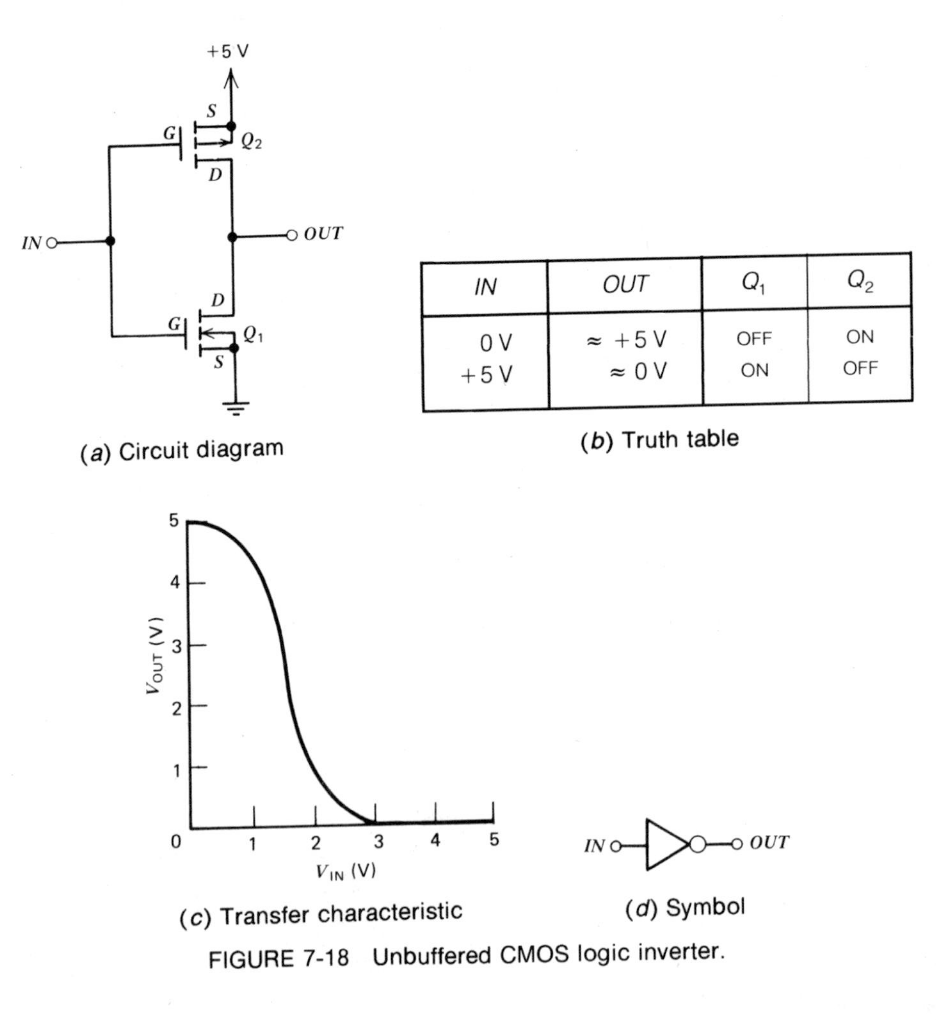

(*a*) Circuit diagram

IN	*OUT*	Q_1	Q_2
0 V	≈ +5 V	OFF	ON
+5 V	≈ 0 V	ON	OFF

(*b*) Truth table

(*c*) Transfer characteristic

(*d*) Symbol

FIGURE 7-18 Unbuffered CMOS logic inverter.

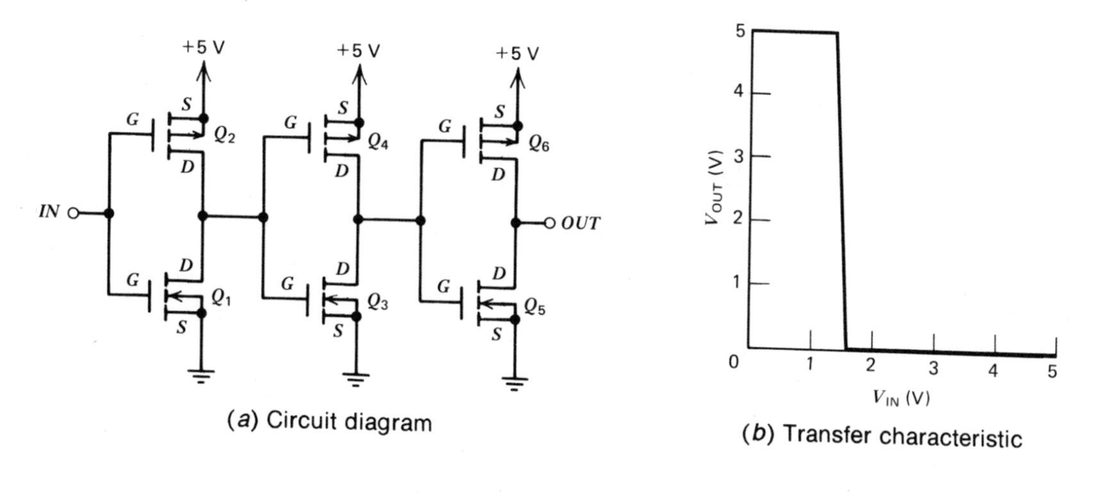

(*a*) Circuit diagram

(*b*) Transfer characteristic

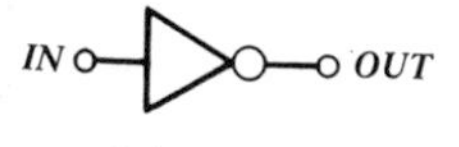

(*c*) Symbol

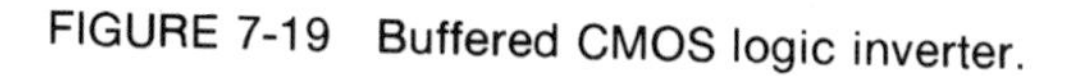

FIGURE 7-19 Buffered CMOS logic inverter.

devices. The transfer characteristic is shown in Fig. 7-18*c*: We can see that the transition is much softer than in a DTL gate (Fig. 2-19*d*, page 19) or in a TTL gate (Fig. 7-1*d*). A symbol is shown in Fig. 7-18*d*.

Note that both *n*-channel device Q_1 and *p*-channel device Q_2 in Fig. 7-18*a* are conducting when the input voltage of the logic inverter (V_{IN} in Fig. 7-18*c*) is in the vicinity of +1.5 V. This results in a "current spike" drawn from the +5 V power supply for a short time during a transition of the input voltage between 0 V and +5 V.

The circuit diagram of a *buffered CMOS logic inverter* is shown in Fig. 7-19*a*: It consists of three unbuffered CMOS logic inverters. The resulting transfer characteristic, shown in Fig. 7-19*b*, exhibits a sharp transition that is comparable to that in TTL. A symbol is shown in Fig. 7-19*c*.

7-6.2 CMOS NAND Gates

The unbuffered logic inverter of Fig. 7-18*a* can be extended to an *unbuffered CMOS NAND gate* as shown in Fig. 7-20*a*. A truth table including the states of the MOS devices is shown in Fig. 7-20*b*. Note that in order for a device to be ON, its gate *G* has to be ON and current to its drain *D* and source *S* must be available. A symbol of the NAND gate is shown in Fig. 7-20*c*.

A *buffered CMOS NAND gate* can be obtained if the circuit of Fig. 7-20*a* is followed by two unbuffered logic inverters, as was similarly done in Fig. 7-19*a*.

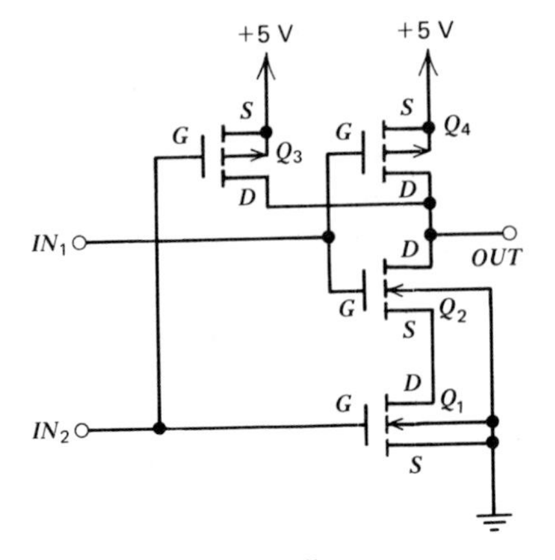

(*a*) Circuit diagram

IN_1	IN_2	OUT	Q_1	Q_2	Q_3	Q_4
0 V	0 V	≈ +5 V	OFF	OFF	ON	ON
0 V	+5 V	≈ +5 V	OFF	OFF	OFF	ON
+5 V	0 V	≈ +5 V	OFF	OFF	ON	OFF
+5 V	+5 V	≈ 0 V	ON	ON	OFF	OFF

(*b*) Truth table including the states of the MOS devices

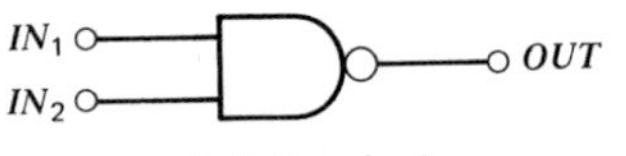

(*c*) Symbol

FIGURE 7-20 Unbuffered CMOS NAND gate.

7-6.3 CMOS NOR Gates

The circuit diagram of an *unbuffered CMOS NOR gate* is shown in Fig. 7-21*a*, a truth table including the states of the MOS devices in Fig. 7-21*b*, and a symbol in Fig. 7-21*c*. A *buffered CMOS NOR gate* can be obtained by following the circuit of Fig. 7-21*a* with two unbuffered logic inverters.

7-6.4 CMOS Bilateral Transmission Gates

In many applications it would be desirable to connect in parallel the outputs of several CMOS logic gates—however, these present ≈ 1 kΩ output impedances both in their LOW and HIGH output states; hence, parallel connection of outputs is not practical. For this reason, CMOS logic gates with 3-state outputs have been developed; these are similar in function to the 3-state TTL gates of Fig. 7-4; also, an alternative is provided by *CMOS bilateral transmission gates*, described below.

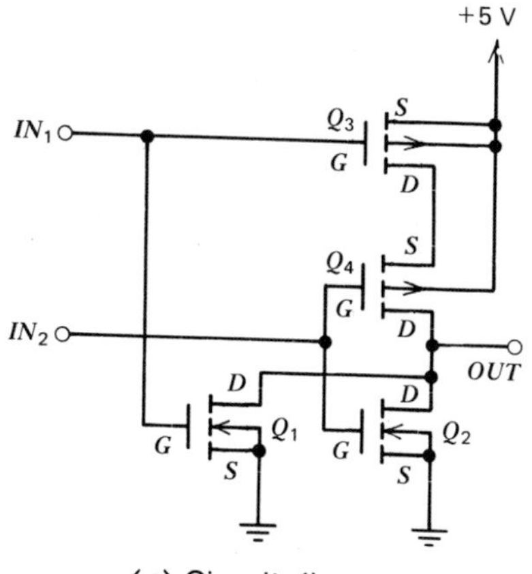

(a) Circuit diagram

IN_1	IN_2	OUT	Q_1	Q_2	Q_3	Q_4
0 V	0 V	$\approx +5$ V	OFF	OFF	ON	ON
0 V	+5 V	≈ 0 V	OFF	ON	OFF	OFF
+5 V	0 V	≈ 0 V	ON	OFF	OFF	OFF
+5 V	+5 V	≈ 0 V	ON	ON	OFF	OFF

(b) Truth table including the states of the MOS devices

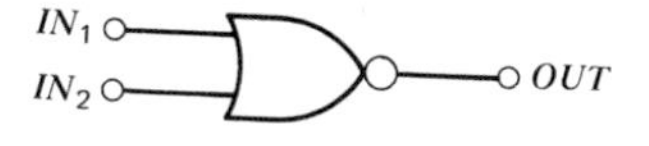

(c) Symbol

FIGURE 7-21 Unbuffered CMOS NOR gate.

Consider the circuit of Fig. 7-22a. It consists of n-channel enhancement-mode device Q_1, p-channel enhancement-mode device Q_2, and a logic inverter that can be either unbuffered or buffered (Fig. 7-18a or Fig. 7-19a). In addition to ground and +5 V, the circuit has three terminals: control input G and signal connections S_1 and S_2. Control input G is either at 0 V or at +5 V, and signal connections S_1 and S_2 are limited to voltages between 0 V and +5 V.

When the voltage at control input G is 0 V, both Q_1 and Q_2 are cut OFF and the circuit approximates an open circuit between signal terminals S_1 and S_2. When the voltage at control input G is +5 V, both Q_1 and Q_2 are turned ON and the circuit presents a low impedance ($\approx 1\ \mathrm{k}\Omega$) between signal terminals S_1 and S_2. Hence, the circuit can be described as a switch between S_1 and S_2: The switch is open when the voltage at G is 0 V and it is closed when the voltage at G is +5 V.

In digital applications, one of the two signal terminals S_1 and S_2 is connected to either 0 V or +5 V. The resulting operating conditions are summarized in the

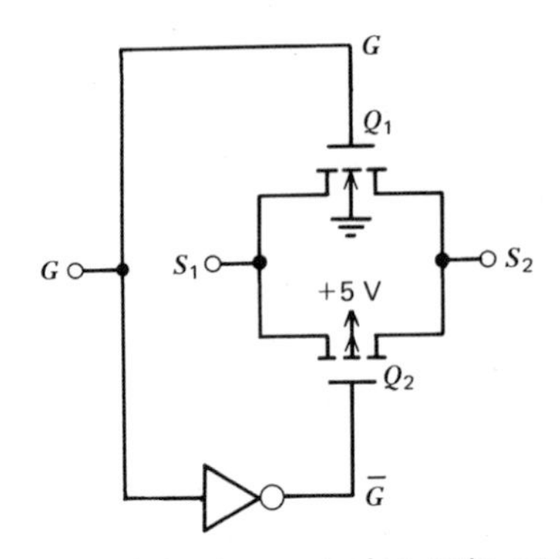

(a) Circuit diagram of the transmission gate and drive circuitry

G	$\overline{G}$	S_1	S_2	Q_1	Q_2
0 V	+5 V	0 V	Open circuit	OFF	OFF
0 V	+5 V	+5 V	Open circuit	OFF	OFF
+5 V	0 V	0 V	≈ 0 V	ON	OFF
+5 V	0 V	+5 V	≈ +5 V	OFF	ON

(b) Truth table including the states of the MOS devices

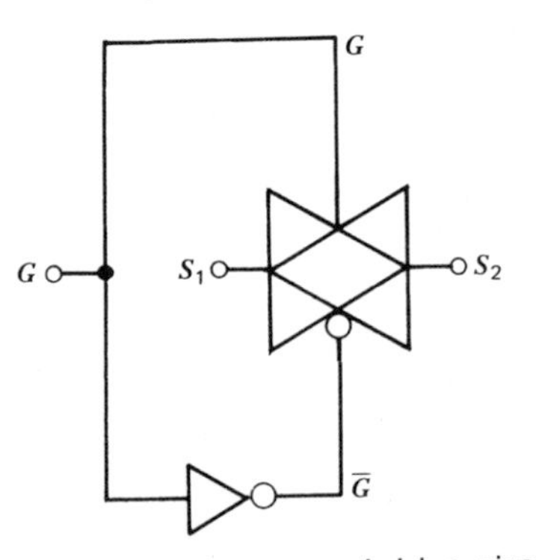

(c) Symbol for the transmission gate and drive circuitry shown in (a)

FIGURE 7-22 CMOS bilateral transmission gate.

truth table of Fig. 7-22b with signal terminal S_1 connected to 0 V or +5 V. We can see that signal terminal S_2 is an open circuit when control input G is at 0 V. On the other hand, the voltage of signal terminal S_2 follows the voltage of signal terminal S_1 when control input G is at +5 V.

A symbol for the transmission gate and its drive circuitry is shown in Fig. 7-22c. An application of CMOS bilateral transmission gates is described in Ex. 7-5, which follows.

EXAMPLE 7-5

Figure 7-23*a* shows two CMOS bilateral transmission gates with their S_2 signal terminals connected in parallel as output *OUT*. What is the truth table of the circuit?

Solution

The truth table is shown in Fig. 7-23*b*. Note that the circuit of Fig. 7-23*a* performs the same function as the 3-state TTL circuit of Fig. 7-5*a*, and that the truth table of Fig. 7-23*b* is identical to the truth table of Fig. 7-5*b* with the sequence of lines rearranged. Thus, for example, in the second to last and third to last lines of the truth table of Fig. 7-23*b*, the two transmission gates are in conflict—one trying to establish an output voltage of ≈ 0 V, the other one an output voltage of $\approx +5$ V; these lines correspond to the second and third lines in the truth table of Fig. 7-5*b*.

We should mention that the transmission gate circuit shown in Fig. 7-22*a* is the simplest possible. When G is at $+5$ V in this circuit, the resistance between S_1 and S_2 is in the 1-kΩ range, but its value depends on the voltages at S_1 and S_2. This is often objectionable in *linear* applications, and for this reason additional MOS devices controlling the substrate voltage of Q_1 are usually included to make the ON resistance between S_1 and S_2 more constant (''*linearized transfer characteristics*'').

7-6.5 Propagation Delays

The magnitudes of propagation delays in a CMOS logic gate depend on the MOS devices it uses, on whether it is unbuffered or buffered, on the power

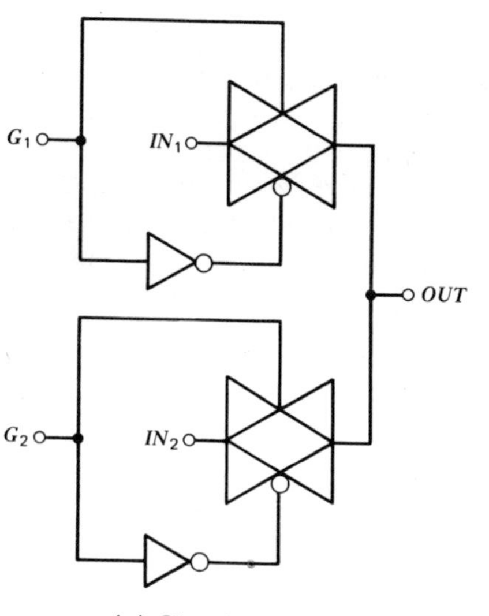

(*a*) Circuit diagram

FIGURE 7-23 Two CMOS bilateral transmission gates with their outputs connected in parallel.

G_1	G_2	IN_1	IN_2	OUT
0 V	0 V	0 V	0 V	Open circuit
0 V	0 V	0 V	+5 V	Open circuit
0 V	0 V	+5 V	0 V	Open circuit
0 V	0 V	+5 V	+5 V	Open circuit
0 V	+5 V	0 V	0 V	≈ 0 V
0 V	+5 V	0 V	+5 V	≈ +5 V
0 V	+5 V	+5 V	0 V	≈ 0 V
0 V	+5 V	+5 V	+5 V	≈ +5 V
+5 V	0 V	0 V	0 V	≈ 0 V
+5 V	0 V	0 V	+5 V	≈ 0 V
+5 V	0 V	+5 V	0 V	≈ +5 V
+5 V	0 V	+5 V	+5 V	≈ +5 V
+5 V	+5 V	0 V	0 V	≈ 0 V
+5 V	+5 V	0 V	+5 V	≈ ?
+5 V	+5 V	+5 V	0 V	≈ ?
+5 V	+5 V	+5 V	+5 V	≈ +5 V

(*b*) Truth table

FIGURE 7-23 (*Continued*)

supply voltage, and on the capacitance loading the output. We first look at the *intrinsic propagation delays*, that is, at propagation delays with unloaded output.

Table 7-2 shows *typical* intrinsic propagation delays $(t_{PLH})_{intrinsic}$ and $(t_{PHL})_{intrinsic}$ of various CMOS logic inverters; propagation delays of other CMOS logic gates are comparable. Note that *high-speed CMOS* logic circuits operate at a maximum power supply voltage of 6 V, they are always buffered, and they use MOS devices with higher current-carrying capabilities than standard CMOS logic circuits. In comparing the intrinsic propagation delays of *standard* CMOS logic inverters shown in Table 7-2, we can see that the intrinsic propagation delays in a standard *buffered* logic inverter are longer than in a standard *unbuffered* logic inverter—due mainly to the additional two logic inverters in the buffered circuit (see Fig. 7-19*a*).

TABLE 7-2
Typical Intrinsic Propagation Delays $(t_{PLH})_{intrinsic} = (t_{PHL})_{intrinsic}$ in CMOS Logic Inverters at Room Temperature

CMOS Logic Inverter	Power Supply Voltage	
	5 V	15 V
Standard unbuffered	50 ns	10 ns
Standard buffered	100 ns	30 ns
High-speed (buffered)	10 ns	—

TABLE 7-3
Typical Equivalent Output Resistance R_{eq} in CMOS Logic Inverters at Room Temperature

CMOS Logic Inverter	Power Supply Voltage	
	5 V	15 V
Standard unbuffered	2500 Ω	750 Ω
Standard buffered	1500 Ω	500 Ω
High-speed (buffered)	150 Ω	—

While the *intrinsic* propagation delays are longer in a buffered logic inverter than in an unbuffered one, the situation is different for large load capacitances because of the lower output resistance of a buffered logic inverter. Since all logic gate circuits are nonlinear, their output resistances vary during a transition between their LOW and HIGH output states; in fact, the output characteristics of a CMOS logic gate during a transition could be somewhat better described by a current source. Instead, in what follows we crudely approximate the output characteristics by an *equivalent output resistance* R_{eq}, which is the voltage-to-current ratio averaged during a transition.

Table 7-3 shows the typical equivalent output resistance R_{eq} of CMOS logic inverters; equivalent output resistances of other CMOS logic gates are comparable. Note that, as was the case for the intrinsic propagation delays, we make no distinction between the transition from the LOW to the HIGH output state and the transition from the HIGH to the LOW output state. These approximations are valid because the logic gates are *designed* to have similar properties for the two transitions.

Given the intrinsic propagation delay $(t_{PLH})_{intrinsic}$ or $(t_{PHL})_{intrinsic}$, the equivalent output resistance R_{eq}, and the load capacitance C at the output, we *approximate* the propagation delays as

$$t_{PLH} \approx (t_{PLH})_{intrinsic} + 0.5\,R_{eq}\,C \tag{7-1}$$

and

$$t_{PHL} \approx (t_{PHL})_{intrinsic} + 0.5\,R_{eq}\,C \tag{7-2}$$

Thus the propagation delays, t_{PLH} and t_{PHL}, are functions of the intrinsic propagation delays shown in Table 7-2, of the equivalent output resistance shown in Table 7-3, and of capacitance C loading the output. We can see in the tables that the intrinsic propagation delays and the equivalent output resistance of a high-speed logic inverter are one tenth of those of a standard buffered logic inverter. Hence, when operated at power supply voltages of 5 V, for identical C the high-speed logic inverter is 10 times faster than the standard buffered logic inverter.

The use of eqs. 7-1 and 7-2 is illustrated in Ex. 7-6, which follows.

EXAMPLE 7-6

Using eqs. 7-1 and 7-2 and the data from Tables 7-2 and 7-3, plot propagation delays of CMOS logic inverters for load capacitances of 0 to 500 pF.

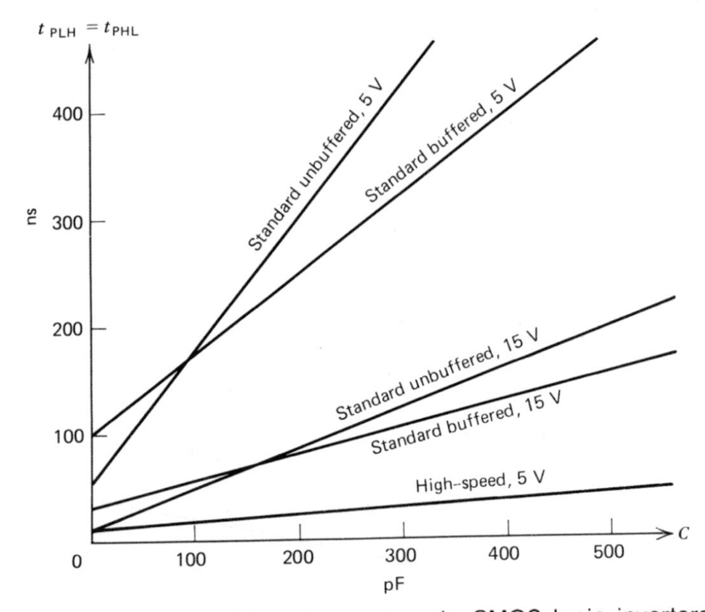

FIGURE 7-24 Propagation delays $t_{PLH} = t_{PHL}$ in CMOS logic inverters as functions of external load capacitance C.

Solution

The propagation delays are plotted in Fig. 7-24. We can see that when $C = 0$, the propagation delays revert to those in Table 7-2. Also, the slope of each line is 0.5 R_{eq}. For example, for the standard buffered logic inverter at 15 V, we get from Table 7-3 an $R_{eq} = 500\ \Omega$ and the slope of the corresponding line in Fig. 7-24 is 0.25 ns/pF = 250 Ω, which indeed equals 0.5 R_{eq}.

Another parameter that influences the propagation delays is the *input capacitance*, C_{IN}. The value of C_{IN} is typically 10 pF for an unbuffered CMOS logic gate and 5 pF for a buffered CMOS logic gate.

EXAMPLE 7-7

A standard unbuffered CMOS logic inverter is operated at a power supply voltage of +5 V. Its output is connected either to the inputs of 20 standard unbuffered logic inverters or to the inputs of 20 standard buffered logic inverters. Find the propagation delays in both cases if the wiring capacitances are negligible.

Solution

The resulting input capacitance of 20 *unbuffered* logic inverters is 20 × 10 pF = 200 pF and the resulting propagation delay from Fig. 7-24 is **300 ns**.

The resulting input capacitance of 20 *buffered* logic inverters is 20 × 5 pF = 100 pF and the resulting propagation delay from Fig. 7-24 is **175 ns**.

7-6.6 Power Dissipation

When a CMOS logic circuit is in one of its quiescent states (HIGH or LOW), the only currents that flow are leakage currents, typically 10 μA or less. Thus, when operated from a power supply of 5 V, the *quiescent (dc) power dissipation* is at most 5 V × 10 μA = 50 μW. Because of this low quiescent (dc) power dissipation, *ac power dissipation* (or *transient power dissipation*) often becomes dominant in CMOS logic.

It is shown by circuit theory that when the voltage across a capacitance C changes by V_c, its stored energy changes by $CV_c^2/2$ and an equal energy is lost in the driving circuit. Thus, each time we switch the voltage across a capacitance C from 0 V to V_c *and* back from V_c to 0 V, we dissipate an energy of CV_c^2 in the driving circuit.

Capacitance C consists of several parts. CMOS logic circuits are characterized by a *no-load power dissipation capacitance*, C_{pd}. This includes the output capacitance, as well as internal capacitances of the circuit that are switched between 0 V and the power supply voltage V_c; it also includes the effects of "current spikes" (see Sec. 7-6.1). In order to find the total capacitance C, we have to add to C_{pd} input capacitances of subsequent circuit and wiring capacitances loading the output of the circuit. This is illustrated in Ex. 7-8 below.

EXAMPLE 7-8

A CMOS logic circuit has a no-load power dissipation capacitance of C_{pd} = 20 pF. Its output is connected to the inputs of three other CMOS logic circuits, each with an input capacitance of C_{IN} = 5 pF; the output is also loaded by a wiring capacitance of C_{WIRING} = 15 pF. The circuit is operated at a power supply voltage of V_c = +5 V. Find

a. The total capacitance C,

b. The energy dissipated in the circuit when its output is switched from V_c = +5 V to 0 V and back from 0 V to V_c = +5 V.

Solution

a. The total capacitance is $C = C_{pd} + (3 \times C_{IN}) + C_{WIRING}$ = 20 pF + (3 × 5 pF) + 15 pF = **50 pF**.

b. The energy dissipated is CV_c^2 = 50 pF × (5 V)^2 = 1.25×10^{-9} joule = **1.25 nJ**.

When the voltage across a capacitance C is switched back and forth between 0 V and V_c at a frequency of f_{CLOCK} ("*clock frequency*"), the resulting ac power dissipation in the circuit driving the capacitance is $CV_c^2 f_{CLOCK}$. This is illustrated in Ex. 7-9 below.

EXAMPLE 7-9

The CMOS logic circuit of Ex. 7-8 is operated at a clock frequency of f_{CLOCK} = 10 MHz. Thus, there is a clock pulse once every 100 ns; also, the output of the circuit is switched back and forth between 0 V and V_c once every 100 ns. Find the ac power dissipation.

Solution

The ac power dissipation is $CV_c^2 f_{CLOCK}$ = 50 pF × (5 V)^2 × 10 MHz = **12.5 mW**. Note that this power dissipation is comparable to that of a TTL gate.

The *total power dissipation* P_t in a CMOS logic circuit is the sum of the quiescent (dc) power dissipation P_{dc} and the ac power dissipation P_{ac}:

$$P_t = P_{dc} + P_{ac} \qquad (7\text{-}3)$$

The use of eq. 7-3 is illustrated in Ex. 7-10, which follows.

EXAMPLE 7-10

Find the total power dissipation in the CMOS logic circuit of Ex. 7-9 if its quiescent power dissipation is $P_{dc} = 50\ \mu W$.

Solution

The ac power dissipation, from Ex. 7-9, is $P_{ac} = 12.5$ mW. The quiescent power dissipation is $P_{dc} = 50\ \mu W = 0.05$ mW. Thus, the total power dissipation is $P_t = P_{dc} + P_{ac} = 0.05\ mW + 12.5\ mW =$ **12.55 mW**. Note that this power dissipation is comparable to that of a TTL gate.

7-6.7 Availability and Use

A large variety of CMOS logic gates, as well as other CMOS circuits discussed later, are available. They are used principally in digital systems where low power dissipation or operation at high voltage (up to 15 V) or both are important. Custom-designed CMOS digital circuits include pocket-size calculators with low standby power ("*continuous memory*"), automotive electronics operating from a 12-V battery, and low-power digital watch circuits operating from a 1.25-V battery.

SUMMARY

The chapter described transistor-transistor logic, TTL; emitter-coupled logic, ECL; integrated-injection logic, I^2L; metal-oxide-semiconductor, MOS, logic; and complementary MOS, CMOS, logic circuits. AND, OR, NAND, NOR, AND-OR-INVERT (AOI), and transmission gates were introduced, and truth tables, transfer characteristics, power dissipation, propagation delays, and loading effects were discussed.

SELF-EVALUATION QUESTIONS

1. Compare the properties of standard TTL, high-speed Schottky TTL, and low-power Schottky TTL.
2. Sketch the circuit diagram of a TTL AND-OR-INVERT gate.
3. Sketch the circuit diagram of a 3-state TTL inverter.
4. Sketch a current-switching pair and describe its operation.
5. Describe the operation of ECL gates in wired-OR circuits.
***6.** Sketch an I^2L circuit.
7. Sketch a 2-input NAND gate using *n*-channel MOS devices.
8. Sketch circuit diagrams of (**a**) an unbuffered and (**b**) a buffered CMOS logic inverter.
9. Sketch the circuit diagram and the symbol of a CMOS bilateral transmission gate.

*Optional material.

10. Compare propagation delays of CMOS logic inverters with various load capacitances.
11. Discuss ac power dissipation in CMOS logic gates.
12. Compare salient operating features of TTL, ECL, and CMOS.

PROBLEMS

7-1. Find the value of the current that flows through the collector-base junction of Q_1 in Fig. 7-1*b*. What is the direction of the current?

7-2. Find the value of the current that flows out of the emitter of Q_1 in Fig. 7-1*a*. To where is this current flowing?

7-3. Estimate the value of the current that flows in the collector-base junction of Q_1 in Fig. 7-2*a* if $IN_1 = +5$ V and $IN_2 = +5$ V.

7-4. Estimate the value of the current that flows through input IN_1 in Fig. 7-2*a* if $IN_1 = 0$ V and $IN_2 = +5$ V.

***7-5.** The output of the TTL inverter shown in Fig. 7-1*a* is loaded by a resistor of 3.7 MΩ to ground. Use the transfer characteristic of Fig. 7-1*d* and find output voltage V_{OUT}.

***7-6.** The output of the TTL inverter shown in Fig. 7-1*a* is loaded by a resistor of 1.2 kΩ to ground. Use the transfer characteristic of Fig. 7-1*d* and find output voltage V_{OUT}.

7-7. The output of the TTL NAND gate of Fig. 7-2*a* is loaded by a capacitance of $C = 100$ pF. The two inputs are simultaneously switched from a voltage of 0 V to a voltage of +5 V. The output impedance during the transition can be approximated as $R_{OUT} = 150\ \Omega$. The intrinsic propagation delay $(t_{PHL})_{intrinsic} = 8$ ns. Estimate propagation delay t_{PHL}.

7-8. The output of the TTL NAND gate of Fig. 7-2*a* is loaded by a capacitance of $C = 15$ pF. The two inputs are simultaneously switched from a voltage of +5 V to a voltage of 0 V. The output impedance during the transition can be approximated as $R_{OUT} = 150\ \Omega$. The intrinsic propagation delay $(t_{PLH})_{intrinsic} = 10$ ns. Estimate propagation delay t_{PLH}.

***7-9.** Find the values of the voltages at the bases of Q_2 and Q_2' and at points x and $\bar{x}$ in Fig. 7-3 if $IN_1 = 0$ V, $IN_2 = 0$ V, $IN_3 = +5$ V, and $IN_4 = +5$ V.

***7-10.** Find the values of the voltages at the bases of Q_2 and Q_2' and at points x and $\bar{x}$ in Fig. 7-3 if $IN_1 = 0$ V, $IN_2 = +5$ V, $IN_3 = 0$ V, and $IN_4 = +5$ V.

*Optional problem.

7-11. The output of a 3-state TTL gate circuit in its high-impedance state exhibits a leakage current of 40 μA to ground. When many outputs are connected in parallel in a wired-OR configuration (illustrated for 2 outputs in Fig. 7-5*a*), a single output in its HIGH ($+2.4$ V to $+5$ V) output state has to be able to carry the sum of the leakage currents of all other outputs connected in parallel. Assume that each output is characterized by Fig. 7-1*d* and find the maximum number of outputs that can be connected in parallel if an output voltage of at least $+2.4$ V is desired in the HIGH output state.

7-12. The output of a 3-state TTL gate circuit in its high-impedance state exhibits a leakage current of 40 μA to ground. When many outputs are connected in parallel in a wired-OR configuration (illustrated for 2 outputs in Fig. 7-5*a*), a single output in its HIGH ($+2.4$ V to $+5$ V) output state has to be able to carry the sum of the leakage currents of all other outputs connected in parallel. Assume that each output in its HIGH state is capable of providing a load current of 5.2 mA ($I_L = -5.2$ mA in Fig. 7-1*a*) at an output voltage of $+2.4$ V and find the maximum number of outputs that can be connected in parallel if an output voltage of at least $+2.4$ V is desired in the HIGH output state.

7-13. Find the collector currents of transistors Q_1, Q_2, and Q_3 in the ECL gate circuit of Fig. 7-8*a* with input voltages of -1.6 V at both inputs.

7-14. Find the collector currents of transistors Q_1, Q_2, and Q_3 in the ECL gate circuit of Fig. 7-8*a* if the voltage at input IN_1 is -0.8 V and if at input IN_2 it is -1.6 V.

7-15. Prepare a truth table for OUT_A and OUT_B in the wired-OR ECL circuit of Fig. 7-9*a*.

7-16. Prepare a truth table for OUT_A and OUT_B in the wired-OR ECL circuit of Fig. 7-10*a*.

***7-17.** The speed-power product of an I^2L gate is 5 picojoules (pJ). Find the current drawn from the 0.8-V power supply if the propagation delay of the logic gate is 10 ns.

***7-18.** The speed-power product of an I^2L gate is 6 pJ. Find the propagation delay if the logic gate draws a current of 0.5 mA from the 0.8-V power supply.

***7-19.** A 1-MΩ resistor is connected between $+5$ V and terminal S_2 in the CMOS bilateral transmission gate of Fig. 7-22*a*. The circuit is operated as described by the first line in the truth table of Fig. 7-22*b*. Which terminal of MOS device Q_1 is the source terminal? (*Note:* In an *n*-channel MOS device the source is defined as the more negative of the source and drain terminals.)

*Optional problem.

***7-20.** A 1-MΩ resistor is connected between +5 V and terminal S_2 in the CMOS bilateral transmission gate of Fig. 7-22*a*. The circuit is operated as described by the first line in the truth table of Fig. 7-22*b*. Which terminal of MOS device Q_2 is the source terminal? (*Note:* In a *p*-channel MOS device the source is defined as the more positive of the source and drain terminals.)

7-21. A CMOS logic gate has an intrinsic propagation delay of 20 ns and an equivalent output resistance of $R_{eq} = 1$ kΩ. Find the propagation delay for a load capacitance of $C = 100$ pF.

7-22. A CMOS logic gate has an intrinsic propagation delay of 75 ns and an equivalent output resistance of $R_{eq} = 800$ Ω. What is the propagation delay if the load capacitance is $C = 200$ pF?

7-23. A high-speed CMOS logic inverter is operated at a power supply voltage of 5 V and its output is loaded by a capacitance of 15 pF. Find the propagation delay at room temperature.

7-24. A high-speed CMOS logic inverter is operated at a power supply voltage of 5 V and its output is loaded by a capacitance of 50 pF. Find the propagation delay at room temperature.

7-25. A CMOS logic gate is operated from a power supply voltage of 10 V, its leakage current is 5 μA, and it is loaded by a capacitance of $C =$ 500 pF. The clock frequency is $f_{CLOCK} = 200$ kHz. Find the maximum total power dissipation.

7-26. A CMOS logic gate is operated from a power supply voltage of 5 V, its leakage current is 5 μA, and it is loaded by a capacitance of $C =$ 200 pF. The clock frequency is $f_{CLOCK} = 500$ kHz. What is the maximum total power dissipation?

*Optional problem.

CHAPTER

8 Flip-Flops (FFs)

Instructional Objectives

This chapter describes latches and flip-flops (FFs), which are principal components of counters, shift registers, computers, and other digital systems, and which recur in later chapters of this book. After you read it you should be able to:

1. Analyze the operation of R-S and D latches and FFs.
2. Follow the timing diagrams of clocked R-S and D latches and FFs.
3. Draw state tables and function tables for R-S, D, J-K, and T FFs.
4. Describe the properties of master–slave FFs.
5. Describe the effects of clock skew.
6. Describe the properties of edge-triggered FFs.
7. Describe the operation of direct preset and clear inputs.
8. Evaluate timing constraints imposed by the maximum clock frequency and by the minimum clock pulse width.
9. Evaluate timing constraints of the direct preset and clear inputs.
10. Explain the terms setup time and hold time.

8-1 OVERVIEW

Thus far we have described logic gates and combinational logic circuits that have no *memory*: after the propagation delays, the outputs of a circuit were independent of previous inputs. In contrast, a *memory element* is capable of indefinitely remembering its input even if this input has changed in the meantime.

The most basic memory element is the *latch*, which has two stable states; transitions between these states can be effected via suitable inputs. A latch can be assembled from logic gates, although this is rarely done in SSI or MSI; however, it is frequently done in LSI gate arrays (Ch. 11).

We start with a discussion of latches, and then assemble a *flip-flop* (*FF*) from two latches. We then describe R-S, D, J-K, and T FFs, as well as master–slave

FFs, master–slave FFs with data lockout, edge-triggered FFs, direct preset and clear inputs, and timing constraints.

8-2 R-S STORAGE LATCHES

The *R-S storage latch* is also known as *R-S storage flip-flop* or as *bistable multivibrator*. We first assemble it from two DTL NAND gates and then discuss its basic properties.

8-2.1 Assembly from Two DTL NAND Gates

Consider the circuit shown in Fig. 8-1*a*, which consists of two DTL NAND gates. One input of each gate is connected to the output of the other gate. The remaining input of each gate is for the time being connected to +5 V; these inputs are used later.

8-2.2 The Two Stable States

We now show that the circuit of Fig. 8-1*a* has two *stable states* and that it can be used as a memory element.

In the circuit of Fig. 8-1*a* the transistors are labeled Q_1 and Q_2, the outputs *OUT* and $\overline{OUT}$, and two junction points x_1 and x_2; also shown are voltages in the circuit. Here we demonstrate that the circuit is stable with the voltages as shown. As a start, we *assume* that output $\overline{OUT}$ is at +5 V; hence the diode between $\overline{OUT}$ and x_1 is cut off (reverse biased). If this is the case, then transistor Q_1 is turned on, output *OUT* is near 0 V, Q_2 is cut off, and output $\overline{OUT}$ is at +5 V. This, however, reinforces our original assumption; hence, the voltages shown in Fig. 8-1*a* represent a *stable state*.

The circuit of Fig. 8-1*a* is shown again in Fig. 8-1*b* but with a different set of voltages. If, as a start, we assume that output $\overline{OUT}$ is at ≈ 0 V, then Q_1 is cut off, output *OUT* is at +5 V, Q_2 is turned on, and output $\overline{OUT}$ is at ≈ 0 V. This, however, reinforces our assumption; hence, the voltages shown in Fig. 8-1*b* also represent a stable state.

8-2.3 The Unstable State

The circuit of Fig. 8-1 may also take up a third state with voltages and currents as shown in Fig. 8-2. This third state is, however, an *unstable state*, as can be shown by considering the small *noise* voltages and currents that are always present in a real circuit. Let us assume, for example, that the noise voltage at the base of Q_1 turns Q_1 on slightly more. Thus the voltage at output *OUT* is lowered slightly below the +1.4 V shown in Fig. 8-2. This, in turn, lowers the voltage at the base of Q_2, leading to a lower collector current in Q_2, and to a voltage at output $\overline{OUT}$ that is slightly higher than the +1.4 V shown in Fig. 8-2. This now further raises the base voltage of Q_1, thus lowering the voltage at

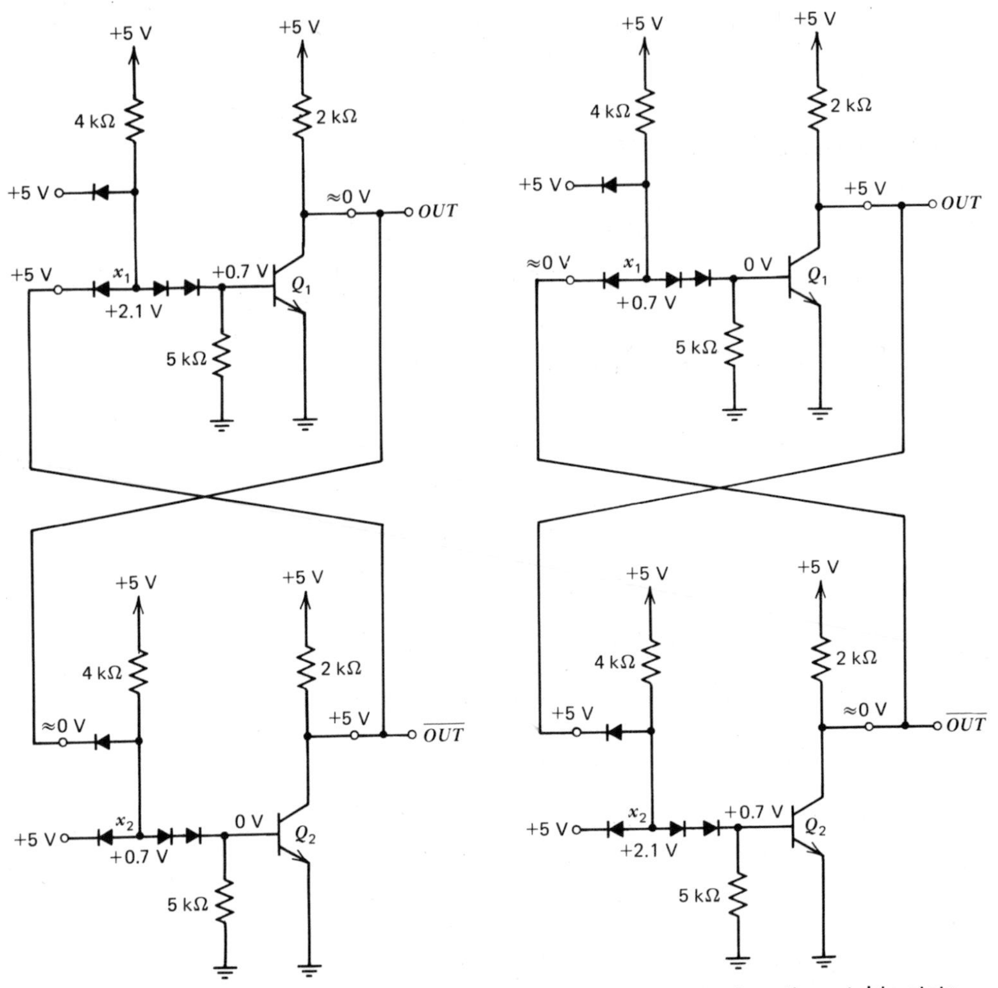

(*a*) Voltages in one stable state (*b*) Voltages in the other stable state

FIGURE 8-1 An R-S storage latch consisting of two cross-connected DTL NAND gates.

output *OUT* even more. This regenerative sequence ("*positive feedback*") drives the circuit to the stable state shown in Fig. 8-1*a*. It can be shown that with different noise conditions the circuit may also end up in the stable state of Fig. 8-1*b*.

8-2.4 Transitions and State Tables

Transitions between the two stable states of the latch shown in Fig. 8-1 can be effected by suitable voltages applied to the inputs that were thus far connected to +5 V. These inputs are marked by $\overline{R}$ and $\overline{S}$ in Fig. 8-3. Note that R and S

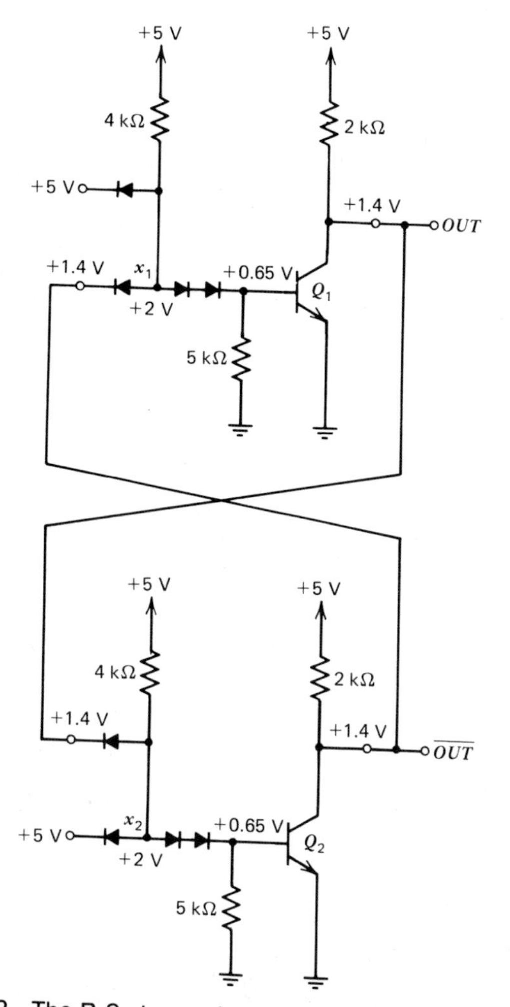

FIGURE 8-2 **The R-S storage latch of Fig. 8-1 in its unstable state.**

stand, respectively, for reset and set. Also note that the outputs are now marked Q and $\overline{Q}$ as is customary in digital circuits.*

We now describe the states of the latch by a *state table*, also known as a *transition table*, which is similar to a truth table and shows all possible states of the latch and of the inputs. Figure 8-4 shows such a state table that describes the states of the latch of Fig. 8-3.

In the first line of the table, inputs $\overline{R}$ and $\overline{S}$ are both at +5 V, reverse biasing the diodes at the inputs. In this case the circuit may assume either of the two stable states described in Fig. 8-1. The second line of the table shows how the

*Unfortunately, the use of Q for an output conflicts somewhat with transistor designations that also use the letter Q. In what follows, however, we use Q and $\overline{Q}$ for outputs, as well as Q_1, $\overline{Q}_1$, Q_2, $\overline{Q}_2$, and so on where more than one circuit is involved.

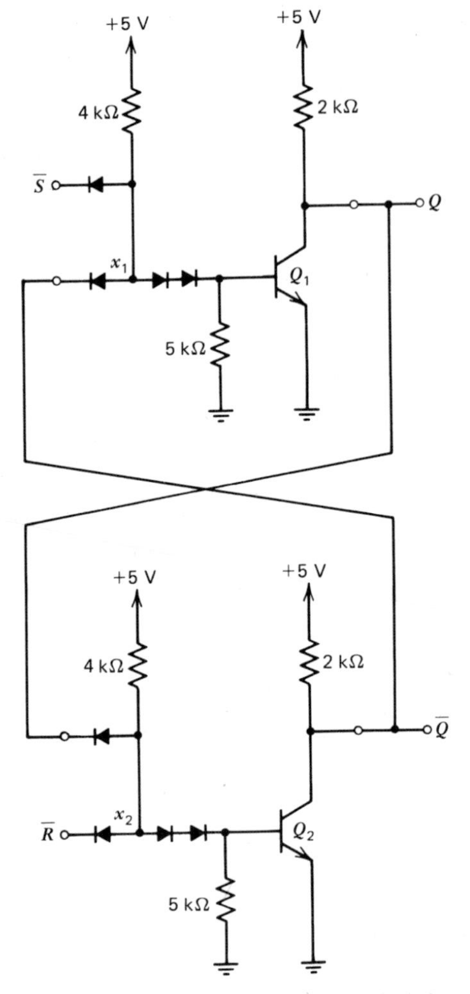

FIGURE 8-3 The R-S storage latch of Fig. 8-1 with inputs and outputs labeled.

latch can be forced into its $Q = +5$ V, $\overline{Q} = 0$ V state by inputs $\overline{R} = +5$ V, $\overline{S} = 0$ V. Note that the latch remains in this state even after input $\overline{S}$ returns to $+5$ V, assuming that $\overline{R}$ remains at $+5$ V. Similarly, the third line of the table shows how the latch can be forced into its $Q = 0$ V, $\overline{Q} = +5$ V state by inputs $\overline{R} = 0$ V, $\overline{S} = +5$ V; the latch remains in this state even after input $\overline{R}$ returns to $+5$ V, assuming that $\overline{S}$ remains at $+5$ V.

In the fourth line of the table, $\overline{R} = 0$ V and $\overline{S} = 0$ V. As the reader can readily ascertain from Fig. 8-3, this results in $Q = +5$ V and $\overline{Q} = +5$ V. Furthermore, in considering the next state for $\overline{R} = +5$ V and $\overline{S} = +5$ V, note that in practice $\overline{R}$ and $\overline{S}$ do not return to $+5$ V at exactly the same instant; thus the resulting state of the latch is undetermined since it can assume either of the two stable states shown in Fig. 8-1. For this reason, the input combination of $\overline{R} = 0$ V and $\overline{S} = 0$ V is usually not allowed.

$\overline{R}$	$\overline{S}$	Q	$\overline{Q}$
+5 V	+5 V	[0 V +5 V]	[+5 V 0 V]
+5 V	0 V	+5 V	0 V
0 V	+5 V	0 V	+5 V
0 V	0 V	+5 V	+5 V

FIGURE 8-4 State table showing voltages in the latch of Fig. 8-3.

8-2.5 Other Circuits

Figure 8-5 describes the R-S storage latch by a *logic diagram*, also known as a *functional block diagram*. Although the latch of Fig. 8-3 used DTL gates, Fig. 8-5 is also applicable to R-S storage latches using gates from any other logic family such as TTL, ECL, I^{2}L, or CMOS.

Inputs $\overline{R}$ and $\overline{S}$ are often preceded by logic inverters as shown in Fig. 8-6*a*. A state table for this latch is shown in Fig. 8-6*b*. Note that since the state table of Fig. 8-6*b* uses logic 0s and logic 1s, it can be used with any logic family. It is also customary to refer to the $Q = 0$ state as simply the "0" state and to the $Q = 1$ state as the "1" state.

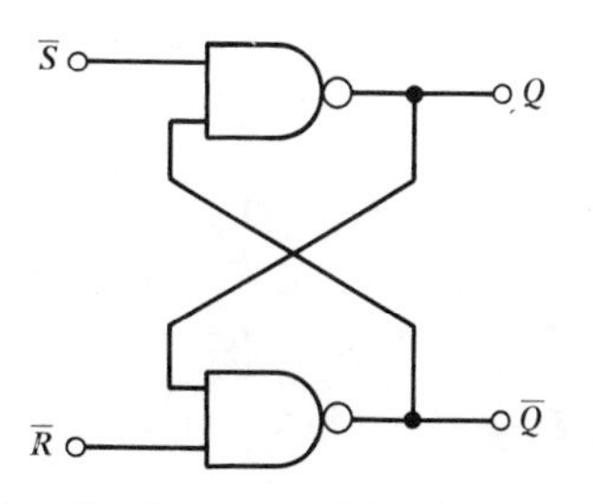

FIGURE 8-5 Logic diagram of the latch shown in Fig. 8-3.

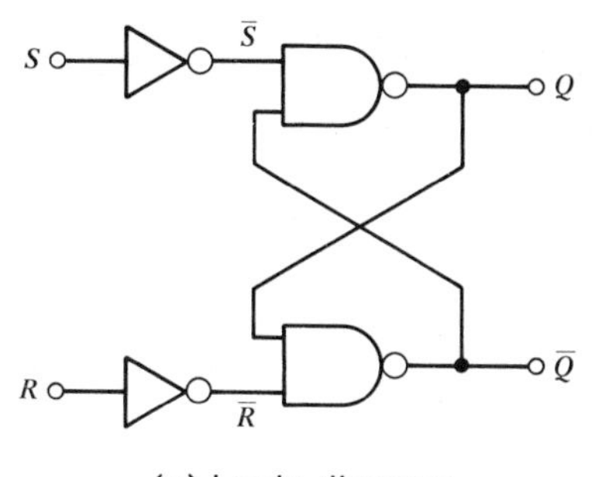

(*a*) Logic diagram

R	S	Q	$\overline{Q}$
0	0	[0 1]	[1 0]
0	1	1	0
1	0	0	1
1	1	1	1

(*b*) State table

FIGURE 8-6 R-S storage latch consisting of two NAND gates and two logic inverters.

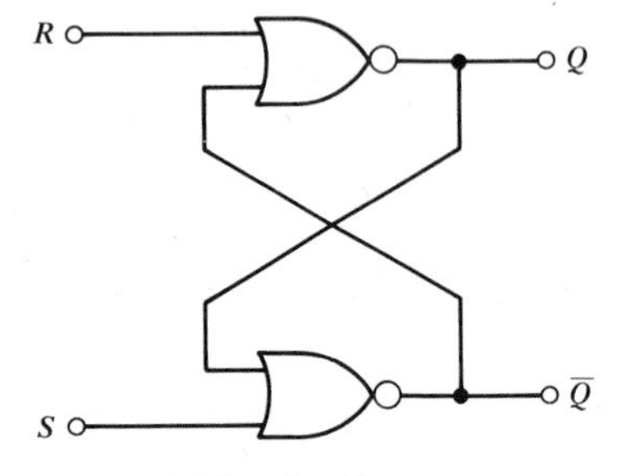

(a) Logic diagram

R	S	Q	$\overline{Q}$
0	0	$\begin{bmatrix}0\\1\end{bmatrix}$	$\begin{bmatrix}1\\0\end{bmatrix}$
0	1	1	0
1	0	0	1
1	1	0	0

(b) State table

FIGURE 8-7 R-S storage latch consisting of two NOR gates.

Figure 8-7*a* shows a realization of the R-S storage latch using two 2-input NOR gates. A state table is shown in Fig. 8-7*b*. Note that the outputs for the "unallowed" input combination of $R = 1$, $S = 1$ differ from the corresponding outputs in the state table of Fig. 8-6*b*, but otherwise the two state tables are identical.

8-2.6 Summary of Operation

We can summarize the operation of R-S storage latches in a single state table, as shown in Fig. 8-8*a*. The first three lines in this table are identical with the first three lines of the state tables of Figs. 8-6*b* and 8-7*b*. The last line, however, shows question mark entries for Q and $\overline{Q}$, indicating that these depend on the details of the implementation; hence, in general they are not known.

A symbol of the R-S storage latch is shown in Fig. 8-8*b*. The locations of S and R in the symbol are often interchanged. Also, in many circuits S and R are replaced by $\overline{S}$ and $\overline{R}$, requiring logic 0 inputs for set and reset.

8-2.7 Availability and Use

R-S storage latches are available in various logic families. In some families both the NAND-gate type of Fig. 8-6 and the NOR-gate type of Fig. 8-7 are provided. Four latches with the $\overline{Q}$ outputs omitted can be packaged in a 16-pin DIP; also, the outputs can be 3-state outputs with their enables in common.

R	S	Q	$\overline{Q}$
0	0	$\begin{bmatrix}0\\1\end{bmatrix}$	$\begin{bmatrix}1\\0\end{bmatrix}$
0	1	1	0
1	0	0	1
1	1	?	?

(a) State table

(b) Symbol

FIGURE 8-8 R-S storage latch.

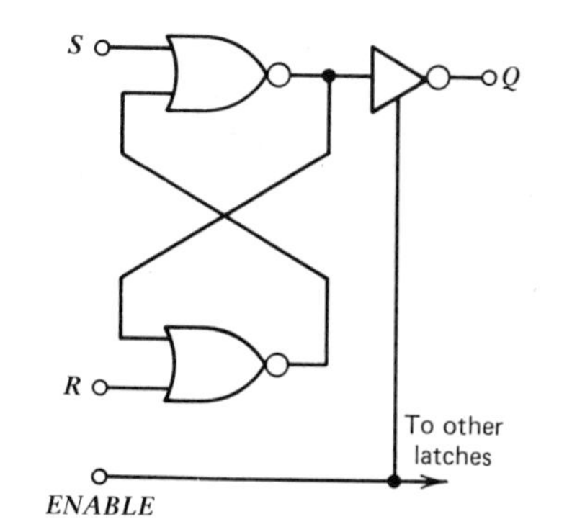

FIGURE 8-9 Logic diagram of one latch in Ex. 8-1.

EXAMPLE 8-1

The Type 4043 CMOS quad "NOR" R-S latch consists of four identical latches with 3-state outputs. A logic diagram of one latch is shown in Fig. 8-9. Find the output waveform for the inputs shown as the first two waveforms in Fig. 8-10, if the initial state of the latch is $Q = 1$ and if propagation delays, rise times, and fall times are negligible.

Solution

The output is shown as the last waveform in Fig. 8-10. The $R = 1$ pulse resets the latch to its $Q = 0$ state and the first $S = 1$ pulse sets it to its $Q = 1$ state. The second $S = 1$ pulse has no effect since the latch is already in its $Q = 1$ state. The simultaneous presence of $R = 1$ and $S = 1$ results in an unknown state. The operation of the 3-state output was described in Sec. 7-2.7.

8-3 CLOCKED R-S LATCHES

The versatility of the R-S storage latch can be enhanced by the addition of a *clock*, or *gate*, input. The result is a *clocked R-S* latch, also known as a *gated R-S latch* or a *transparent R-S latch*. In this section we show two implementa-

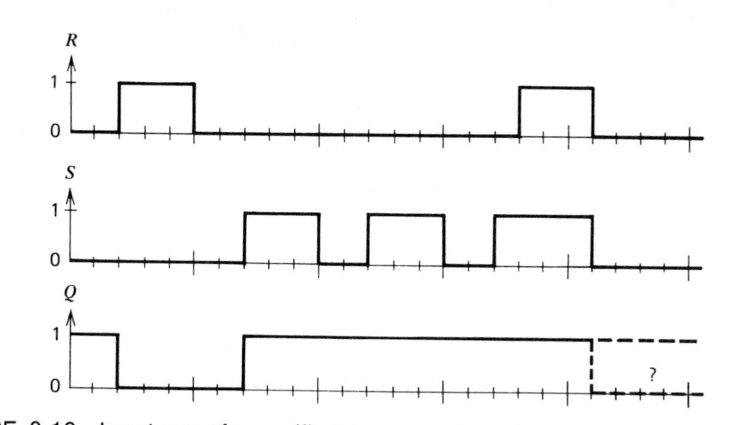

FIGURE 8-10 Input waveforms (first two waveforms) and output waveform (last waveform) in Ex. 8-1.

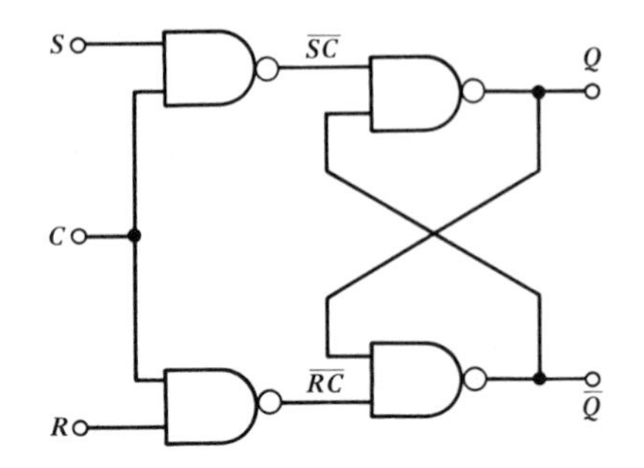

FIGURE 8-11 Extension of the R-S storage latch of Fig. 8-6*a* to a clocked R-S latch.

tions and describe their basic properties. The operation of the clocked R-S latch is further discussed later in this chapter.

8-3.1 Development from the R-S Storage Latch

Figure 8-11 shows the R-S storage latch of Fig. 8-6*a* with the input inverters replaced by NAND gates. In addition to *control inputs S* and *R*, it also has a *clock input C*, and the circuit is called a clocked *R-S* latch. Figure 8-12 shows another implementation that was developed from Fig. 8-7*a*.

8-3.2 Operation

Clock input C is ANDed with both control inputs S and R, which are thus ineffective when $C = 0$. The control inputs are, however, active when $C = 1$. As a precaution to ensure correct operation, the control inputs are not allowed to change *while* clock $C = 1$. Applications where this requirement is relaxed are discussed in Sec. 8-6. The operation of the clocked R-S latch of Fig. 8-11 is illustrated in Ex. 8-2, which follows.

EXAMPLE 8-2

Find the output waveform in the clocked R-S latch of Fig. 8-11 for the inputs shown as the first three waveforms in Fig. 8-13. Assume that the initial state of

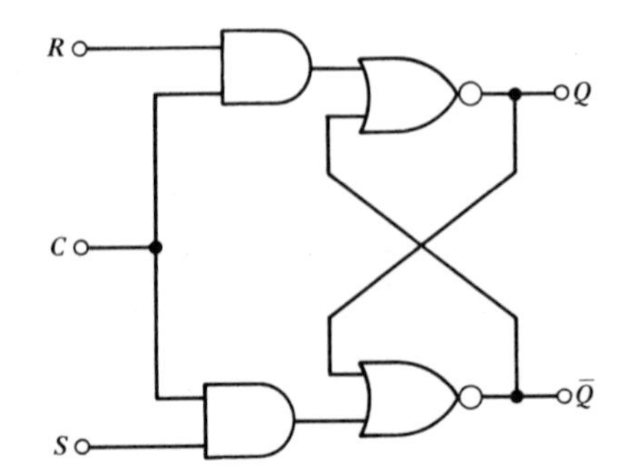

FIGURE 8-12 Extension of the R-S storage latch of Fig. 8-7*a* to a clocked R-S latch.

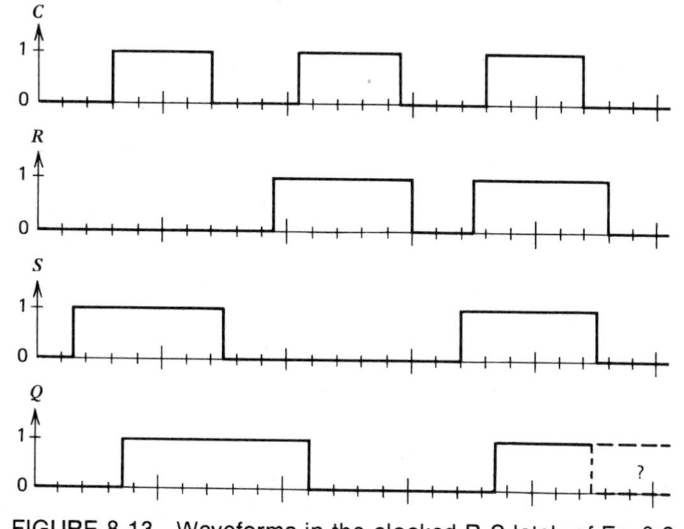

FIGURE 8-13 Waveforms in the clocked R-S latch of Ex. 8-2.

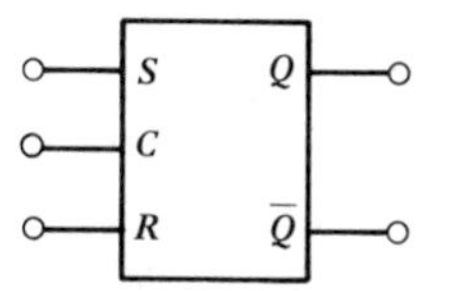

FIGURE 8-14 A symbol of the clocked R-S latch.

the latch is $Q = 0$ and that the propagation delay of output Q from the leading edge of the clock pulse equals one-half of a small division on the time scale in the figure.

Solution

The output is shown as the last waveform in Fig. 8-13. An $S = 1$ and $C = 1$ sets the latch to its 1 state, an $R = 1$ and $C = 1$ to its 0 state. A simultaneous $S = 1$, $R = 1$, and $C = 1$ leads to an undetermined state.

A symbol for the clocked R-S latch is shown in Fig. 8-14. The locations of S and R are often interchanged. Output $\overline{Q}$ is frequently omitted; also, clock input C may be made common to several latches within a package. Both of these measures are aimed at reducing the number of pins required, thus increasing the number of latches in a package.

8-4 STATE TABLES AND FUNCTION TABLES OF CLOCKED LATCHES

In our description of the operation of clocked latches, the state of Q before a clock pulse ($C = 1$) is usually called the *present state*, Q_n, and the state of Q

Q_n	S_n	R_n	Q_{n+1}
0	0	0	0
0	0	1	0
0	1	0	1
0	1	1	1?
1	0	0	1
1	0	1	0
1	1	0	1
1	1	1	?

(*a*) State table

S_n	R_n	Q_{n+1}
0	0	Q_n
0	1	0
1	0	1
1	1	?

(*b*) Function table

FIGURE 8-15 Clocked R-S latch.

after a clock pulse is called the *next state*, Q_{n+1}. Transitions of a latch can be described by a state table listing the next states for all combinations of present states and present control inputs.

Figure 8-15*a* shows a state table for the clocked R-S latch. Symbols S_n and R_n are the *present control inputs* representing control inputs S and R before and during a clock pulse ($C = 1$); frequently these are shown simply as S and R. The question mark entries for the next states in the fourth and last lines indicate undetermined next states for simultaneous $S_n = 1$ and $R_n = 1$. The reader can readily ascertain that this state table indeed describes the operation of a clocked R-S latch.

Transitions between the states of a latch can also be described by a *function table*, which is an abbreviated state table. A function table for the R-S latch is shown in Fig. 8-15*b*. In the first line of the state table of Fig. 8-15*a*, $Q_n = 0$ and $Q_{n+1} = 0$, in the fifth line $Q_n = 1$ and $Q_{n+1} = 1$, and $S_n = 0$ and $R_n = 0$ in both lines. These two lines of the state table are combined as $Q_{n+1} = Q_n$ in the first line of the function table. The second line in the function table combines the second and sixth lines of the state table, which are both characterized by $S_n = 0$, $R_n = 1$, and $Q_{n+1} = 0$. Similarly, the third line in the function table combines the third and the seventh lines of the state table, which are both characterized by $S_n = 1$, $R_n = 0$, and $Q_{n+1} = 1$. Finally, the last line in the function table combines the fourth and the eighth (last) line of the state table, which are both characterized by $S_n = 1$ and $R_n = 1$, hence by undetermined next states as indicated by question mark entries for Q_{n+1}.

8-5 CLOCKED D LATCHES

The *clocked D latch* has one control input D in addition to the clock input C. In its simplest form it can be obtained from the clocked R-S latch as shown in Fig. 8-16*a*, although this approach is rarely followed in practical circuits. Also, for simplicity we ignore the propagation delay of the logic inverter at the input.

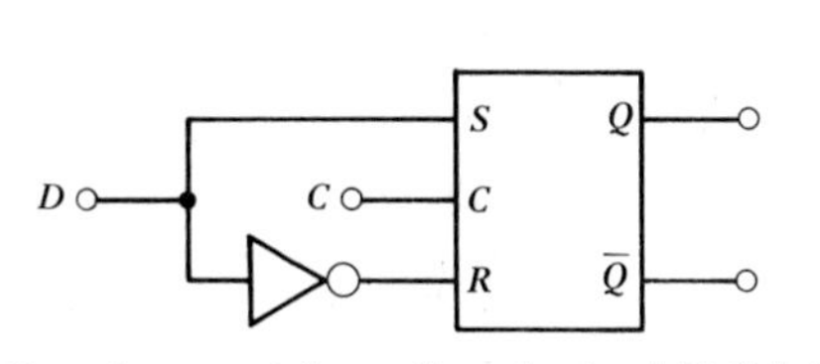

(*a*) Development from the clocked R-S latch

Q_n	D_n	Q_{n+1}
0	0	0
0	1	1
1	0	0
1	1	1

(*b*) State table

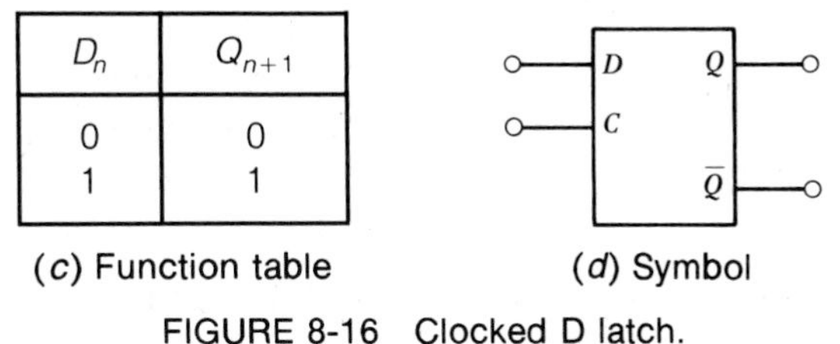

D_n	Q_{n+1}
0	0
1	1

(*c*) Function table (*d*) Symbol

FIGURE 8-16 Clocked D latch.

We can see in Fig. 8-16*a* that $S = \overline{R} = D$ at all times. Thus the second line in the function table of the clocked R-S latch shown in Fig. 8-15*b* is applicable when $D_n = 0$ and the third line when $D_n = 1$. Hence, in a clocked D latch the next state $Q_{n+1} = D_n$, which is independent of the present state Q_n.

A state table of the clocked D latch is shown in Fig. 8-16*b*. Since there is only one control input, *D*, there are only four combinations of Q_n and D_n, hence only four lines in the state table. We can see that, as stated before, $Q_{n+1} = D_n$, irrespective of Q_n. This can be also seen in the function table of the clocked D latch, which is shown in Fig. 8-16*c*. A symbol of the clocked D latch is shown in Fig. 8-16*d*.

Note that control input *D* is not allowed to change during a clock pulse ($C = 1$): this restriction ensures that the latch is left in a definite state after each clock pulse.

EXAMPLE 8-3

A clocked D latch is used for storing the state of control input *D*. Clock input *C* is shown as the first waveform in Fig. 8-17, and control input *D* in the second

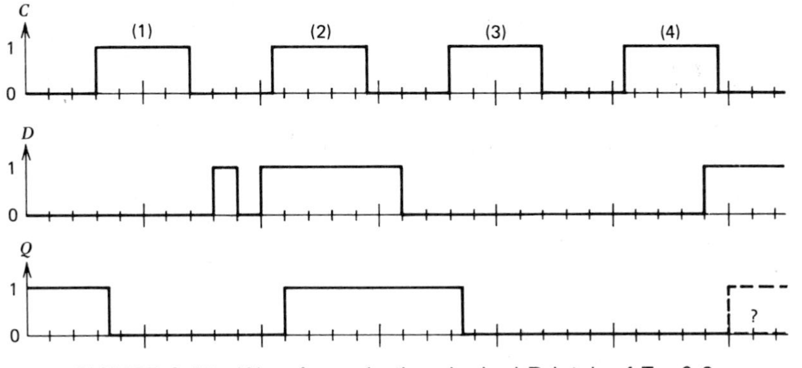

FIGURE 8-17 Waveforms in the clocked D latch of Ex. 8-3.

line. Find the output waveform if the initial state of the latch is $Q = 1$, and if the propagation delay of the output from the leading edge of the clock pulse equals one-half of a small division on the time scale in the figure.

Solution

Output Q is shown as the third waveform in Fig. 8-17. Since control input D is 0 immediately before and during clock pulse (1), output Q is 0 after this clock pulse. Though D is changing between clock pulses (1) and (2), it is 1 immediately before and during clock pulse (2); hence Q becomes 1. Control input D is 0 immediately before and during clock pulse (3); hence, Q becomes 0. Finally, control input D changes from 0 to 1 *during* clock pulse (4). Depending upon the details of circuitry, output Q may or may not change from 0 to 1, or it may *teeter* between 0 and 1. Hence, output Q is undetermined for times following clock pulse (4).

8-6 USE OF CLOCKED LATCHES FOR STORAGE

Thus far the control inputs of clocked latches were not allowed to change while the clock was $C = 1$. This constraint can be relaxed when a latch is used for storage, which is an important use.

Consider the clocked R-S latch shown in Fig. 8-11, (similar considerations hold for Fig. 8-12). Assume that the propagation delay of each gate is t_P and control inputs S and R do not change during the last 2 t_P before the negative-going edge of the clock. If $S = 1$ and $R = 0$ during this time, then the state of the latch is $Q = 1$ after C becomes 0. Similarly, if $S = 0$ and $R = 1$ during this time, then the state of the latch is $Q = 0$ after C becomes 0. If $S = 0$ and $R = 0$ during this time, then we need previous history to find the resulting state of Q; if $S = 1$ and $R = 1$ during this time, then the resulting state of Q is undetermined.

Thus, the clocked R-S latch of Fig. 8-11 *stores* a logic 1 if $S = 1$ and $R = 0$ during the last 2 t_P before the negative-going edge of the clock. Similarly, it stores a logic 0 if $S = 0$ and $R = 1$ during the last 2 t_P before the negative-going edge of the clock.

A clocked D latch stores a logic 1 if its control input $D = 1$ during the last 2 t_P before the negative-going edge of the clock. Similarly, it stores a logic 0 if $D = 0$ during the last 2 t_P before the negative-going edge of the clock.

8-7 MASTER–SLAVE FFs

In the latches shown thus far, the control inputs were not allowed to change while the clock input was $C = 1$. This restriction ensured that the latch was left in a definite state after each clock pulse. Also as illustrated in Fig. 8-13 and Fig. 8-17, the *outputs* of a latch may change state while the clock input is still $C = 1$. This causes no problem as long as the latch is used only for storage. However, there are other applications, for example counters, where the control inputs are Boolean functions of outputs. In such applications we would like to delay changes of the outputs until the control inputs are no longer sensitive to

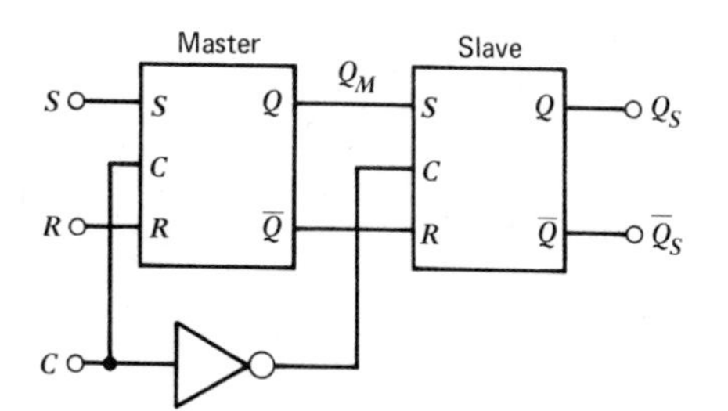

FIGURE 8-18 Simple master–slave R-S FF.

changes. Thus, in Figs. 8-13 and 8-17 we would like to delay changes of the outputs until the clock has returned to its $C = 0$ state. One way to delay changes in the outputs until C becomes 0 is by use of *master–slave flip-flops* (*master–slave FFs*).

The essential difference between a latch and a flip-flop is that *a flip-flop is assembled from two clocked latches.* Also, the clock signals of the two clocked latches are *nonoverlapping*, that is, at no time do they activate the control inputs of *both* clocked latches simultaneously.

This section introduces the simplest master–slave FFs and shows the structure of master–slave R-S and D FFs. Sections 8-8 and 8-9 introduce more complex FFs with master–slave structures: the master–slave FF with data lockout and the edge-triggered FF. Sections 8-10 and 8-11 introduce two new FF types: the J-K FF and the T FF; these, as well as all other FF types, can be implemented as master–slave FFs, master–slave FFs with data lockout, or edge-triggered FFs.

8-7.1 A Simple Circuit

A simple master–slave R-S FF is shown in Fig. 8-18. It consists of two clocked R-S latches marked *master* and *slave* and a logic inverter. Each clocked R-S latch may be implemented by the circuit of either Fig. 8-11 or Fig. 8-12.

When clock input $C = 0$, the clock input of the master latch is 0; hence, its control inputs S and R are ineffective. However, the clock input of the slave latch is 1; hence, the state of the master latch is forced into the slave latch via its control inputs S and R.

When clock input $C = 1$, control inputs S and R of the master latch are active and the control inputs of the slave latch are ineffective. Hence the slave latch is, in effect, isolated from the master latch. The slave latch continues holding the state it had preceding the $C = 1$ clock pulse until the clock input again becomes $C = 0$.

EXAMPLE 8-4

Find output waveforms Q_M and Q_S in the master–slave R-S FF of Fig. 8-18 for the inputs shown as the first three waveforms of Fig. 8-13. Assume that the initial state of the FF is $Q_M = Q_S = 0$ and that the propagation delay of output Q_S from the *trailing edge* (negative-going transition) of the clock pulse equals one-half of a small division on the time scale of the figure.

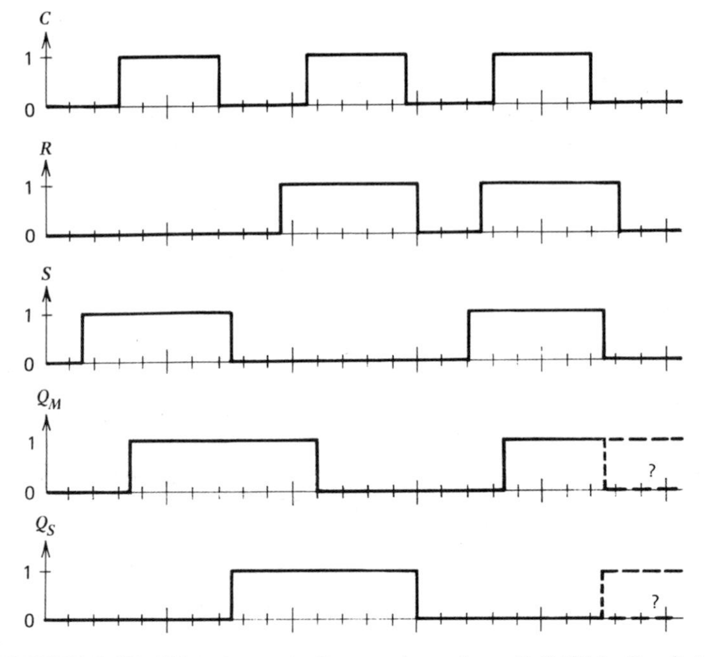

FIGURE 8-19 Waveforms in the master–slave R-S FF in Ex. 8-4.

Solution

Output waveforms Q_M and Q_S are shown in Fig. 8-19, together with the waveforms of C, R, and S copied from Fig. 8-13. The waveform of Q_M is identical to the waveform of output Q in Fig. 8-13, and it changes following the *leading edge* (positive-going transition) of clock pulse C. Output Q_S is shown as the last waveform in Fig. 8-19, and it changes following the *trailing edge* (negative-going transition) of clock pulse C.

8-7.2 An Improved Master–Slave R-S FF

The simple master–slave R-S FF of Fig. 8-18 operates correctly with clock pulses that have fast transitions. However, its operation may become marginal if the rise time of clock input C becomes too long and the threshold voltage of the logic inverter is higher (closer to logic 1) than the threshold voltage of the master latch.* Under such conditions there is a time interval during the slow rise time of the clock when the control inputs of both the master latch and the slave latch are active. For clock rise times longer than about 1 μs, this time interval may become long enough to set the slave latch to an erroneous state. Similar problems arise if the fall time of the clock is longer than about 1 μs.

A solution to these problems is outlined in Fig. 8-20. The voltage of the battery is in the vicinity of 0.7 V in TTL (lower in ECL, higher in CMOS) and in reality it is always replaced by components that are easy to include in an

*Note that, as before, we assume positive logic (active-high logic). Thus logic 0 is equivalent to LOW or "L" and logic 1 is equivalent to HIGH or "H."

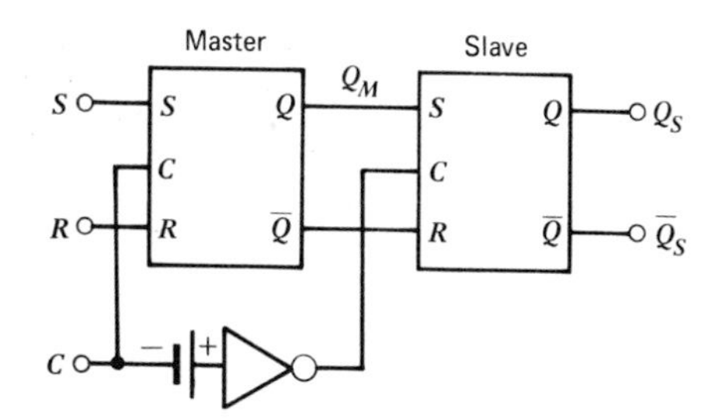

FIGURE 8-20 An improved master–slave R-S FF.

integrated circuit. This added voltage dominates any initial difference in threshold voltages; hence the resulting threshold voltage of the logic inverter with the battery is always lower than the threshold voltage of the clock input of the master latch.

The operation of the improved master–slave FF of Fig. 8-20 is described here for a clock input C that starts at logic 0, makes a transition to logic 1, and then returns to logic 0. While the clock is at $C = 0$, the control inputs of the master latch are ineffective. Also, the control inputs of the slave latch are active; hence, the state of the slave latch is identical to the state of the master latch.

As the clock rises, the threshold voltage of the logic inverter is reached first, the slave latch is isolated from the master latch, and the control inputs of the master latch remain ineffective. Next, the threshold voltage of the master latch clock input is exceeded, which activates its control inputs S and R; the slave latch remains isolated from the master latch. The situation remains unchanged while clock input C continues to rise, reaches the full logic 1 level, stays there, and starts descending at the end of the clock pulse.

As clock input C descends from logic 1, it first drops below the threshold voltage of the clock input of the master latch. This makes the control inputs of the master latch ineffective; the slave latch still remains isolated from the master latch. Finally clock input C descends below the threshold voltage of the logic inverter with the battery, reactivating the control inputs of the slave latch, thus forcing the slave latch to follow the state of the master latch.

The above sequence of events guarantees that the clock signals of the master latch and the slave latch are nonoverlapping, that is, they are never both at logic 1 at the same time. Because of this, the control inputs of the master latch and the control inputs of the slave latch are never active at the same time: the control inputs of the master latch are active during $C = 1$, the control inputs of the slave latch are active during $C = 0$, and all control inputs are inactive for a short time during each transition of the clock.

8-7.3 A Symbol of the Master-Slave R-S FF

Figure 8-21 shows a symbol for the master-slave FF of Fig. 8-20. Inputs S, C, and R of Fig. 8-21 are connected, respectively, to inputs S, C, and R of Fig. 8-20; outputs Q and $\overline{Q}$ of Fig. 8-21 are connected, respectively, to outputs Q_S and $\overline{Q}_S$ of Fig. 8-20. Note the *triangle* at clock input C in Fig. 8-21: it

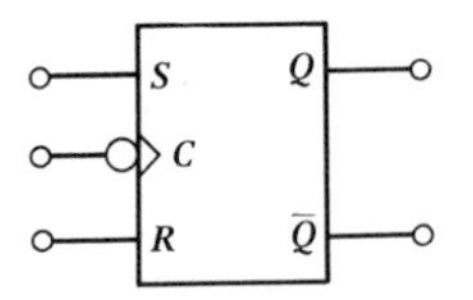

FIGURE 8-21 Symbol of the master–slave R-S FF.

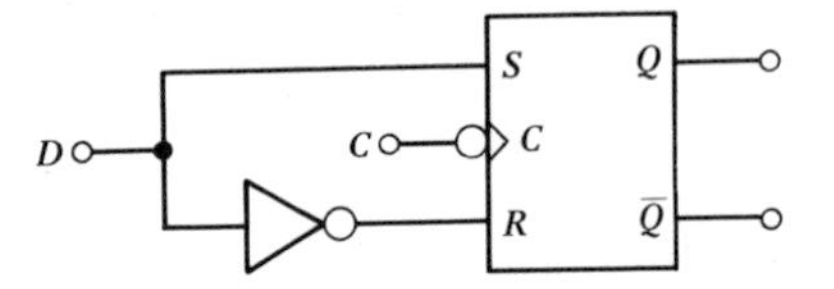

FIGURE 8-22 Development of the master–slave D FF.

indicates that the symbol represents a FF and not a latch. The circle at clock input C in Fig. 8-21 does *not* represent a logic inversion as it would in a combinational logic circuit, but it indicates that the outputs of the FF change states following the *negative-going edge* of the clock. Unfortunately, the symbol of Fig. 8-21 does not unambiguously describe a master–slave FF; as we see later, it is also used for other FF types that change output states following the negative-going edge of the clock. Note that the state table and the function table of the R-S FF are identical to those of the clocked R-S latch shown in Fig. 8-15.

8-7.4 Master-Slave D FFs

A master–slave D FF can be developed from a master–slave R-S FF as shown in Fig. 8-22. A symbol of the resulting FF is shown in Fig. 8-23. Note that the state table and the function table of the D FF are identical to those of the clocked D latch shown in Fig. 8-16*b* and 8-16*c*.

8-8 MASTER–SLAVE FFs WITH DATA LOCKOUT

We have seen that the control inputs in a master–slave FF are sensitive to changes until the end of the $C = 1$ clock pulse. Also, changes in the output of the FF are delayed from this time by a propagation delay (this was equal to one-half of a small time division in Ex. 8-4). This propagation delay provides a safety margin that ensures correct operation when the output of one FF is used as a control input of another FF (either directly of via logic gates).

8-8.1 Clock Skew

The safety margin provided by the propagation delay of the FF is usually adequate in small systems where delays in the interconnecting wiring are negligible. However, circuit malfunction may result if the propagation delay in the interconnecting wiring, called *clock skew*, exceeds the propagation delay of the FF. This is illustrated in Ex. 8-5, which follows.

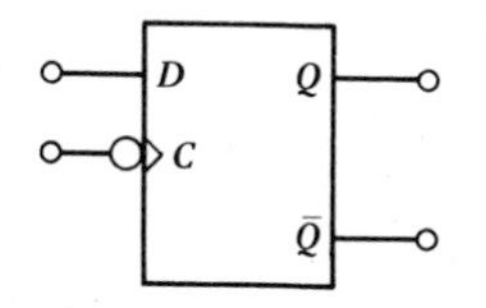

FIGURE 8-23 Symbol of the master – slave D FF.

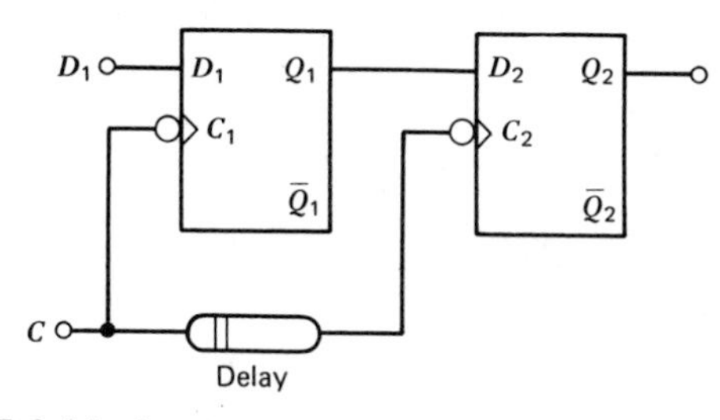

FIGURE 8-24 Logic circuit using two master – slave D FFs.

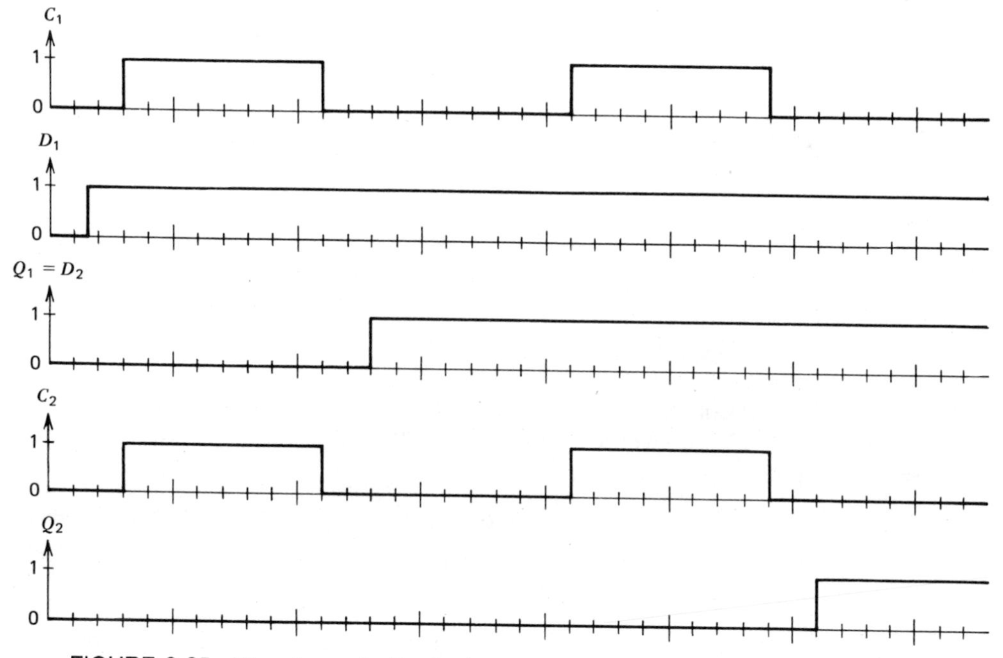

FIGURE 8-25 Waveforms in the logic circuit of Fig. 8-24 with zero delay between clocks C_1 and C_2.

***EXAMPLE 8-5**

Figure 8-24 shows a logic circuit consisting of two master–slave D FFs. The circuit is a *shift-register* and is discussed in more detail later. We now consider the effects of the delay between the clock of the first FF, C_1, and the clock of

*Optional example.

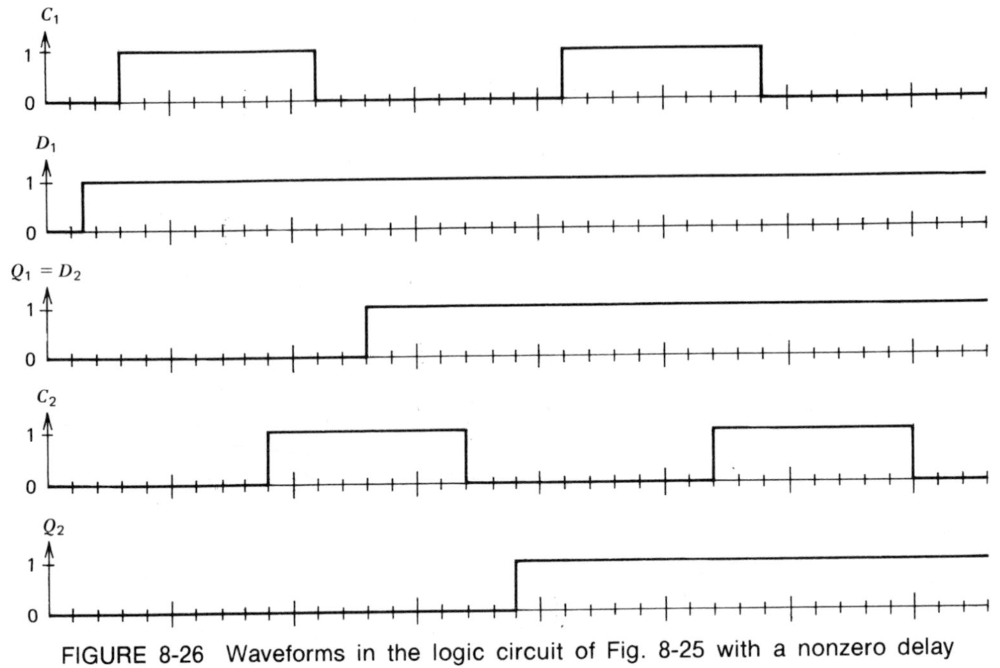

FIGURE 8-26 Waveforms in the logic circuit of Fig. 8-25 with a nonzero delay between clocks C_1 and C_2.

the second FF, C_2. We assume initial states of $Q_1 = 0$ and $Q_2 = 0$ throughout; we also assume that the propagation delay of each FF output from the trailing (negative-going) edge of the clock pulse equals 2 small divisions on the time scale of Fig. 8-25, as well as in subsequent figures of Ex. 8-5 and Ex. 8-6.

Figure 8-25 shows the case when the delay between C_1 and C_2 is zero and the circuit operates correctly. Find how the waveforms of Fig. 8-25 change if the delay between C_1 and C_2 equals 6 small divisions on the time scale of the figure.

Solution

The resulting waveforms are shown in Fig. 8-26. Of particular interest is that control input D_2 now changes from 0 to 1 *during* clock pulse C_2 because of the delay between C_1 and C_2. The time remaining between this change and the end of clock pulse C_2 is *probably* sufficient for the $D_2 = 1$ level to be recognized, as shown in the figure. A comparison with the correct Fig. 8-25 shows that the waveform of FF output Q_2 in Fig. 8-26 is *erroneous*, since Q_2 is supposed to remain 0 between the first and the second clock pulses and change to 1 only after the second clock pulse.

The effects of clock skew can be alleviated by use of *master–slave FFs with data lockout*. The principle is illustrated in Fig. 8-27 for an R-S FF and in Fig. 8-28 for a D FF.

The master–slave FF with data lockout shown in Fig. 8-27 differs from the FF of Fig. 8-20 by the circuitry that is added at the clock input of the master latch

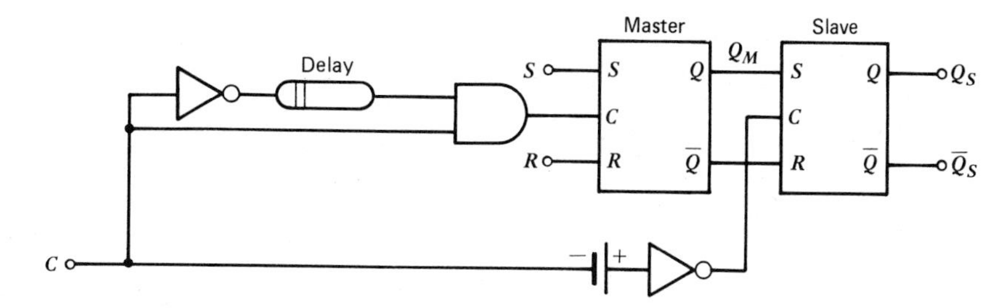

FIGURE 8-27 Master–slave R-S FF with data lockout.

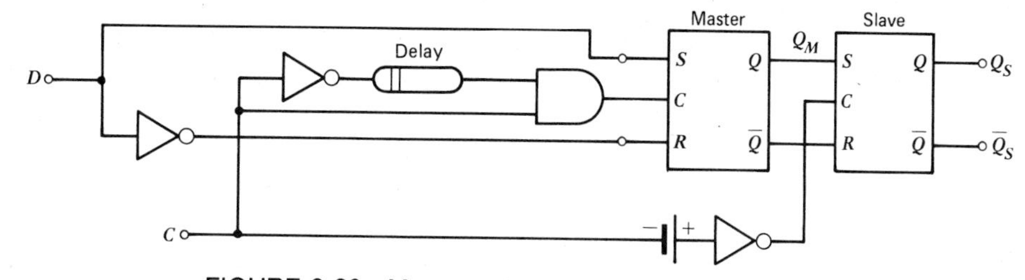

FIGURE 8-28 Master–slave D FF with data lockout.

and that is usually combined with circuitry inside the master latch. This added circuitry reduces the width of the $C = 1$ clock pulse at the clock input of the master latch; the resulting width equals the sum of the "Delay" and the propagation delay of the logic inverter driving it. As a result, control inputs S and R become ineffective ("*locked out*") at the end of this shortened clock pulse and remain so until the next clock pulse.

The situation is similar in the master–slave D FF with data lockout shown in Fig. 8-28, where control input D becomes ineffective (locked out) at the end of the shortened clock pulse and remains so until the next clock pulse. The use of the master–slave D FF with data lockout is illustrated in Ex. 8-6, which follows.

***EXAMPLE 8-6**

Each of the two master–slave D FFs in Fig. 8-24 is replaced by the master–slave D FF with data lockout shown in Fig. 8-28. The waveforms of control input D_1 and of clocks C_1 and C_2 are the same as in Fig. 8-26. The delay inserted at the clock input of the master latch of Fig. 8-28 equals 3 small divisions on the time scale of Fig. 8-26. The propagation delay of the logic inverter driving the delay element is negligible, as is the propagation delay of the AND gate following the delay element. Also, as before, the propagation delay of each FF output from the trailing (negative-going) edge of the clock pulse equals 2 small time divisions on the time scale of Fig. 8-26.

Find the waveforms of Q_1 and Q_2 if the initial states are $Q_1 = 0$ and $Q_2 = 0$.

Solution

Figure 8-29 shows the waveforms of Q_1 and Q_2, as well as the waveforms of C_1, C_2, and D_1, which were copied from Fig. 8-26. The control input of the

*Optional example.

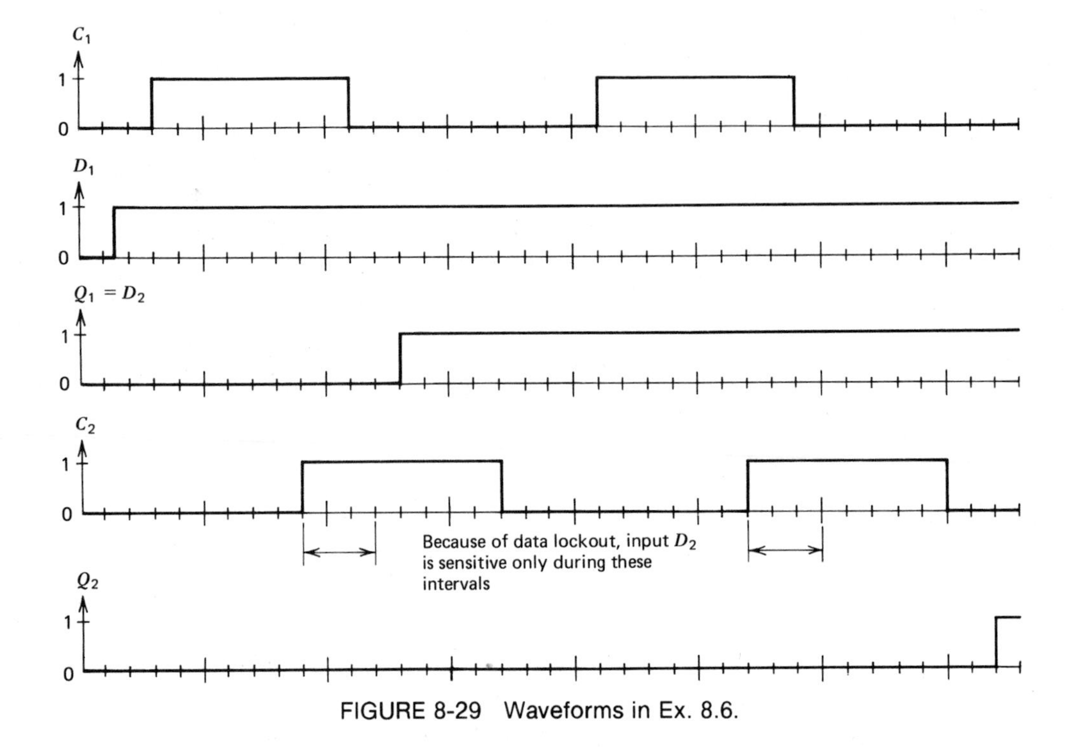

FIGURE 8-29 Waveforms in Ex. 8.6.

second FF, D_2, is sensitive only for an interval with a duration of 3 small time divisions. This interval starts at the leading edge of clock pulse C_2. (See note on the waveform of C_2 in Fig. 8-29.)

Figure 8-29 shows that the circuit operates correctly, as can be seen by comparison with the correct Fig. 8-25.

8-8.2 Advantages and Disadvantages

The use of master–slave FFs with data lockout significantly reduces the detrimental effects of clock skew. The cost in added circuitry is quite small, and the number of pins is not increased.

A disadvantage of the data lockout is that in some circuits the FF may malfunction if the transition times of the clock are too slow and the threshold voltages of the added inverter and the AND gate are not matched. Since perfectly matched threshold voltages cannot be guaranteed, the use of master–slave FFs with data lockout is hazardous for clock transition times in excess of about 1 μs. (It is shown in Ch. 16 how the situation may often be improved by use of a *Schmitt-trigger circuit.*)

We should note that the symbol of a master–slave FF with data lockout is identical to the symbol of a master–slave FF. Thus, a symbol of the FF of Fig. 8-27 is given by Fig. 8-21 and a symbol of the FF of Fig. 8-28 by Fig. 8-23.

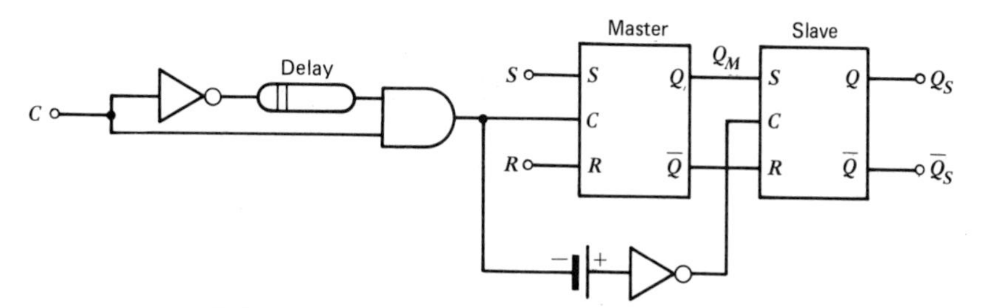

FIGURE 8-30 Positive edge-triggered R-S FF.

8-9 EDGE-TRIGGERED FFs

We have seen that in a master–slave FF with data lockout, the control inputs are effective only during a short interval that starts at the leading (positive-going) edge of the $C = 1$ clock pulse. Also, the outputs of the FF did not change until after the trailing (negative-going) edge of the clock pulse.

We may want to further modify these properties for use in small systems where the clock pulses have fast transitions and clock skew is small and in which we desire fast operation. We want to retain the property that the control inputs are effective only during a short interval. However, in the interest of faster operation, we also want the outputs of the FF to change shortly after this interval instead of waiting until the clock pulse is over. These characteristics are satisfied by *edge-triggered FFs*.

In a *positive edge-triggered FF*, the control inputs are sensitive for a short interval following the *positive-going* transition of the clock and the outputs of the FF change shortly after this interval. In a *negative edge-triggered* FF, the control inputs are sensitive for a short interval following the *negative-going* transition of the clock, and the outputs of the FF change shortly after this interval.

Figure 8-30 shows an implementation of a positive edge-triggered R-S FF. As was the case in a master–slave FF with data lockout (Fig. 8-27), control inputs S and R in Fig. 8-30 are effective only during a short time interval that starts at the leading (positive-going) edge of the $C = 1$ clock input pulse. However, unlike in the master–slave FF with data lockout of Fig. 8-27, the control inputs of the slave latch in Fig. 8-30 are activated *immediately* at the end of this short time interval; hence, FF outputs Q_S and $\overline{Q}_S$ also change shortly thereafter—instead of after the trailing (negative-going) edge of the clock input pulse as was the case in the master–slave FF with data lockout in Fig. 8-27.

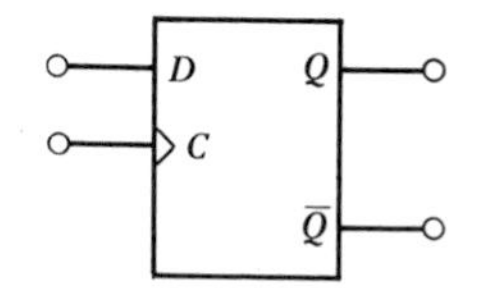

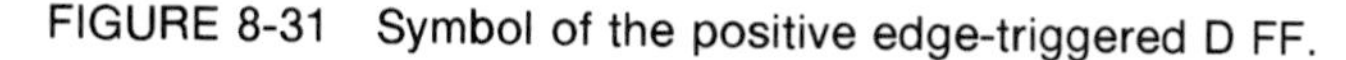
FIGURE 8-31 Symbol of the positive edge-triggered D FF.

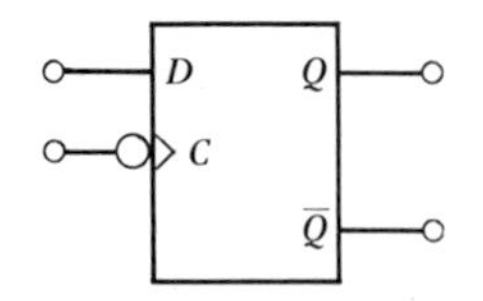

FIGURE 8-32 Symbol of the negative edge-triggered D FF.

A symbol of the positive edge-triggered D FF is shown in Fig. 8-31. A symbol of the negative edge-triggered D FF is shown in Fig. 8-32; it is identical to the master–slave D FF symbol shown in Fig. 8-23. Thus flip-flop symbols do not completely describe a FF, and additional specifications are necessary.

8-10 J-K FFs

We now have three different FF implementations with output transitions that occur after the control inputs ceased to be active: the master–slave FF, the master–slave FF with data lockout, and the edge-triggered FF. Any one of these three can be used to implement the J-K FF that we shortly specify; however, we start the discussion by assuming the use of master–slave FFs.

Consider the circuit of Fig. 8-33*a*. It consists of a master–slave R-S FF shown in Fig. 8-21 and of two AND gates. (In practical circuits these AND gates are often combined with the AND gates at the control inputs inside the R-S FF.) The two AND gates realize two Boolean equations: $S = J\overline{Q}$ and $R = KQ$. By applying these equations to the present states, we get $S_n = J_n\overline{Q}_n$ and $R_n = K_nQ_n$. Figure 8-33*b* shows a state table listing the eight possible combinations of Q_n, J_n, and K_n, the Boolean values of S_n and R_n obtained from the above two equations, and the resulting next states Q_{n+1} as given by the state table of the clocked R-S latch of Fig. 8-15*a*, which is also applicable to R-S FFs.

As far as the operation of the circuit at its terminals J, K, C, Q, and $\overline{Q}$ is concerned, we do not need the states of S and R, and the reduced state table shown in Fig. 8-33*c* is adequate. This state table fully specifies the behavior of the J-K FF developed in Fig. 8-33*a*, which has two control inputs, J and K, a clock input C, and outputs Q and $\overline{Q}$. The information contained in the state table of Fig. 8-33*c* can also be given in an abbreviated form as shown in the function table of Fig. 8-33*d*.

We now compare the function table of the J-K FF shown in Fig. 8-33*d* with the function table of the clocked R-S latch shown in Fig. 8-15*b*, which is also applicable to R-S FFs. Both tables describe a circuit with two control inputs. Also, the next states, Q_{n+1}, are identical in the first three lines of the two tables. However, the undetermined next state in the last line of Fig. 18-15*b* is replaced by a definite next state in Fig. 8-33*d*. This shows the superior performance of the J-K FF as compared to the R-S latch and the R-S FF.

Two symbols of the J-K FF are shown in Fig. 8-33*e*. The first symbol represents a master–slave J-K FF, a master–slave J-K FF with data lockout, or a negative edge-triggered J-K FF, all of which change output states following

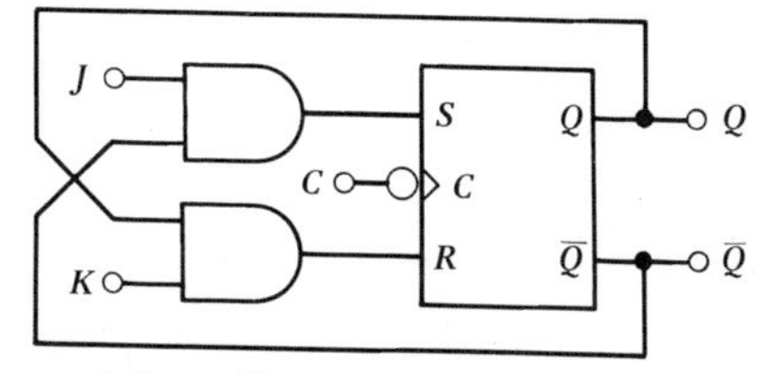

(a) Development from the master-slave R-S FF of Fig. 8-21

Q_n	J_n	K_n	S_n	R_n	Q_{n+1}
0	0	0	0	0	0
0	0	1	0	0	0
0	1	0	1	0	1
0	1	1	1	0	1
1	0	0	0	0	1
1	0	1	0	1	0
1	1	0	0	0	1
1	1	1	0	1	0

(b) State table

Q_n	J_n	K_n	Q_{n+1}
0	0	0	0
0	0	1	0
0	1	0	1
0	1	1	1
1	0	0	1
1	0	1	0
1	1	0	1
1	1	1	0

(c) State table listing only Q_n, J_n, K_n, and Q_{n+1}

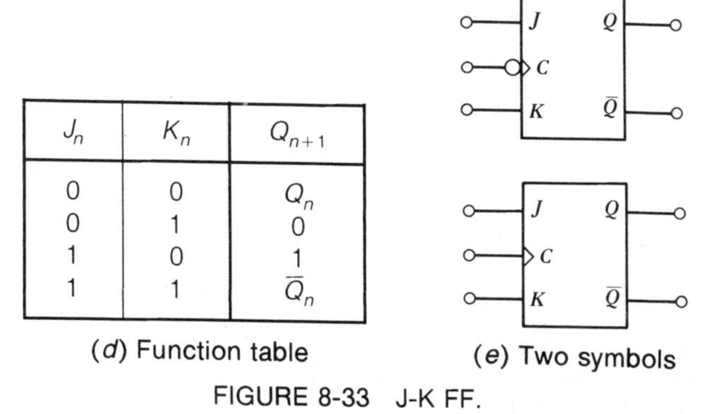

J_n	K_n	Q_{n+1}
0	0	Q_n
0	1	0
1	0	1
1	1	$\overline{Q}_n$

(d) Function table

(e) Two symbols

FIGURE 8-33 J-K FF.

negative-going clock transitions. The second symbol in Fig. 8-33e represents a positive edge-triggered J-K FF, which changes output states following positive-going clock transitions.

8-11 T FFs

T FFs can be developed from J-K FFs as shown in Fig. 8-34a. A state table for these FFs is shown in Fig. 8-34b and a function table in Fig. 8-34c. We can see that the state of the FF is complemented with each application of a clock pulse when $T_n = 1$; hence the name *toggle*, or *T*, *FF*.

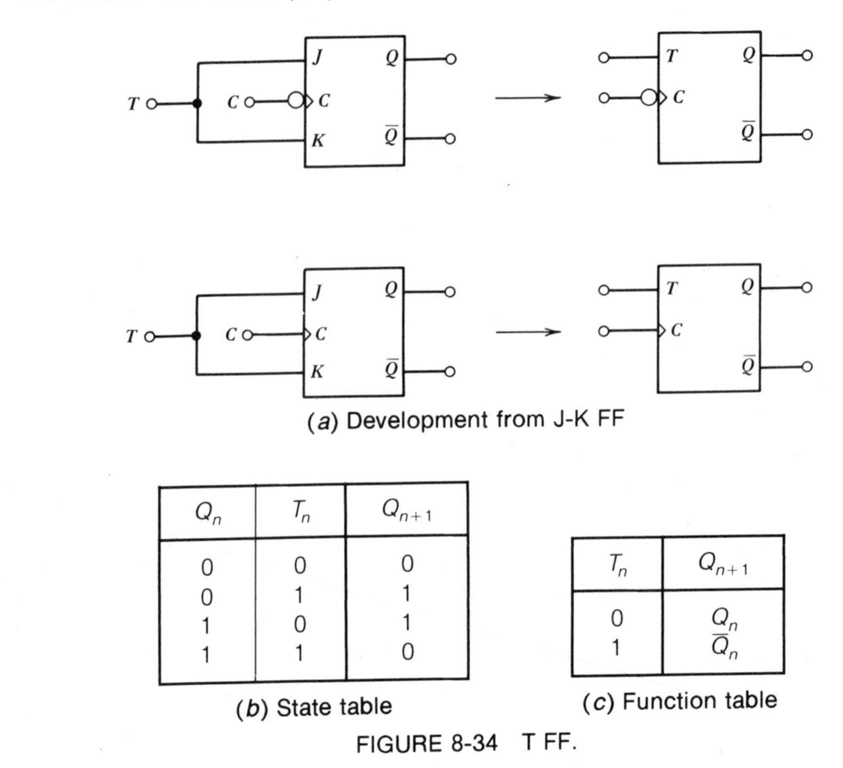

(a) Development from J-K FF

Q_n	T_n	Q_{n+1}
0	0	0
0	1	1
1	0	1
1	1	0

(b) State table

T_n	Q_{n+1}
0	Q_n
1	$\overline{Q}_n$

(c) Function table

FIGURE 8-34 T FF.

8-12 DIRECT PRESET AND CLEAR INPUTS

In addition to control inputs, clock inputs, and outputs, most practical FFs are also equipped with *direct preset* inputs, which are also known as *direct set* inputs, as well as with *direct clear* inputs, which are also known as *direct reset* inputs. When the direct preset and direct clear inputs are both at logic 0, they have no influence on the operation of the FF. When the direct preset input is at logic 1 and the direct clear input is at logic 0, the FF is set to its 1 state irrespective of the control inputs and the clock. Similarly, when the direct clear input is at logic 1 and the direct preset input is at logic 0, the FF is set to its 0 state irrespective of the control inputs and the clock. When both the direct preset and the direct clear inputs are at logic 1, the resulting state of the FF is usually undetermined.

Information on direct preset and direct clear inputs can be included in the function table of the FF as illustrated in Ex. 8-7 below.

EXAMPLE 8-7

The Type 10131 IC package consists of two ECL master–slave D FFs. A function table for each FF is shown in Fig. 8-35, where the first two lines describe clocked transitions. Entries of "H" correspond to -0.8 V, entries of "L" to -1.6 V, and the question mark entry to a next state that is not defined. Find the next state Q_{n+1} if the clear input is at -1.6 V, and if the preset input and the D input are as follows:

CLEAR	*PRESET*	D_n	Q_{n+1}
L	L	L	L
L	L	H	H
L	H	x	H
H	L	x	L
H	H	x	?

FIGURE 8-35 Function table of the Type 10131 ECL master–slave D FF.

a. The preset input is at −1.6 V and the D input is at −1.6 V.
b. The preset input is at −1.6 V and the D input is at −0.8 V.
c. The preset input is at −0.8 V and the D input is at −1.6 V.
d. The preset input is at −0.8 V and the D input is at −0.8 V.

Solution

In all four cases we have a clear input that is at −1.6 V, corresponding to "L." Thus the answers are contained in the first three lines of the function table.

a. Here we have a preset input of "L" and a D input of "L"; hence from the first line of the function table, Q_{n+1} = "L" = **−1.6 V**.
b. Here we have a preset input of "L" and a D input of "H"; hence from the second line of the function table, Q_{n+1} = "H" = **−0.8 V**.
c and d. For these conditions we have, in addition to a clear input of "L," a preset input of "H." According to the third line of the function table, the next state is Q_{n+1} = "H" irrespective of D. Thus the solutions of both **c** and **d** are Q_{n+1} = "H" = **−0.8 V**.

The direct preset and direct clear inputs are often shared by FFs in the same package. We should also note that frequently the *complements* of the direct preset and clear inputs are required: a logic H at these complement inputs permits normal operation and a logic L effects preset or clear.

EXAMPLE 8-8

Figure 8-36 shows a function table of the Type 7470 TTL positive edge-triggered J-K FF. Entries of "H" correspond to a voltage between +2.4 V and +5 V, entries of "L" to ≈ 0 V, and the question mark entry to a next state that

CLEAR	*PRESET*	J_n	K_n	Q_{n+1}
L	L	x	x	?
L	H	x	x	L
H	L	x	x	H
H	H	L	L	Q_n
·	·			
H	H	L	H	L
H	H	H	L	H
H	H	H	H	$\overline{Q}_n$

FIGURE 8-36 Function table of the Type 7470 TTL positive edge-triggered J-K FF.

is not defined. Find the next state Q_{n+1} if the clear input is "L" and the preset input is "H."

Solution

From the second line of the function table, we see that the next state is Q_{n+1} = "L," irrespective of the states of J and K.

8-13 TIMING

In this section we describe timing properties of FFs as follows: (1) timing constraints on the clock; (2) timing constraints on the direct preset and direct clear inputs; (3) propagation delays of the clock; (4) propagation delays of the direct preset and direct clear inputs; (5) timing constraints on the control inputs as specified by the required *setup time* and *hold time*.

8-13.1 Timing Constraints on the Clock

Consider the clocked R-S FF of Fig. 8-12. If we want to change its state from $Q = 0$ to $Q = 1$, we must have $S = 1$, $R = 0$, and $C = 1$; furthermore, we must have $C = 1$ long enough to permit setting of the FF. It can be shown that for correct operation of the circuit of Fig. 8-12 the minimum required duration of the $C = 1$ time interval is $2t_P$, where t_P is the propagation delay of each logic gate. The situation is somewhat more complicated in a master–slave FF where we also have a requirement on the minimum required duration of the $C = 0$ time interval, since the slave latch is set to its new state during $C = 0$.

Thus, timing constraints of the clock in a FF can be specified by the minimum required duration of the $C = 1$ time interval, $t_{1_{min}}$, and by the minimum required duration of the $C = 0$ time interval, $t_{0_{min}}$. We may also specify the *maximum clock frequency*, f_{max}, which is given as

$$f_{max} = \frac{1}{t_{1_{min}} + t_{0_{min}}} \tag{8-1}$$

EXAMPLE 8-9

The Type 74107 IC package consists of two TTL master–slave J-K FFs. The minimum required duration of the $C = 1$ time interval is $t_{1_{min}} = 20$ ns and the minimum required duration of the $C = 0$ time interval is $t_{0_{min}} = 47$ ns. Find the maximum clock frequency, f_{max}.

Solution

According to eq. 8-1,

$$f_{max} = \frac{1}{t_{1_{min}} + t_{0_{min}}} = \frac{1}{20\text{ ns} + 47\text{ ns}} \approx \mathbf{15\ MHz}$$

In many FFs the maximum clock frequency, f_{max}, is specified, but the minimum required duration of the $C = 0$ time interval, $t_{0_{min}}$, is not given. In such a case we can find $t_{0_{min}}$ by expressing it from eq. 8-1 as

$$t_{0_{min}} = \frac{1}{f_{max}} - t_{1_{min}} \tag{8-2}$$

EXAMPLE 8-10

The Type 4027B IC package consists of two CMOS positive edge-triggered J-K FFs. In this example, they are operated at a power supply voltage of 5 V. At this supply voltage the maximum clock frequency is specified as f_{max} = 1.5 MHz and the minimum required duration of the $C = 1$ time interval as $t_{1_{min}} = 0.33\ \mu s$. Find the minimum required duration of the $C = 0$ time interval, $t_{0_{min}}$.

Solution

By use of eq. 8-2

$$t_{0_{min}} = \frac{1}{f_{max}} - t_{1_{min}} = \frac{1}{1.5\text{ MHz}} - 0.33\ \mu s \approx \mathbf{0.33\ \mu s}$$

Thus a *symmetrical* squarewave with a frequency of 1.5 MHz properly operates the clock input of the FF.

8-13.2 Timing Constraints on the Direct Preset and Clear Inputs

The only timing constraint on the direct preset input is a minimum time interval it has to spend in its logic 1 state to ensure a FF state of $Q = 1$ (''*minimum preset pulse width*''). Similarly, the only timing constraint on the direct clear input is a minimum time interval it has to spend in its logic 1 state to ensure a FF state of $Q = 0$ (''*minimum clear pulse width*'').

EXAMPLE 8-11

Timing constraints of the direct preset and direct clear inputs in a Type 7470 TTL positive edge-triggered J-K FF are specified as follows: ''Pulse width, preset or clear low: 25 ns min.'' A function table of the FF is shown in Fig. 8-36. Describe a direct clear input that will set the FF to its $Q = 0$ state.

Solution

From the specification we conclude that preset or clear is attained by a *low* logic level, that is, the *complements* of the preset and clear inputs are required. This is confirmed by the function table of Fig. 8-36. Thus an acceptable direct clear input would stay at +2.4 V to +5 V for normal FF operation, and would have to make a transition to $\approx$ 0 V and stay there for a duration of at least 25 ns to ensure a FF state of $Q = 0$.

8-13.3 Propagation Delay of the Clock

Whether or not the state of a FF is changed by a clock pulse is determined by its state before the clock pulse and by the control inputs. The timing of such a change, however, is determined by the clock pulse.

In a positive edge-triggered FF, the change of the output state follows, after a delay, the positive-going edge of the clock. This delay is the *propagation delay from the clock input to the output*, $t_{P_{CLOCK}}$. It is referenced to the positive-going edge of the clock waveform, which is usually abbreviated by an arrow pointing upward: ↑.

EXAMPLE 8-12

In a Type 7470 TTL positive edge-triggered J-K FF, the maximum propagation delay from the positive-going edge of the clock waveform to a change of state at either output Q or $\overline{Q}$ is 50 ns. Describe this information by abbreviated notation.

Solution

$$t_{P_{\mathrm{CLOCK}}} = \mathbf{50\ ns}\uparrow$$

The propagation delay from the clock to the output of the FF is usually referenced to the negative-going edge of the clock waveform in negative edge-triggered FFs, in most master–slave FFs, and in master–slave FFs with data lockout. In such cases the abbreviation used is an arrow pointing downward: ↓.

Thus far we assumed that the propagation delay from the relevant clock transition to a change in the output state was the same when the output state changed from 0 to 1 as when it changed from 1 to 0. This is frequently *not* the case, as shown in Ex. 8-13 below.

EXAMPLE 8-13

In a Type 7472 TTL master–slave J-K FF, the maximum propagation delay from the negative-going edge of the clock waveform to a change of state at either output is $t_{PLH_{\mathrm{CLOCK}}} = 25$ ns when the state changes from LOW to HIGH, and it is $t_{PHL_{\mathrm{CLOCK}}} = 40$ ns when the state changes from HIGH to LOW. Describe this information by abbreviated notation.

Solution

$$t_{PLH_{\mathrm{CLOCK}}} = \mathbf{25\ ns}\downarrow \qquad t_{PHL_{\mathrm{CLOCK}}} = \mathbf{40\ ns}\downarrow$$

8-13.4 Propagation Delays of the Direct Preset and Clear Inputs

We have seen in Sec. 8-12 that a FF can be forced to its 1 state by the application of a logic 1 at the direct preset input while the direct clear input is held at logic 0. Similarly, the FF can be forced to its 0 state by the application of a logic 1 at the direct clear input while the direct preset input is held at logic 0. The changes at the FF output, however, take place only some time after the application of the logic 1 at the direct preset or direct clear inputs. We designate the propagation delay from the direct preset input to the outputs by $t_{P_{\mathrm{PRESET}}}$ and the progagation delay from the direct clear input to the outputs by $t_{P_{\mathrm{CLEAR}}}$. These propagation delays are often different for negative-going and positive-going output transitions. Thus we may have four different propagation delays: $t_{PLH_{\mathrm{PRESET}}}$, $t_{PHL_{\mathrm{PRESET}}}$, $t_{PLH_{\mathrm{CLEAR}}}$, and $t_{PHL_{\mathrm{CLEAR}}}$, where the subscript PLH refers to positive-going transitions at a FF output and the subscript PHL refers to negative-going transitions at a FF output. In reality we usually have $t_{PLH_{\mathrm{PRESET}}} = t_{PLH_{\mathrm{CLEAR}}}$ and $t_{PHL_{\mathrm{PRESET}}} = t_{PHL_{\mathrm{CLEAR}}}$.

EXAMPLE 8-14

In a Type 7472 TTL master–slave J-K FF, the maximum propagation delays of the direct preset and direct clear inputs are given as $t_{PLH_{PRESET}} = t_{PLH_{CLEAR}} =$ 25 ns and as $t_{PHL_{PRESET}} = t_{PHL_{CLEAR}} =$ 40 ns. What is the maximum propagation delay of the preset input to the $\overline{Q}$ output of the FF?

Solution

The preset input sets the FF into its $Q =$ "H," $\overline{Q} =$ "L" state. Hence output $\overline{Q}$ makes a transition to its "L" state. Thus the propagation delay of the preset input for a negative-going transient at the FF output, $t_{PHL_{PRESET}}$, is applicable; thus the solution is $t_{PHL_{PRESET}} =$ **40 ns.**

8-13.5 Setup Time and Hold Time

We have seen that for predictable operation the control inputs of a master–slave FF are not allowed to change while the clock input is $C = 1$. We have also seen that a master–slave FF with data lockout is sensitive to the control inputs only during a short interval that starts at the positive-going edge of the clock waveform.

The time interval when the control inputs are not allowed to change is specified by the *setup time* and the *hold time* of the FF. Each of these two times is referenced to an edge in the clock waveform, but not necessarily to the same edge. A reference to a positive-going clock edge is abbreviated by an arrow pointing upward: ↑; a reference to a negative-going clock edge is abbreviated by an arrow pointing downward: ↓. The control inputs are not allowed to change during a time interval that starts a setup time *before* the clock edge to which the setup time is referenced and that ends a hold time *after* the clock edge to which the hold time is referenced.

EXAMPLE 8-15

The minimum setup time of a FF is specified as $t_{SETUP} = 20$ ns ↑ and the minimum hold time as $t_{HOLD} = 5$ ns ↓. The width of clock pulse ($C = 1$) is 50 ns. Specify the time interval when the control inputs are not allowed to change.

Solution

The setup time requirement specifies that the time interval when the control inputs are not allowed to change starts 20 ns before the positive-going edge of the clock pulse. The hold time requirement specifies that this interval ends 5 ns after the negative-going edge of the clock pulse. Since the clock pulse width is 50 ns, the duration of the interval is 20 ns + 50 ns + 5 ns = 75 ns. The 75 ns long interval starts 20 ns before the positive-going edge of the clock pulse and it ends 5 ns after the negative-going edge of the clock pulse, as shown in Fig. 8-37.

Thus far we discussed setup time and hold time specifications applicable to master–slave FFs and master–slave FFs with data lockout. The specifications are similar in *edge-triggered* FFs. In a *positive* edge-triggered FF the setup time and the hold time are referenced to the *positive*-going edge of the clock, and in a *negative* edge-triggered FF to the *negative*-going edge (see Probs. 8-9 and 8-10).

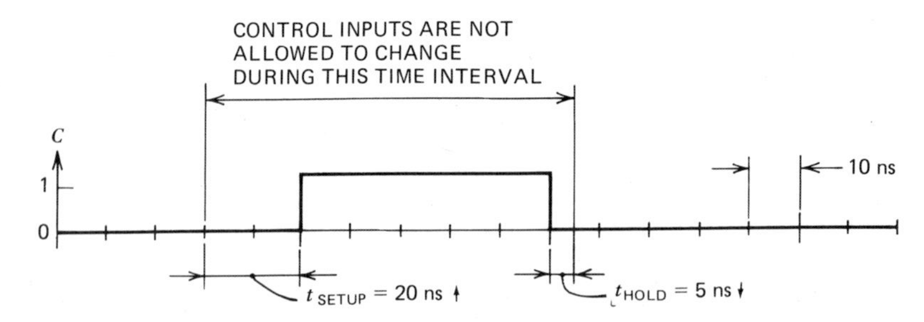

FIGURE 8-37 Illustration of the required setup and hold times in Ex. 8-15.

8-13.6 Summary of Timing Properties

Timing properties of FFs used in the examples and problems are summarized in Table 8-1. Note that the table is incomplete since packaging information is not given, direct preset and clear timing are not included, and clock rise and fall times, loading, and other operating conditions are not given.

TABLE 8-1
Timing Properties of FFs Used in Examples and Problems

Type	Family	Description	f_{max}	$t_{1_{min}}$	$t_{0_{min}}$	$t_{PLH_{CLOCK}}$	$t_{PHL_{CLOCK}}$	t_{SETUP}	t_{HOLD}
7470	TTL	Positive edge-triggered J-K		20 ns	30 ns	50 ns↑	50 ns↑	20 ns↑	5 ns↑
7472	TTL	Master-slave J-K		20 ns	47 ns	25 ns↓	40 ns↓	0 ↑	0 ↓
74107	TTL	Master-slave J-K		20 ns	47 ns	25 ns↓	40 ns↓	0 ↑	0 ↓
74LS107	Low-power Schottky TTL	Negative edge-triggered J-K	30 MHz	20 ns		20 ns↓	30 ns↓	20 ns↓	0 ↓
74109	TTL	Positive edge-triggered J-$\overline{K}$		20 ns	20 ns	16 ns↑	28 ns↑	10 ns↑	6 ns↑
74110	TTL	Master-slave J-K with data lockout		25 ns	25 ns	30 ns↓	20 ns↓	20 ns↑	5 ns↑
74111	TTL	Master-slave J-K with data lockout		25 ns	25 ns	17 ns↓	30 ns↓	0 ↑	30 ns↑
1670	ECL	Positive edge-triggered D	300 MHz	1.66 ns		2.5 ns↑	2.5 ns↑	0.5 ns↑	0.5 ns↑
1690	ECL	Positive edge-triggered D	500 MHz	1 ns		1.5 ns↑	1.5 ns↑	0.3 ns↑	0.3 ns↑
10131	ECL	Positive edge-triggered D	125 MHz	4 ns		4.5 ns↑	4.5 ns↑	2.5 ns↑	1.5 ns↑
10135	ECL	Positive edge-triggered J-K	125 MHz	4 ns		4.5 ns↑	4.5 ns↑	2.5 ns↑	1.5 ns↑
4027B	CMOS	Positive edge-triggered J-K	1.5 MHz	0.33 μs		0.35 μs↑	0.35 μs↑	0.14 μs↑	0 ↑

SUMMARY

After a short overview, an R-S storage latch was assembled from two DTL NAND gates and its properties were described. Transitions between the two stable states were described by a state table, and alternative realizations and use of R-S storage latches were discussed. The clocked R-S latch was introduced next, followed by a description of state tables and function tables for clocked latches, and by the introduction of clocked D latches. Master–slave FFs, master–slave FFs with data lockout, and edge-triggered FFs were introduced, as well as properties of R-S, D, J-K and T FFs, and of direct preset and direct clear inputs. Finally, timing properties of FFs were described.

SELF-EVALUATION QUESTIONS

1. What is a state table?
2. What is a control input of a FF?
3. Explain the terms present state and next state.
4. What is a function table?
5. What is a master–slave FF?
6. What is a master–slave FF with data lockout?
7. What is an edge-triggered FF?
8. What is a J-K FF?
9. Describe the operation of direct preset and direct clear inputs.
10. Summarize timing properties of FFs.

PROBLEMS

8-1. Draw a detailed logic diagram showing each gate in the clocked D latch of Fig. 8-16*a*. Use the clocked R-S latch of Fig. 8-11.

8-2. Draw a detailed logic diagram showing each gate in the clocked D latch of Fig. 8-16*a*. Use the clocked R-S latch of Fig. 8-12.

8-3. Draw a detailed logic diagram showing each gate in the master–slave R-S FF of Fig. 8-18. Use the clocked R-S latch of Fig. 8-12.

8-4. Draw a detailed logic diagram showing each gate in the master–slave R-S FF of Fig. 8-18. Use the clocked R-S latch of Fig. 8-11.

8-5. Draw a detailed logic diagram showing each gate in the J-K FF of Fig. 8-33*a*. Use the clocked R-S latch of Fig. 8-12.

8-6. Reduce the number of gates by 2 gates in the logic circuit obtained in the preceding problem.

8-7. The Type 74109 IC package consists of two positive edge-triggered TTL J-$\overline{K}$ FFs (i.e., control inputs J and $\overline{K}$ are available). Clock requirements are specified by a minimum $C = 1$ interval of $t_{1_{min}} = 20$ ns and by a minimum $C = 0$ interval of $t_{0_{min}} = 20$ ns. Find the maximum clock frequency, f_{max}.

8-8. The Type 74LS107 IC consists of two low-power Schottky TTL negative edge-triggered J-K FFs. Clock requirements are specified by a maximum clock frequency of $f_{max} = 30$ MHz and by a minimum $C = 1$ interval of $t_{1_{min}} = 20$ ns. Find the minimum required duration of the $C = 0$ time interval, $t_{0_{min}}$.

8-9. The Type 74109 IC package consists of two TTL positive edge-triggered FFs. The minimum required setup time of each FF is specified as $t_{SETUP} = 10$ ns ↑ and the minimum required hold time as $t_{HOLD} = 6$ ns ↑. The width of the clock pulse ($C = 1$) is 20 ns. Specify the time interval when the control inputs are not allowed to change.

8-10. The minimum required setup time in each FF of the Type 74LS107 low-power Schottky TTL negative edge-triggered dual J-K FFs is specified as $t_{SETUP} = 20$ ns ↓ and the minimum required hold time as $t_{HOLD} = 0$ ns ↓. The width of the clock pulse ($C = 1$) is 20 ns. Specify the time interval when the control inputs are not allowed to change.

8-11. The *excitation table of the D FF*, shown in Fig. P8-1, is used in the *design* of sequential digital circuits. It consists of three columns and four rows. The left two columns show the four combinations of present state Q_n and the *desired* next state Q_{n+1}; the rightmost column shows present control input D_n required to effect the desired transition. Demonstrate that Fig. P8-1 is consistent with the state table of the D FF given in Fig. 8-16*b*. (Fig. 8-16*b* is also a state table for the clocked D latch.)

Q_n	Q_{n+1}	D_n
0	0	0
0	1	1
1	0	0
1	1	1

FIGURE P8-1.

8-12. The *excitation table of the T FF*, shown in Fig. P8-2, is used in the *design* of sequential digital circuits. It consists of three columns and four rows. The left two columns show the four combinations of present state Q_n and the *desired* next state Q_{n+1}; the rightmost column shows present control input T_n required to effect the desired transition. Demonstrate that Fig. P8-2 is consistent with the state table of the T FF given in Fig. 8-34*b*.

Q_n	Q_{n+1}	T_n
0	0	0
0	1	1
1	0	1
1	1	0

FIGURE P8-2.

8-13. The *excitation table of the J-K FF*, shown in Fig. P8-3, is used in the *design* of sequential digital circuits. It consists of four columns and four rows. The left two columns show the four combinations of present state Q_n and the *desired* next state Q_{n+1}; the right two columns show present control inputs J_n and K_n required to effect the desired transition. Demonstrate that Fig. P8-3 is consistent with the state table of the J-K FF given in Fig. 8-33*c*.

Q_n	Q_{n+1}	J_n	K_n
0	0	0	x
0	1	1	x
1	0	x	1
1	1	x	0

FIGURE P8-3.

8-14. A D-E FF has two control inputs D and E, and is characterized by a next state that equals the present state when $E_n = 0$ and that equals D_n when $E_n = 1$, that is, $Q_{n+1} = \bar{E}_n Q_n + E_n D_n$. Prepare a state table for the D-E FF.

8-15. The *excitation table of the R-S FF,* shown in Fig. P8-4, is used in the *design* of sequential digital circuits. It consists of four columns and four rows. The left two columns show the four combinations of present state Q_n and the *desired* next state Q_{n+1}; the right two columns show present control inputs S_n and R_n required to effect the desired transition. Demonstrate that Fig. P8-4 is consistent with the state table of the R-S FF given in Fig. 8-15*a*. (Fig 8-15*a* is also a state table for the clocked R-S latch.)

Q_n	Q_{n+1}	S_n	R_n
0	0	0	x
0	1	1	0
1	0	0	1
1	1	x	0

FIGURE P8-4.

8-16. The state table of the R-S FF, Fig. 8-15*a*, includes two question mark entries. How do these effect the excitation table of the R-S FF, Fig. P8-4 above?

CHAPTER 9 Counters

Instructional Objectives

This chapter describes basic counter circuits and their operation. After you read it you should be able to:

1. Describe the properties of ripple counters.
2. Discuss the use and limitations of decoding in ripple counters.
3. Draw state diagrams for counters.
4. Describe synchronous counters, including programmable counters.
5. Discuss divide-by-N counters.
6. Describe hybrid counters.
7. Evaluate operating speed limitations of counters.

9-1 OVERVIEW

Counters are simple digital systems that can be used for *dividing* the frequency of an input signal as well as for *counting*, or totalizing, the number of input pulses. This chapter concentrates on basic counter circuits; counters based on shift registers are described in Ch. 10.

Logic diagrams in this chapter, as well as in later chapters, use mainly T, J-K, and D FFs. In reality, these FFs are implemented as governed by the technology utilized: by J-K and J-$\overline{\text{K}}$ FFs in TTL, and by D FFs in ECL, n-channel and p-channel MOS, and CMOS.

9-2 DIVIDE-BY-2 COUNTERS

We saw in Sec. 8-11 (page 197) that a clock pulse had no effect on the state of a T FF when its control input was $T = 0$ and it complemented the state of a T FF when its control input was $T = 1$. In many counter applications we permanently connect control input T as $T = 1$, resulting in a FF whose state is complemented on every clock pulse, that is, the next state Q_{n+1} is the complement of the present state Q_n: $Q_{n+1} = \overline{Q}_n$. The resulting FF is a $T = 1$ FF, also known as a *toggle FF*. (Unfortunately the name "toggle FF" is also used to describe T FFs.)

$T = 1$ FFs are shown in Figs. 9-1*a* and 9-1*b*, where we introduced *IN* for clock C, *OUT* for Q, and $\overline{OUT}$ for $\overline{Q}$. Since the T input is permanently

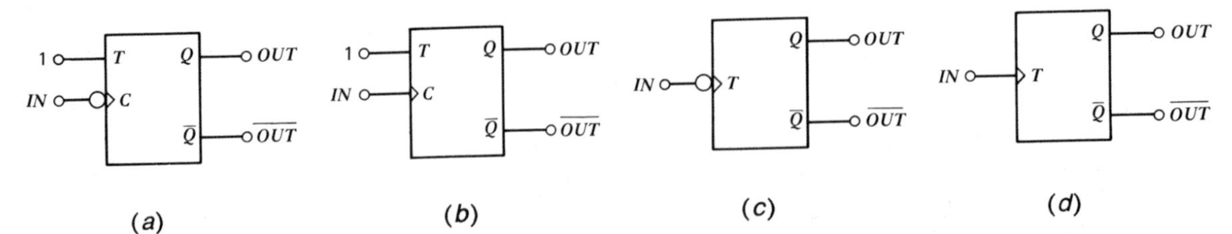

FIGURE 9-1 $T = 1$ FFs. (*a*) Master – slave and negative edge-triggered FFs (*b*) Positive edge-triggered FF (*c*) Symbol for master – slave and negative edge-triggered FFs (*d*) Symbol for positive edge-triggered FF

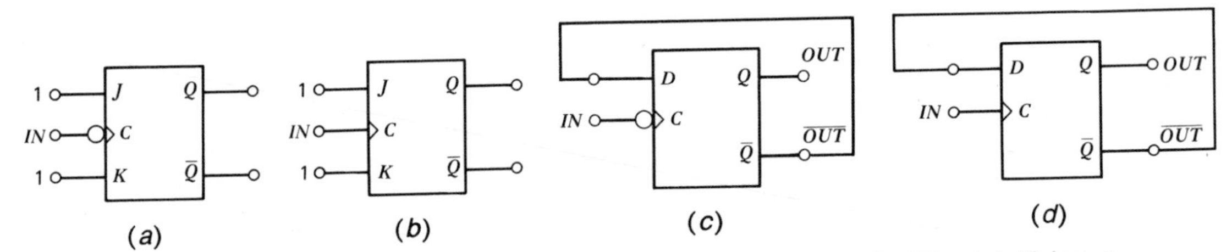

FIGURE 9-2 Implementation of $T = 1$ FFs using J-K and D FFs. (*a*) Using a master – slave or a negative edge-triggered J-K FF (*b*) Using a positive edge-triggered J-K FF (*c*) Using a master – slave or a negative edge-triggered D FF (*d*) Using a positive edge-triggered D FF

connected to logic 1, a $T = 1$ FF has only one input, input *IN*. $T = 1$ FFs can be described by the symbols shown in Figs. 9-1*c* and 9-1*d*.

In Fig. 9-1 we implemented $T = 1$ FFs using T FFs. However, because present technology does not favor T FFs, we also have to look at other possibilities. Based on Fig. 8-34*a* (page 198), we can implement $T = 1$ FFs using J-K FFs as shown in Figs. 9-2*a* and 9-2*b*. $T = 1$ FFs can be also implemented using D FFs as shown in Figs. 9-2*c* and 9-2*d* where $Q_{n+1} = D_n$ and $D_n = \overline{Q}_n$, hence $Q_{n+1} = \overline{Q}_n$ as required in a $T = 1$ FF.

Figure 9-3*a* shows a timing diagram that is applicable to master–slave $T = 1$ FFs as well as to negative edge-triggered $T = 1$ FFs; the propagation delays from the negative-going edge of input *IN* to outputs *OUT* and $\overline{OUT}$ are assumed to be one small time division. We can see that outputs *OUT* and $\overline{OUT}$ are square waves with frequencies that are half of the frequency of input *IN*; for this reason the $T = 1$ FF is also called a *divide-by-2 counter*. Figure 9-2*b* shows a timing diagram applicable to positive edge-triggered $T = 1$ FFs, again with propagation delays equaling one small time division.

9-3 RIPPLE COUNTERS

We saw in the preceding section that the output frequency of a $T = 1$ FF was half of the frequency of its input; hence the name divide-by-2 counter. We also

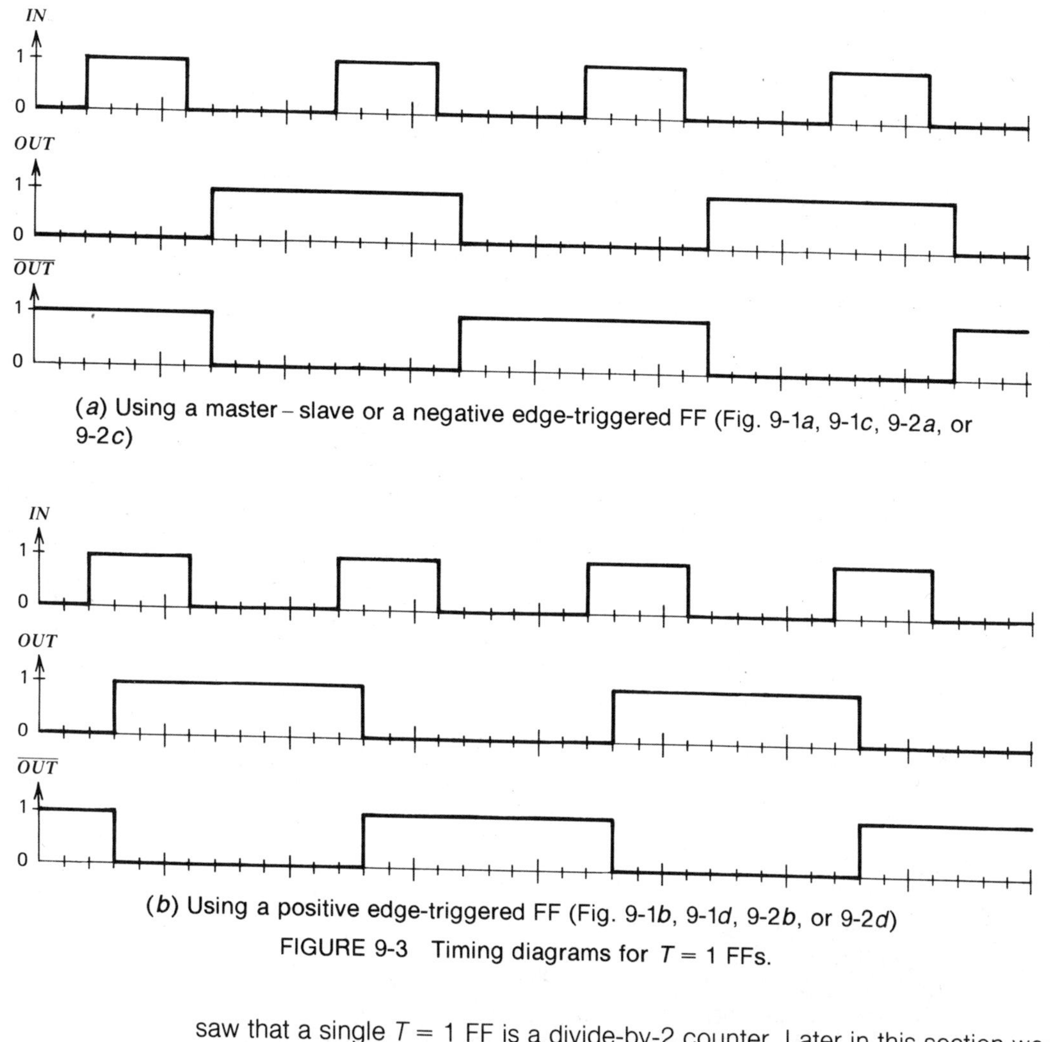

(*a*) Using a master–slave or a negative edge-triggered FF (Fig. 9-1*a*, 9-1*c*, 9-2*a*, or 9-2*c*)

(*b*) Using a positive edge-triggered FF (Fig. 9-1*b*, 9-1*d*, 9-2*b*, or 9-2*d*)

FIGURE 9-3 Timing diagrams for $T = 1$ FFs.

saw that a single $T = 1$ FF is a divide-by-2 counter. Later in this section we see that cascading two $T = 1$ FFs results in a divide-by-4 *binary ripple counter*. Similarly, cascading three $T = 1$ FFs results in a divide-by-8 binary ripple counter. In general, when we cascade m flip-flops that are $T = 1$ FFs, the result is a divide-by-n binary ripple counter, where

$$n = 2^m \tag{9-1}$$

EXAMPLE 9-1

Figure 9-4 shows a simplified logic diagram of the CMOS Type 4024B 7-stage binary ripple counter. It consists of 7 cascaded master–slave $T = 1$ FFs. Find n.

Solution

For a 7-stage binary ripple counter $m = 7$. Hence, by use of eq. 9-1, $n = 2^m = 2^7 =$ **128**. Thus the Type 4024B is a divide-by-128 binary ripple counter.

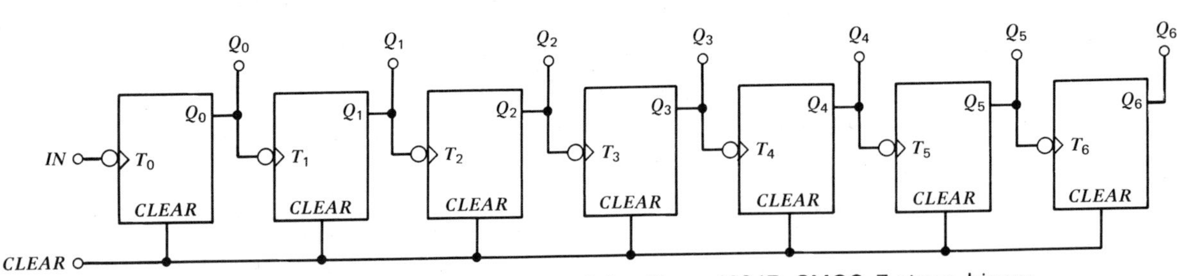

FIGURE 9-4 Simplified logic diagram of the Type 4024B CMOS 7-stage binary ripple counter.

9-3.1 Divide-by-4 Ripple Counters

Figure 9-5*a* shows a 2-stage binary ripple counter using two master–slave or negative edge-triggered $T = 1$ FFs. The operation is illustrated in the timing diagram shown in Fig. 9-6*a* with initial states of 0 for both Q_0 and Q_1 and with the propagation delay of each FF equaling one small time division.

The first waveform of Fig. 9-6*a* shows the input; since this is connected to input T_0 of the first FF, we also have $IN = T_0$. The second waveform of Fig. 9-6*a* shows output Q_0 of the first FF; since this is connected to input T_1 of the second FF, we also have $Q_0 = T_1$. The third waveform of Fig. 9-6*a* shows output Q_1 of the second FF.

We can see in Fig. 9-6*a* that Q_0 complements at times $t_{P_{CLOCK}}$ after each negative-going edge of T_0; similarly Q_1 complements at times $t_{P_{CLOCK}}$ after each negative-going edge of T_1. Also, from input *IN* to output *OUT* we have a divide-by-4 counter.

A 2-stage binary ripple counter can also be implemented using positive edge-triggered FFs as shown in Fig. 9-5*b*. Note that in Fig. 9-5*b* the *complement* output $\overline{Q}_0$ of the first FF is connected to the input of the second FF: $T_1 = \overline{Q}_0$, *not* $T_1 = Q_0$ as in Fig. 9-5*a*.

Since Fig. 9-5*b* uses positive edge-triggered FFs, the second FF complements after each positive-going edge of T_1, that is, after each positive-going edge of $\overline{Q}_0$. This is shown in the timing diagram of Fig. 9-6*b* with $t_{P_{CLOCK}}$ of each FF equaling one small time division.

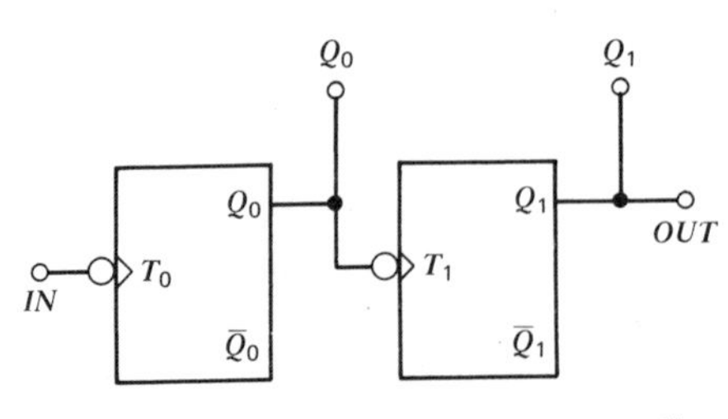

(*a*) Using master–slave or negative edge-triggered $T = 1$ FFs

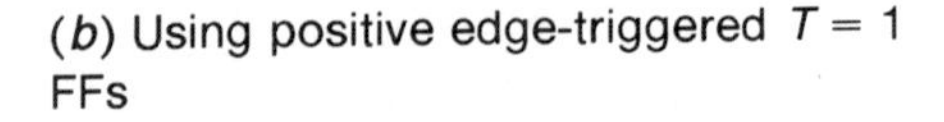

(*b*) Using positive edge-triggered $T = 1$ FFs

FIGURE 9-5 Two-stage binary ripple counters.

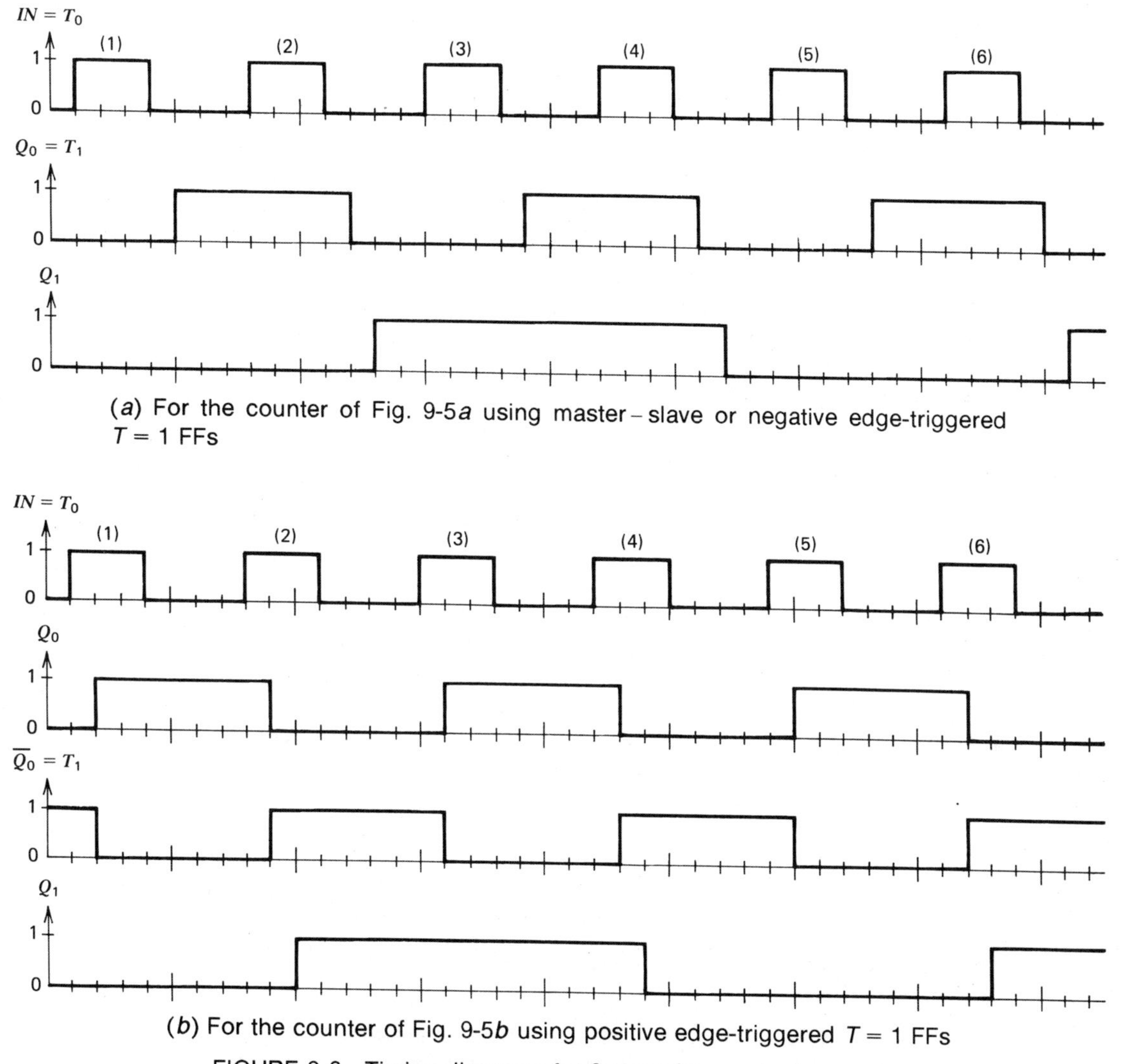

(*a*) For the counter of Fig. 9-5*a* using master–slave or negative edge-triggered $T = 1$ FFs

(*b*) For the counter of Fig. 9-5*b* using positive edge-triggered $T = 1$ FFs

FIGURE 9-6 Timing diagrams for 2-stage binary ripple counters.

Positive-going edges of $\overline{Q}_0$, however, coincide with negative-going edges of Q_0; hence, the operation of the ripple counter in Fig. 9-5*b* is similar to that of Fig. 9-5*a*. We can also see from the timing diagram of Fig. 9-6*b* that from input *IN* to output *OUT* we have a divide-by-4 counter.

9-3.2 Other Binary Ripple Counters

The 2-stage binary ripple counters shown in Figs. 9-5*a* and 9-5*b* can be readily extended to more stages. Figure 9-7 shows a 4-stage binary ripple counter using master–slave $T = 1$ FFs.

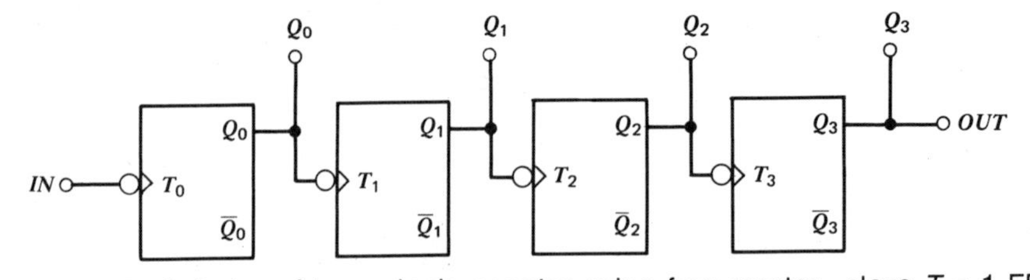

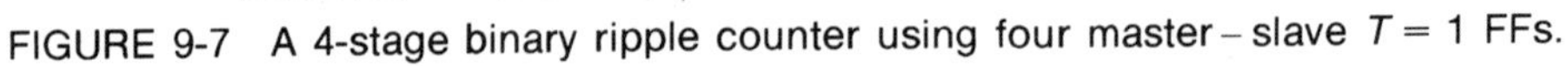
FIGURE 9-7 A 4-stage binary ripple counter using four master–slave $T = 1$ FFs.

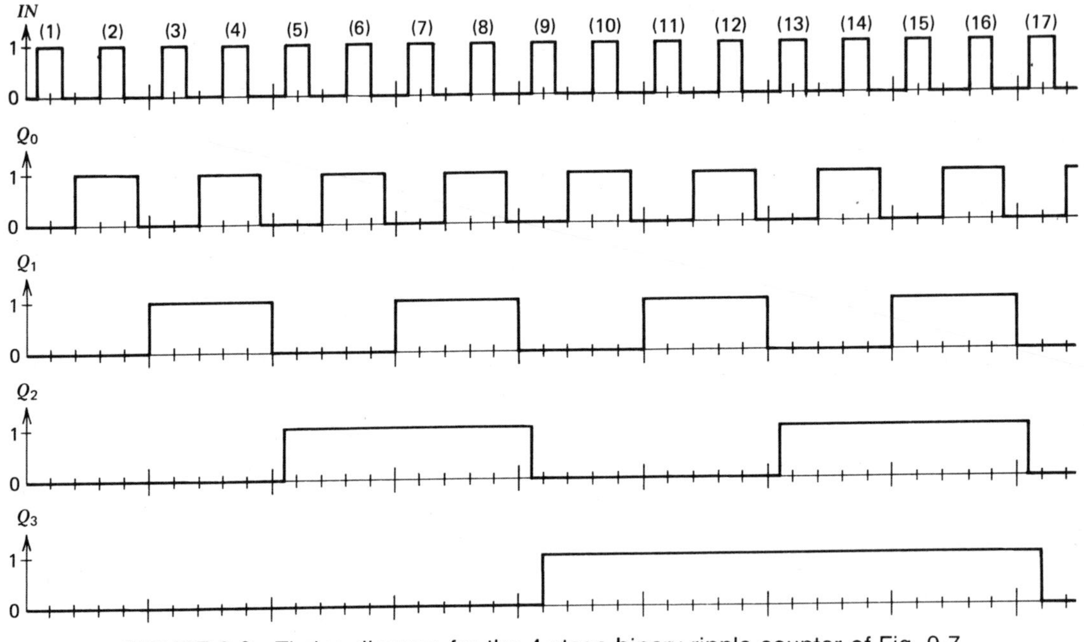

FIGURE 9-8 Timing diagram for the 4-stage binary ripple counter of Fig. 9-7.

Figure 9-8 shows a timing diagram for the 4-stage binary ripple counter of Fig. 9-7; the propagation delay of each FF equals one-half of a small time division. We can see that the counter is a divide-by-16 counter. This is in accordance with eq. 9-1, that is, for a divide-by-n-counter consisting of $m = 4$ stages $n = 2^m = 2^4 = 16$.

9-3.3 Decimal Ripple Counters

In the preceding sections we assembled binary ripple counters from divide-by-2 counters. Because of the common use of the decimal number system, we also have a need for *decimal ripple counters*; such a counter can be assembled from a divide-by-2 ripple counter and a divide-by-5 ripple counter.

The logic diagram of a divide-by-5 ripple counter is shown in Fig. 9-9. It consists of three master–slave (or negative edge-triggered) FFs: a T FF, a $T = 1$

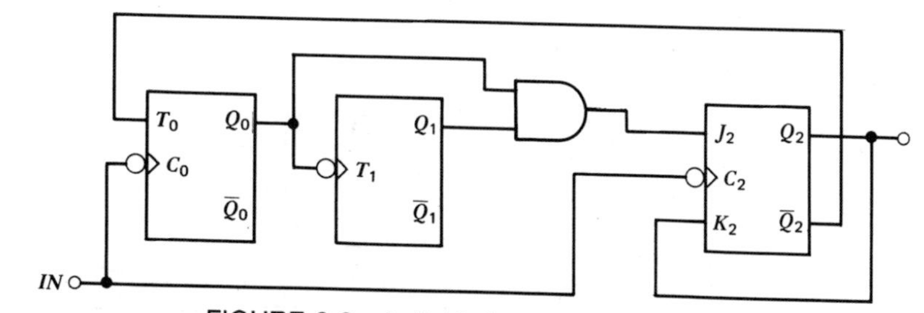

FIGURE 9-9 A divide-by-5 ripple counter.

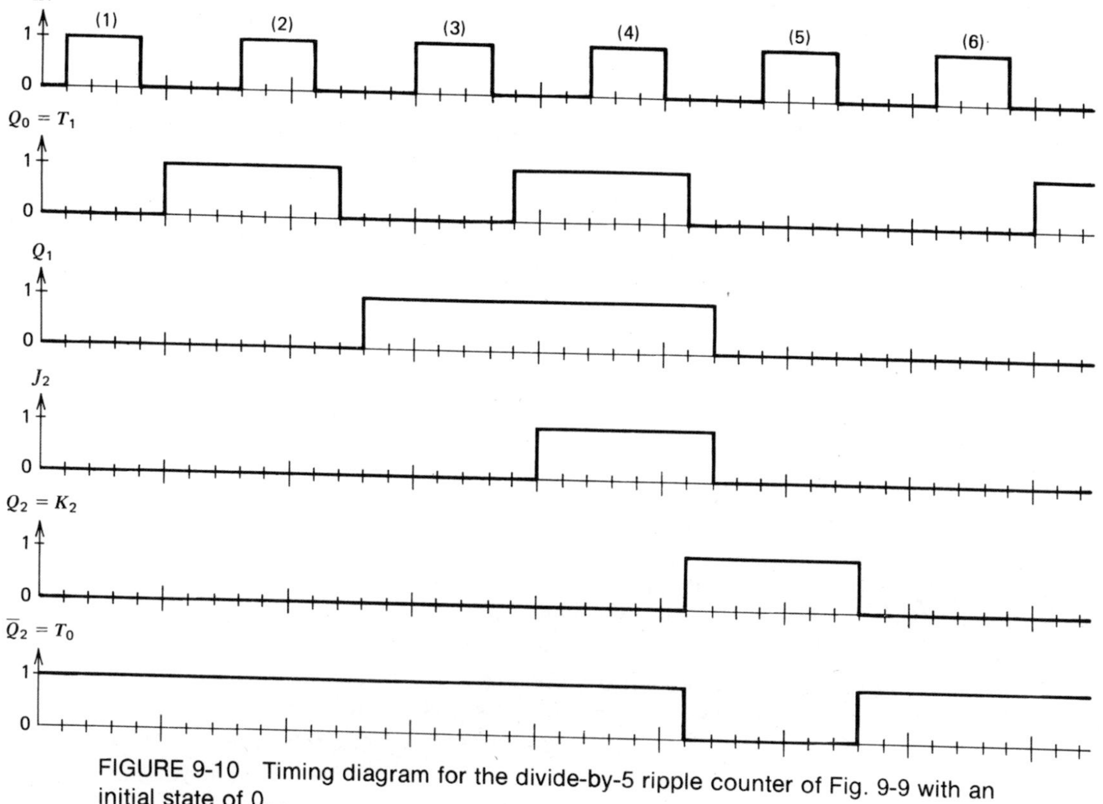

FIGURE 9-10 Timing diagram for the divide-by-5 ripple counter of Fig. 9-9 with an initial state of 0.

FF, and a J-K FF. The operation is illustrated in the timing diagram shown in Fig. 9-10 with propagation delays of each FF and the AND gate equaling one small time division. We can see that the circuit of Fig. 9-9 is indeed a divide-by-5 counter.

The operation of the counter of Fig. 9-9 is described in the state table shown in Fig. 9-11. It is divided into three major parts: Present States, Present Control Inputs, and Next States. Three FFs have eight states; these are identified as the

Present States				Present Control Inputs				Next States			
Decimal	Q_{2_n}	Q_{1_n}	Q_{0_n}	J_{2_n}	K_{2_n}	T_{1_n}	T_{0_n}	$Q_{2_{n+1}}$	$Q_{1_{n+1}}$	$Q_{0_{n+1}}$	Decimal
0	0	0	0	0	0	0	1	0	0	1	1
1	0	0	1	0	0	1	1	0	1	0	2
2	0	1	0	0	0	0	1	0	1	1	3
3	0	1	1	1	0	1	1	1	0	0	4
4	1	0	0	0	1	0	0	0	0	0	0
5	1	0	1	0	1	1	0	0	0	1	1
6	1	1	0	0	1	0	0	0	1	0	2
7	1	1	1	1	1	1	0	0	1	1	3

FIGURE 9-11 State table for the divide-by-5 ripple counter of Fig. 9-9.

eight Present States and are marked by decimal numerals 0 through 7. Each decimal numeral is the decimal equivalent of the binary number $Q_{2_n}Q_{1_n}Q_{0_n}$.

We can see in the logic diagram of Fig. 9-9 that the control inputs of the FFs are described by the equations $J_2 = Q_1 \cdot Q_0$, $K_2 = Q_2$, $T_1 = Q_0$, and $T_0 = \overline{Q}_2$. The present control inputs obtained from these equations are shown as Present Control Inputs in Fig. 9-11.

FF Q_2 is a J-K FF; given its present state Q_{2_n} and its present control inputs J_{2_n} and K_{2_n}, its next state $Q_{2_{n+1}}$ can be found by use of the function table in Fig. 8-33*d* (page 197). The next states thus obtained are shown in the first column under Next States in Fig. 9-11. FF Q_1 is a master–slave (or negative edge-triggered) $T = 1$ FF; it *complements on each negative-going transition of its input.* FF Q_0 is a T FF; given its present states and present control inputs, its next states can be found by use of the function table in Fig. 8-34*c* (page 198).

The next states thus obtained are shown in the second and third columns under Next States in Fig. 9-11. The last column under Next States in Fig. 9-11 shows the decimal equivalents of the binary numbers $Q_{2_{n+1}}Q_{1_{n+1}}Q_{0_{n+1}}$.

The first five rows in Fig. 9-11 describe the 5 states that comprise the regular divide-by-5 operation of counting through the states 0-1-2-3-4-0. . . . The last three rows describe *unallowed states*, 5, 6, and 7, also known as *unused states*, *undefined states*, *improper states*, or *illegal states*. These states are not used during counting, but the counter may enter them following power turn-on or as a result of improper input.

We can see that the transitions from the three unallowed states lead to states 1, 2, and 3. Thus the counter will eventually count correctly, even if it starts in an unallowed state. Such counters are called *self-starting counters*.

A decimal ripple counter can be assembled in two different ways: the divide-by-2 counter may precede or may follow the divide-by-5 counter. The former choice provides faster operation but results in an output waveform that is *asymmetric*, that is, the output waveform spends unequal intervals at logic 0 and logic 1. The latter choice operates more slowly but provides a *symmetric* square wave output.

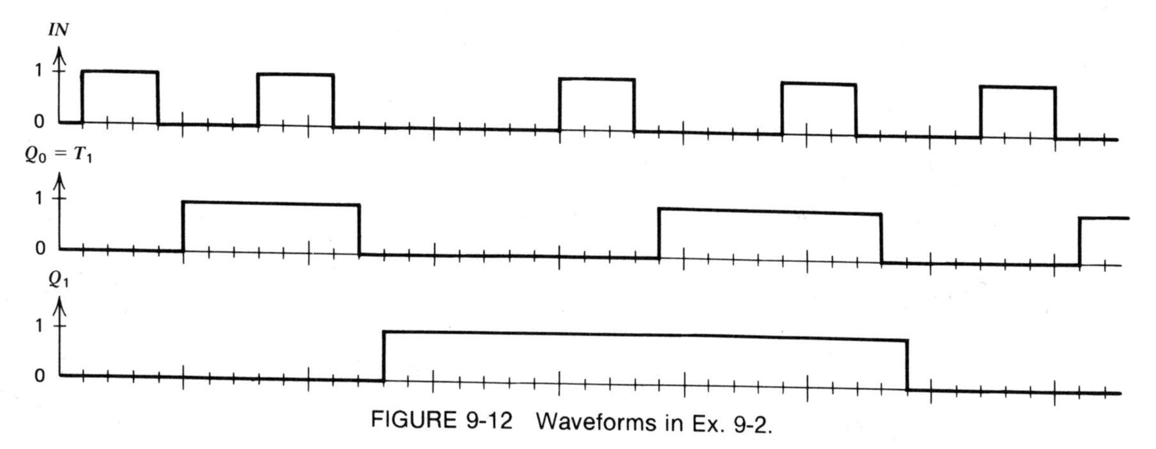

FIGURE 9-12 Waveforms in Ex. 9-2.

9-3.4 Aperiodic Operation

Thus far all ripple counters in this chapter were operated with input waveforms that consisted of identical and evenly spaced pulses, that is, the input waveforms were *periodic*. Ripple counters can also operate with *aperiodic* input waveforms, as illustrated in Ex. 9.2 below.

EXAMPLE 9-2

The ripple counter shown in Fig. 9-5*a* is used for counting pulses that arrive at randomly spaced time intervals. The input of the counter is shown as the first waveform in Fig. 9-12. Draw the waveforms of outputs Q_0 and Q_1 in addition to the input waveform. Assume that propagation delays equal one small time division.

Solution

Outputs Q_0 and Q_1 are shown as the last two waveforms in Fig. 9-12. Since Fig. 9-5*a* uses master–slave FFs, transitions take place following negative-going edges of $T_0 = IN$ and of T_1.

9-3.5 Decoding

Thus far we described how ripple counters *count*, but we did not show how we can *read out* the *contents* of the counter. The simplest way to read out the contents of a binary counter is bit by bit, as illustrated in Ex. 9-3 below.

EXAMPLE 9-3

The CMOS Type 4024B 7-stage binary ripple counter is used for counting randomly spaced pulses as shown in Fig. 9-13. Each of the 7 FF outputs drives a display light via a *lamp driver* circuit; a light is ON when the corresponding FF output is $Q = 1$. All FFs in the counter are initially reset to 0 via the direct *CLEAR* input. Next the pulse source is turned on for an interval of 1 minute. During this interval 89 pulses are sent to the input of the counter. At the end of the 1-minute interval the pulse source is turned OFF and the lights are read.

Find which lights are ON and which lights are OFF.

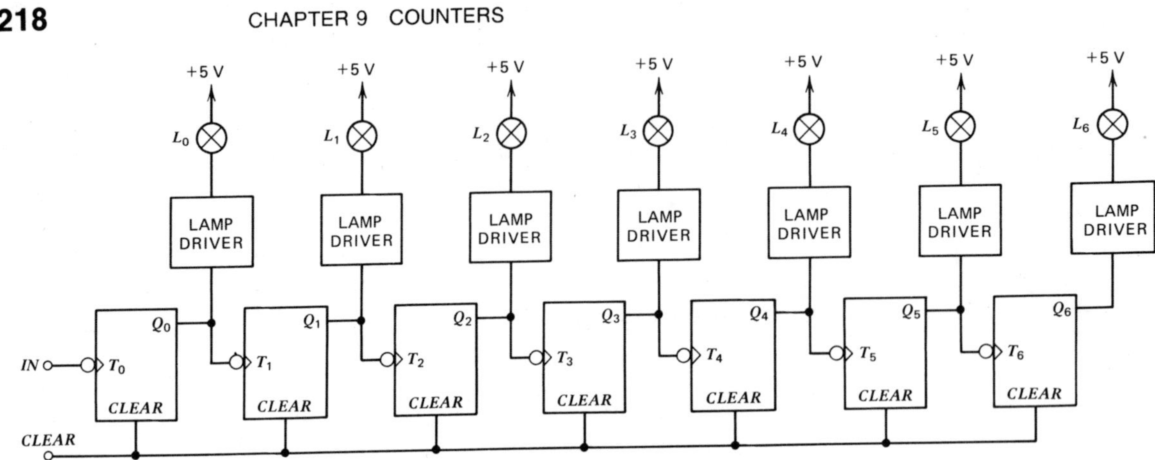

FIGURE 9-13 Binary ripple counter in Ex. 9-3.

Solution

The binary equivalent of 89_{10} is 1011001_2; hence, the lights are lit as follows:

L_6	L_5	L_4	L_3	L_2	L_1	L_0
ON	OFF	ON	ON	OFF	OFF	ON

In Ex. 9-3 the contents of the counter were read out *after* the counting was completed. However, in many applications we want to monitor the contents of the counter *during* counting and generate an output signal when a certain count is reached. For example, a traffic control system turns the light red after 20 vehicles go by, or a computer activates the line feed of a printer after 80 characters in a line. Such applications often require *decoding* of the contents of the counter.

Decoders were discussed in Sec. 6-8. Figure 9-14 shows the 2-stage ripple counter of Fig. 9-5*a* connected to a 2-line to 4-line decoder similar to the one shown in Fig. 6-25 (page 130). Output $OUT_0 = 1$ when the content of the counter is 0; otherwise $OUT_0 = 0$. Similarly, output $OUT_1 = 1$ when the content of the counter is 1, and $OUT_1 = 0$ otherwise. Outputs OUT_2 and OUT_3 operate likewise.

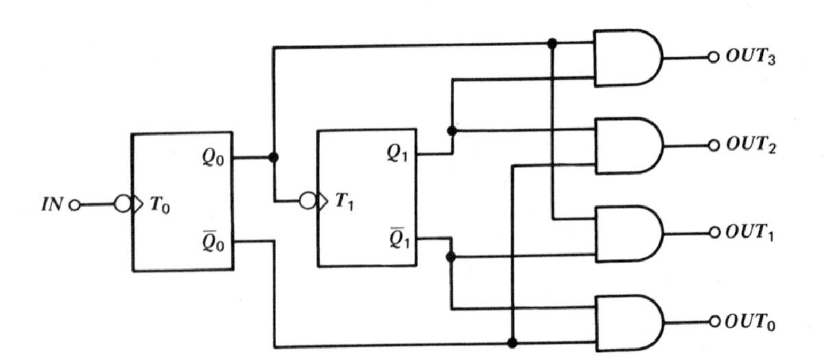

FIGURE 9-14 The ripple counter of Fig. 9-5*a* and a 2-line to 4-line decoder.

9-3.6 Operating Speed

There are two basic limitations on the maximum operating speed of ripple counters. One of these is the maximum *counting* speed that arises from limitations of the individual FFs as given by f_{max}, $t_{1_{min}}$, $t_{P_{CLOCK}}$, t_{SETUP}, and t_{HOLD} (see Sec. 8-13)—this limitation is discussed in Sec. 9-3.6.1. The other limitation arises from ambiguities in the decoded waveforms when the contents of the counter are read out *during* counting—this limitation is discussed in Sec. 9-3.6.2.

9-3.6.1 Maximum Counting Speed

When the contents of a ripple counter are read out *after* counting is completed as was the case in Ex. 9-3, its operating speed is limited only by its *maximum counting speed.* We can state that as long as f_{max}, $t_{1_{min}}$, t_{SETUP}, and t_{HOLD} requirements of *each* FF in a ripple counter are satisfied, then the counter will operate correctly and its maximum counting speed limitation will not be exceeded. In what follows, these limitations are first illustrated for a single FF in Ex. 9-4, and are then extended to a 2-stage ripple counter in Ex. 9-5.

EXAMPLE 9-4

The ECL Type 1690 is a positive edge-triggered D FF. It has the specifications $f_{max} = 500$ MHz, $t_{1_{min}} = 1$ ns, $t_{P_{CLOCK}} = 1.5$ ns, $t_{SETUP} = 0.3$ ns, and $t_{HOLD} = 0.3$ ns. It is used as a $T = 1$ FF, according to Fig. 9-2*d*. Is the maximum counting speed of the FF exceeded if input *IN* is a 500-MHz symmetrical square wave?

Solution

The input waveform is a 500-MHz square wave, thus the $f_{max} = 500$ MHz specification of the FF is satisfied. Also, the input waveform is a *symmetrical* square wave; hence, it spends equal intervals at logic 0 and at logic 1. Since the period of a 500-MHz square wave is $1/500$ MHz $= 2$ ns, the input square wave spends 1 ns at logic 0 and 1 ns at logic 1. Consequently, the $t_{1_{min}} = 1$ ns specification is not violated.

Next we use the $t_{P_{CLOCK}} = 1.5$ ns specification, and for the time being we *assume* that t_{SETUP} and t_{HOLD} are satisfied. If this happens, then we can draw the timing diagram shown in Fig. 9-15. Note that since $t_{P_{CLOCK}} = 1.5$ ns, the

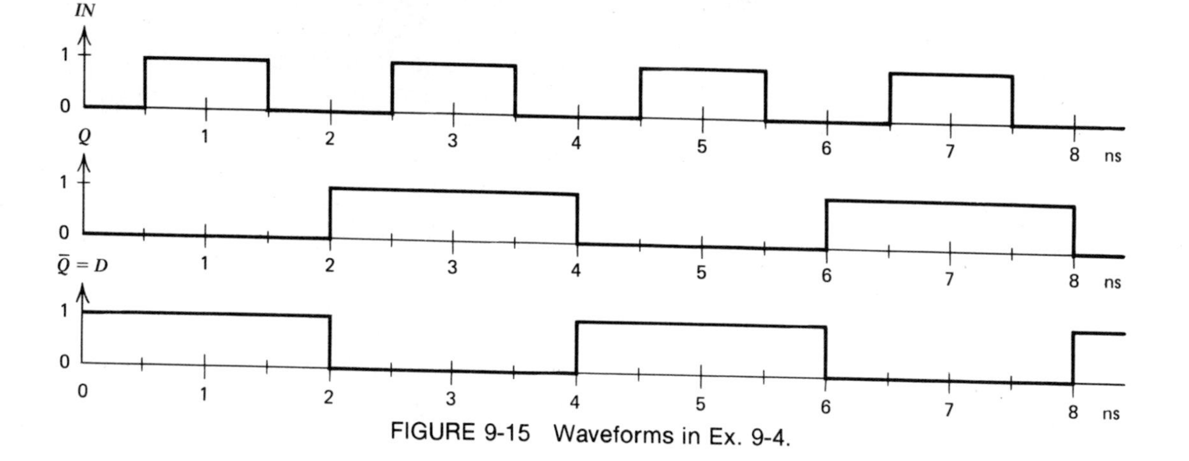

FIGURE 9-15 Waveforms in Ex. 9-4.

outputs of the FF complement 1.5 ns after the positive-going edge of the input waveform at *IN*; this turns out to be 0.5 ns after the negative-going edge of the input waveform.

According to the setup time specification, control input D has to settle no later than 0.3 ns before a positive-going edge of the input waveform. In Fig. 9-15 this settling takes place 0.5 ns before each positive-going edge of the input waveform; hence, the setup time specification is satisfied.

According to the hold time specification, control input D is not allowed to change until 0.3 ns after a positive-going edge of the input waveform. In Fig. 9-15 such changes take place 1.5 ns after positive-going edges of the input waveform; hence, the hold time specification is satisfied.

Thus, we first established that the f_{max} and $t_{1_{min}}$ specifications are satisfied. Next, we used $t_{P_{CLOCK}}$ and drew a timing diagram. We saw that this timing diagram satisfied the t_{SETUP} and t_{HOLD} specifications. Hence, we conclude that the FF operates correctly with a 500-MHz symmetrical square wave input, its maximum counting speed limitation is not exceeded, and the timing diagram of Fig. 9-15 is correct.

When several FFs are used in a ripple counter, we have to verify that *all* specifications of *every* FF are satisfied. The procedure is outlined for a 2-stage ripple counter in Ex. 9-5.

EXAMPLE 9-5

An ECL Type 1690 FF described in Ex. 9-4 is used as the first stage in a 2-stage ripple counter as shown in Fig. 9-16. The second FF is an ECL Type 1670 positive edge-triggered D FF with $f_{max} = 300$ MHz, $t_{1_{min}} = 1.66$ ns, $t_{P_{CLOCK}} = 2.5$ ns, $t_{SETUP} = 0.5$ ns, and $t_{HOLD} = 0.5$ ns. Verify that the maximum counting speed limitation of the counter is not exceeded if the input waveform is a 500-MHz symmetrical square wave.

Solution

The correct operation of the first FF was verified in Ex. 9-4. Its output Q_0 is a 250-MHz symmetrical square wave spending 2 ns at logic 0 and 2 ns at logic 1. Since this waveform is the input waveform of the second FF, the $f_{max} =$ 300 MHz and $t_{1_{min}} = 1.66$ ns specifications of the second FF are satisfied. By *assuming* that the second FF of the counter also operates correctly, we can draw the timing diagram shown in Fig. 9-17. We can see that the t_{SETUP} and t_{HOLD} specifications of the second FF are satisfied; this means that the counter indeed operates correctly, its maximum counting speed limitation is not exceeded, and the timing diagram of Fig. 9-17 is correct.

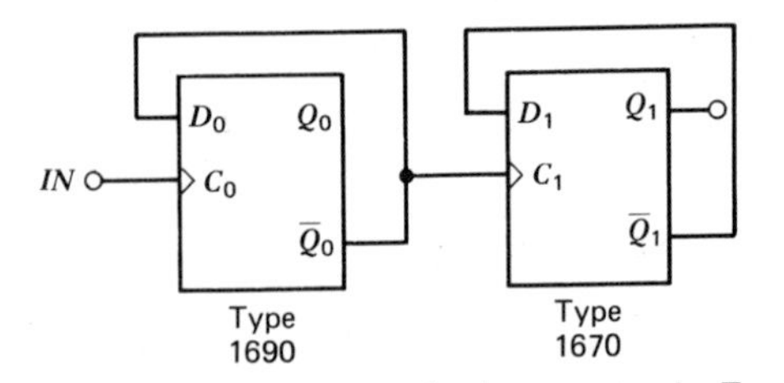

FIGURE 9-16 Two-stage ripple counter in Ex. 9-5.

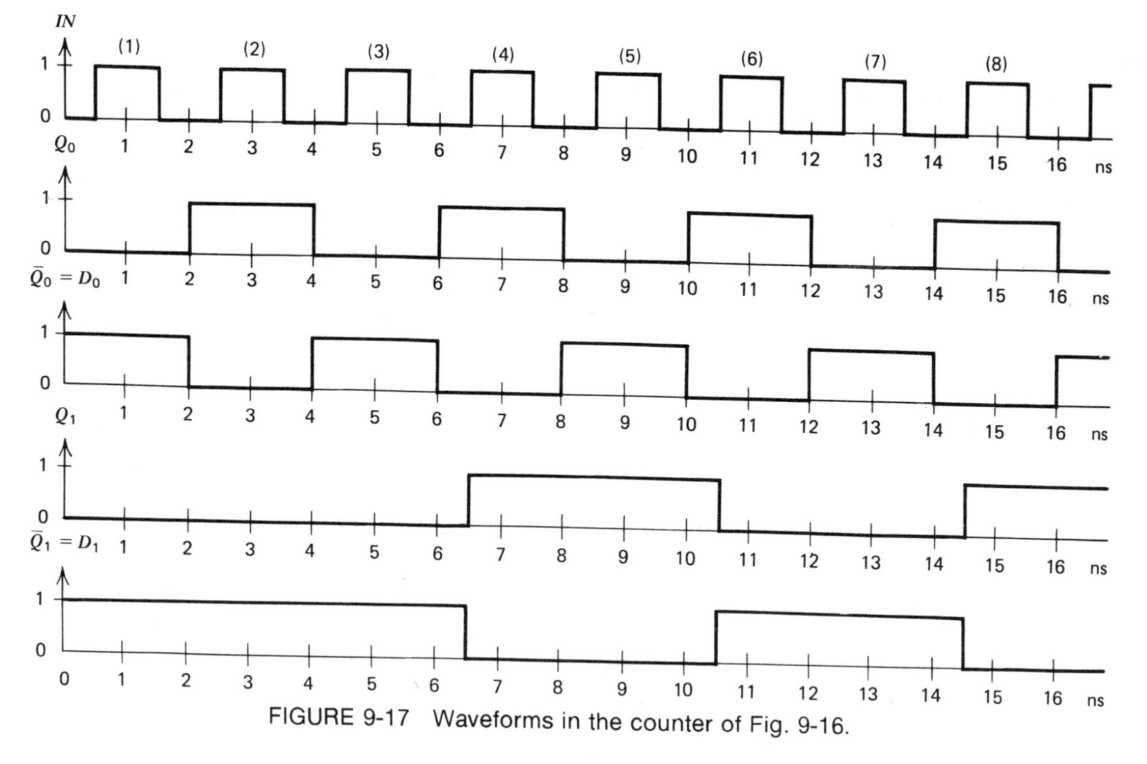

FIGURE 9-17 Waveforms in the counter of Fig. 9-16.

9-3.6.2 Speed Limitations of Decoded Outputs

When the contents of a ripple counter are read out *during* counting, additional speed limitations result. Consider the ripple counter and the 2-line to 4-line decoder shown in Fig. 9-14, and assume for simplicity that propagation delays through the AND gates of the decoder are zero. We can see in the timing diagram shown in Fig. 9-18 that the transitions of the outputs of the two FFs in the two-stage counter are not simultaneous. As a result, "spikes" or "glitches" appear at the decoder outputs as shown.

Whether the glitches in Fig. 9-18 are troublesome or not depends on the application. For example, if the decoder outputs drive lights, the short glitches probably will not be noticed. However, if some action such as the line feed of a printer is initiated by the decoded output, the glitches may lead to erroneous operation.

Erroneous operation due to the glitches can often be corrected by "deglitching" the outputs. This can be attained by ANDing the decoded outputs in Fig. 9-14 with input *IN*. The circuit of Fig. 6-25 (page 130) with its $\bar{E}$ input connected to the complement of input *IN* can be used for this purpose. It can be shown that a criterion for such deglitching to work in a binary ripple counter consisting of master–slave FFs is as follows: The sum of all FF propagation delays must be less than the duration of a logic 0 interval between two input pulses of logic 1. If the input is aperiodic, then the shortest logic 0 interval has to be taken into account.

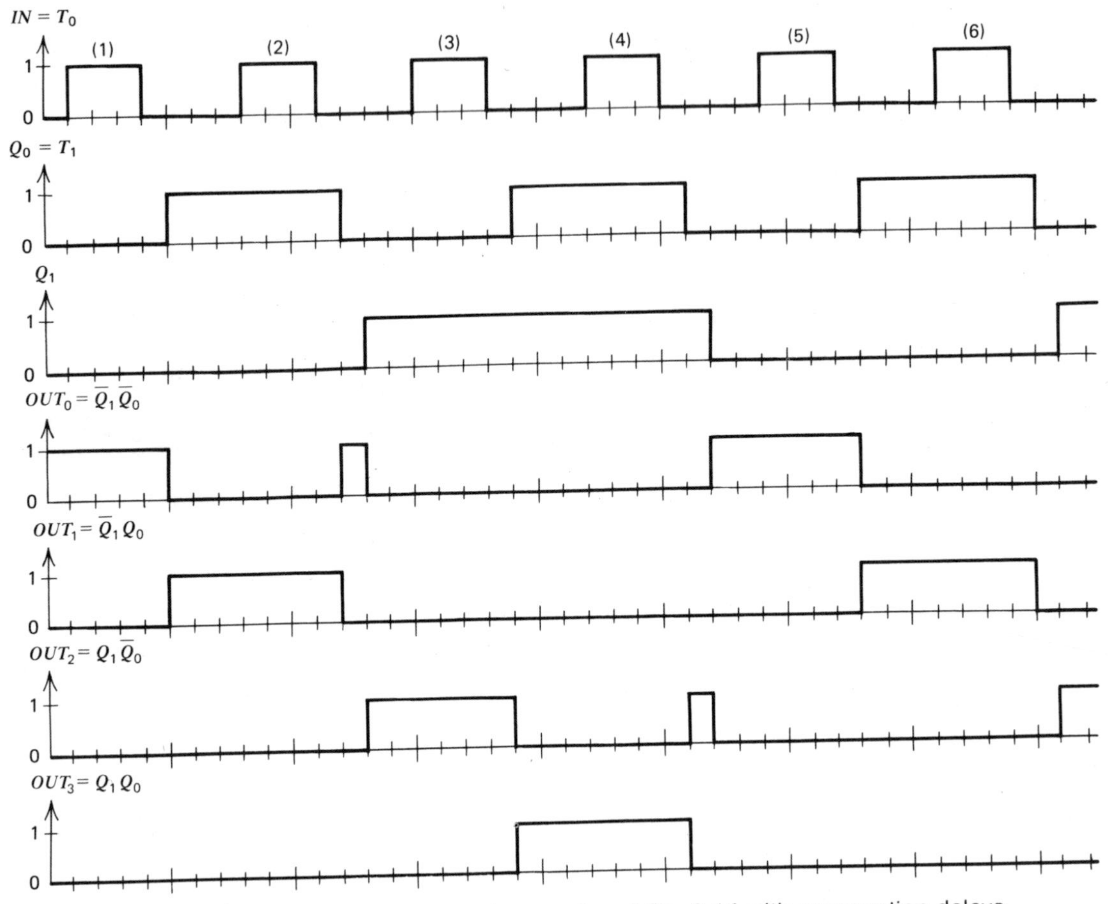

FIGURE 9-18 Timing diagram for the counter of Fig. 9-14 with propagation delays of the AND gates equaling zero.

9-4 ONE-BIT COUNTERS

Section 9-2 introduced the $T = 1$ FF by permanently connecting the T input of a T FF to logic 1. The resulting $T = 1$ FF is a divide-by-2 counter with only one input. We have also seen that the state of a $T = 1$ FF is complemented by each input pulse.

This section introduces more complex divide-by-2 counters. The importance of these counters is that they form the basis of synchronous counters to be described in a later section. Master–slave FFs are used throughout this section; however, the circuits may also be built using edge-triggered FFs.

9-4.1 One-Bit Counter with CARRY-OUT

A 1-bit counter with a *CARRY-OUT* output is shown in Fig. 9-19*a*. It is identical to the $T = 1$ FF of Fig. 9-1*a* except that the labeling is different. We consider the

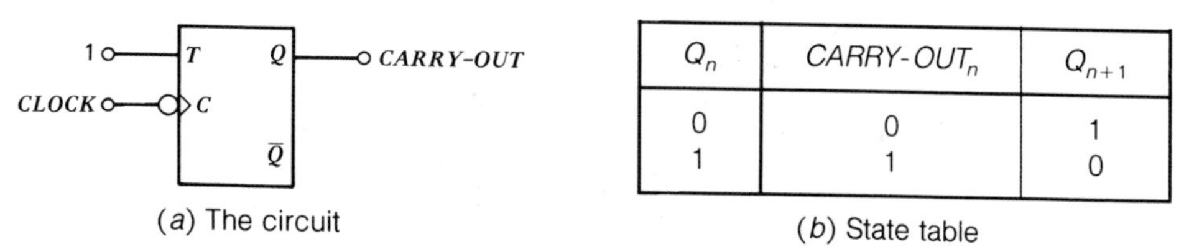

(a) The circuit

Q_n	$CARRY\text{-}OUT_n$	Q_{n+1}
0	0	1
1	1	0

(b) State table

FIGURE 9-19 One-bit counter with *CARRY-OUT*.

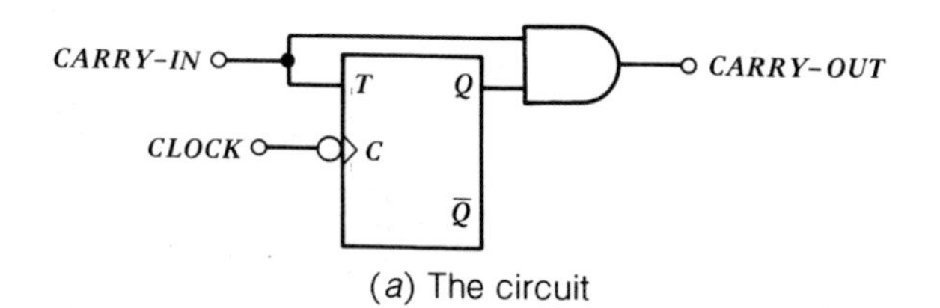

(a) The circuit

$CARRY\text{-}IN_n$	Q_n	$CARRY\text{-}OUT_n$	Q_{n+1}
0	0	0	0
0	1	0	1
1	0	0	1
1	1	1	0

(b) State table

FIGURE 9-20 One-bit counter with *CARRY-IN* and *CARRY-OUT*.

circuit of Fig. 9-19*a* as a 1-bit counter that adds 1 to its content on each *CLOCK* pulse; also it generates a *CARRY-OUT* = 1 when it is full, that is, when its content $Q = 1$. The operation of the 1-bit counter with *CARRY-OUT* is summarized in the state table of Fig. 9-19*b*.

9-4.2 One-Bit Counter with CARRY-IN and CARRY-OUT

Figure 9-20*a* shows a 1-bit counter with *CARRY-IN* and *CLOCK* inputs, and a *CARRY-OUT* output. Each *CLOCK* pulse adds 1 to the content of the counter if *CARRY-IN* = 1; *CARRY-OUT* = 1 when *CARRY-IN* = 1 and $Q = 1$. The operation is summarized in the state table of Fig. 9-20*b*.

The counter of Fig. 9-20*a* uses a T FF. A J-K FF may be used instead by connecting the *J* and *K* inputs in parallel to provide the *T* input. A D FF may also be used as shown in Fig. 9-21.

9-4.3 One-Bit Counter with ENABLE Input

In many applications it is desirable to be able to stop counting without influencing the generation of the *CARRY-OUT*. This can be done by the application of a logic 0 at an additional *ENABLE* input. The 1-bit counter shown in Fig. 9-22*a* has such an input in addition to the *CARRY-IN* and *CLOCK* inputs.

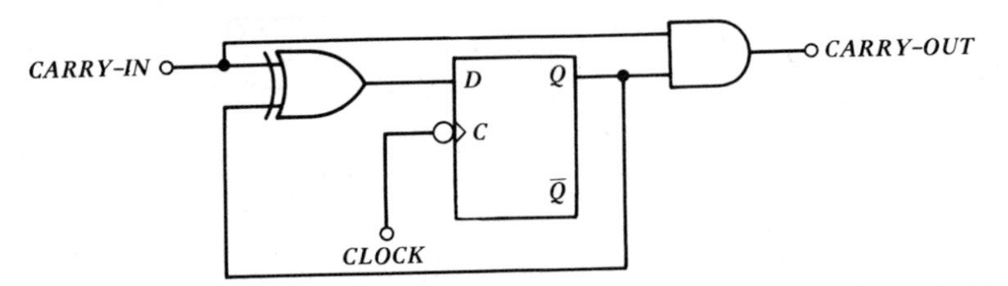

FIGURE 9-21 Realization of the 1-bit counter of Fig. 9-20*a* by a D FF.

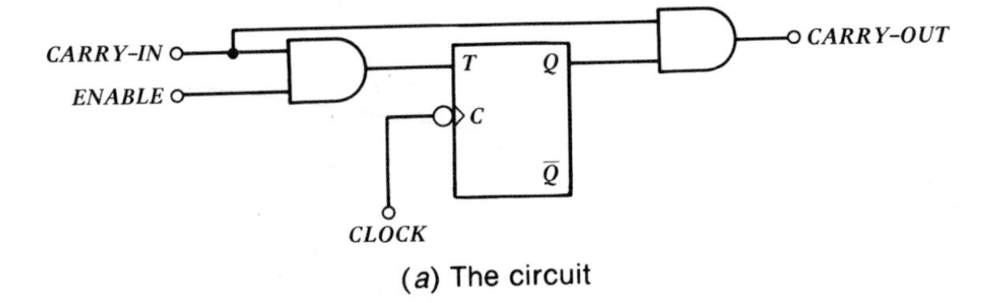

(*a*) The circuit

$ENABLE_n$	$CARRY\text{-}IN_n$	Q_n	$CARRY\text{-}OUT_n$	Q_{n+1}
0	0	0	0	0
0	0	1	0	1
0	1	0	0	0
0	1	1	1	1
1	0	0	0	0
1	0	1	0	1
1	1	0	0	1
1	1	1	1	0

(*b*) State table

$ENABLE_n$	$CARRY\text{-}IN_n$	Q_n	Q_{n+1}
0	x	x	Q_n
1	0	0	0
1	0	1	1
1	1	0	1
1	1	1	0

(*c*) Function table without the *CARRY-OUT*

FIGURE 9-22 One-bit counter with *ENABLE* input.

When *ENABLE* = 1, the counter counts the same way as the circuit of Fig. 9-20*a*. When *ENABLE* = 0 the counter holds its contents. This can also be seen in the state table shown in Fig. 9-22*b* and in the function table of 9-22*c*.

9-5 STATE DIAGRAMS

Thus far we described the operation of digital circuits by a state table (see, e.g., Fig. 9-22*b*), or by a function table, which is an abbreviated state table (see, e.g., Fig. 9-22*c*). These tables list all possible states, as well as outputs and next states for the various input combinations. The states of FF control inputs may also be included (see, e.g., Fig. 9-11).

The states and transitions described by a state table may also be presented in a *state diagram*, also known as a *transition diagram*, which is a pictorial representation. This is illustrated here in a step-by-step introduction of a state diagram for the 1-bit counter that was described in Fig. 9-20. State diagrams with more than two states are introduced in later sections of this chapter.

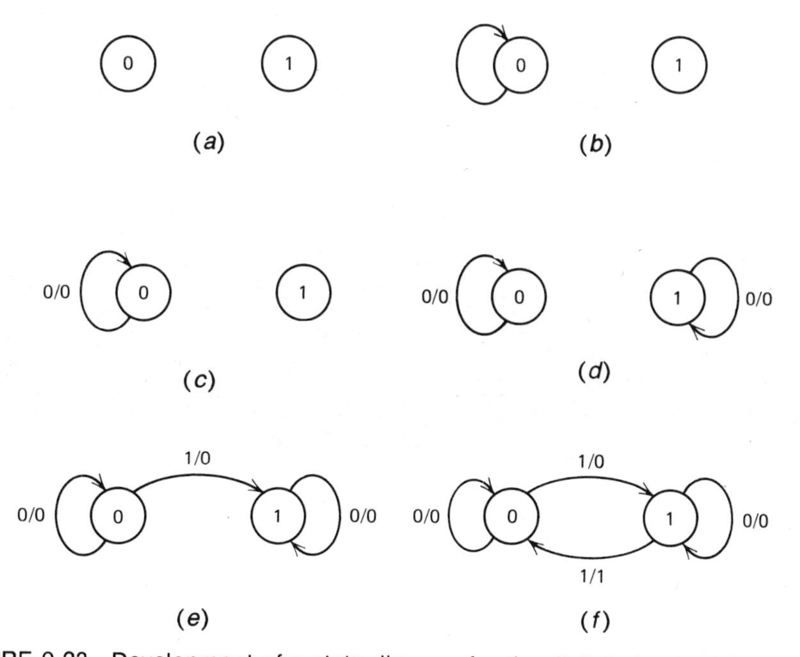

FIGURE 9-23 Development of a state diagram for the digital circuit of Fig. 9-20*a*. (*a*) The two states (*b*) Addition of a branch corresponding to the first row in the state table of Fig. 9-20*b* (*c*) Labeling of the branch corresponding to the first row in the state table (*d*) Addition of a labeled branch corresponding to the second row in the state table (*e*) Addition of a labeled branch corresponding to the third row in the state table (*f*) Addition of a labeled branch corresponding to the last row in the state table.

The digital circuit of Fig. 9-20*a* includes only one FF; hence, there are two possible states: $Q = 0$ and $Q = 1$. These two states are represented by two circles in Fig. 9-23*a*: the circle identified by 0 represents the $Q = 0$ state, the circle identified by 1 the $Q = 1$ state.

Next we look at the first row in the state table of Fig. 9-20*b*. It shows a transition from state 0 to state 0. In the state diagram we represent this transition by a *branch*, which is a line with an arrow. This branch originates from state 0 and returns to state 0, as shown in Fig. 9-23*b*.

The first row in the state table of Fig. 9-20*b* shows two additional pieces of information: $CARRY\text{-}IN_n = 0$ and $CARRY\text{-}OUT_n = 0$. We enter these in the state diagram by a *label* placed on the branch. The format of the label is $CARRY\text{-}IN_n/CARRY\text{-}OUT_n$. For the branch corresponding to the first row in the state table we thus have a label of 0/0 as shown in Fig. 9-23*c*. The first 0 in this label represents a $CARRY\text{-}IN_n = 0$, and the second 0 a $CARRY\text{-}OUT_n = 0$.

In Fig. 9-23*d* we add the branch corresponding to the second row in the state table of Fig. 9-20*b*. This branch describes a transition from state 1 to state 1 with $CARRY\text{-}IN_n = 0$ and $CARRY\text{-}OUT_n = 0$.

In Fig. 9-23*e* we add the branch corresponding to the third row in the state table of Fig. 9-20*b*. This branch describes a transition from state 0 to state 1 with $CARRY\text{-}IN_n = 1$ and $CARRY\text{-}OUT_n = 0$.

As a last step, shown in Fig. 9-23*f*, we add the branch corresponding to the last row in the state table of Fig. 9-20*b*. This branch describes a transition from state 1 to state 0 with $CARRY\text{-}IN_n = 1$ and $CARRY\text{-}OUT_n = 1$.

The result, Fig. 9-23*f*, is a complete state diagram for the digital circuit of Fig. 9-20*a*. It shows the same information as the state table of Fig. 9-20*b*, but in a different form.

A state diagram with two states can have at most four branches. However, a branch may represent *more than one transition*, as illustrated in Ex. 9-6, which follows.

EXAMPLE 9-6

The figure below shows a state diagram for the digital circuit of Fig. 9-22*a*. Each branch is labeled $ENABLE_n$, $CARRY\text{-}IN_n$ / $CARRY\text{-}OUT_n$ with the comma omitted in the state diagram. Two of the branches have more than one label. Each of the eight labels corresponds to a row in the state table of Fig. 9-22*b*. Find which label in the state diagram corresponds to which row in the state table of Fig. 9-22*b*.

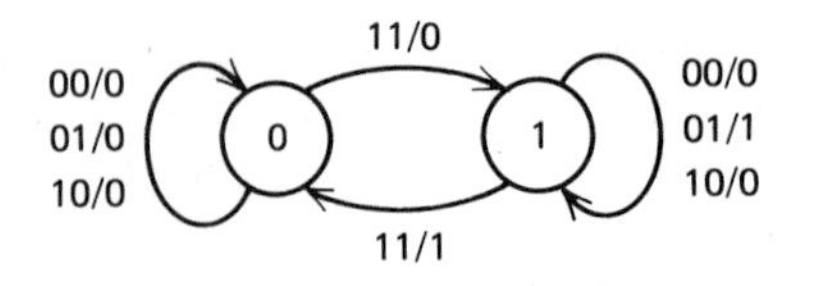

State diagram in Ex. 9-6.

Solution

The branch originating from state 1 and ending at state 0 corresponds to the last row, and the branch originating from state 0 and ending at state 1 corresponds to the next to last (seventh) row.

The branch originating from state 0 and returning to state 0 corresponds to three transitions. These are, in sequence, in the first, third, and fifth rows of the state table. The branch originating from state 1 and returning to state 1 corresponds to the three transitions in the second, fourth, and sixth rows of the state table.

9-6 SYNCHRONOUS COUNTERS

We have seen that in a ripple counter the transitions of each FF are delayed by $t_{P_{CLOCK}}$ from the transitions of the preceding FF. This led to glitches in the decoded outputs, or to slower operation when the glitches were removed by additional gating. These problems are greatly alleviated in *synchronous counters* in which the states of all FFs change approximately *simultaneously*. The circuitry of synchronous counters is more complex than that of ripple counters, but this disadvantage is becoming less significant with the evolution of integrated circuit technology.

9-6.1 Two-Bit Counters

Two-bit counters are divide-by-3 or divide-by-4 counters constructed from two FFs. In what follows we first describe divide-by-4 synchronous counters and then divide-by-3 synchronous counters.

9-6.1.1 Divide-by-4 Up Counter

Figure 9-24 shows the interconnection of two 1-bit counters of Fig. 9-20*a*. The resulting circuit is a 2-bit counter with a *CLOCK* input, a *CARRY-IN* input, and a *CARRY-OUT* output.

We return to Fig. 9-24 later, but for now we simplify it by making input *CARRY-IN* = 1. In this case the AND gate between the two FFs is not needed and the circuit of Fig. 9-25*a* results.

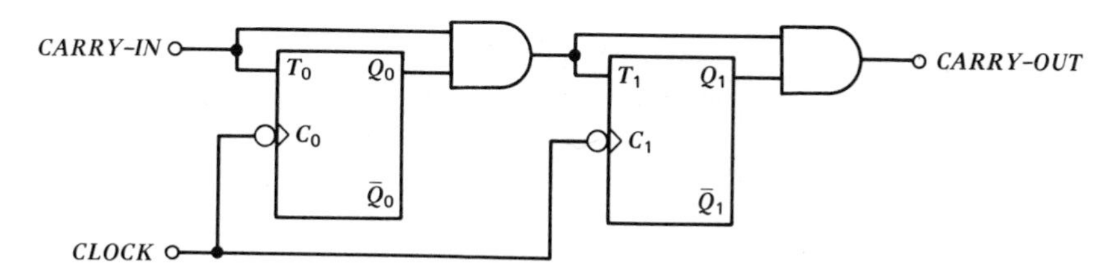

FIGURE 9-24 Interconnection of two 1-bit counters from Fig. 9-20*a*.

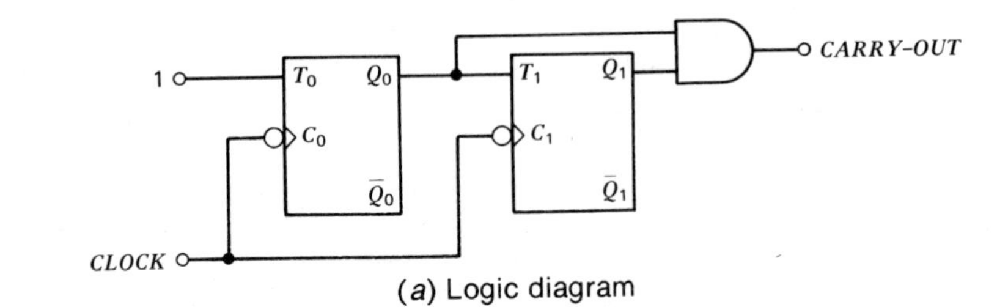

(a) Logic diagram

Q_{1_n}	Q_{0_n}	T_{1_n}	T_{0_n}	$Q_{1_{n+1}}$	$Q_{0_{n+1}}$	$CARRY\text{-}OUT_n$
0	0	0	1	0	1	0
0	1	1	1	1	0	0
1	0	0	1	1	1	0
1	1	1	1	0	0	1

(b) State table

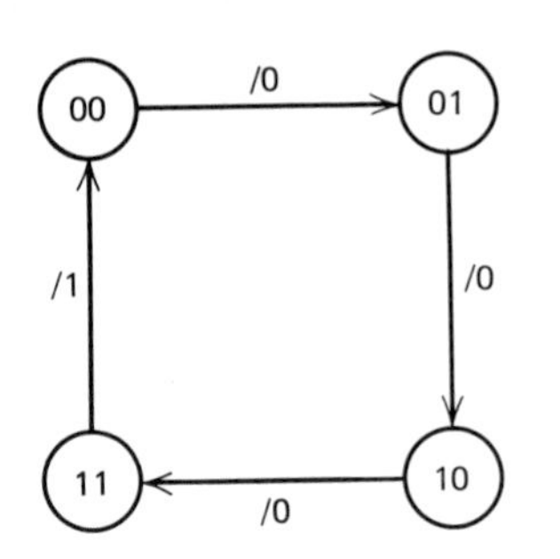

(c) State diagram with the branches labeled $/CARRY\text{-}OUT_n$

FIGURE 9-25 Divide-by-4 synchronous up counter.

The states of the two FFs in a divide-by-4 up counter change in the sequence 00-01-10-11-00 . . . ; hence the name *up* counter. In what follows we demonstrate that the circuit shown in Fig. 9-25*a* is a divide-by-4 up counter.

Figure 9-25*b* shows a state table for the circuit of Fig. 9-25*a*. It lists the four combinations of present states Q_{1_n} and Q_{0_n} in the leftmost two columns. The next two columns list present control inputs T_{1_n} and T_{0_n} in accordance with the circuit of Fig. 9-25*a* ($T_1 = Q_0$, $T_0 = 1$). The last two columns show next states $Q_{1_{n+1}}$ and $Q_{0_{n+1}}$. Next state $Q_{1_{n+1}}$ is obtained from Q_{1_n} and T_{1_n} in accordance with the function table of the T FF shown in Fig. 8-34*c* (page 198). Next state $Q_{0_{n+1}}$ is obtained similarly from Q_{0_n} and T_{0_n}.

Figure 9-25*c* shows a state diagram. The counter has no input other than the *CLOCK*; thus the branches are labeled $/CARRY\text{-}OUT_n$. The circle with 00 represents the $Q_1 = 0$, $Q_0 = 0$ state; the circle with 01 the $Q_1 = 0$, $Q_0 = 1$ state; the circle with 10 the $Q_1 = 1$, $Q_0 = 0$ state; and the circle with 11 the $Q_1 = 1$, $Q_0 = 1$ state. We can see that the counter counts up in the sequence 00-01-10-11-00 . . . ; hence, it is a divide-by-4 up counter.

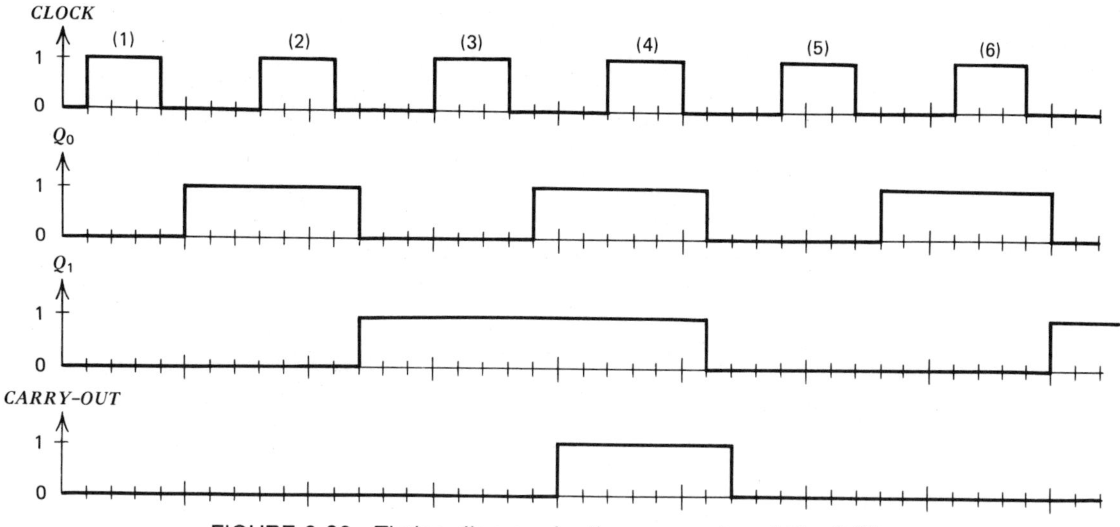

FIGURE 9-26 Timing diagram for the up counter of Fig. 9-25*a*.

Figure 9-26 shows a timing diagram for the counter of Fig. 9-25*a*; the propagation delay of each FF and each AND gate equals one small time division. We can see that, unlike in a ripple counter, the states of Q_0 and Q_1 change simultaneously following the negative-going transition of the common *CLOCK*. It is for this reason that the counters of Figs. 9-24 and 9-25*a*, as well as similar counters described later, are called synchronous counters.

9-6.1.2 Divide-by-4 Gray Counter

The states of the two FFs in a divide-by-4 Gray counter change in the sequence 00-01-11-10-00 . . . , that is, they follow a Gray code (see Sec. 4-3.1). Figure 9-27*a* shows a logic diagram for this counter.

A state table is shown in Fig. 9-27*b*. The leftmost two columns list the four present states in binary sequence. The next two columns show the control inputs: from Fig. 9-27*a* these are $T_{1_n} = Q_{1_n} \oplus Q_{0_n}$ and $T_{0_n} = Q_{1_n} \oplus \overline{Q}_{0_n}$. Next state $Q_{1_{n+1}}$ is obtained from Q_{1_n} and T_{1_n} in accordance with the function table of the T FF shown in Fig. 8-34*c* (page 198). Next state $Q_{0_{n+1}}$ is obtained similarly from Q_{0_n} and T_{0_n}. Output $CARRY\text{-}OUT_n = Q_{1_n} \cdot \overline{Q}_{0_n}$ according to Fig. 9-27*a*.

A state diagram is shown in Fig. 9-27*c* with the states identified as $Q_{1_n}Q_{0_n}$. The counter has no input other than the *CLOCK*; thus the branches are labeled $/CARRY\text{-}OUT_n$. The state diagram shows that the counter counts in the sequence 00-01-11-10-00 . . . , and it provides a *CARRY-OUT* = 1 in its 10 state.

The rows in the state table of Fig. 9-27*b* are arranged in the binary sequence 00-01-10-11 for the present states. However, the counter counts in the Gray code 00-01-11-10. Thus it is also useful to show a state table with the rows in this sequence, as in Fig. 9-27*d*.

Figure 9-27*e* shows a timing diagram for the Gray counter of Fig. 9-27*a*. The propagation delay of each FF and each gate equals one small time division.

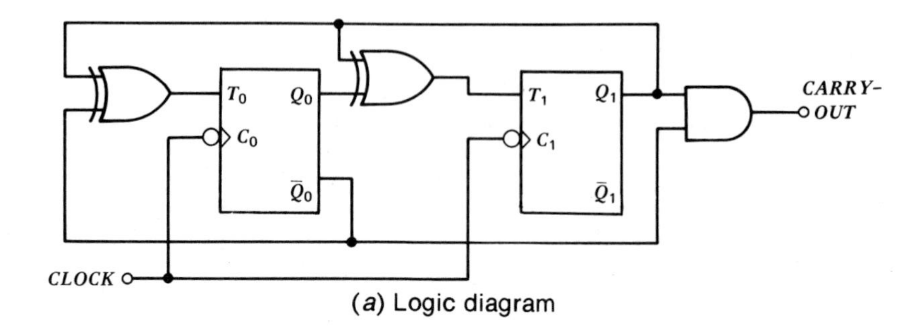

(*a*) Logic diagram

Q_{1_n}	Q_{0_n}	T_{1_n}	T_{0_n}	$Q_{1_{n+1}}$	$Q_{0_{n+1}}$	$CARRY\text{-}OUT_n$
0	0	0	1	0	1	0
0	1	1	0	1	1	0
1	0	1	0	0	0	1
1	1	0	1	1	0	0

(*b*) State table

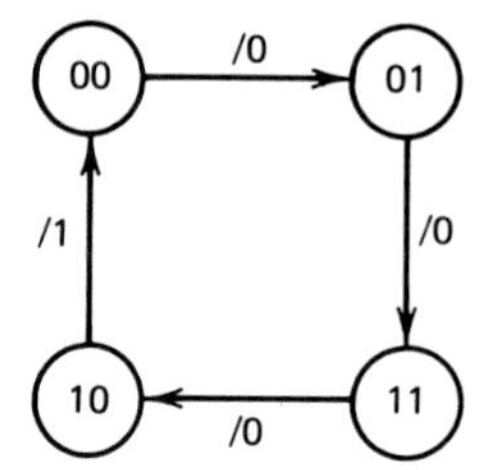

(*c*) State diagram with the branches labeled $/CARRY\text{-}OUT_n$

Q_{1_n}	Q_{0_n}	T_{1_n}	T_{0_n}	$Q_{1_{n+1}}$	$Q_{0_{n+1}}$	$CARRY\text{-}OUT_n$
0	0	0	1	0	1	0
0	1	1	0	1	1	0
1	1	0	1	1	0	0
1	0	1	0	0	0	1

(*d*) The state table of (*b*) with the sequence of rows rearranged

FIGURE 9-27 Divide-by-4 synchronous Gray counter.

9-6.1.3 Divide-by-4 Down Counter

The states of the two FFs in a divide-by-4 down counter change in the sequence 11-10-01-00-11 . . . , hence the name *down* counter. Figure 9-28*a* shows the circuit, and Fig. 9-28*b* a state table without control inputs $T_{1_n} = \overline{Q}_{0_n}$ and $T_{0_n} = 1$. Figure 9-28*c* shows a state diagram with the states identified as $Q_{1_n}Q_{0_n}$ and the branches labeled $/BORROW\text{-}OUT_n$. We can see that the

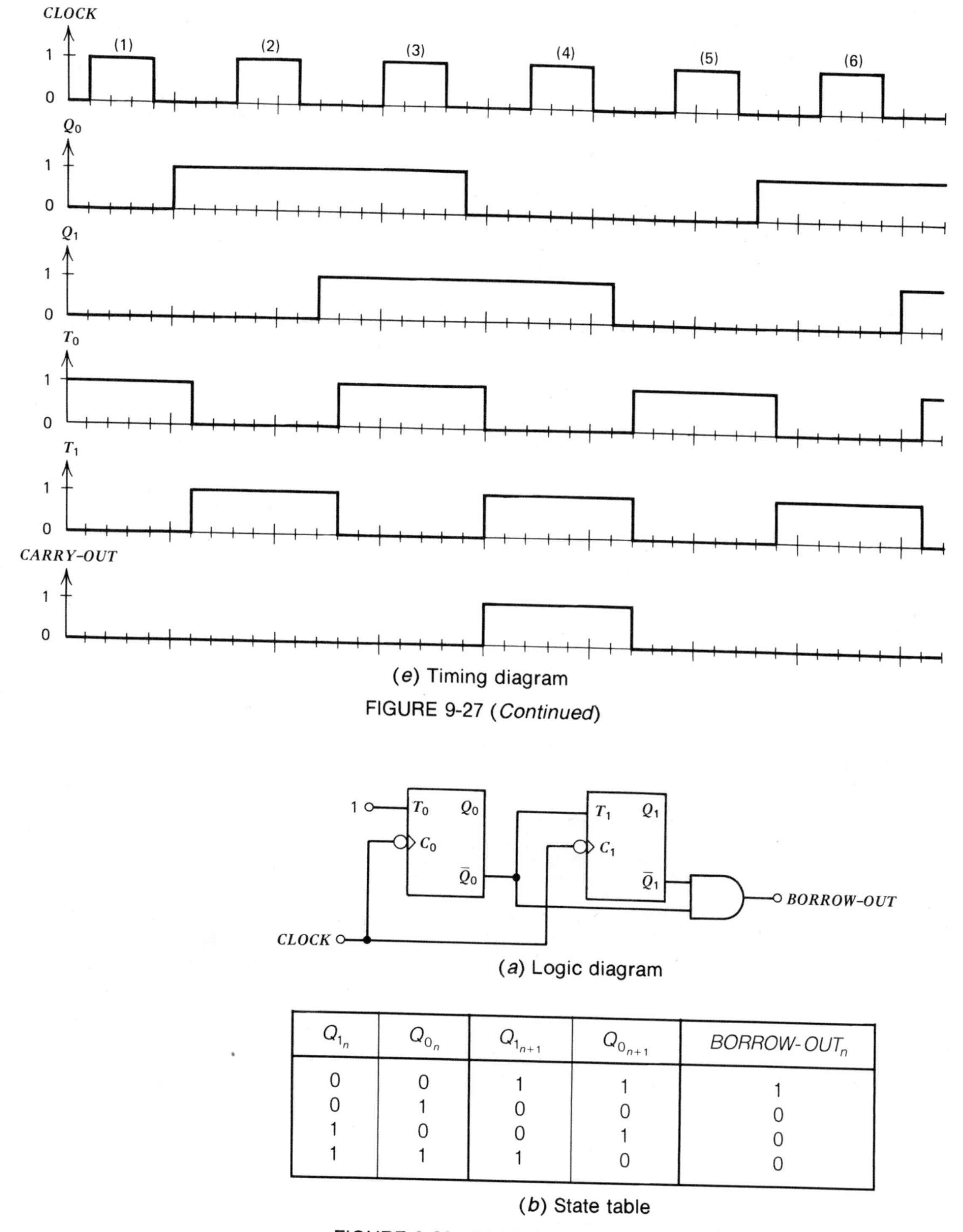

(*e*) Timing diagram

FIGURE 9-27 (*Continued*)

(*a*) Logic diagram

Q_{1_n}	Q_{0_n}	$Q_{1_{n+1}}$	$Q_{0_{n+1}}$	$BORROW\text{-}OUT_n$
0	0	1	1	1
0	1	0	0	0
1	0	0	1	0
1	1	1	0	0

(*b*) State table

FIGURE 9-28 Divide-by-4 synchronous down counter.

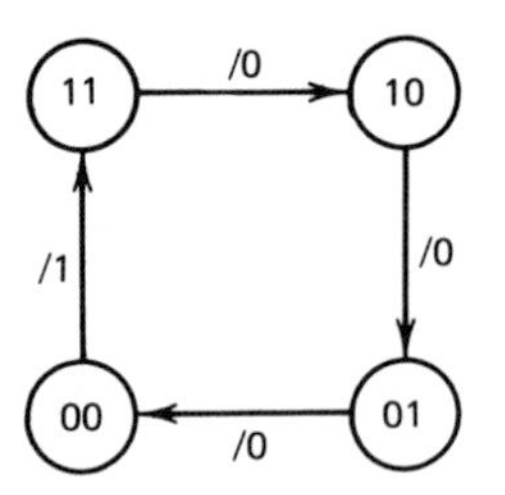

(c) State diagram with the branches labeled $/BORROW\text{-}OUT_n$

Q_{1_n}	Q_{0_n}	$Q_{1_{n+1}}$	$Q_{0_{n+1}}$	$BORROW\text{-}OUT_n$
1	1	1	0	0
1	0	0	1	0
0	1	0	0	0
0	0	1	1	1

(d) The state table of (b) with the sequence of rows rearranged

FIGURE 9-28 (*Continued*)

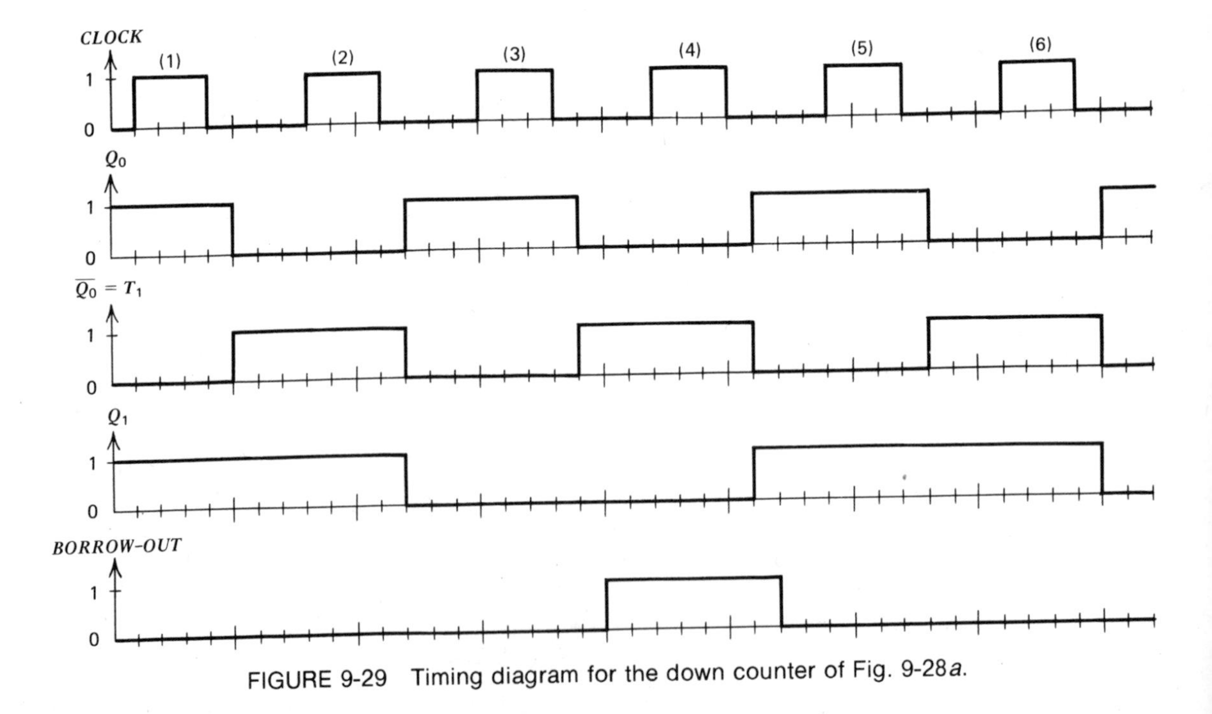

FIGURE 9-29 Timing diagram for the down counter of Fig. 9-28a.

counter indeed counts in the sequence 11-10-01-00-11 . . . ; also, it provides a *BORROW-OUT* = 1 in its 00 state. Figure 9-28*d* shows a state table with the sequence of the rows rearranged.

The operation of the down counter is illustrated in the timing diagram of Fig. 9-29. The propagation delay of each flip-flop and of the AND gate equals one small time division.

9-6.1.4 Divide-by-4 Up-Down Counter

Figure 9-30*a* shows a divide-by-4 *up-down counter*. The control inputs of the FFs are $T_0 = 1$ and $T_1 = UP \oplus \overline{Q}_0$. When input $UP = \overline{DOWN} = 1$, then control input $T_1 = \overline{Q}_0 \oplus 1 = \overline{Q}_0 \cdot 0 + Q_0 \cdot 1 = Q_0$, and the circuit of Fig. 9-30*a* re-

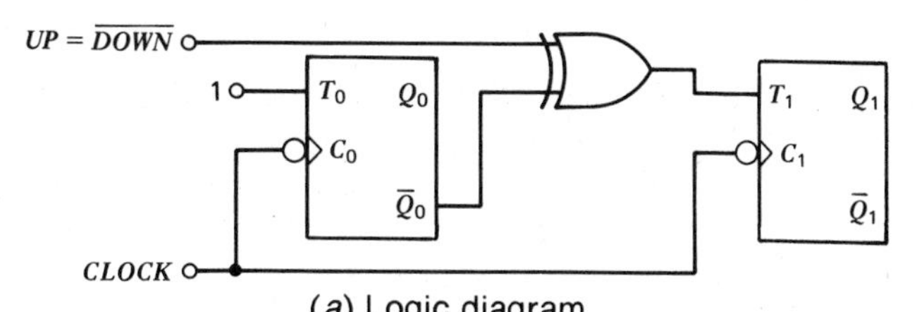

(*a*) Logic diagram

$UP_n = \overline{DOWN_n}$	Q_{1_n}	Q_{0_n}	T_{1_n}	$Q_{1_{n+1}}$	$Q_{0_{n+1}}$
0	0	0	1	1	1
0	0	1	0	0	0
0	1	0	1	0	1
0	1	1	0	1	0
1	0	0	0	0	1
1	0	1	1	1	0
1	1	0	0	1	1
1	1	1	1	0	0

(*b*) State table

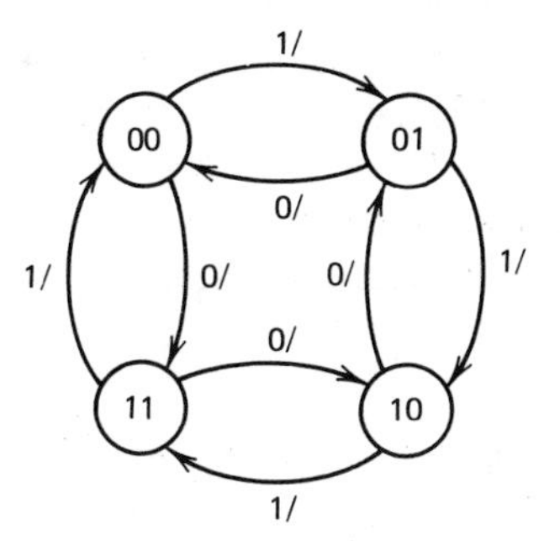

(*c*) State diagram with the branches labeled as UP_n /

FIGURE 9-30 Divide-by-4 synchronous up-down counter.

verts to the up counter of Fig. 9-25*a* with the *CARRY-OUT* omitted. When input $UP = \overline{DOWN} = 0$ (that is, $DOWN = 1$), then control input $T_1 = \overline{Q}_0 \oplus 0 = \overline{Q}_0 \cdot 1 + Q_0 \cdot 0 = \overline{Q}_0$ and the circuit of Fig. 9-30*a* reverts to the down counter of Fig. 9-28*a* with the *BORROW-OUT* omitted.

A state table is shown in Fig. 9-30*b*. It lists the eight combinations of the present states $UP_n = \overline{DOWN}_n$, Q_{1_n}, and Q_{0_n}, as well as the resulting T_{1_n} and next states $Q_{1_{n+1}}$ and $Q_{0_{n+1}}$.

Figure 9-30*c* shows a state diagram with the states identified as $Q_{1_n}Q_{0_n}$ and the branches labeled as $UP_n/$. When $UP_n = \overline{DOWN}_n = 1$, the counter counts *up* along the outer branches in accordance with the timing diagram of Fig. 9-26. When $UP_n = \overline{DOWN}_n = 0$, that is, when $DOWN_n = 1$, the counter counts *down* along the inner branches in accordance with the timing diagram of Fig. 9-29.

9-6.1.5 Self-Starting Divide-by-3 Counter

The states of the two FFs in a divide-by-3 counter change in the sequence 00-01-10-00. . . . The fourth state, state 11, is not used in the counting. In a *self-starting* divide-by-3 counter, there is a transition from the unused state 11 to at least one of the other states. In a non-self-starting divide-by-3 counter, there is no transition from the unused state 11 to any of the other states. A 2-bit synchronous counter is shown in Fig. 9-31*a*; in what follows we demonstrate that it is a self-starting divide-by-3 counter.

Figure 9-31*b* shows a state table for the counter of Fig. 9-31*a*. The control inputs of the two FFs are $J_1 = Q_0$, $K_1 = 1$, $J_0 = \overline{Q}_1$, and $K_0 = 1$, in accordance with Fig. 9-31*a*. Next, states $Q_{1_{n+1}}$ and $Q_{0_{n+1}}$ are found with the aid of the J-K FF function table, Fig. 8-33*d* (page 197).

The first row in the state table of Fig. 9-31*b* shows a transition from state 00 to state 01, the second row from state 01 to state 10, and the third row from state 10 to state 00. Thus we indeed have a divide-by-3 counter. Furthermore, the fourth row in the state table shows a transition from the unused state 11 to state 00. Since state 00 is part of the normal counting sequence, we conclude that the counter is self-starting.

The information of the state table of Fig. 9-31*b* is also shown in the state diagram of Fig. 9-31*c*, where the states are identified as $Q_{1_n}Q_{0_n}$. There is no input to the circuit other than the clock, and no output other than FF outputs Q_0 and Q_1. Thus there are no labels on the branches in the state diagram.

Figure 9-31*d* shows a timing diagram for the counter of Fig. 9-31*a* with the propagation delay of each FF equal to one small time division. The counter is initially in unallowed state 11, from which it makes a transition to state 00, and from there on it counts in the sequence 00-01-10-00. . . .

9-6.1.6 Non-Self-Starting Divide-by-3 Counter

Figure 9-32*a* shows a 2-bit synchronous counter. In what follows we demonstrate that it is a non-self-starting divide-by-3 counter with a normal counting sequence 00-01-10-00. . . , and with an unused state 11 that has no transition to any of the other states.

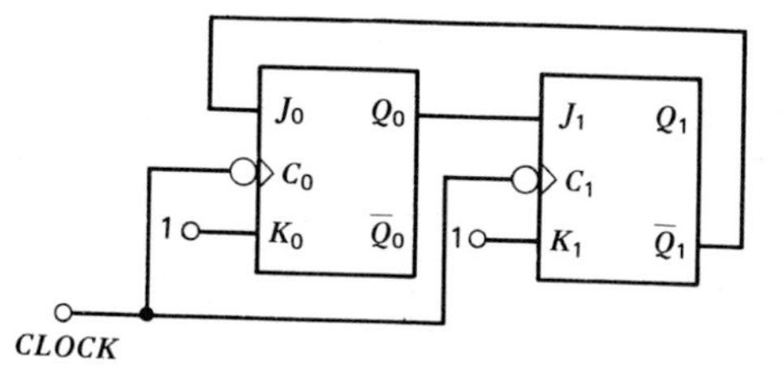

(*a*) Logic diagram

Q_{1_n}	Q_{0_n}	J_{1_n}	K_{1_n}	J_{0_n}	K_{0_n}	$Q_{1_{n+1}}$	$Q_{0_{n+1}}$
0	0	0	1	1	1	0	1
0	1	1	1	1	1	1	0
1	0	0	1	0	1	0	0
1	1	1	1	0	1	0	0

(*b*) State table

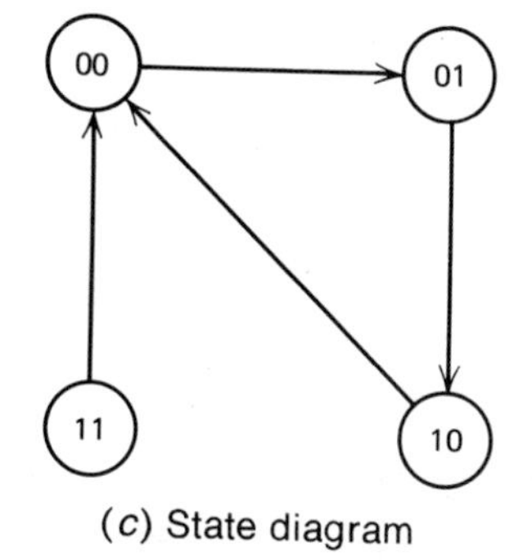

(*c*) State diagram

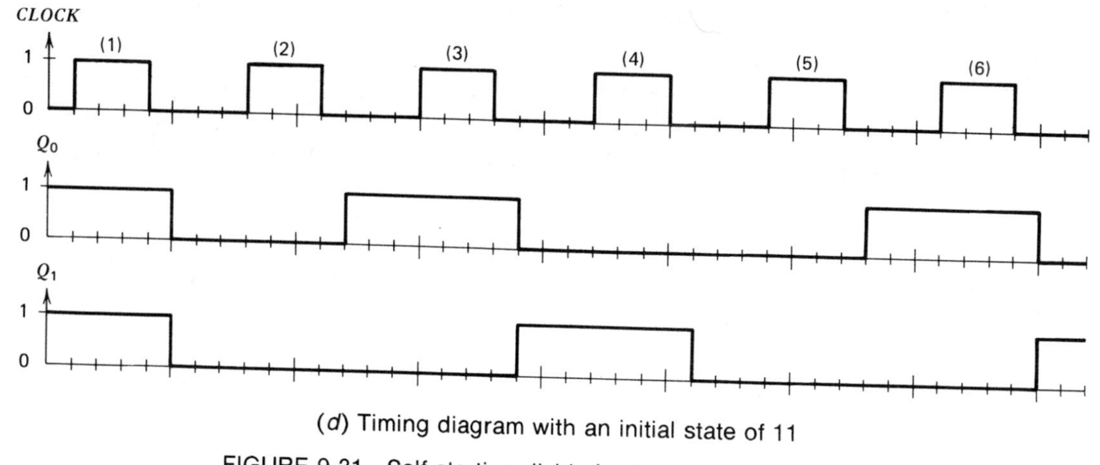

(*d*) Timing diagram with an initial state of 11

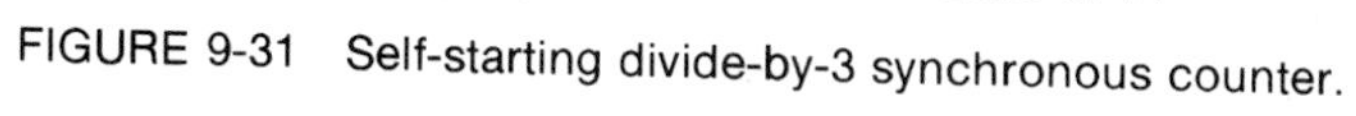
FIGURE 9-31 Self-starting divide-by-3 synchronous counter.

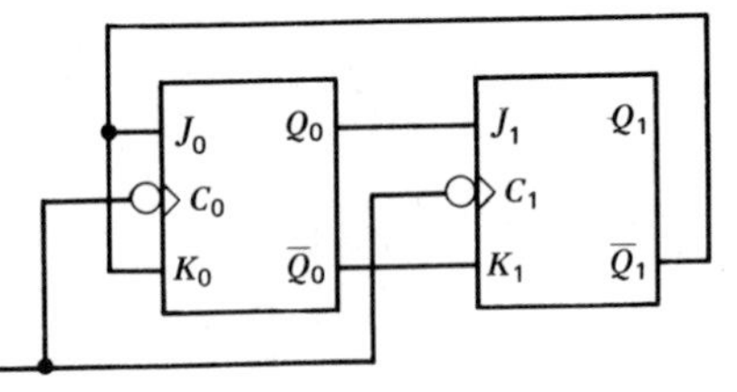

(*a*) Logic diagram

Q_{1_n}	Q_{0_n}	J_{1_n}	K_{1_n}	J_{0_n}	K_{0_n}	$Q_{1_{n+1}}$	$Q_{0_{n+1}}$
0	0	0	1	1	1	0	1
0	1	1	0	1	1	1	0
1	0	0	1	0	0	0	0
1	1	1	0	0	0	1	1

(*b*) State table

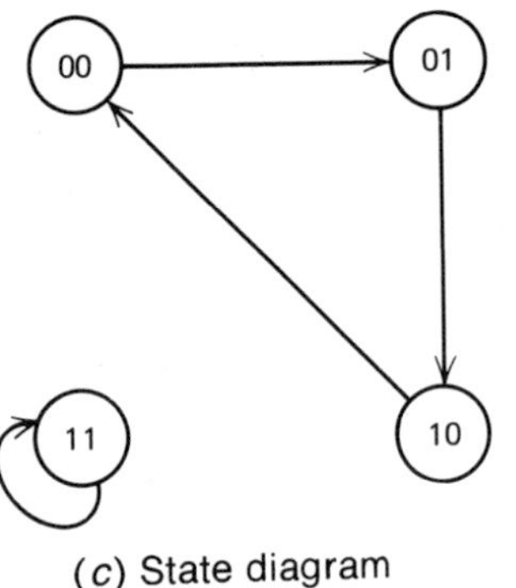

(*c*) State diagram

FIGURE 9-32 Non-self-starting divide-by-3 synchronous counter illustrating a poor design.

Figure 9-32*b* shows a state table for the counter of Fig. 9-32*a*. The first three rows in the state table show that the counting sequence is indeed 00-01-10-00. . . . However, the fourth row in the state table shows a transition from the unused 11 state back to itself. Thus once the counter enters state 11, it will remain there. Therefore the counter is not self-starting.

The information in the state table of Fig. 9-32*b* is also shown in the state diagram of Fig. 9-32*c*, where the states are identified as $Q_{1_n}Q_{0_n}$. There are no labels on the branches, since the only input to the circuit is the clock and the only outputs are FF outputs Q_0 and Q_1.

Consider next what happens in the case of an intermittent loss of power. When the power returns, the counter may enter any state including the unused state 11. Since there is no way out of state 11 ("*hang-up state*"), the counter remains there no matter how many pulses it receives at its *CLOCK* input. Such apparent malfunction may be unacceptable in many applications.

A commonly used remedy for this problem is to clear all FFs via their direct *CLEAR* inputs after power turn-on. When this can be done *manually*, it safely starts the counter in its zero state, and no unallowed state is entered. However, when such start-up systems are automated, they do not always work. This is because such a system may get confused and may not reset the counter to its zero state when the power is off only for a short time.

Thus we conclude that the use of a non-self-starting counter is poor practice. The same can be said of any other digital system that has no transition from an unused state leading back to the normal operating states.

9-6.2 Three-Bit Counters

Three-bit counters can be divide-by-5, divide-by-6, divide-by-7, or divide-by-8 counters. In what follows here, we describe divide-by-8 and divide-by-5 synchronous up counters. Logic diagrams of divide-by-6 and divide-by-7 synchronous up counters are given, respectively, in Probs. 9-11 and 9-12.

9-6.2.1 Divide-by-8 Up Counters

Figure 9-33*a* shows the interconnection of three 1-bit counters from Fig. 9-20*a*. It is a 3-bit counter with a *CLOCK* input, a *CARRY-IN* input, and a *CARRY-OUT* output.

We make two changes in the circuit of Fig. 9-33*a*: we permanently connect the *CARRY-IN* to logic 1, and we rearrange the logic gates to reduce propagation delays from the FFs to *CARRY-OUT*. The resulting circuit is shown in Fig.

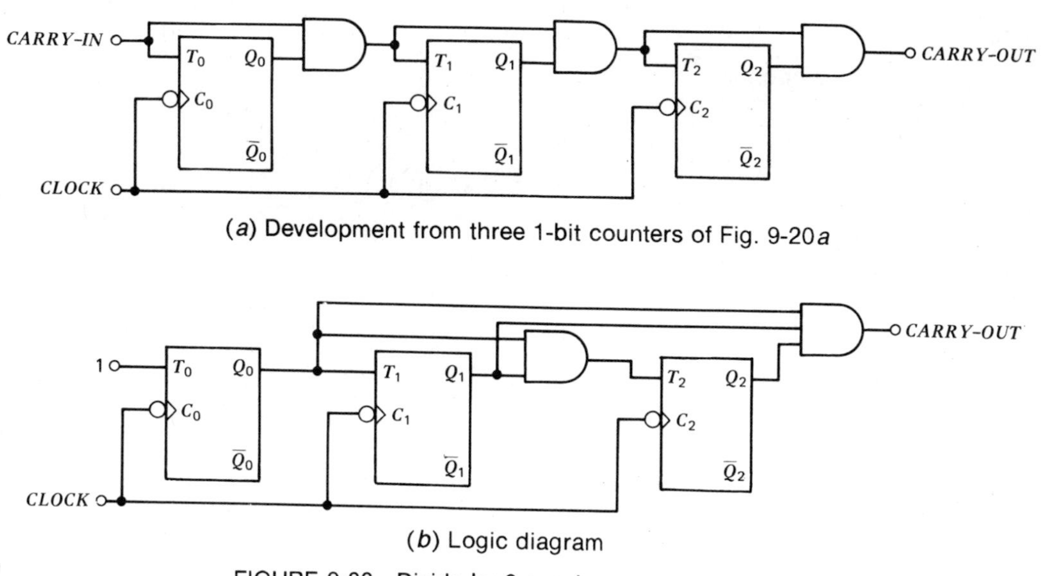

(*a*) Development from three 1-bit counters of Fig. 9-20*a*

(*b*) Logic diagram

FIGURE 9-33 Divide-by-8 synchronous up counter.

Present States			Present Control Inputs			Next States			Present Output
Q_{2_n}	Q_{1_n}	Q_{0_n}	T_{2_n}	T_{1_n}	T_{0_n}	$Q_{2_{n+1}}$	$Q_{1_{n+1}}$	$Q_{0_{n+1}}$	$CARRY\text{-}OUT_n$
0	0	0	0	0	1	0	0	1	0
0	0	1	0	1	1	0	1	0	0
0	1	0	0	0	1	0	1	1	0
0	1	1	1	1	1	1	0	0	0
1	0	0	0	0	1	1	0	1	0
1	0	1	0	1	1	1	1	0	0
1	1	0	0	0	1	1	1	1	0
1	1	1	1	1	1	0	0	0	1

FIGURE 9-34 State table for the counter of Fig. 9-33*b*.

9-33*b*. We now demonstrate that the counter of Fig. 9-33*b* is a divide-by-8 synchronous up counter.

Figure 9-34 shows a state table for the counter of Fig. 9-33*b*. The three present states Q_{2_n}, Q_{1_n}, and Q_{0_n} have eight possible combinations. These are listed in increasing binary sequence under Present States in Fig. 9-34. The next three columns list Present Control Inputs T_{2_n}, T_{1_n}, and T_{0_n} in accordance with Fig. 9-33*b* ($T_2 = Q_1Q_0$, $T_1 = Q_0$, and $T_0 = 1$). The columns $Q_{2_{n+1}}$, $Q_{1_{n+1}}$, and $Q_{0_{n+1}}$ list the Next States obtained with the function table of the T FF shown in Fig. 8-34*c* (page 198).

Figure 9-35 shows a state diagram deduced from the state table of Fig. 9-34. The states are identified by the decimal equivalents of $Q_{2_n}Q_{1_n}Q_{0_n}$; the branches are labeled $/CARRY\text{-}OUT_n$. We can see that the counter counts *up* in a sequence of 0-1-2-3-4-5-6-7-0 . . . , that is, in a binary sequence of 000-001-010-011-100-101-110-111-000 . . . . Thus the counter is indeed a divide-by-8 up counter.

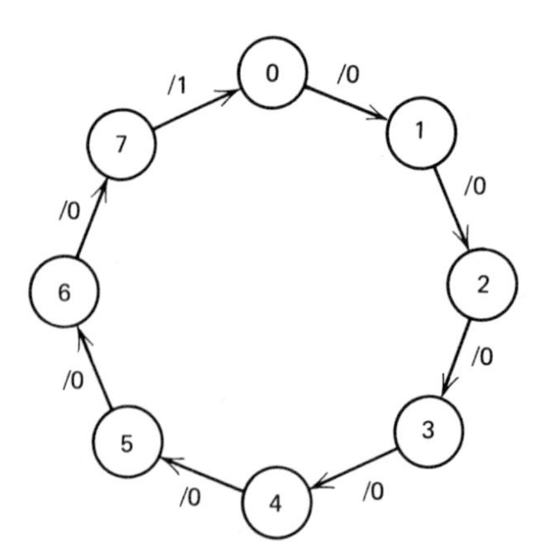

FIGURE 9-35 State diagram for the counter of Fig. 9-33*b*.

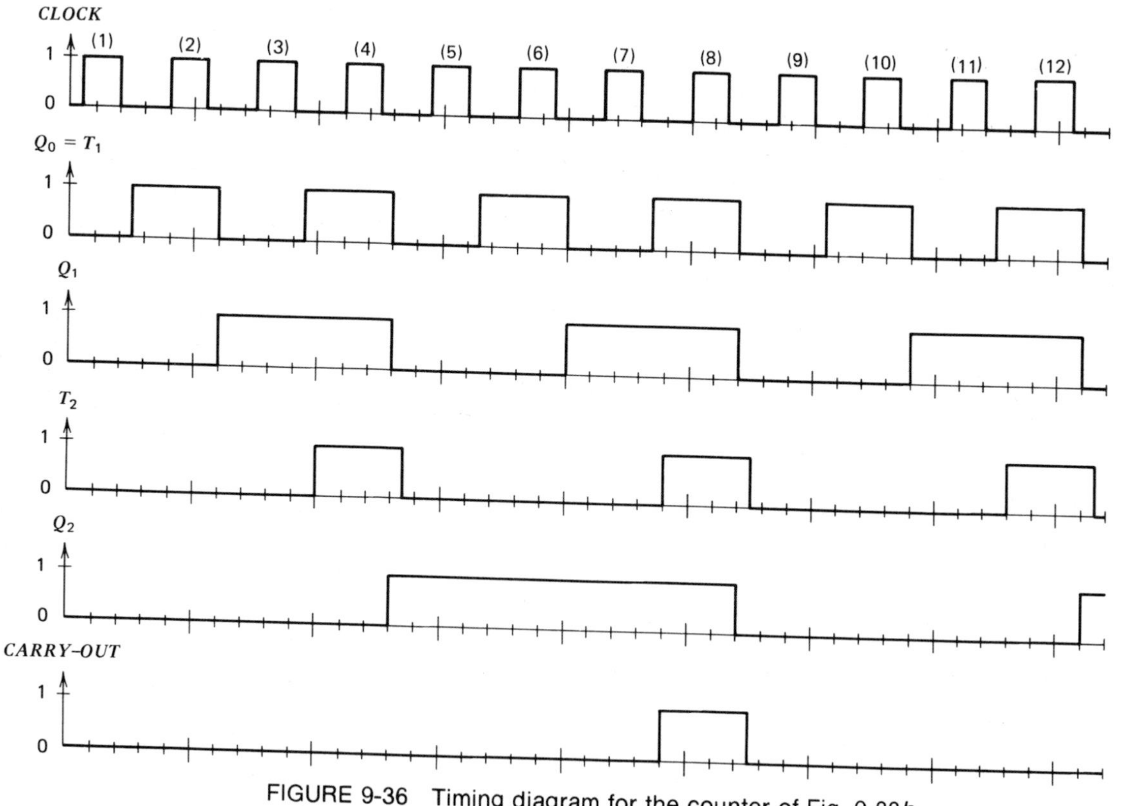

FIGURE 9-36 Timing diagram for the counter of Fig. 9-33*b*.

Figure 9-36 shows a timing diagram for the counter of Fig. 9-33*b*. The initial states of the FFs are arbitrarily set as $Q_2 = 0$, $Q_1 = 0$, and $Q_0 = 0$. The propagation delay of each FF and each gate equals one-half of a small time division.

9-6.2.2 Divide-by-5 Up Counters

The importance of divide-by-5 counters is due to their use as a building block in divide-by-10 counters. The simplest divide-by-5 synchronous up counter can be built by use of J-K FFs as shown in Fig. 9-37*a*. The first two FFs can be either T FFs as shown, or each T FF can be substituted by a J-K FF with control inputs *J* and *K* connected in parallel to form a *T* input. However, the rightmost FF has to be J-K.

Figure 9-37*b* shows a state table for the counter of Fig. 9-37*a*. The Present States list the eight possible present states: the five states participating in the normal counting sequence above the broken line, and the three unused states below the broken line. The Present Control Inputs list the control inputs as obtained from Fig. 9-37*a* ($J_2 = Q_1Q_0$, $K_2 = 1$, $T_1 = Q_0$, $T_0 = \overline{Q}_2$). The Next States list $Q_{2_{n+1}}$, $Q_{1_{n+1}}$, $Q_{0_{n+1}}$ and the resulting decimal equivalents. These next states were obtained using the function tables of the J-K and T FFs from Fig. 8-33*d* (page 197) and Fig. 8-34*c* (page 198), respectively.

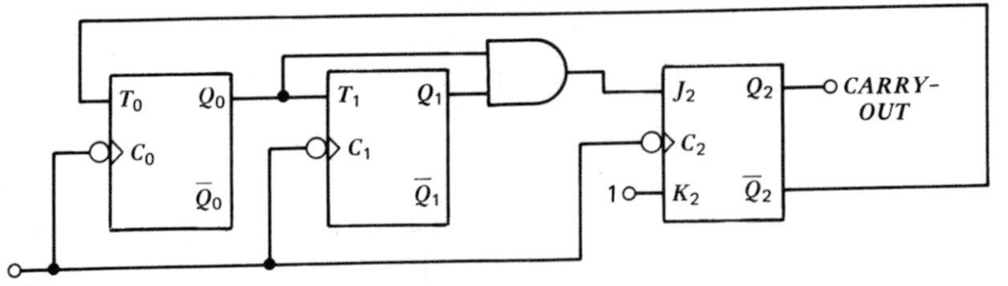

(*a*) Logic diagram

Present States				Present Control Inputs				Next States			
Decimal	Q_{2_n}	Q_{1_n}	Q_{0_n}	J_{2_n}	K_{2_n}	T_{1_n}	T_{0_n}	$Q_{2_{n+1}}$	$Q_{1_{n+1}}$	$Q_{0_{n+1}}$	Decimal
0	0	0	0	0	1	0	1	0	0	1	1
1	0	0	1	0	1	1	1	0	1	0	2
2	0	1	0	0	1	0	1	0	1	1	3
3	0	1	1	1	1	1	1	1	0	0	4
4	1	0	0	0	1	0	0	0	0	0	0
5	1	0	1	0	1	1	0	0	1	1	3
6	1	1	0	0	1	0	0	0	1	0	2
7	1	1	1	1	1	1	0	0	0	1	1

(*b*) State table

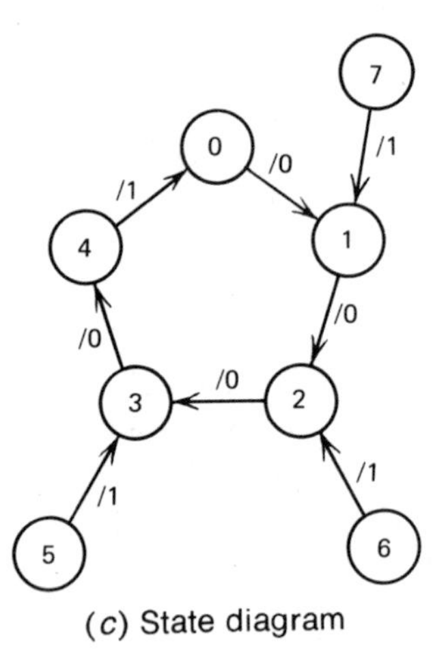

(*c*) State diagram

FIGURE 9-37 Divide-by-5 synchronous up counter using two T FFs and one J-K FF.

Figure 9-37*c* shows a state diagram based on the state table of Fig. 9-37*b*. As before, the states are identified by the decimal equivalents of $Q_{2_n}Q_{1_n}Q_{0_n}$, and the branches are labeled $/CARRY\text{-}OUT_n$. We can see that we indeed have a divide-by-5 up counter. Also, it is a self-starting counter, since all unused states have transitions back to the normal counting sequence.

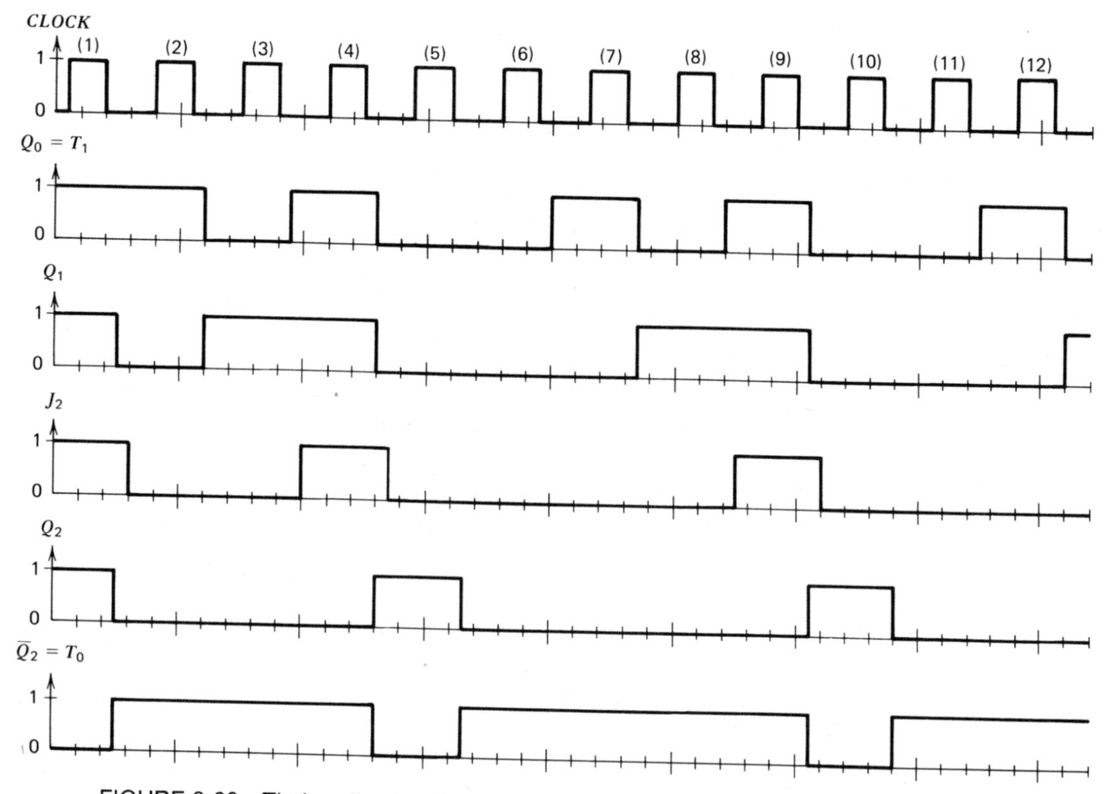

FIGURE 9-38 Timing diagram for the counter of Fig. 9-37*a* with an initial state of 7.

Figure 9-38 shows a timing diagram for the counter of Fig. 9-37*a*. The initial state of the counter is arbitrarily set to unallowed state 7, corresponding to $Q_2 = 1$, $Q_1 = 1$, and $Q_0 = 1$. The propagation delay of each FF and each gate equals one-half of a small time division.

9-6.3 Four-Bit Counters

This section describes basic divide-by-16 and divide-by-10 synchronous counters. Additional 4-bit counters are described in Secs. 9-6.4 through 9-7.

9-6.3.1 Divide-by-16 Up Counters

A divide-by-16 synchronous up counter can be assembled from four 1-bit counters of Fig. 9-20*a*, as shown in Fig. 9-39*a*. The same counter is shown in Fig. 9-39*b* with *CARRY-IN* = 1 and the logic gates rearranged to obtain shorter propagation delays from the FFs to *CARRY-OUT*. A state diagram for the counter of Fig. 9-39*b* is shown in Fig. 9-39*c*.

The T FFs in Fig. 9-39*b* can be directly replaced J-K FFs with their *J* and *K* inputs connected together as a *T* input. However, the configuration is somewhat different with D FFs, as illustrated in Ex. 9-7, which follows.

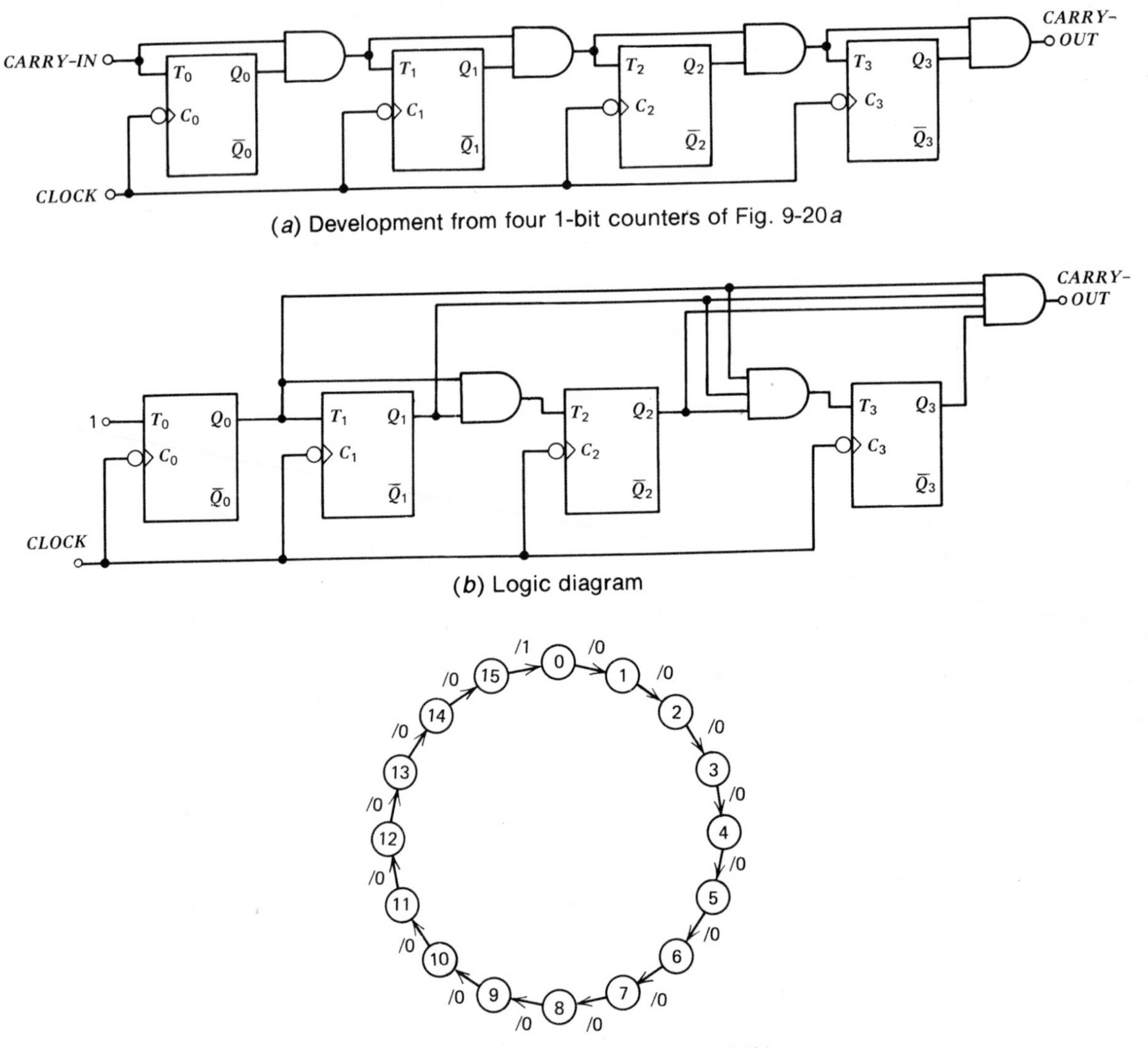

(*a*) Development from four 1-bit counters of Fig. 9-20*a*

(*b*) Logic diagram

(*c*) State diagram for the counter of (*b*)

FIGURE 9-39 Divide-by-16 up counter.

EXAMPLE 9-7

The CMOS Type 4520B IC package consists of two identical divide-by-16 synchronous up counters. Figure 9-40 shows a simplified logic diagram of one counter (in a real counter the *CLOCK* input is preceded by a logic inverter). The counter consists of four negative edge-triggered D FFs and has a *CLOCK* input, an *ENABLE* input, and a direct *CLEAR* input. The four FF outputs Q_0, Q_1, Q_2, and Q_3 are also available. However, no *CARRY-OUT* is available, because of the limitations in the number of available pins.

Show that the circuit of Fig. 9-40 is indeed a divide-by-16 synchronous up counter.

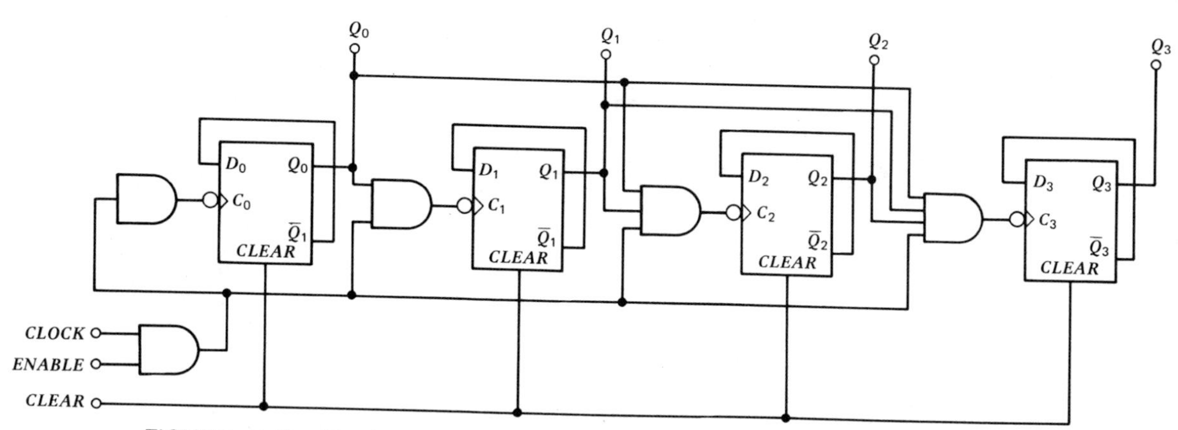

FIGURE 9-40 Simplified logic diagram of one-half of a CMOS Type 4520B divide-by-16 synchronous up counter. (In the real counter the *CLOCK* input is preceded by a logic inverter.)

Solution

The states of all four FFs change at the same time; this happens two gate propagation delays after the negative-going edge of the *CLOCK* waveform. Thus the counter is synchronous.

Each D FF is connected as a $T = 1$ FF in a manner shown in Fig. 9-2*c*. Each FF is activated on the return of its clock input from logic 1 to logic 0. The logic equations for the clock inputs of the FFs can be written as

$$C_0 = CLOCK \cdot T_0 \qquad C_1 = CLOCK \cdot T_1$$
$$C_2 = CLOCK \cdot T_2 \qquad C_3 = CLOCK \cdot T_3$$

where

$$T_0 = ENABLE \qquad T_1 = ENABLE \cdot Q_0$$
$$T_2 = ENABLE \cdot Q_0 \cdot Q_1, \qquad T_3 = ENABLE \cdot Q_0 \cdot Q_1 \cdot Q_2$$

Note that when $ENABLE = 1$, then the *T logic functions* T_0, T_1, T_2, and T_3 are the same as in Fig. 9-39*b*.

Thus, for example, clock input C_3 of the rightmost FF becomes logic 1 when $ENABLE = 1$, $T_3 = 1$, and $CLOCK = 1$, and it subsequently changes to logic 0 when the *CLOCK* input becomes logic 0. Since the FF is negative edge-triggered, it complements on this negative-going transition of the *CLOCK*. Similar considerations apply to the other FFs.

Hence, each FF complements following the negative-going transition of the *CLOCK* input if, and only if, its *T* logic function (T_0, T_1, T_2, or T_3) is at logic 1. These *T* logic functions are the correct logic functions for operation as a divide-by-16 up counter. Therefore, the counter of Fig. 9-40 is indeed a divide-by-16 synchronous up counter.

9-6.3.2 Divide-by-16 Up-Down Counter

The logic diagram of a divide-by-4 synchronous up-down counter was shown in Fig. 9-30*a*. The logic diagram of a divide-by-16 synchronous up-down counter

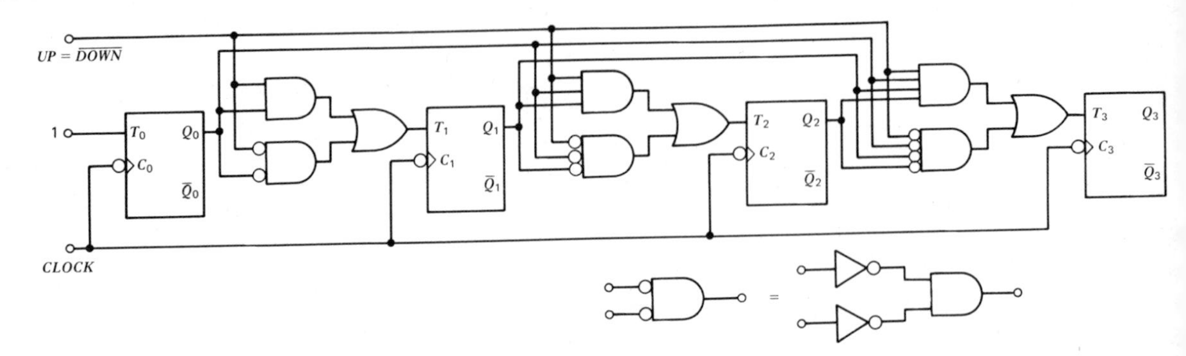

(*a*) Logic diagram

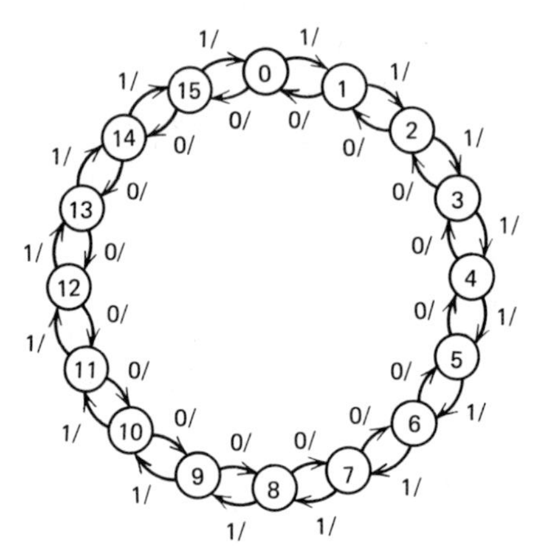

(*b*) State diagram

FIGURE 9-41 Divide-by-16 synchronous up-down counter.

is shown in Fig. 9-41*a*. We can see that when counting up, control input $UP = 1$, and the resulting control inputs of the FFs are $T_0 = 1$, $T_1 = Q_0$, $T_2 = Q_1Q_0$, and $T_3 = Q_2Q_1Q_0$, which are the same as in an up counter. Also, when counting down, $UP = \overline{DOWN} = 0$, and we have $T_0 = 1$, $T_1 = \overline{Q}_0$, $T_2 = \overline{Q}_1\overline{Q}_0$, and $T_3 = \overline{Q}_2\overline{Q}_1\overline{Q}_0$, which are the same as in a down counter.

A state diagram is shown in Fig. 9-41*b*. The states are identified by the decimal equivalents of $Q_{3_n}Q_{2_n}Q_{1_n}Q_{0_n}$ and the branches are labeled $UP_n/$.

9-6.3.3 Divide-by-10 Up Counters

A divide-by-10 synchronous up counter can be assembled from a divide-by-5 synchronous counter and a divide-by-2 synchronous counter. This can be done in two different ways: the divide-by-5 counter can precede or it can follow the divide-by-2 counter. The former is a *biquinary* counter, the latter is a *binary-coded-decimal*, or *BCD* counter.

Figure 9-42*a* shows a synchronous biquinary counter assembled from the divide-by-5 counter of Fig. 9-37*a* followed by the divide-by-2 counter of Fig.

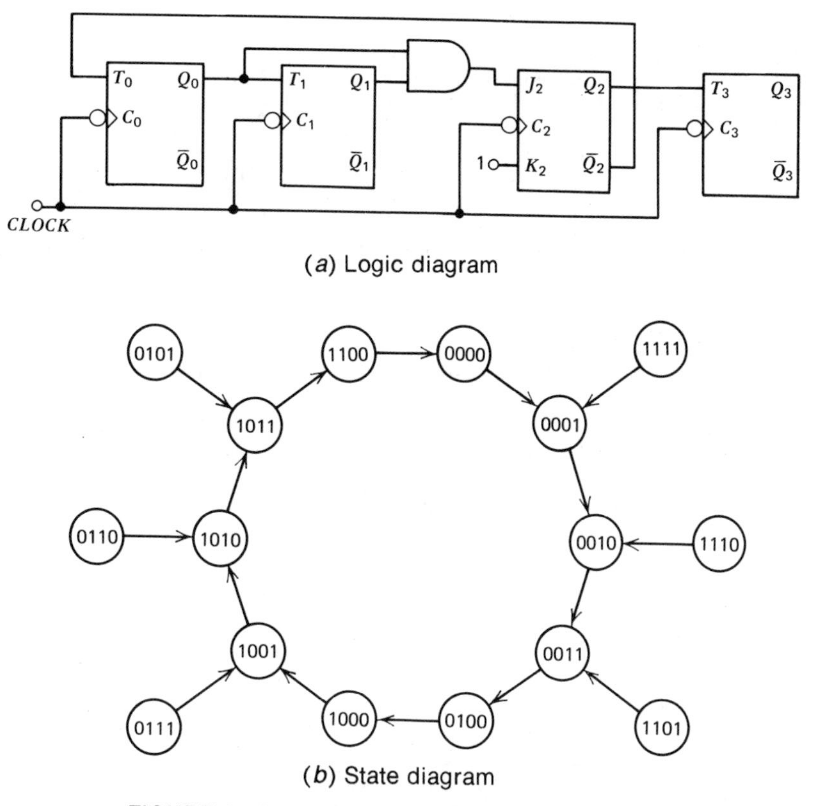

(*a*) Logic diagram

(*b*) State diagram

FIGURE 9-42 Synchronous biquinary up counter.

9-20*a* without the *CARRY-OUT*. Figure 9-42*b* shows a state diagram with the states identified as $Q_{3_n}Q_{2_n}Q_{1_n}Q_{0_n}$. We can see that the counting sequence is 0000-0001-0010-0011-0100-1000-1001-1010-1011-1100-0000. . . . This is the biquinary counting sequence; it is different from the BCD counting sequence. The state diagram also shows that all unused states have transitions back to the main counting sequence; thus the counter is self-starting.

Figure 9-43 shows a timing diagram for the counter of Fig. 9-42*a* with an initial state of 0000. We can see that the output waveform at Q_3 is a *symmetrical* square wave, which is advantageous in many applications. A disadvantage of the biquinary counter is its coding, which is weighted as 5-4-2-1.

Next we assemble a synchronous *BCD* up counter by preceding a divide-by-5 counter that consists of three T FFs (not discussed previously) by the divide-by-2 counter of Fig. 9-19*a*. The result is shown in Fig. 9-44. In what follows here, we present without proof a state diagram and a timing diagram for the counter of Fig. 9-44.

Figure 9-45 shows a state diagram for the counter of Fig. 9-44: it describes a self-starting BCD up counter. A timing diagram is shown in Fig. 9-46. It assumes a 0 initial state for each FF and invisibly short propagation delays. We can see that the output waveform at Q_3 is an asymmetric square wave.

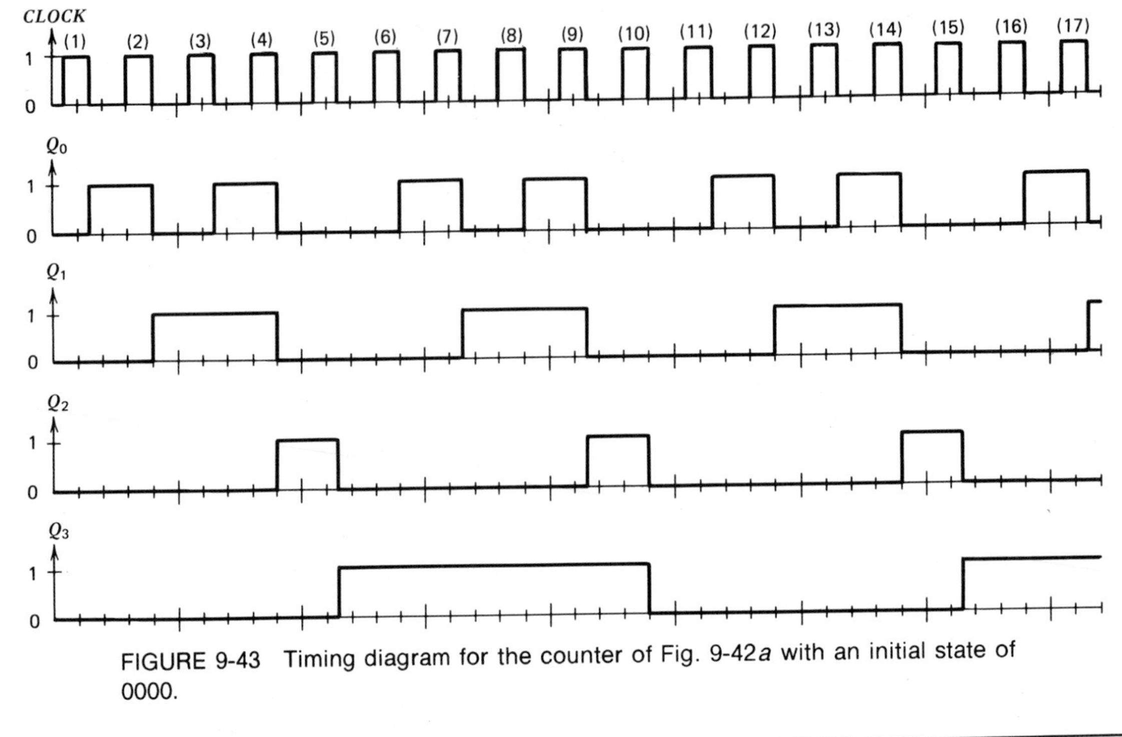

FIGURE 9-43 Timing diagram for the counter of Fig. 9-42*a* with an initial state of 0000.

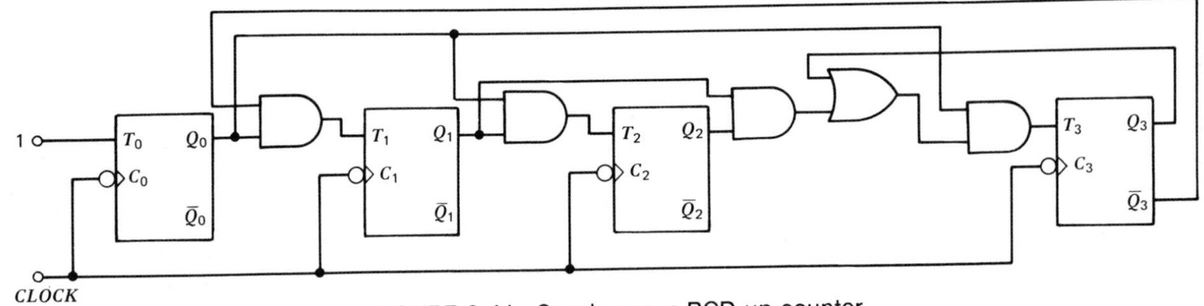

FIGURE 9-44 Synchronous BCD up counter.

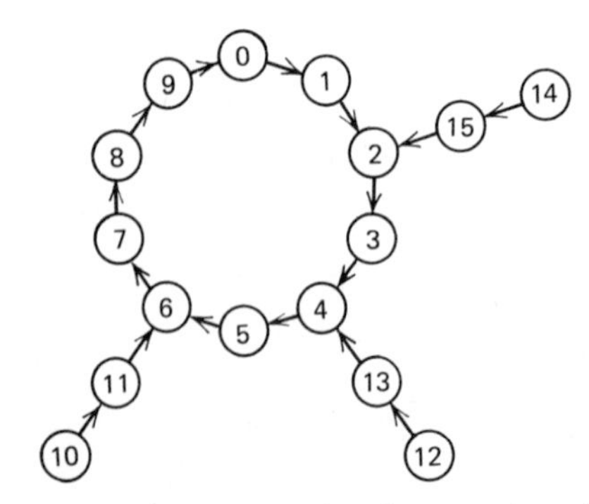

FIGURE 9-45 State diagram for the counter of Fig. 9-44.

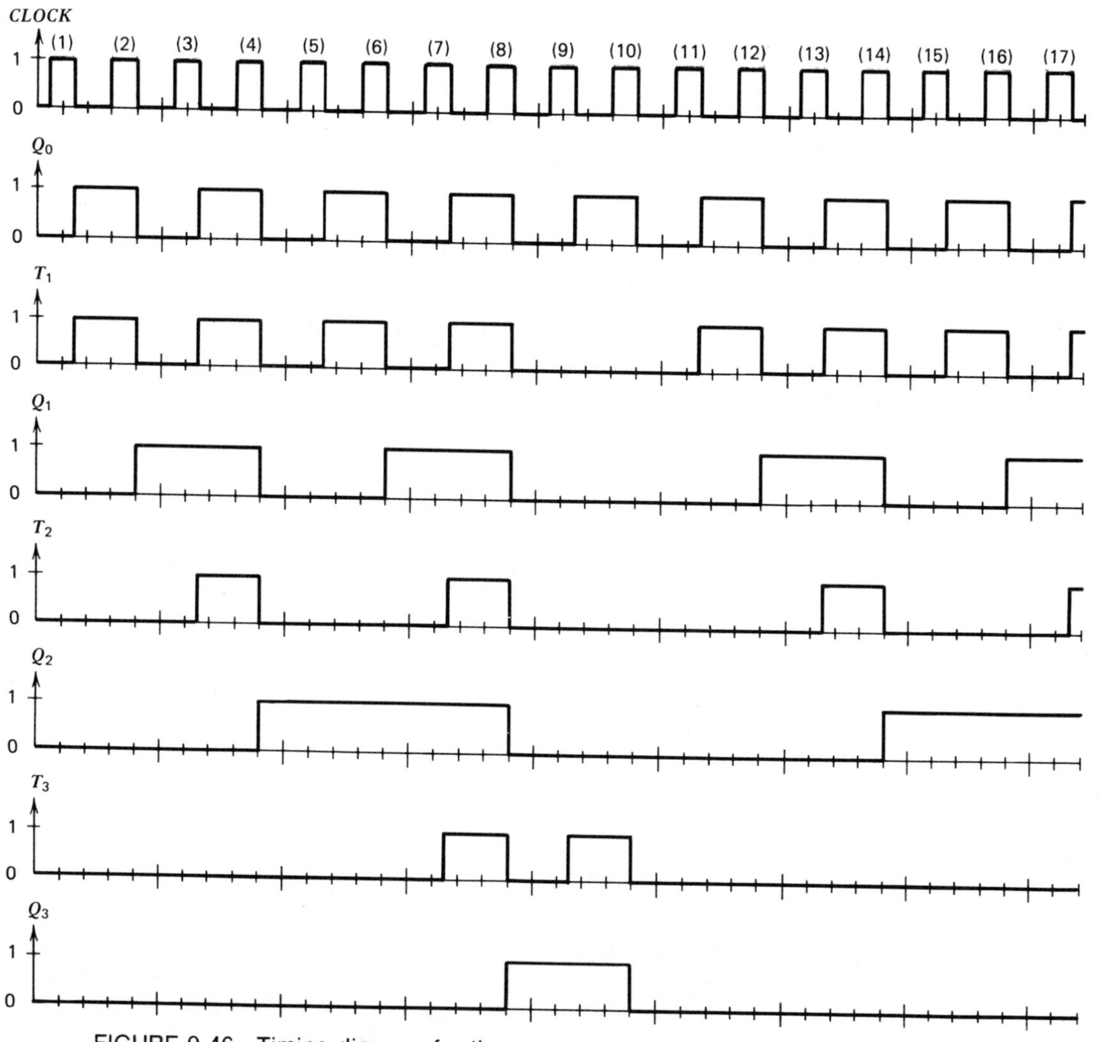

FIGURE 9-46 Timing diagram for the counter of Fig. 9-44 with an initial state of 0.

Note that today's manufacturing technology realizes T FFs by J-K FFs in TTL, and by D FFs in ECL, *n*-channel and *p*-channel MOS, and CMOS.

EXAMPLE 9-8

The CMOS Type 4518B IC package consists of two identical divide-by-10 synchronous up counters. Figure 9-47 shows a simplified logic diagram of one counter (in a real counter the *CLOCK* input is preceded by a logic inverter). Show that the counter of Fig. 9-47 is indeed a divide-by-10 synchronous up counter, and that it is self-starting.

Solution

Each D FF is connected as a T FF in accordance with Fig. 9-2*c*. Also, the clock input of each FF in Fig. 9-47 is driven by the same logic function as the corresponding *T* input in Fig. 9-44, but with *ENABLE* · *CLOCK* ANDed. These

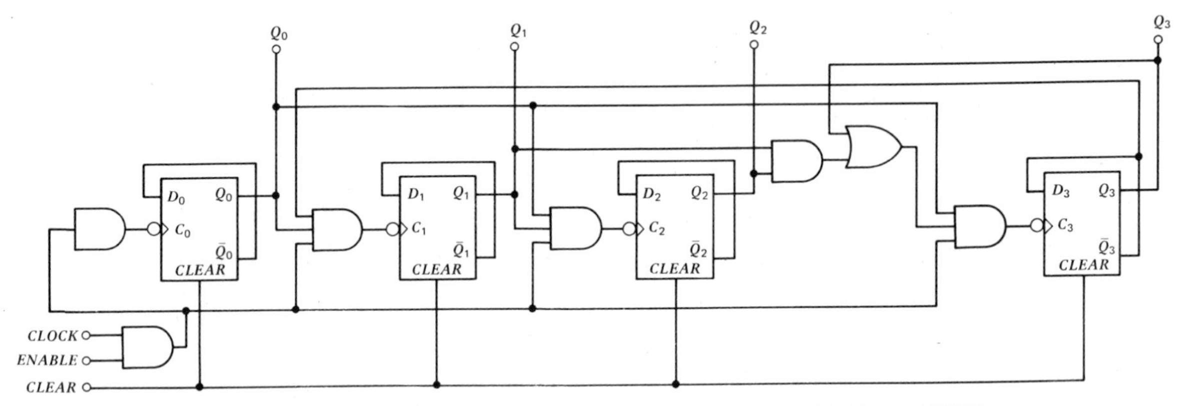

FIGURE 9-47 Simplified logic diagram of one-half of a CMOS Type 4518B synchronous BCD up counter.

T logic functions are the correct ones for a BCD up counter; also we saw in Ex. 9-7 that the ANDing of *ENABLE* · *CLOCK* results in a negative edge-triggered counter. Thus, Fig. 9-47 is a negative edge-triggered divide-by-10 synchronous up counter.

Because of the identical *T* logic functions in Figs. 9-44 and 9-47, the state diagram of Fig. 9-45 is also applicable to Fig. 9-47. Thus the counter of Fig. 9-47 is self-starting.

9-6.4 Programmable Counters

In the counters described thus far, the states of the FFs changed sequentially according to a specified counting sequence. In a *programmable counter* the state of each FF can also be *programmed*, or *preset*, to 0 or 1. This can be done in two different ways: the programming can be via the direct *PRESET* and *CLEAR* inputs (''*asynchronous programming*''), or via the control inputs of the FFs (''*synchronous programming*''). Since the direct *PRESET* and *CLEAR* inputs of a FF override the clock and the control inputs, asynchronous (direct) programming forces the FF into the desired state. When the programming is synchronous via the control input or inputs, additional circuitry is required.

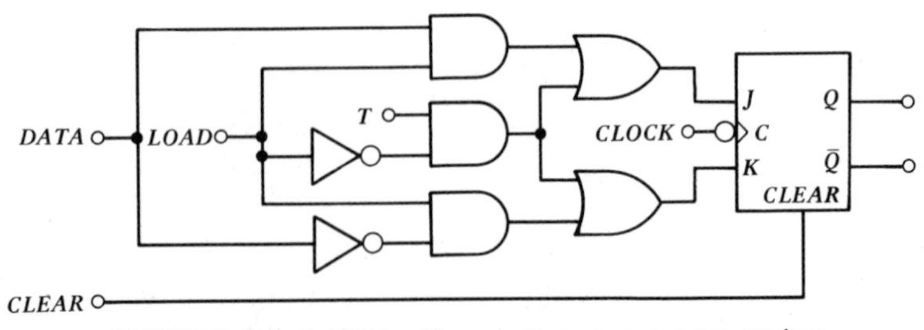

FIGURE 9-48 A T FF with synchronous programming.

TABLE 9-1
Function Table for the FF of Fig. 9-48

$LOAD$	T	Q_{n+1}
0	0	Q_n
0	1	$\overline{Q}_n$
1	x	$DATA$

EXAMPLE 9-9

Figure 9-48 shows a T FF with synchronous programming and with direct *CLEAR*. It is built using a J-K FF and seven logic gates.

a. Show that it acts as a T FF at terminals T and *CLOCK* when $LOAD = 0$.

b. Show that it is loaded by the binary value of *DATA* on a *CLOCK* pulse when $LOAD = 1$.

c. Summarize its operation in a function table.

Solution

a. When $LOAD = 0$, $J = T$ and $K = T$; these are the correct control inputs for realizing a T FF by a J-K FF.

b. When $LOAD = 1$, $J = DATA$ and $K = \overline{DATA}$; hence the next state of the FF equals *DATA*.

c. The operation can be summarized by the function table of Table 9-1.

The programming feature is available in up counters and in up-down counters, both binary and BCD. A programmable BCD up-down counter is described in Ex. 9-10 below.

EXAMPLE 9-10

The ECL Type 10137 IC is a 4-bit divide-by-10 programmable synchronous up-down counter. The following inputs to the circuit are provided: *CLOCK*, control lines S_1 and S_2, $DATA_3$, $DATA_2$, $DATA_1$, $DATA_0$, and *CARRY-IN*. The following outputs are available: Q_3, Q_2, Q_1, Q_0, and $CARRY\text{-}OUT = Q_3Q_2Q_1Q_0CARRY\text{-}IN$. Summarize the operation of the counter in a function table, if control lines S_1 and S_2 operate as shown in Table 9-2.

Solution

The function table is shown in Table 9-3. When $S_1 = 0$ and $S_2 = 0$ (first row), the counter is preset to the data word irrespective of *CARRY-IN*. When $S_1 = 1$

TABLE 9-2
Operation of Control Inputs S_1 and S_2 in Ex. 9-10

S_1	S_2	FUNCTION
0	0	Preset (program)
0	1	Count up in BCD
1	0	Count down in BCD
1	1	Hold contents

TABLE 9-3
Function Table for the ECL Type 10137 Counter in Ex. 9-10

S_1	S_2	*CARRY-IN*	$Q_{3_{n+1}}$	$Q_{2_{n+1}}$	$Q_{1_{n+1}}$	$Q_{0_{n+1}}$
0	0	x	$DATA_3$	$DATA_2$	$DATA_1$	$DATA_0$
0	1	0	Q_{3_n}	Q_{2_n}	Q_{1_n}	Q_{0_n}
0	1	1	Count up in BCD			
1	0	0	Q_{3_n}	Q_{2_n}	Q_{1_n}	Q_{0_n}
1	0	1	Count down in BCD			
1	1	x	Q_{3_n}	Q_{2_n}	Q_{1_n}	Q_{0_n}

and $S_2 = 1$ (last row), the counter holds its contents irrespective of *CARRY-IN*. When $S_1 = 0$ and $S_2 = 1$ (second and third rows), the counter counts up in BCD, but only if *CARRY-IN* = 1. When $S_1 = 1$ and $S_2 = 0$ (fourth and fifth rows), the counter counts down in BCD, but only if *CARRY-IN* = 1.

9-6.5 Asynchronous and Synchronous Clear

Many counters provide a separate *CLEAR* input in addition to the *LOAD* input and the *DATA* inputs. A *direct CLEAR* or *asynchronous CLEAR* input is simply a parallel connection of the direct *CLEAR* inputs of the individual FFs in the counter.

EXAMPLE 9-11

The TTL Type 74161 IC is a 4-bit divide-by-16 programmable synchronous up counter. The following inputs are provided: *CLOCK*, $\overline{LOAD}$, $DATA_3$, $DATA_2$, $DATA_1$, $DATA_0$, *ENABLE* (or "*ENABLE P*"), *CARRY-IN* (or "*ENABLE T*"), and asynchronous (direct) $\overline{CLEAR}$. The following outputs are available: Q_3, Q_2, Q_1, Q_0, and $\overline{CARRY\text{-}OUT} = \overline{Q_3Q_2Q_1Q_0\ CARRY\text{-}IN}$. Summarize the operation of the counter in a function table.

Solution

When $\overline{CLEAR}$ = H, the operation of the counter can be described by the function table of Table 9-4, which shows only clocked transitions. When

TABLE 9-4
Function Table Showing Clocked Transitions in the TTL Type 74161 Counter of Ex. 9-11

$\overline{LOAD}$	*ENABLE*	*CARRY-IN*	$Q_{3_{n+1}}$	$Q_{2_{n+1}}$	$Q_{1_{n+1}}$	$Q_{0_{n+1}}$
H	L	x	Q_{3_n}	Q_{2_n}	Q_{1_n}	Q_{0_n}
H	x	L	Q_{3_n}	Q_{2_n}	Q_{1_n}	Q_{0_n}
H	H	H	Count up in binary			
L	x	x	$DATA_3$	$DATA_2$	$DATA_1$	$DATA_0$

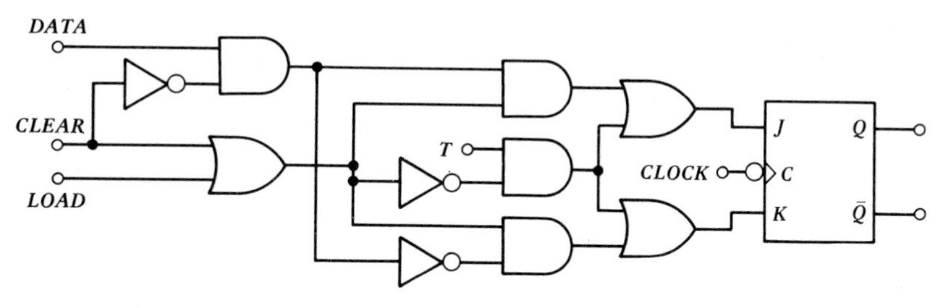

(*a*) Logic diagram

CLEAR	*LOAD*	*T*	Q_{n+1}
0	0	0	Q_n
0	0	1	$\overline{Q}_n$
0	1	x	*DATA*
1	x	x	0

(*b*) Function table

FIGURE 9-49 The T FF of Fig. 9-48 with synchronous *CLEAR* added and with the direct *CLEAR* removed.

$\overline{CLEAR}$ = L, then Q_3, Q_2, Q_1, and Q_0 become L irrespective of the *CLOCK* or of any other input.

An alternative to a direct (asynchronous) *CLEAR* input is a *synchronous CLEAR* input, shown for a T FF in Fig. 9-49. The synchronous *CLEAR* is entered via two logic gates that are added to the gating network of the FF control inputs in Fig. 9-48.

When *CLEAR* = 1 in Fig. 9-49*a*, it forces a *LOAD* = 1 via the added OR gate and forces a *DATA* = 0 via the lower input of the added AND gate. A function table is shown in Fig. 9-49*b*. Note that the synchronous *CLEAR* does *not* override the clock, but is effected by the appropriate transition of the clock waveform: positive-going transition in positive edge-triggered FFs, negative-going transition in negative edge-triggered or master–slave FFs.

EXAMPLE 9-12

The TTL Type 74163 IC is a 4-bit divide-by-16 programmable synchronous up counter with synchronous $\overline{CLEAR}$. Its input and output connections are the same as in the Type 74161 described in Ex. 9-11, except that the direct clear is replaced by a synchronous clear. Summarize the operation of the counter in a function table.

Solution

The operation of the counter is summarized in the function table of Table 9-5. When the synchronous $\overline{CLEAR}$ = H, the function table reverts to the simpler Table 9-4. When the synchronous $\overline{CLEAR}$ = L, the states of all four FFs become L on the next clock pulse.

TABLE 9-5
Function Table Showing Clocked Transitions in the TTL Type 74163 Counter of Ex. 9-12

$\overline{CLEAR}$	$\overline{LOAD}$	*ENABLE*	*CARRY-IN*	$Q_{3_{n+1}}$	$Q_{2_{n+1}}$	$Q_{1_{n+1}}$	$Q_{0_{n+1}}$
H	H	L	x	Q_{3_n}	Q_{2_n}	Q_{1_n}	Q_{0_n}
H	H	x	L	Q_{3_n}	Q_{2_n}	Q_{1_n}	Q_{0_n}
H	H	H	H	Count up in binary			
H	L	x	x	$DATA_3$	$DATA_2$	$DATA_1$	$DATA_0$
L	x	x	x	L	L	L	L

Programmable counters with asynchronous (direct) *CLEAR* and with synchronous *CLEAR* are also available as BCD counters. This is illustrated in the examples that follow: in Ex. 9-13 with asynchronous (direct) *CLEAR* and in Ex. 9-14 with synchronous *CLEAR*.

EXAMPLE 9-13

The TTL Type 74160 IC is a 4-bit divide-by-10 programmable synchronous BCD up counter with an asynchronous (direct) $\overline{CLEAR}$. Describe a function table for this counter.

Solution

The function table of this counter is identical to Table 9-4, except the phrase "Count up in binary" has to be replaced by "Count up in BCD."

EXAMPLE 9-14

The TTL Type 74162 IC is a 4-bit divide-by-10 programmable synchronous BCD up counter with synchronous $\overline{CLEAR}$. Describe a function table for this counter.

Solution

The function table of this counter is identical to Table 9-5, except the phrase "Count up in binary" has to be replaced by "Count up in BCD."

9-6.6 Divide-by-*N* Counters

In many applications we may require a divide-by-*N* counter, where *N* is an integer, $2 \leq N \leq 16$. We have already discussed counters that divided by 2, 3, 4, 5, 8, 10, and 16, and counters with other *N* can be similarly built. Here we explore two techniques that use standard divide-by-16 counters for a divide-by-*N* counter of any *N* between 2 and 16.

One method uses a programmable counter (Sec. 9-6.4)—this approach is described in Sec. 9-6.6.1. The other possibility is to use a decoder—this approach is described in Sec. 9-6.6.2.

9-6.6.1 Use of Programmable Counters

A programmable counter can be programmed (preset) to any number that is presented at its *DATA* inputs. Also, the counter provides a *CARRY-OUT* =

$Q_3Q_2Q_1Q_0$*CARRY-IN*. Such a counter can be used as a divide-by-N counter by connecting the *CARRY-OUT* to *LOAD*, connecting the *DATA* inputs to the 2's complement of N, and setting *CARRY-IN* = 1 and *ENABLE* = 1.

EXAMPLE 9-15

The TTL Type 74161 4-bit divide-by-16 programmable synchronous up counter described in Ex. 9-11 is used as a divide-by-9 counter. The *CARRY-IN* and *ENABLE* inputs are connected to logic 1. The $\overline{CARRY\text{-}OUT}$ is connected to $\overline{LOAD}$. Since $9_{10} = 1001$, its 2's complement is $0110 + 1 = 0111$; hence the *DATA* inputs are connected as $DATA_3 = 0$, $DATA_2 = 1$, $DATA_1 = 1$, and $DATA_0 = 1$.

Present States					Present Outputs	Next States				
Decimal	Q_{3_n}	Q_{2_n}	Q_{1_n}	Q_{0_n}	$CARRY\text{-}OUT_n = LOAD_n$	$Q_{3_{n+1}}$	$Q_{2_{n+1}}$	$Q_{1_{n+1}}$	$Q_{0_{n+1}}$	Decimal
0	0	0	0	0	0	0	0	0	1	1
1	0	0	0	1	0	0	0	1	0	2
2	0	0	1	0	0	0	0	1	1	3
3	0	0	1	1	0	0	1	0	0	4
4	0	1	0	0	0	0	1	0	1	5
5	0	1	0	1	0	0	1	1	0	6
6	0	1	1	0	0	0	1	1	1	7
7	0	1	1	1	0	1	0	0	0	8
8	1	0	0	0	0	1	0	0	1	9
9	1	0	0	1	0	1	0	1	0	10
10	1	0	1	0	0	1	0	1	1	11
11	1	0	1	1	0	1	1	0	0	12
12	1	1	0	0	0	1	1	0	1	13
13	1	1	0	1	0	1	1	1	0	14
14	1	1	1	0	0	1	1	1	1	15
15	1	1	1	1	1	0	1	1	1	7

(*a*) State table

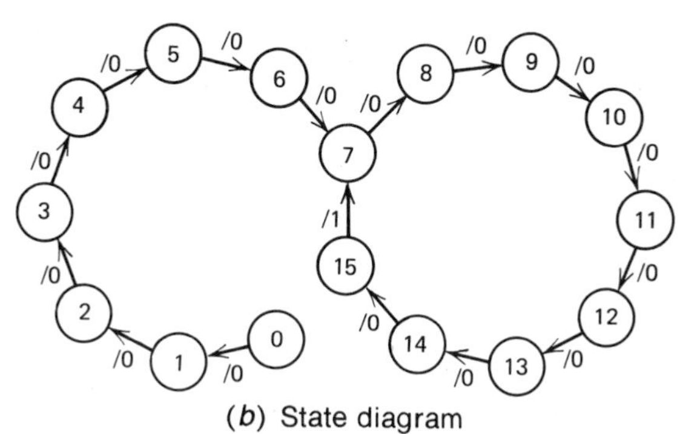

(*b*) State diagram

FIGURE 9-50 Divide-by-9 counter in Ex. 9-15.

Describe the operation of the counter by a state table and a state diagram.

Solution

A state table is shown in Fig. 9-50*a*. The first 15 rows describe the regular binary counting sequence. However, in the last row $CARRY\text{-}OUT_n = Q_{3_n}Q_{2_n}Q_{1_n}Q_{0_n}CARRY\text{-}IN = 1$; thus $LOAD_n = 1$, and the *DATA* word 0111 is loaded into the counter, whereby the next state is $0111_2 = 7_{10}$. Subsequently, the counter will count in the sequence of 7-8-9-. . . 14-15-7-8. . . .

A state diagram is shown in Fig. 9-50*b*. The states are identified by the decimal equivalents of $Q_{3_n}Q_{2_n}Q_{1_n}Q_{0_n}$ and the branches are labeled $/CARRY\text{-}OUT_n$. We can see that the state diagram represents a self-starting divide-by-9 counter that provides a $CARRY\text{-}OUT_n = 1$ in present state 15.

9-6.6.2 Counters Using Decoders

Another method of constructing a divide-by-*N* counter is by use of a decoder. In this scheme, the inputs of the decoder are connected to the outputs of the FFs in the counter, and the decoder provides a logic 1 output when the content of the counter is $N - 1$. The output of the decoder is connected to the synchronous *CLEAR* input of the counter. The resulting counting sequence is 0-1-. . . -($N - 1$)-0-1 . . . . This is illustrated in Ex. 9-16, which follows.

EXAMPLE 9-16

A divide-by-16 synchronous counter has the following connections available: Q_3, $\overline{Q}_3$, Q_2, $\overline{Q}_2$, Q_1, $\overline{Q}_1$, Q_0, $\overline{Q}_0$, *CLOCK*, and synchronous *CLEAR*. A 4-input AND gate is also available. Assemble a divide-by-9 counter, and provide a logic diagram and a state diagram.

Solution

Since $9_{10} = 1001_2$, $N = 9$, $N - 1 = 8 = 1000_2$, and the inputs of the AND gate are connected to Q_3, $\overline{Q}_2$, $\overline{Q}_1$, $\overline{Q}_0$. The output of the AND gate is connected to the synchronous *CLEAR* input of the counter as shown in the logic diagram of Fig. 9-51*a*.

A state diagram is shown in Fig. 9-51*b*. The states are identified by the decimal equivalents of $Q_{3_n}Q_{2_n}Q_{1_n}Q_{0_n}$, and the branches are labeled /CARRY-

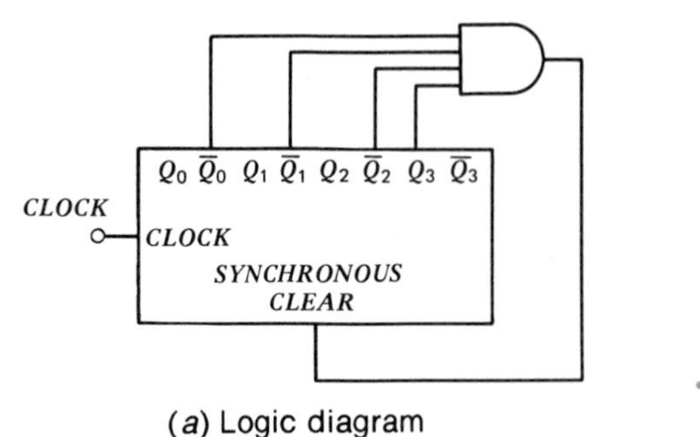

(*a*) Logic diagram

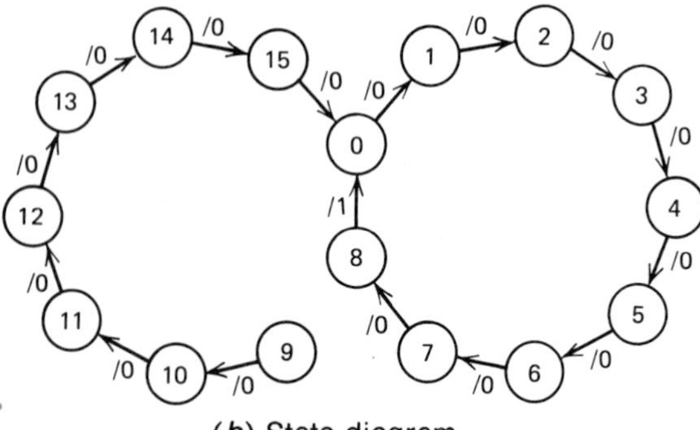

(*b*) State diagram

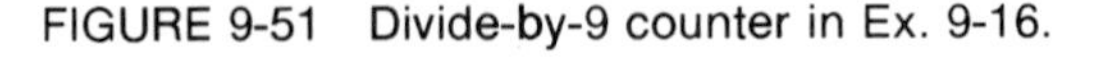

FIGURE 9-51 Divide-by-9 counter in Ex. 9-16.

OUT_n. We can see that the counter is a self-starting divide-by-9 counter that provides a $CARRY\text{-}OUT = 1$ in state 8.

When a synchronous *CLEAR* is not available, an asynchronous *CLEAR* may be used instead—at least in principle. The difficulties arise in the relationships of the propagation delay of the direct *CLEAR* through a FF, $t_{P_{CLEAR}}$, the propagation delay of the AND gate, $t_{P_{AND}}$, and the minimum required *CLEAR* pulse width. We state here without proof that for reliable operation $t_{P_{CLEAR}} + t_{P_{AND}}$ must be *greater* than the minimum required *CLEAR* pulse width. This condition is not easy to guarantee, because *minimum* values of $t_{P_{CLEAR}}$ and $t_{P_{AND}}$ are rarely specified.

9-6.7 Operating Speed

In general, a synchronous counter operates correctly as long as the timing requirements of *each* FF in it are satisfied. These timing requirements include f_{max}, $t_{1_{min}}$, t_{SETUP}, t_{HOLD}, $t_{CLEAR_{min}}$, and so on. Because connections to the FF inputs are made via various external logic gates, the evaluation of timing constraints can often be difficult. In what follows, we restrict our attention to timing constraints that arise from FFs and from delays of logic gates that are part of the counter. Thus, for example, we do not discuss timing constraints of divide-by-N counters (see, e.g., Fig. 9-51*a*) since these involve external interconnections that influence timing.

The *CLOCK* input in a synchronous counter is connected to the clock inputs of the individual FFs. Thus the *CLOCK* waveform has to satisfy f_{max} and $t_{1_{min}}$ of each FF. Further limitations are imposed by t_{SETUP} of the FFs.

EXAMPLE 9-17

The divide-by-5 synchronous counter of Fig. 9-37*a* is built using three TTL Type 7470 *positive* edge-triggered J-K FFs and an AND gate. Each of the two leftmost FFs is realized by parallel connection of the J and K inputs to form a T input.

The TTL Type 7470 J-K FF is characterized by $t_{1_{min}} = 20$ ns, $t_{0_{min}} = 30$ ns; hence, also by an $f_{max} = 1/(t_{1_{min}} + t_{0_{min}}) = 1/(20 \text{ ns} + 30 \text{ ns}) = 20$ MHz. Also $t_{PLH_{CLOCK}} = t_{PHL_{CLOCK}} = t_{P_{CLOCK}} = 50$ ns $\uparrow$, $t_{SETUP} = 20$ ns $\uparrow$, and $t_{HOLD} = 5$ ns $\uparrow$. The propagation delay of the AND gate is 20 ns.

Does the counter operate correctly if the *CLOCK* waveform is a symmetrical square wave with a period of (**a**) 100 ns, (**b**) 80 ns, (**c**) 60 ns?

Solution

First we check whether $t_{1_{min}}$, $t_{0_{min}}$, and f_{max} are satisfied. In the most critical case (**c**) we have a symmetrical square wave with a period of 60 ns. Thus, the *CLOCK* waveform is at logic 1 for 30 ns and at logic 0 for 30 ns. Therefore, $t_{1_{min}}$, $t_{0_{min}}$, and hence also f_{max} are satisfied. It follows that they are also satisfied in the less critical cases (**a**) and (**b**).

Next we look at the setup times in case (**a**), when the positive-going transitions of the *CLOCK* waveform are separated by 100 ns. The outputs of the leftmost two FFs, Q_0 and Q_1, change $t_{P_{CLOCK}} = 50$ ns after a positive-going transition of the *CLOCK* waveform. The AND gate adds a delay of 20 ns; hence J_2 changes 70 ns after a positive-going transition of the *CLOCK* waveform. This is 30 ns *before* the *next* positive-going *CLOCK* transition. Thus the $t_{SETUP} = 20$ ns ↑ is satisfied for J_2. It can be shown that J_2 is the most critical control input in the counter; thus all setup times are satisfied in case (**a**).

In case (**b**), the positive-going transitions of the *CLOCK* waveform are separated by 80 ns. As before, J_2 changes 70 ns after a positive-going transition of the *CLOCK* waveform. Now, however, this is only 10 ns before the next positive-going *CLOCK* transition. Thus the $t_{SETUP} = 20$ ns ↑ is *not* satisfied for J_2, and the counter will probably *not* operate correctly. Note, however, that the counter *would* operate correctly if the AND gate had a propagation delay of only 10 ns instead of 20 ns.

In case (**c**), the positive-going transitions of the *CLOCK* waveform are separated by 60 ns. As before, J_2 changes 70 ns after a positive-going transition of the *CLOCK* waveform. This certainly comes too late, and the counter will *not* operate correctly. Note that in case (**c**) the counter would not operate correctly even if the AND gate had zero propagation delay.

Finally, we would have to check whether the hold times are satisfied. It can be shown that the hold times are satisfied in all three cases; hence, they provide no limitation in this counter.

Note that timing diagrams for cases (**a**) and (**b**) are the subject of Prob. 9-15 and Prob. 9-16, respectively.

9-7 HYBRID COUNTERS

When two or more synchronous counters are cascaded in a ripple-counter configuration, a *hybrid counter* results.

EXAMPLE 9-18

The TTL Type 7492A IC consists of two independent parts, as shown in Fig. 9-52. One part is a divide-by-2 counter; the other part is a divide-by-6 hybrid

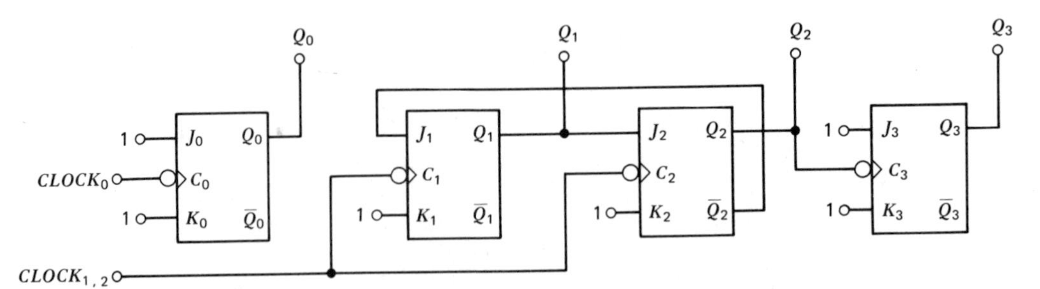

FIGURE 9-52 The TTL Type 7492A counter. (Direct *PRESET* and *CLEAR* are not shown.)

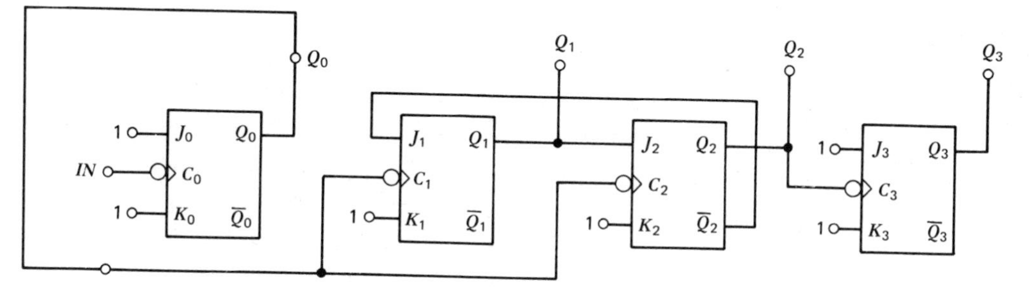

FIGURE 9-53 Divide-by-12 hybrid counter in Ex. 9-18.

counter consisting of a divide-by-3 synchronous counter (Fig. 9-31*a*) followed by a divide-by-2 counter (Fig. 9-2*a*) in a ripple counter configuration. Show connections for a divide-by-12 counter.

Solution

One possibility is shown in Fig. 9-53 where the divide-by-6 counter follows the divide-by-2 counter.

SUMMARY

The chapter started with the simplest counter: the divide-by-2 counter consisting of a single FF. This was followed by a description of ripple counters and their operation, including decoding and speed limitations.

One-bit counters with *CARRY-OUT*, with *CARRY-IN* and *CARRY-OUT*, and with *ENABLE* were described. State diagrams were introduced and used as a tool to describe counter operation.

Structure and operation of synchronous counters were described, including programmable counters, counters with synchronous clear, and divide-by-*N* counters. A brief discussion of hybrid counters was also presented.

SELF-EVALUATION QUESTIONS

1. What is a toggle FF?
2. What is a divide-by-2 counter?
3. Sketch a divide-by-16 ripple counter.
4. What is a self-starting counter?
5. Explain the term "maximum counting speed."
6. Describe the reason for glitches in the decoded outputs of ripple counters.
7. What is a state diagram?
8. What is a synchronous counter?
9. Sketch a divide-by-16 synchronous counter.
10. Explain the operation of programmable counters.

PROBLEMS

9-1. Which figure in this chapter can be used to describe the timing diagram of a 2-stage binary ripple counter using negative edge-triggered FFs?

9-2. Which figure in this chapter can be used to describe the timing diagram of a 4-stage binary ripple counter using negative edge-triggered FFs?

9-3. Figure 9-10 shows a timing diagram with an initial state of 0. Sketch a similar diagram with an initial state of 6.

9-4. Figure 9-10 shows a timing diagram with an initial state of 0. Sketch a similar diagram with an initial state of 7.

9-5. Sketch a state diagram for a J-K FF. Label the branches $J_nK_n/$.

9-6. Sketch a state diagram for a D FF. Label the branches $D_n/$.

9-7. Sketch a state diagram for the counter shown in Fig. 9-24. Identify the states as $Q_{1_n}Q_{0_n}$ and label the branches $CARRY\text{-}IN_n/CARRY\text{-}OUT_n$.

9-8. Sketch a state diagram for the counter shown in Fig. 9-20*a*. Identify the states by the decimal equivalents of $Q_{2_n}Q_{1_n}Q_{0_n}$ and label the branches $CARRY\text{-}IN_n/CARRY\text{-}OUT_n$.

9-9. Figure 9-38 shows a timing diagram with an initial state of 7. Sketch a similar diagram with an initial state of 6.

9-10. Figure 9-38 shows a timing diagram with an initial state of 7. Sketch a similar diagram with an initial state of 5.

9-11. Draw a state table and a state diagram for the counter shown in Fig. P9-1. Use no labels on the branches of the state diagram. Show that the counter is a self-starting divide-by-6 synchronous counter.

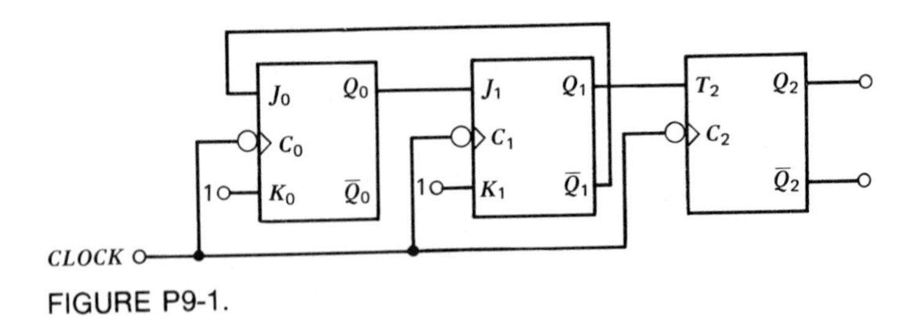

FIGURE P9-1.

9-12. Draw a state table and a state diagram for the counter shown in Fig. P9-2. Use no labels on the branches of the state diagram. Show that the counter is a self-starting divide-by-7 synchronous counter.

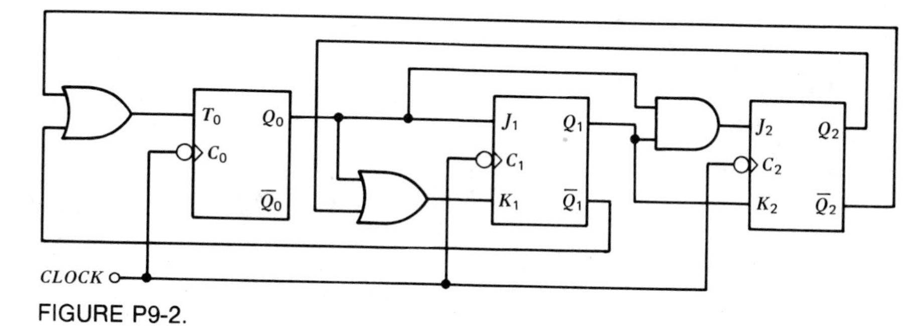

FIGURE P9-2.

9-13. Sketch a state diagram for the counter shown in Fig. 9-39*a*. Identify the states by the numerals 0 through 15 and label the branches as $CARRY\text{-}IN_n/CARRY\text{-}OUT_n$.

9-14. Draw a state table for the counter shown in Fig. 9-44. Show the present states and the next states of all four FFs and the present states of the four FF control inputs.

9-15. Sketch a timing diagram for case (**a**) in Ex. 9-17. Show the waveforms of the *CLOCK*, $Q_0 = T_1$, Q_1, J_2, Q_2, and the setup time interval with respect to J_2.

9-16. Sketch a timing diagram for case (**b**) in Ex. 9-17. Show the waveforms of the *CLOCK*, $Q_0 = T_1$, Q_1, J_2, Q_2, and the setup time interval with respect to J_2.

9-17. Use the excitation table of the D FF (Fig. P8-1, page 206) to *design* an implementation of the $T = 1$ FF using a D FF, and arrive at the circuit shown in Fig. 9-2*c* or Fig. 9-2*d*.

9-18. Use the excitation table of the J-K FF (Fig. P8-3, page 207) to *design* an implementation of the $T = 1$ FF using a J-K FF, and arrive at the circuit shown in Fig. 9-2*a* or Fig. 9-2*b*.

9-19. Use the excitation table of the T FF (Fig. P8-2, page 207) to *design* an implementation of the state table given in Fig. 9-20*b* using a T FF, and arrive at the circuit shown in Fig. 9-20*a*. (*Hint:* Add a column showing T_n to the state table of Fig. 9-20*b*.)

9-20. Use the excitation table of the D FF (Fig. P8-1, page 206) to *design* an implementation of the state table given in Fig. 9-20*b* using a D FF, and arrive at the circuit shown in Fig. 9-21. (*Hint:* Add a column showing D_n to the state table of Fig. 9-20*b*.)

CHAPTER

10 Shift Registers and Shift-Register Counters

Instructional Objectives

This chapter describes shift registers and shift-register counters. After you read it, you should be able to:

1. Describe the structure of shift registers.
2. Use shift registers for data storage and data manipulation.
3. Describe the operation of bidirectional shift registers with synchronous parallel loading.
4. Compare the properties of shift-register counters.
5. Describe the operation or ring counters.
6. Describe the operation of Johnson counters.
7. Evaluate the operating speed limitations of shift registers and shift-register counters.

10-1 OVERVIEW

The preceding chapter described various counters and their applications. The subject of this chapter is another widely used digital circuit: the shift register. While a counter is characterized by a *counting loop* in its *state diagram*, the principal characteristic of a shift register is that its basic *hardware structure* consists of a *chain of FFs*.

A 2-stage shift register was already introduced in Fig. 8-24 (page 191); in a well-designed circuit the delay between C_1 and C_2 is negligible, as shown in the timing diagram of Fig. 8-25 (page 191). In the following discussion, shift registers and their application as counters are described in detail.

Section 10-2 describes shift registers and their application for data storage and for data manipulation such as shift-right, shift-left, parallel load, and parallel retrieval. Section 10-3 uses shift registers for counters. Section 10-4 summarizes operating speed limitations of shift registers and shift-register counters. Applications of MOS shift registers as memory elements are further discussed in Chapter 11.

10-2 SHIFT REGISTERS

The basic shift-register structure is illustrated in Fig. 10-1, which shows a 3-stage shift register using master–slave D FFs. In what follows, most shift registers are shown with master–slave FFs; they could also be built using edge-triggered FFs. Also, each D FF can be replaced by a J-K FF and an inverter, or by a J-$\overline{K}$ FF.

In a shift register, all FF clock inputs (C_1, C_2, and C_3 in Fig. 10-1) are activated simultaneously—or at least nearly so (see discussion in Ex. 8-5, page 191). In a D FF, the next state Q_{n+1} of output Q equals the present state D_n of input D. Thus in Fig. 10-1, $Q_{1_{n+1}} = D_{1_n} = IN_n$, $Q_{2_{n+1}} = D_{2_n} = Q_{1_n}$, and $Q_{3_{n+1}} = D_{3_n} = Q_{2_n}$. Therefore, data entered at input IN are *shifted right* by one stage on each clock pulse.

EXAMPLE 10-1

The initial state of each FF in Fig. 10-1 is 0. Input IN is 0 during the first two clock pulses, it is 1 during the next clock pulse, and it is 0 during the next four clock pulses.

Assume that all setup time and hold time requirements are satisfied, and summarize the operation of the circuit as follows:

a. Prepare a state table showing the states of IN, D_1, Q_1, D_2, Q_2, D_3, and Q_3 during seven clock pulses.

b. Sketch a timing diagram with invisibly short propagation delays.

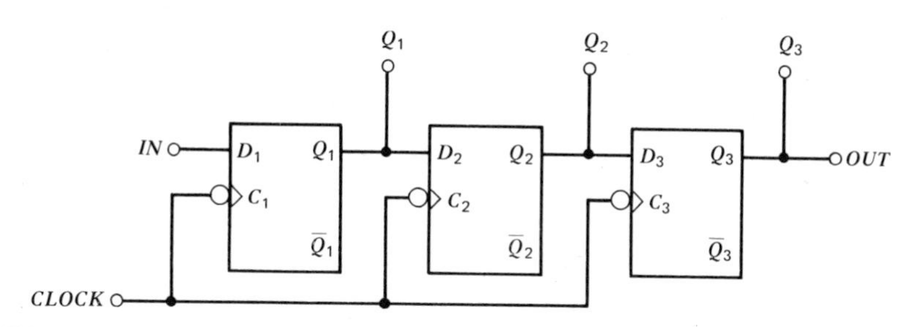

FIGURE 10-1 Logic diagram of a 3-stage shift register using master–slave D FFs.

TABLE 10-1
States in Ex. 10-1

Clock Pulse	1	2	3	4	5	6	7
IN	0	0	1	0	0	0	0
D_1	0	0	1	0	0	0	0
Q_1	0	0	0	1	0	0	0
D_2	0	0	0	1	0	0	0
Q_2	0	0	0	0	1	0	0
D_3	0	0	0	0	1	0	0
Q_3	0	0	0	0	0	1	0

Solution

a. The state table is shown in Table 10-1. The states are in accordance with the logic equations $Q_{1_{n+1}} = D_{1_n} = IN_n$, $Q_{2_{n+1}} = D_{2_n} = Q_{1_n}$, and $Q_{3_{n+1}} = D_{3_n} = Q_{2_n}$.

b. Figure 10-2 shows the timing diagram, derived from Table 10-1.

Data in the shift register of Fig. 10-1 are shifted from the left to the right; for this reason such shift registers are called *shift-right shift registers*. Also, data are *entered serially* at input IN, which is therefore called a *serial input*.

Output from the shift register is available *serially* at the output of the last FF, Q_3: this is called *serial output*. *Parallel outputs* Q_1, Q_2, and Q_3 are also

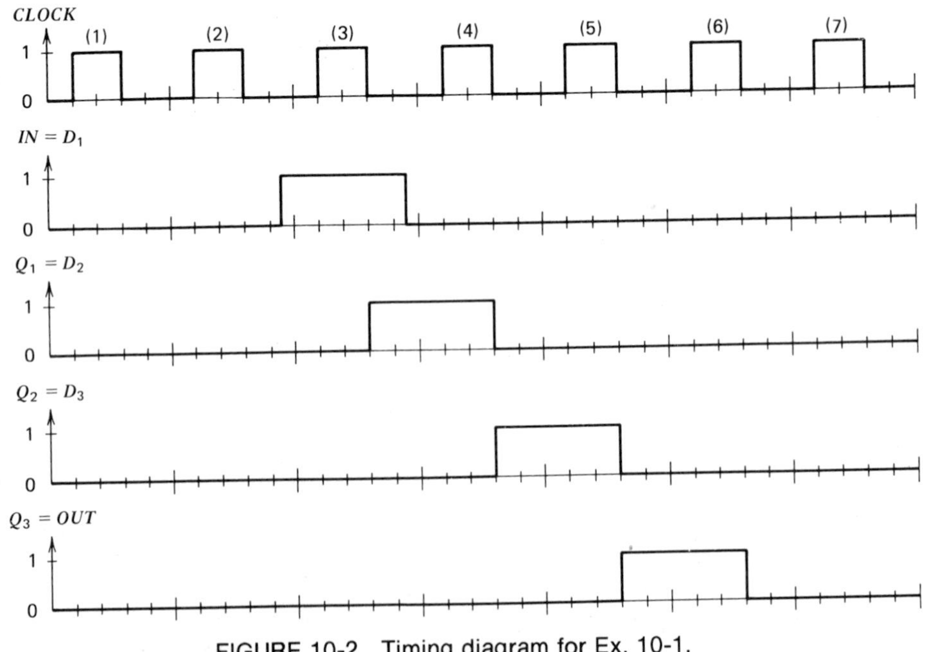

FIGURE 10-2 Timing diagram for Ex. 10-1.

available in Fig. 10-1—such outputs are often omitted because of IC pin limitations; however, the output of the last FF, the serial output, is always available.

Figure 10-3 outlines the various operating modes of a 4-stage shift register. Figure 10-3*a* shows the connections of the four FFs for a *shift-right* shift register, and Fig. 10-3*b* for a *shift-left* shift register. Figure 10-3*c* shows the connections of the four FFs for *synchronous parallel loading.*

The shift-right, shift-left, and parallel load operations of Fig. 10-3 may also be implemented using the *same* four FFs; this is done by use of suitable interconnection via logic gates. Thus in a *shift-right shift-left shift register*, which is also known as a *bidirectional shift register*, data are shifted right or left depending on

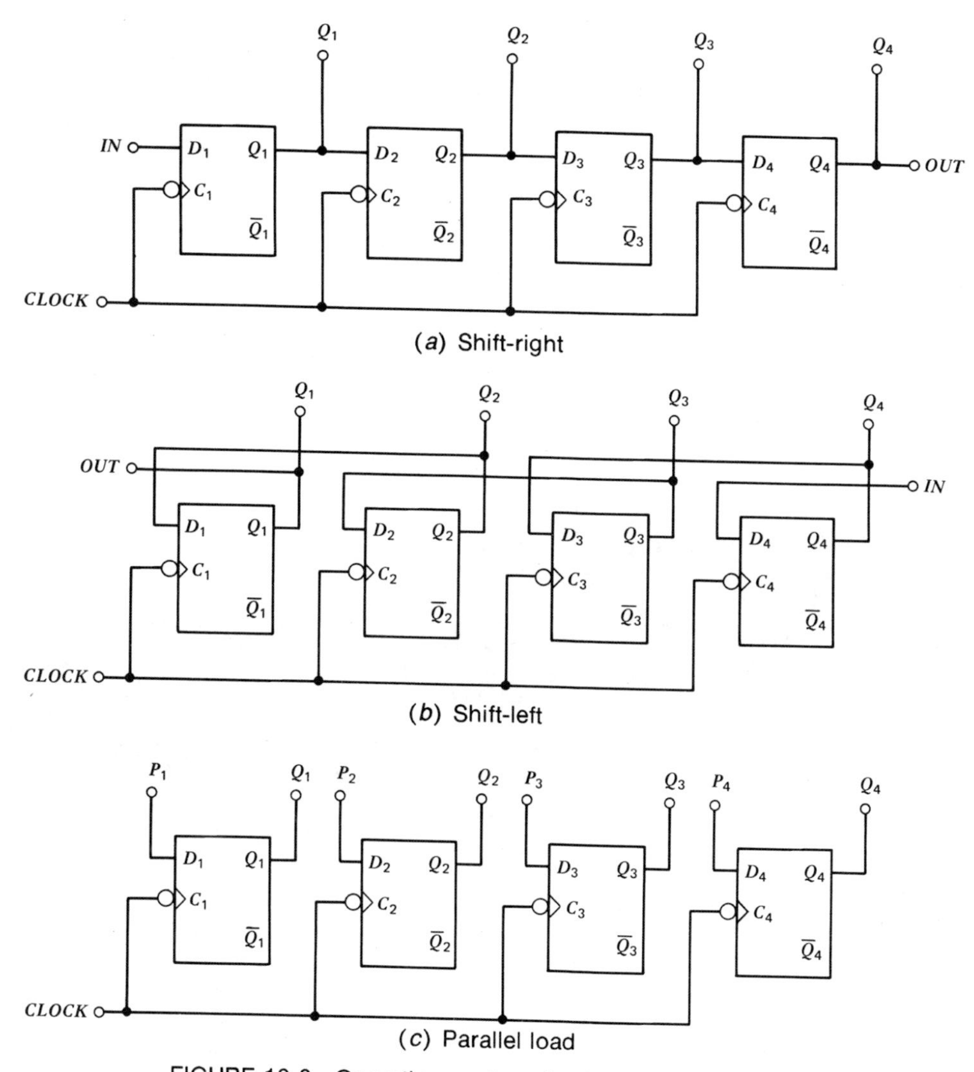

(*a*) Shift-right

(*b*) Shift-left

(*c*) Parallel load

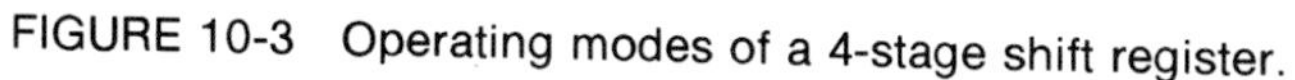
FIGURE 10-3 Operating modes of a 4-stage shift register.

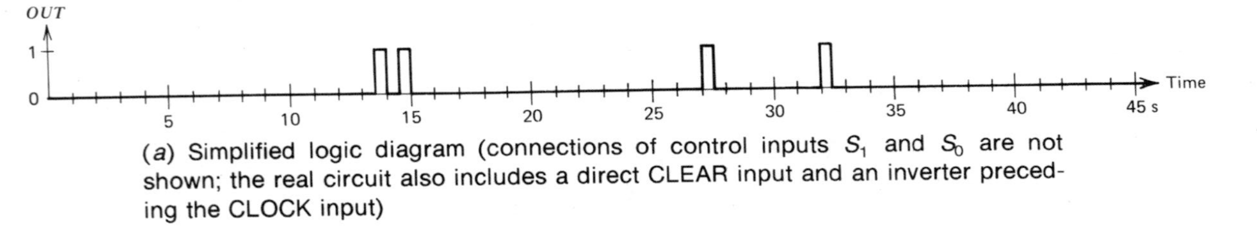

(a) Simplified logic diagram (connections of control inputs S_1 and S_0 are not shown; the real circuit also includes a direct CLEAR input and an inverter preceding the CLOCK input)

S_1	S_0	Operating Mode
0	0	Do nothing
0	1	Shift right
1	0	Shift left
1	1	Parallel load

(b) Operating modes

FIGURE 10-4 The TTL Type 74194 4-stage bidirectional shift register with synchronous parallel loading.

the state of a direction-control input signal. Some shift registers also have provisions for *synchronous parallel loading* whereby external data can be jammed into all FFs in a single clock pulse. This is illustrated in Ex. 10-2, which follows.

EXAMPLE 10-2

Figure 10-4*a* shows a simplified logic diagram of the TTL Type 74194 4-stage bidirectional shift register with parallel loading. It consists of four master-slave D FFs and of gating networks.

Shift direction and parallel loading are controlled by control inputs S_1 and S_0 in accordance with Fig. 10-4*b*. When $S_1 = 0$ and $S_0 = 0$, the *CLOCK* input in Fig. 10-4*a* is inhibited and the FFs do not change state (''*Do nothing*''). When $S_1 = 0$ and $S_0 = 1$, then also $\bar{S}_0 = 0$, $S_1S_0 = 0$, and $\bar{S}_1 = 1$. Therefore, the lowermost four AND gates in Fig. 10-4*a* are activated. Thus, in this mode of operation, D_1 is connected to *S.R.IN* (shift-right serial input), D_2 to Q_1, D_3 to Q_2, and D_4 to Q_3. This is the *shift-right* mode of operation.

When $S_1 = 1$ and $S_0 = 0$, then also $\bar{S}_0 = 1$, $S_1S_0 = 0$, and $\bar{S}_1 = 0$. Therefore, the uppermost four AND gates in Fig. 10-4*a* are activated. Thus, in this mode of operation, D_4 is connected to *S.L.IN* (shift-left serial input), D_3 to Q_4, D_2 to Q_3, and D_1 to Q_2. This is the *shift-left* mode of operation.

When $S_1 = 1$ and $S_0 = 1$, then also $\bar{S}_0 = 0$, $S_1S_0 = 1$, and $\bar{S}_1 = 0$. Therefore, the center four AND gates in Fig. 10-4*a* are activated. Thus, in this mode of operation, D_1 is connected to P_1, D_2 to P_2, D_3 to P_3, and D_4 to P_4. As a result, the states of *parallel inputs* P_1, P_2, P_3, and P_4 are loaded into FFs Q_1, Q_2, Q_3, and Q_4, respectively.

Prepare a function table that shows the states of S_{1_n}, S_{0_n}, $(S.R.IN)_n$, $(S.L.IN)_n$, $Q_{1_{n+1}}$, $Q_{2_{n+1}}$, $Q_{3_{n+1}}$, and $Q_{4_{n+1}}$.

Solution

The function table is shown in Fig. 10-5. When $S_{1_n} = 0$ and $S_{0_n} = 0$ (first row), the operating mode is ''Do nothing''; thus the states of the FFs do not change, no matter what the states of the other inputs are. When $S_{1_n} = 0$ and $S_{0_n} = 1$ (second and third rows), the operating mode is ''shift right''; thus $Q_{1_{n+1}} = (S.R.IN)_n$, $Q_{2_{n+1}} = Q_{1_n}$, $Q_{3_{n+1}} = Q_{2_n}$, and $Q_{4_{n+1}} = Q_{3_n}$, irrespective of $(S.L.IN)_n$. When $S_{1_n} = 1$ and $S_{0_n} = 0$ (fourth and fifth rows), the operating mode is ''shift left''; thus $Q_{1_{n+1}} = Q_{2_n}$, $Q_{2_{n+1}} = Q_{3_n}$, $Q_{3_{n+1}} = Q_{4_n}$, and $Q_{4_{n+1}} =$

S_{1_n}	S_{0_n}	$(S.R.IN)_n$	$(S.L.IN)_n$	$Q_{1_{n+1}}$	$Q_{2_{n+1}}$	$Q_{3_{n+1}}$	$Q_{4_{n+1}}$
0	0	x	x	Q_{1_n}	Q_{2_n}	Q_{3_n}	Q_{4_n}
0	1	0	x	0	Q_{1_n}	Q_{2_n}	Q_{3_n}
0	1	1	x	1	Q_{1_n}	Q_{2_n}	Q_{3_n}
1	0	x	0	Q_{2_n}	Q_{3_n}	Q_{4_n}	0
1	0	x	1	Q_{2_n}	Q_{3_n}	Q_{4_n}	1
1	1	x	x	P_{1_n}	P_{2_n}	P_{3_n}	P_{4_n}

FIGURE 10-5 Function table for Ex. 10-2.

$(S.L.IN)_n$, irrespective of $(S.R.IN)_n$. Finally, when $S_{1_n} = 1$ and $S_{0_n} = 1$ (last row), the operating mode is "parallel load"; thus $Q_{1_{n+1}} = P_{1_n}$, $Q_{2_{n+1}} = P_{2_n}$, $Q_{3_{n+1}} = P_{3_n}$, and $Q_{4_{n+1}} = P_{4_n}$, irrespective of $(S.R.IN)_n$ and $(S.L.IN)_n$.

10-3 SHIFT-REGISTER COUNTERS

We saw in Ch. 9 that a divide-by-n counter can be built by the use of m FFs as long as n was at most 2^m, that is, $n \leq 2^m$ held. The circuits in Ch. 9 used the minimum possible number of FFs that satisfied the $n \leq 2^m$ limit; thus, the counter circuits in Ch. 9 were optimal in the utilization of FFs. However, when *decoded* outputs were required, these counters used decoders that often produced undesirable "glitches." These glitches were prominent in ripple counters but could be eliminated in synchronous counters with well-matched propagation delays in the counter and in the decoder.

Two types of *shift-register counters* are described in this section: the *ring counter* and the *Johnson counter* (also known as *twisted-ring counter*). Both of these counters use more FFs than required by the $n \leq 2^m$ limit and both provide completely glitch-free decoded outputs. A divide-by-n ring counter uses n FFs —but requires no decoder at all. A divide-by-n Johnson counter uses $n/2$ FFs and counts in a unit-distance code (see Sec. 4-3).

10-3.1 Three-Stage Non-Self-Starting Ring Counter

Figure 10-6*a* shows a 3-stage ring counter. It is a 3-stage shift register with output Q_3 of the last FF connected to input D_1 of the first FF. Consider the operation of the circuit with initial states of $Q_1Q_2Q_3 = 100$. After one clock pulse the states become $Q_1Q_2Q_3 = 010$, and after a second clock pulse $Q_1Q_2Q_3 = 001$. However, since $D_1 = Q_3$, after the third clock pulse the states will be $Q_1Q_2Q_3 = 100$. Therefore, the digital circuit of Fig. 10-6*a* counts in the sequence of 100-010-001-100-010 . . . .

A state table is shown in Fig. 10-6*b* and a state diagram in Fig. 10-6*c*. These describe a divide-by-3 counter. However, in addition to the main counting loop of 100-010-001, the state diagram also shows 5 additional states (3 FFs have a total of $2^3 = 8$ states). These 5 states have no transitions back to the main counting loop. Hence, the counter is not self-starting; for proper operation, it has to be initialized into its main counting loop by direct *PRESET* and *CLEAR* inputs, which are not shown.

When operating in its main counting loop, the divide-by-3 counter of Fig. 10-6*a* provides decoded outputs on Q_1, Q_2, and Q_3; hence no additional decoding circuitry is required. Also, outputs Q_1, Q_2, and Q_3 are completely glitch-free.

10-3.2 Three-Stage Self-Starting Ring Counter

Figure 10-7*a* shows a digital circuit consisting of 3 FFs and a NOR gate. We now show that it is a self-starting divide-by-3 ring counter.

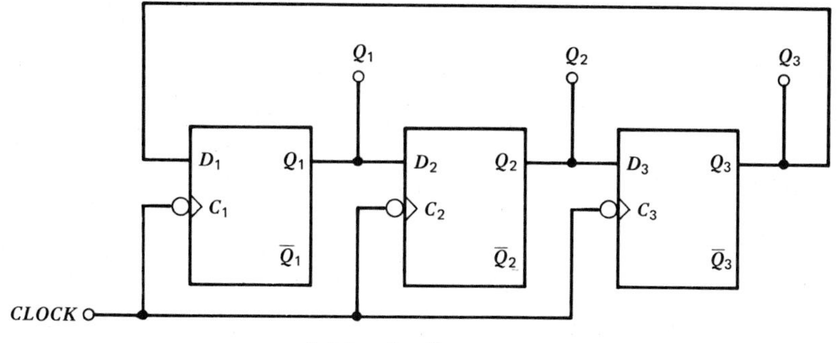

(a) Logic diagram

Present States			Present Control Inputs			Next States		
Q_{1_n}	Q_{2_n}	Q_{3_n}	D_{1_n}	D_{2_n}	D_{3_n}	$Q_{1_{n+1}}$	$Q_{2_{n+1}}$	$Q_{3_{n+1}}$
0	0	0	0	0	0	0	0	0
0	0	1	1	0	0	1	0	0
0	1	0	0	0	1	0	0	1
0	1	1	1	0	1	1	0	1
1	0	0	0	1	0	0	1	0
1	0	1	1	1	0	1	1	0
1	1	0	0	1	1	0	1	1
1	1	1	1	1	1	1	1	1

(b) State table

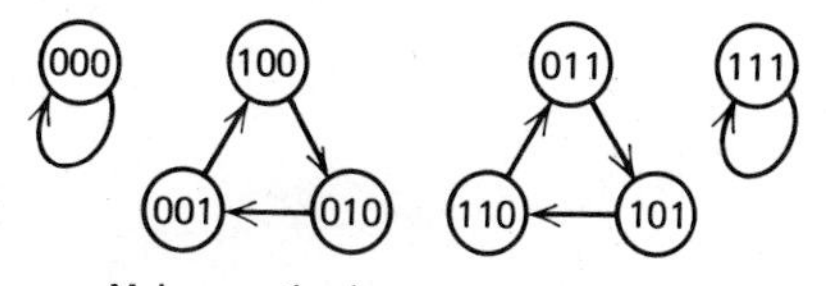

(c) State diagram with the states identified as $Q_{1_n}Q_{2_n}Q_{3_n}$

FIGURE 10-6 Three-stage non-self-starting divide-by-3 ring counter.

A state table is shown in Fig. 10-7*b*. The entries of the Present Control Inputs are obtained in accordance with Fig. 10-7*a* as $D_1 = \overline{Q_1 + Q_2}$, $D_2 = Q_1$, and $D_3 = Q_2$. Figure 10-7*c* shows a state diagram that was derived from the state table of Fig. 10-7*b*. We can see that Fig. 10-7*a* is indeed a self-starting divide-by-3 counter. Also, the three states in the counting loop are coded as 100-010-001; hence the counter is a ring counter with completely glitch-free outputs.

10-3.3 Three-Stage Non-Self-Starting Johnson Counter

Figure 10-8*a* shows a 3-stage Johnson counter. It is a 3-stage shift register with complementary output $\overline{Q}_3$ of the last FF connected to input D_1 of the first FF.

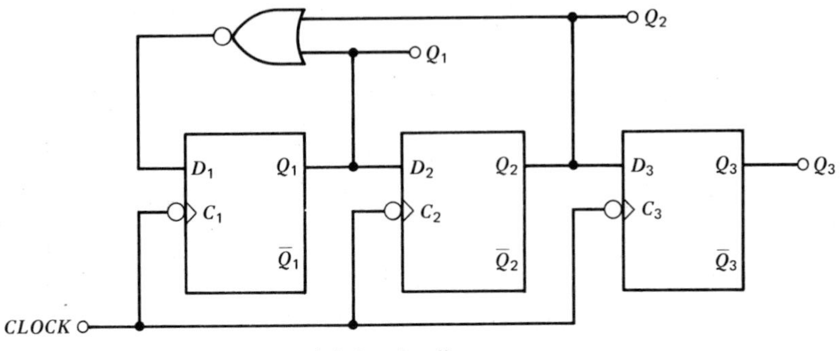

(*a*) Logic diagram

Present States			Present Control Inputs			Next States		
Q_{1_n}	Q_{2_n}	Q_{3_n}	D_{1_n}	D_{2_n}	D_{3_n}	$Q_{1_{n+1}}$	$Q_{2_{n+1}}$	$Q_{3_{n+1}}$
0	0	0	1	0	0	1	0	0
0	0	1	1	0	0	1	0	0
0	1	0	0	0	1	0	0	1
0	1	1	0	0	1	0	0	1
1	0	0	0	1	0	0	1	0
1	0	1	0	1	0	0	1	0
1	1	0	0	1	1	0	1	1
1	1	1	0	1	1	0	1	1

(*b*) State table

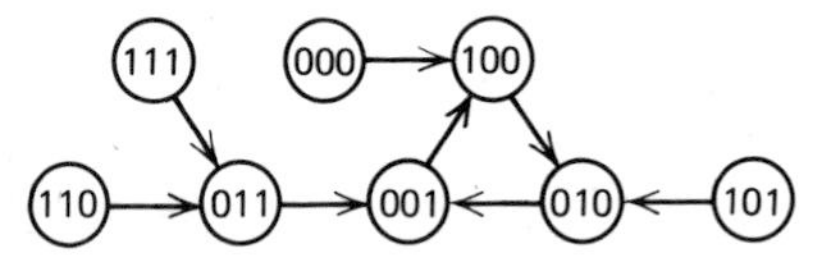

(*c*) State diagram with the states identified as $Q_{1_n}Q_{2_n}Q_{3_n}$

FIGURE 10-7 Three-stage self-starting divide-by-3 ring counter.

Figure 10-8*b* shows a state table, Fig. 10-8*c* a state diagram. These show a divide-by-6 counter that counts in the sequence of 000-100-110-111-011-001-000-100. . . . However, the state diagram also shows the two additional states of 010 and 101. There is no transition from these states back to the main counting loop; hence, the counter is not self-starting. The outputs are again completely glitch-free.

10-3.4 Three-Stage Self-Starting Johnson Counter

Figure 10-9*a* shows a digital circuit consisting of 3 FFs and two logic gates. We now show that it is a self-starting divide-by-6 Johnson counter.

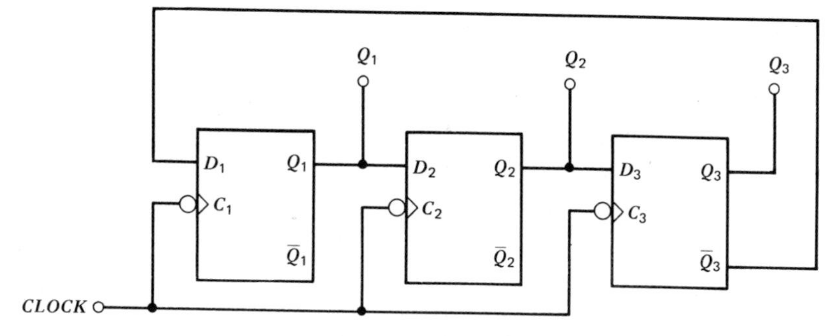

(*a*) Logic diagram

Present States			Present Control Inputs			Next States		
Q_{1_n}	Q_{2_n}	Q_{3_n}	D_{1_n}	D_{2_n}	D_{3_n}	$Q_{1_{n+1}}$	$Q_{2_{n+1}}$	$Q_{3_{n+1}}$
0	0	0	1	0	0	1	0	0
1	0	0	1	1	0	1	1	0
1	1	0	1	1	1	1	1	1
1	1	1	0	1	1	0	1	1
0	1	1	0	0	1	0	0	1
0	0	1	0	0	0	0	0	0
0	1	0	1	0	1	1	0	1
1	0	1	0	1	0	0	1	0

(*b*) State table

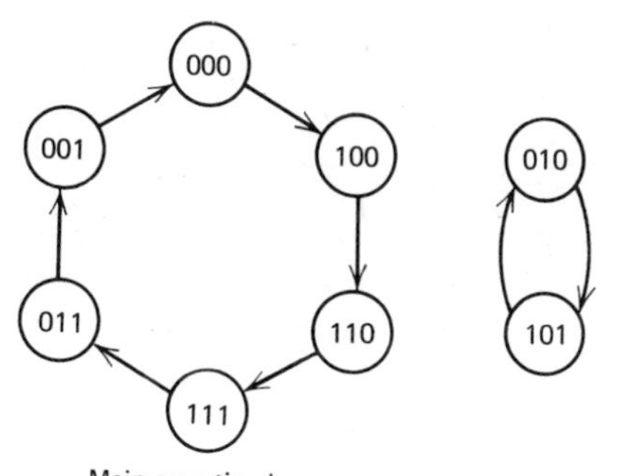

(*c*) State diagram with the states identified as $Q_{1_n}Q_{2_n}Q_{3_n}$

FIGURE 10-8 Three-stage non-self-starting divide-by-6 Johnson counter.

A state table is shown in Fig. 10-9*b*. The entries of the Present Control Inputs are obtained in accordance with Fig. 10-9*a* as $D_1 = \overline{Q}_3$, $D_2 = Q_1$, and $D_3 = Q_2 \cdot (Q_1 + Q_3)$. Figure 10-9*c* shows a state diagram that was derived from the state table of Fig. 10-9*b*. We can see that Fig. 10-9*a* is indeed a self-starting divide-by-6 counter. Also, in its counting loop the counter counts in the

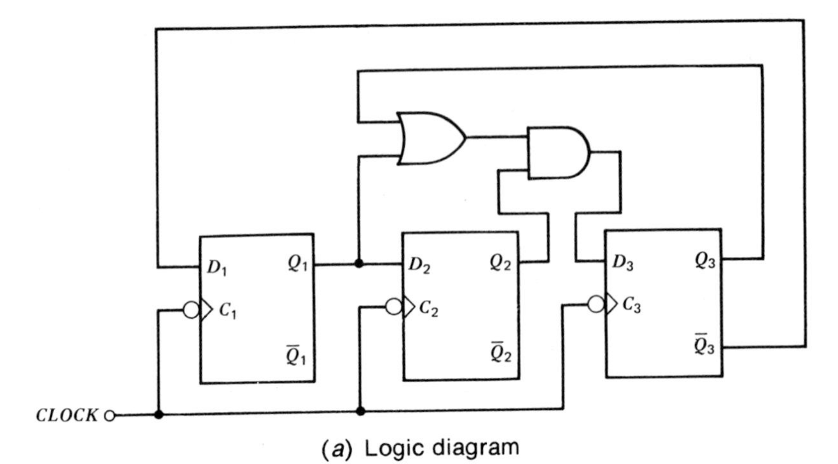

(*a*) Logic diagram

Present States			Present Control Inputs			Next States		
Q_{1_n}	Q_{2_n}	Q_{3_n}	D_{1_n}	D_{2_n}	D_{3_n}	$Q_{1_{n+1}}$	$Q_{2_{n+1}}$	$Q_{3_{n+1}}$
0	0	0	1	0	0	1	0	0
1	0	0	1	1	0	1	1	0
1	1	0	1	1	1	1	1	1
1	1	1	0	1	1	0	1	1
0	1	1	0	0	1	0	0	1
0	0	1	0	0	0	0	0	0
0	1	0	1	0	0	1	0	0
1	0	1	0	1	0	0	1	0

(*b*) State table

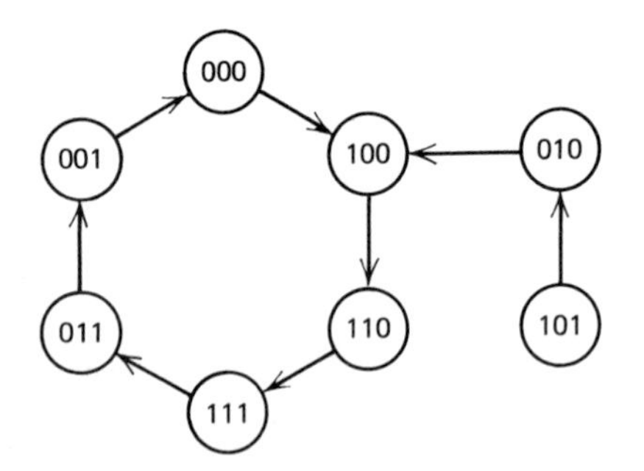

(*c*) State diagram with the states identified as $Q_{1_n}Q_{2_n}Q_{3_n}$

FIGURE 10-9 Three-stage self-starting divide-by-6 Johnson counter.

sequence of 000-100-110-111-011-001-000-100 . . . ; hence it is a Johnson counter with completely glitch-free outputs.

10-3.5 Decoding of Johnson Counters

Figure 10-10*a* shows a decoder for the Johnson counter of Fig. 10-9*a*. When interconnected with Fig. 10-9*a*, the operation of the decoder can be described by the state table shown in Fig. 10-10*b*. This state table was derived from Fig.

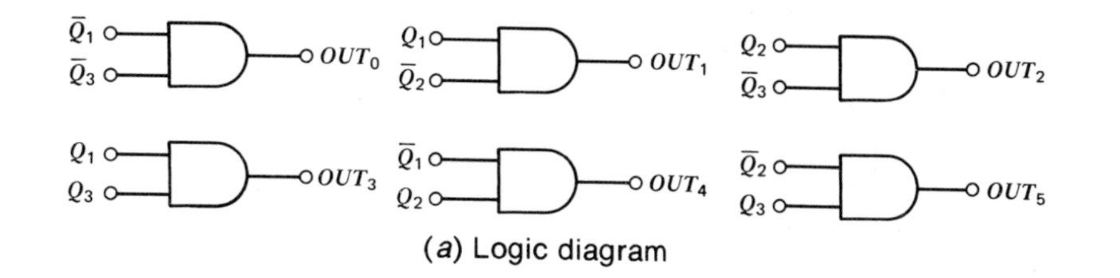

(*a*) Logic diagram

Q_{1_n}	Q_{2_n}	Q_{3_n}	D_{1_n}	D_{2_n}	D_{3_n}	OUT_{0_n}	OUT_{1_n}	OUT_{2_n}	OUT_{3_n}	OUT_{4_n}	OUT_{5_n}	$Q_{1_{n+1}}$	$Q_{2_{n+1}}$	$Q_{3_{n+1}}$
0	0	0	1	0	0	1	0	0	0	0	0	1	0	0
1	0	0	1	1	0	0	1	0	0	0	0	1	1	0
1	1	0	1	1	1	0	0	1	0	0	0	1	1	1
1	1	1	0	1	1	0	0	0	1	0	0	0	1	1
0	1	1	0	0	1	0	0	0	0	1	0	0	0	1
0	0	1	0	0	0	0	0	0	0	0	1	0	0	0
0	1	0	1	0	0	1	0	1	0	1	0	1	0	0
1	0	1	0	1	0	0	1	0	1	0	1	0	1	0

(*b*) State table

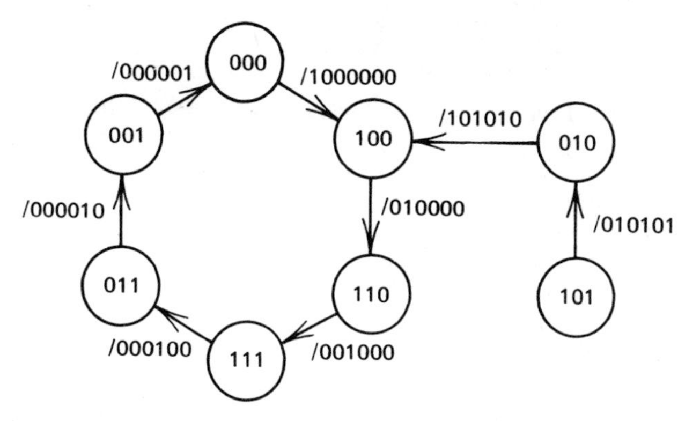

(*c*) State diagram with the states identified as $Q_{1_n}Q_{2_n}Q_{3_n}$ and the branches labeled as $/OUT_0OUT_1OUT_2OUT_3OUT_4OUT_5$

FIGURE 10-10 Decoder for the Johnson counter of Fig. 10-9*a*.

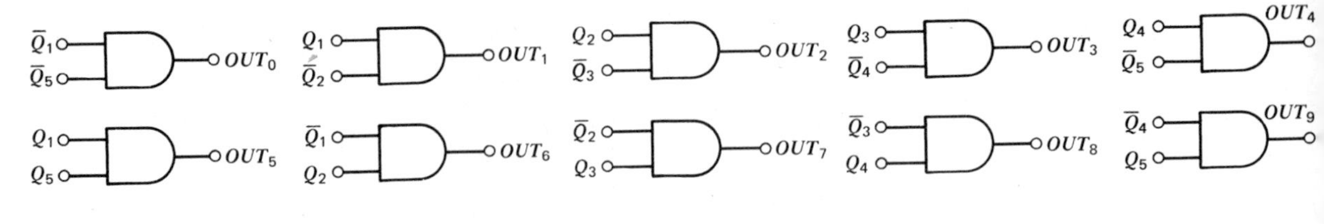

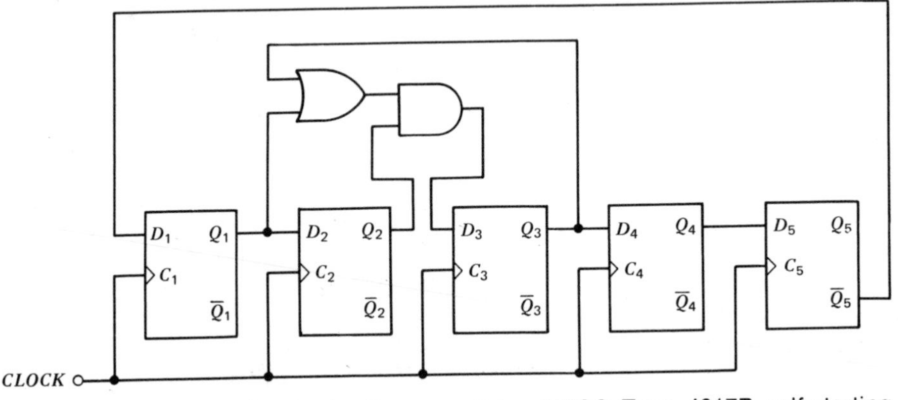

FIGURE 10-11 Simplified logic diagram of the CMOS Type 4017B self-starting divide-by-10 Johnson counter. (The real circuit also includes an *ENABLE* and a direct *CLEAR* input, as well as complementary output $\overline{Q}_5$.) Interconnections of the FF outputs to the decoding AND gates are not shown.

10-9*b* by adding columns OUT_0 through OUT_5. The entries in these columns were obtained from Fig. 10-10*a* as $OUT_0 = \overline{Q}_1\overline{Q}_3$, $OUT_1 = Q_1\overline{Q}_2$, $OUT_2 = Q_2\overline{Q}_3$, $OUT_3 = Q_1Q_3$, $OUT_4 = \overline{Q}_1Q_2$, and $OUT_5 = \overline{Q}_2Q_3$.

Figure 10-10*c* shows a state diagram with the decoded outputs, derived from Fig. 10-10*b*. We can see that the decoded outputs become logic 1 in the sequence of $OUT_0 - OUT_1 - OUT_2 - OUT_3 - OUT_4 - OUT_5 - OUT_0 \ldots$; hence the decoder operates correctly.

10-3.6 Five-Stage Self-Starting Johnson Counter with Decoder

Figure 10-11 shows a simplified logic diagram of the CMOS Type 4017B self-starting divide-by-10 Johnson counter. It consists of 5 FFs and two logic gates. The derivation of a state table and a state diagram is left to the reader as an exercise (Probs. 10-11 and 10-12).

10-4 OPERATING SPEED

The operating speed limitations of shift registers and shift-register counters are similar to the limitations described for synchronous counters in Sec. 10-6.7. Thus, the circuit operates correctly as long as the timing requirements of *each*

FF are satisfied. These timing requirements include f_{max}, $t_{1_{min}}$, t_{SETUP}, t_{HOLD}, $t_{CLEAR_{min}}$, and other FF timing specifications.

SUMMARY

This chapter presented shift registers and shift-register counters. After a short introduction, it described the structure and operation of shift registers, including bidirectional shift registers with parallel loading. This was followed by a description of ring counters and Johnson counters. Operating speed limitations were also outlined.

SELF-EVALUATION QUESTIONS

1. What is a shift register?
2. What is a bidirectional shift register?
3. Describe the operation of a shift register with synchronous parallel loading.
4. Compare the properties of ring counters and Johnson counters.
5. Describe a 3-stage non-self-starting ring counter.
6. Describe a 3-stage self-starting ring counter.
7. Describe a 3-stage non-self-starting Johnson counter.
8. Describe a 3-stage self-starting Johnson counter.
9. Describe a 5-stage self-starting Johnson counter.
10. Describe a decoder for a 3-stage Johnson counter.
11. What limits the operating speed of a shift register?

PROBLEMS

10-1. Sketch a logic diagram of a 2-stage shift register and draw a state table. Include all four states and show the present states of Q_1, Q_2, and IN, the present states of D_1 and D_2, and the next states of Q_1 and Q_2.

10-2. Draw a state table for the 3-stage shift register of Fig. 10-1. Include all eight states and show the present states of Q_1, Q_2, Q_3, and IN, the present states of D_1, D_2, and D_3, and the next states of Q_1, Q_2, and Q_3.

10-3. Draw a state diagram for the 2-stage shift register of Prob. 10-1. Identify the states as $Q_{1_n}Q_{2_n}$ and label the branches as $IN_n/$.

***10-4.** Draw a state diagram for the 3-stage shift register of Fig. 10-1. Identify the states as $Q_{1_n}Q_{2_n}Q_{3_n}$ and label the branches as $IN_n/$. Arrange the states such that no two branches cross each other.

*Optional problem.

10-5. The initial state of each FF in Fig. 10-1 is 0. Input *IN* is 1 during the first clock pulse, and it is 0 during the next four clock pulses. Assume that all setup time and hold time requirements are satisfied and summarize the operation as follows:

a. Sketch a table similar to Table 10-1 showing the states of *IN*, D_1, Q_1, D_2, Q_2, D_3, and Q_3 during 5 clock pulses.

b. Sketch a timing diagram with invisibly short propagation delays.

10-6. The initial state of each FF in Fig. 10-1 is 0. Input *IN* is 1 during the first clock pulse, it is 0 during the next clock pulse, it is 1 during the third clock pulse, and it is 0 during the next four clock pulses. Assume that all setup time and hold time requirements are satisfied and summarize the operation as follows:

a. Sketch a table similar to Table 10-1 showing the states of *IN*, D_1, Q_1, D_2, Q_2, D_3, and Q_3 during seven clock pulses.

b. Sketch a timing diagram with invisibly short propagation delays.

10-7. Figure P10-1 shows the logic diagram of a 4-stage self-starting ring counter. Draw a state table and a state diagram.

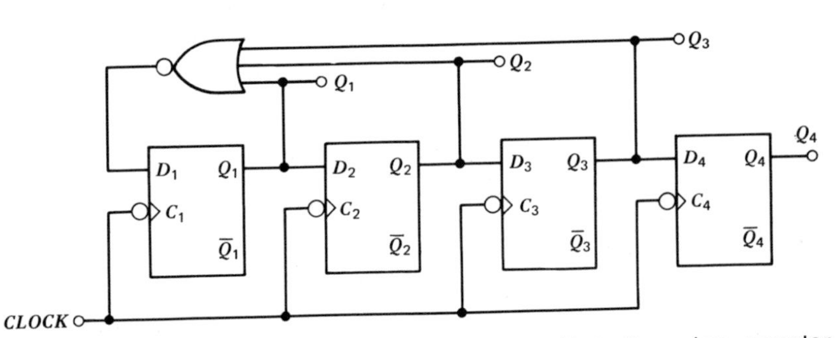

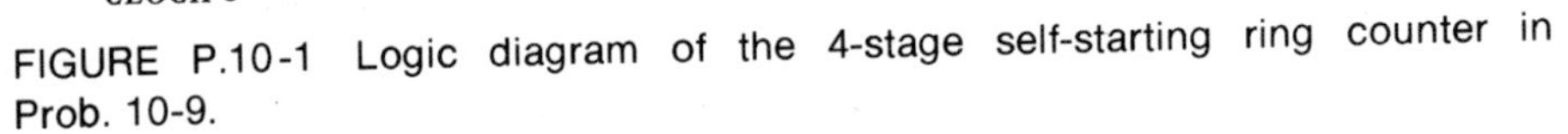
FIGURE P.10-1 Logic diagram of the 4-stage self-starting ring counter in Prob. 10-9.

10-8. Draw a logic diagram, a state table, and a state diagram for a 4-stage non-self-starting ring counter. Identify the main counting loop in the state diagram.

10-9. Draw a logic diagram, a state table, and a state diagram for a 4-stage non-self-starting Johnson counter. Identify the main counting loop in the state diagram.

10-10. Figure P10-2 shows the logic diagram of a 4-stage self-starting Johnson counter. Draw a state table and a state diagram.

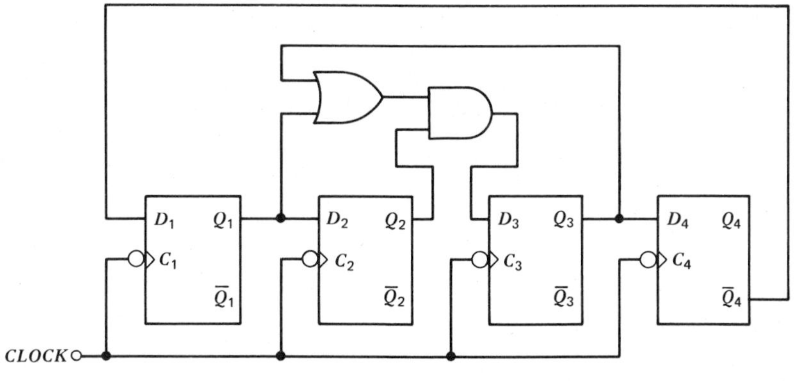

FIGURE P.10-2 Logic diagram of the 4-stage self-starting Johnson counter in Prob. 10-10.

10-11. **a.** Draw a state table for the counter of Fig. 10-11. Include only the 10 states of the main counting loop. Show the present states of the FFs, the present states of the FF control inputs, and the next states of the FFs. Do not show the decoded outputs.

b. Draw a state diagram for the counter of Fig. 10-11. Show only the 10 states of the main counting loop. Identify the states as $Q_{1_n}Q_{2_n}Q_{3_n}Q_{4_n}Q_{5_n}$. Do not label the branches.

***10-12.** **a.** Draw a state table for the counter of Fig. 10-11. Include all 32 states. Show the present states of the FFs, the present states of the FF control inputs, and the next states of the FFs. Do not show the decoded outputs.

b. Draw a state diagram for the counter of Fig. 10-11. Show all 32 states. Identify the states as $Q_{1_n}Q_{2_n}Q_{3_n}Q_{4_n}Q_{5_n}$. Do not label the branches.

10-13. The FFs in the circuit of Fig. 10-6*a* are replaced by three Type 74109 positive edge-triggered J-$\overline{K}$ FFs (see Table 8-1, page 204). The J and $\overline{K}$ inputs of each FF are connected in parallel as a D input. What is the maximum counting frequency of the circuit?

10-14. The FFs in the circuit of Fig. 10-8*a* are replaced by three Type 74109 positive edge-triggered J-$\overline{K}$ FFs (see Table 8-1, page 204). The J and $\overline{K}$ inputs of each FF are connected in parallel as a D input. What is the maximum counting frequency of the circuit?

10-15. The FFs in the circuit of Fig. 10-7*a* are replaced by three Type 74109 positive edge-triggered J-$\overline{K}$ FFs (see Table 8-1, page 204). The

*Optional problem.

propagation delay of the NOR gate is 12 ns. What is the maximum counting frequency of the circuit?

10-16. The FFs in the circuit of Fig. 10-9*a* are replaced by three Type 74109 positive edge-triggered J-$\overline{K}$ FFs (see Table 8-1, page 204). The propagation delay of the OR gate is 12 ns and the propagation delay of the AND gate is 12 ns. What is the maximum counting frequency of the circuit?

CHAPTER 11 LSI and VLSI

Instructional Objectives

This chapter describes large-scale integrated circuits (LSI) and very-large-scale integrated circuits (VLSI), as well as their applications, with additional circuits and applications discussed in later chapters. After you read the chapter you should be able to:

1. Explain the operation of dynamic MOS logic circuits.
2. Discuss the operation of MOS shift registers.
3. Describe the structure of random-access memory (RAM) cells.
4. Explain the operation of read-only memories (ROMs).
5. Describe programmable logic arrays (PLAs).
***6.** Outline the operation of associative memories.
7. Describe the structure of gate arrays.

11-1 OVERVIEW

Large-scale integration (LSI) and very-large-scale integration (VLSI) are terms describing the level of complexity of an integrated circuit (''*IC*'' or ''*chip*''). Three technologies are used in manufacturing LSI and VLSI circuits: (1) bipolar, which includes primarily ECL and I^2L, (2) *n*-channel and *p*-channel MOS, characterized by the smallest area and a low power dissipation per logic gate, and (3) complementary MOS (CMOS), characterized by the lowest dc power dissipation per logic gate, but requiring larger area than *n*-channel or *p*-channel MOS, as well as more complex manufacturing technology.

LSI chips usually exceed a size of 100 × 100 mils, VLSI chips a size of 300 × 300 mils. In such a large size there is a significant probability of *defects* such as, for example, pinholes that reduce the production *yield.* Thus one limitation on the complexity of LSI and VLSI is the size of the chip, which limits the number of logic gates that can be included; maximum allowable power dissipation of the chip constitutes another limitation. The semiconductor in-

*Optional material.

dustry has been very inventive in reducing size and power dissipation per logic gate, resulting in increasingly complex circuits.

The advent of LSI and VLSI circuits led to improved performance in several areas of application: (1) a drastic reduction in the volume required by electronic equipment such as digital computers, electronic instrumentation, and communication equipment; (2) a considerable decrease in the power required to implement a given function; (3) improved reliability, partly as a result of better semiconductor manufacturing techniques and also because many interconnections are now internal to the IC and are thus more reliable than external wiring; (4) higher speeds because the short interconnections between components inside a chip have low stray capacitances to ground.

This chapter describes first the operation of various MOS logic circuits and shift registers. This is followed by descriptions of random-access memory (RAM) circuits and organizations, read-only memories (ROMs), programmable logic arrays (PLAs), and associative memories. The chapter concludes with a discussion of gate arrays.

11-2 DYNAMIC MOS LOGIC CIRCUITS

Fully static MOS logic inverters and gates were discussed in Sec. 7-5. The fully static logic inverter of Fig. 7-15*a* (page 158) consists of a *driver* (pull-down) enhancement-mode device Q_1 and of a *load* (pull-up) depletion-mode device Q_2. In order to obtain the transfer characteristic of Fig. 7-15*b*, Q_2 must have a resistance that is large compared to the resistance of Q_1; this results in a device with a large area—hence in a limitation on the number of logic gates that can be included on an IC chip. Also, both Q_1 and Q_2 conduct when the input is HIGH, resulting in a continuously dissipated power—hence in another limitation on the number of logic gates that can be included on an IC chip. Thus the two parameters of area and power dissipation that are essential for high-density LSI and VLSI are not favorable in a fully static MOS logic circuit.

Dynamic MOS logic circuits are made possible by the high resistance of MOS devices in their cutoff state, making capacitive charge storage practical for durations of milliseconds. The power dissipation is lower in dynamic MOS logic circuits than in static MOS logic circuits; also, in dynamic MOS logic circuits all devices are small and of identical size. Thus dynamic MOS logic circuits permit higher circuit density on an IC chip.

The disadvantages of dynamic MOS logic circuits are the somewhat more complex circuitry to implement a given function and the requirement for two or more *clock* signals that must be accurately timed with respect to each other. However, the advantages of dynamic operation usually outweigh the disadvantages when LSI and VLSI circuits with the highest logic gate density are desired.

Figure 11-1*a* shows a circuit diagram of a dynamic 2-phase MOS logic inverter. It consists of two parts: A *sampling circuit* consisting of driver device

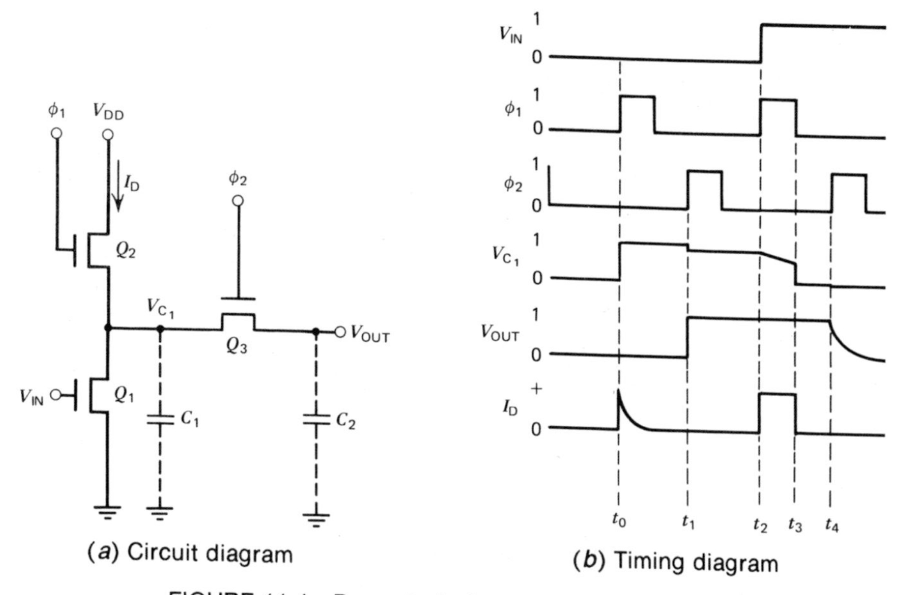

(*a*) Circuit diagram (*b*) Timing diagram

FIGURE 11-1 Dynamic 2-phase MOS logic inverter.

Q_1, load device Q_2, and stray capacitance C_1; and of an *output circuit* consisting of *transfer gate* Q_3 and stray capacitance C_2.

Figure 11-1*b* shows a timing diagram. The timing sequence is governed by the two *clock phases* ϕ_1 and ϕ_2, which are *nonoverlapping*, that is, they are never both HIGH at the same time.

The gate terminal of load device Q_2 in Fig. 11-1*a* is HIGH only while *clock phase* ϕ_1 is HIGH. Thus power can be drawn from the power supply only during the ϕ_1 = HIGH periods, irrespective of the state of input voltage V_{IN}—as can be seen in power supply current I_D shown as the last waveform in Fig. 11-1*b*.

When V_{IN} is LOW (times $t < t_2$ in Fig. 11-1*b*), Q_1 is OFF and V_{C_1} charges to near V_{DD} via Q_2 while ϕ_1 is HIGH. However, when V_{IN} is HIGH (times $t > t_2$), a V_{C_1} of close to zero is reached after the ϕ_1 = HIGH interval. A later clock phase, ϕ_2, turns Q_3 ON and transfers charge from C_1 to C_2. Since C_1 is much larger than C_2, V_{OUT} becomes almost V_{DD} when V_{IN} is LOW ($t < t_2$) and it becomes near zero when V_{IN} is HIGH ($t > t_2$).

Output V_{OUT} thus becomes the complement of input V_{IN}. The clock signals provide three functions: (1) ϕ_1 samples input V_{IN}; (2) ϕ_2 transfers the complement of V_{IN} to the output; and (3) both clocks reinforce the charge states of capacitors C_1 and C_2 through periodic recharging ("*refresh*"). A *minimum clock frequency* is, therefore, specified for dynamic MOS circuits to prevent loss of data because of charge or discharge of capacitors by leakage currents; this minimum clock frequency is in the vicinity of 1 kHz, that is, the maximum permissible time interval between consecutive refreshes is in the vicinity of a millisecond.

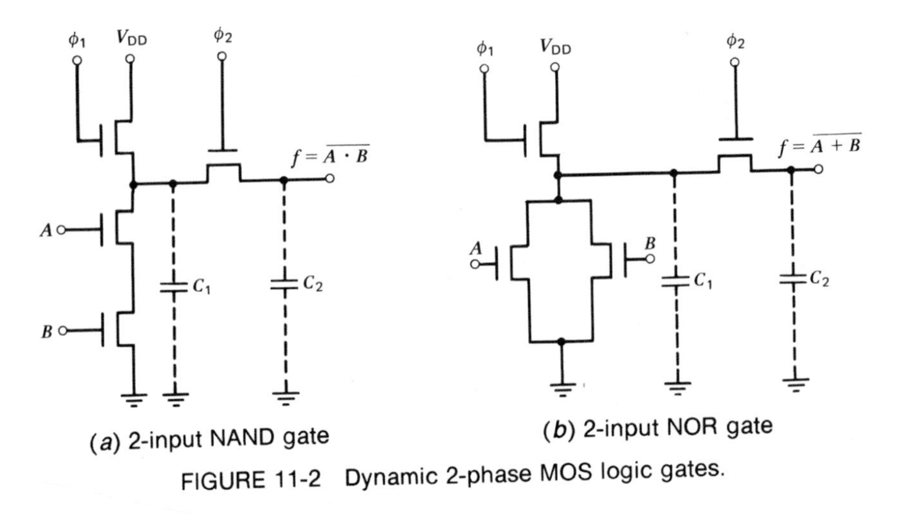

(*a*) 2-input NAND gate (*b*) 2-input NOR gate

FIGURE 11-2 Dynamic 2-phase MOS logic gates.

The dynamic 2-phase logic inverter described above may also be extended to other MOS logic gates. Figure 11-2 shows schematic diagrams of a dynamic 2-phase MOS NAND gate and of a dynamic 2-phase MOS NOR gate.

11-3 DYNAMIC MOS SHIFT REGISTERS

Figure 11-3*a* shows a circuit diagram of one stage of dynamic 2-phase MOS shift register. It consists of two *fully static logic inverters* (Fig. 7-15a, page 158), Q_2-Q_3 and Q_5-Q_6, and of two transfer gates Q_1 and Q_4. The timing sequence of Fig. 11-3*b* shows two nonoverlapping clock phases ϕ_1 and ϕ_2. The state of the *DATA IN* terminal is transferred to C_1 when ϕ_1 is HIGH, and its complement gets established at point B shortly thereafter. The state of point B is transferred to C_2 when ϕ_2 is HIGH and its complement gets established at the *DATA OUT*

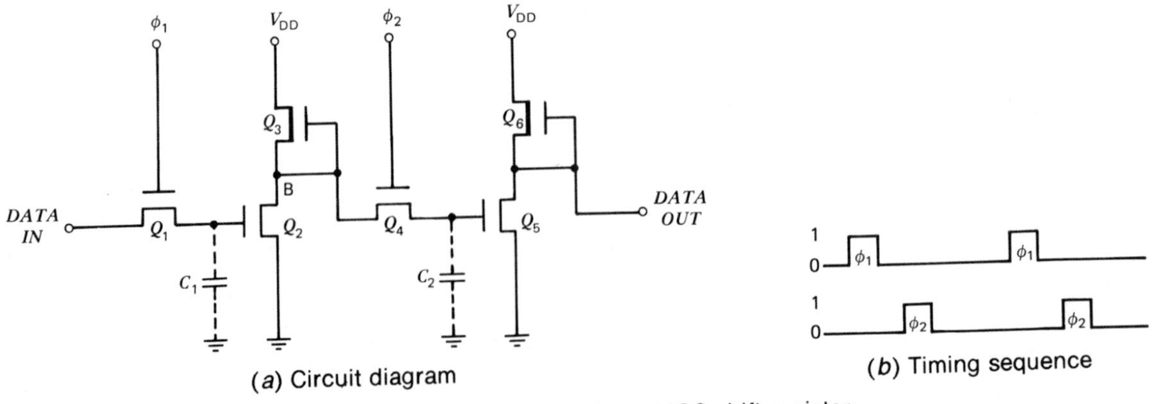

(*a*) Circuit diagram (*b*) Timing sequence

FIGURE 11-3 Dynamic 2-phase MOS shift register.

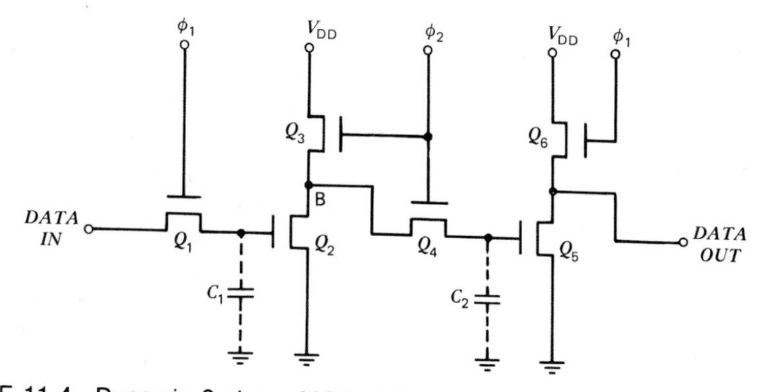

FIGURE 11-4 Dynamic 2-phase MOS shift register with clocked load devices Q_3 and Q_6.

terminal shortly thereafter. Note that there is no logic inversion from *DATA IN* to *DATA OUT*.

Thus the circuit acts as a master–slave D FF. Transfer gate Q_1, capacitance C_1, and logic inverter Q_2-Q_3 constitute the master latch; transfer gate Q_4, capacitance C_2, and logic inverter Q_5-Q_6 the slave latch.

In Fig. 11-3*a*, depletion-mode load devices Q_3 and Q_6 are continuously on. Thus, when *DATA IN* is HIGH, Q_2 and Q_3 conduct following the application of ϕ_1 = HIGH, since capacitor C_1 retains its logic HIGH level even after ϕ_1 becomes LOW and turns Q_1 off. Similarly, when *DATA IN* is LOW, point B becomes HIGH, and Q_5 and Q_6 conduct following the application of ϕ_2 = HIGH. As a result, unlike the dynamic MOS logic circuits of Figs. 11-1 and 11-2, the dynamic MOS shift register of Fig. 11-3 may dissipate power continuously.

The power dissipation in the circuit of Fig. 11-3 may be reduced considerably by *clocking* load devices Q_3 and Q_6 by ϕ_2 and ϕ_1, respectively. This is shown in Fig. 11-4, where the timing sequence of Fig. 11-3*b* is again applicable. Since logic inverters Q_2-Q_3 and Q_5-Q_6 in Fig. 11-4 draw current from the power supply only when their input is HIGH *and* the corresponding clock phase is HIGH, power dissipation is reduced considerably from that of Fig. 11-3*a*.

The disadvantage of the clocked load devices in Fig. 11-4 is that they increase the capacitive loading of the ϕ_1 and ϕ_2 clock lines to approximately twice of what they are in Fig. 11-3*a* and they also somewhat increase the ac power dissipation. Still, a decreased overall power dissipation is attained at the cost of decreased operating speed.

Dynamic MOS shift registers are used primarily as digital delays and as serial memories. In such an application, the input of the shift register is connected by logic gates to the data input during data entry and to the output of the shift register during storage, thus recirculating the contents to ensure refreshing, while a separate counter keeps track of the location of the data. For example, the Type 9401 IC consists of two dynamic MOS 1024-bit shift registers with a maximum clock frequency of 2 MHz; the two shift registers may also be used as a single 2048-bit memory with an 11-bit counter keeping track of the location of the data.

11-4 STATIC MOS CIRCUITS

Figure 8-20 (page 189) showed a master–slave FF where the nonoverlapping clocks of the master and slave latches were derived from a *single* clock input by means of a battery and a logic inverter; thus only one clock signal had to be distributed. In contrast, as seen in Figs. 11-1*b* and 11-3*b*, in MOS circuits we often prefer a separate distribution of the two clock phases in order to maintain a more accurate control of their timing. In fact, ϕ_1 and ϕ_2 of Fig. 11-1*b* or Fig. 11-3*b* could also be used as clocks for the master and slave latches in Fig. 8-20.

Note that the two clocks are nonoverlapping in Fig. 8-20, as well as in Figs. 11-1*b* and 11-3*b*, that is, they cannot both be HIGH at the same time. Thus there remain three possibilities: One clock is HIGH, the other clock is HIGH, or both clocks are LOW.

The master–slave FF of Fig. 8-20 is *fully static*, that is, it can preserve stored data indefinitely. This is true during all three allowed combinations of the two clocks: when the clock of the master latch is held HIGH, when the clock of the slave latch is held HIGH, or when both clocks are held LOW. This is not true in the *dynamic* MOS circuits that we have discussed thus far: Data are lost if we stop the clocks for much longer than a millisecond, no matter whether we stop while ϕ_1 is HIGH, while ϕ_2 is HIGH, or while both ϕ_1 and ϕ_2 are LOW.

In many systems it would be desirable to be able to stop the clocks without loss of data—for example, for diagnostic purposes. This can be attained by use of *fully static* MOS circuits; such circuits were introduced in Sec. 7-5 and are further discussed later in this chapter. In what follows here, we describe an MOS circuit that permits stopping the clocks without loss of data, but only at certain times—for example, while ϕ_1 is HIGH. Such circuits are called *quasi-static* MOS circuits, or simply *static* MOS circuits.

Figure 11-5 shows a circuit diagram of a *static 2-phase MOS clocked D-latch*, where the timing sequence of Fig. 11-3*b* is applicable. It consists of two fully static logic inverters, Q_2-Q_3 and Q_4-Q_5, and of two transfer gates, Q_1 and Q_6. The state of *DATA IN* is entered via transfer gate Q_1 by a ϕ_1 = HIGH, held temporarily on stray capacitance *C* while ϕ_1 and ϕ_2 are LOW, and is latched as a stable state via transfer gate Q_6 by a ϕ_2 = HIGH.

Note that there is no logic inversion from *DATA IN* to *DATA OUT* and that complementary output $\overline{DATA\ OUT}$ is available at point A in the circuit. Also, a static D FF can be built using the latch of Fig. 11-5 as a master latch, and an additional, identical latch that has the connections of ϕ_1 and ϕ_2 interchanged, as a slave latch.

In the static MOS circuit of Fig. 11-5 there is no loss of data if we stop for any length of time while *either* ϕ_1 *or* ϕ_2 is HIGH. This is in contrast with the dynamic circuits of Figs. 11-3*a* and 11-4, where such stopping is not possible at any time without loss of data. However, even in the circuit of Fig. 11-5 there is loss of data if we stop the clocks while *both* ϕ_1 and ϕ_2 are LOW—if we stop for much longer than a millisecond. Hence, the circuit of Fig. 11-5 is *not fully static*, but it is only *static* (or *quasi-static*).

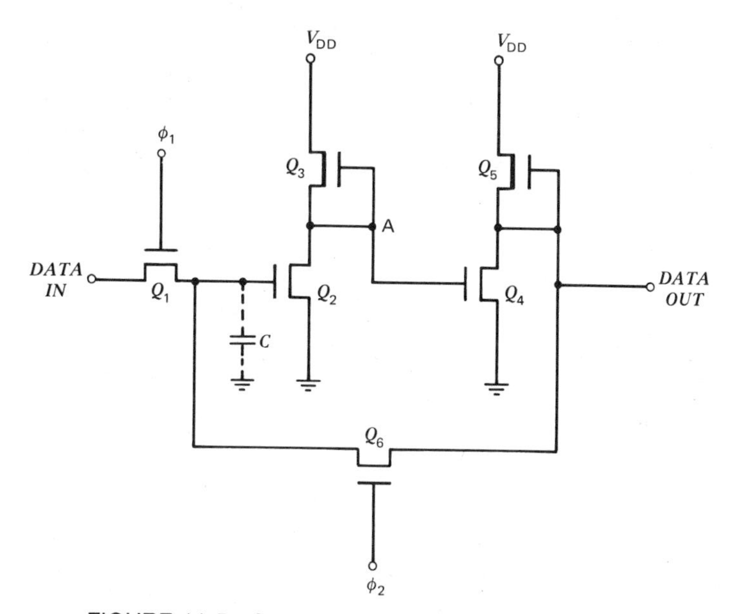

FIGURE 11-5 Static 2-phase MOS clocked D latch.

11-5 RANDOM-ACCESS MEMORIES (RAMs)

In a *random-access memory* (*RAM*) we have a direct, or *random*, read and write access to any bit, or *cell*, of data storage. Also, in a RAM read and write *access times* are comparable; when this is not the case, the memory is not considered a RAM (we return to such memories later in this chapter). In what follows here, we describe dynamic MOS RAMs, fully static MOS RAMs, bipolar RAMs, organization of RAM ICs, and memory systems.

11-5.1 Dynamic MOS RAMs

The most widely used dynamic MOS RAM cells consist of either 1 or 3 MOS devices. Figure 11-6 shows a 1-transistor dynamic MOS RAM cell. Information is retained in the cell by means of charge stored on stray capacitance C, which must be *refreshed* by a write or by a read-rewrite operation at least once every millisecond.

The cells are laid out in a rectangular array with each horizontal *WORD SELECT LINE* shared by a row of cells and each vertical *BIT LINE* shared by a column of cells. When the *WORD SELECT LINE* of a cell is LOW, MOS device Q_1 is cut off and the cell is inactive. A cell is *selected* for a read or a write operation by raising its *WORD SELECT LINE* to HIGH, thus turning on Q_1. In a write operation capacitance C is charged via Q_1 to the logic level present on the *BIT LINE*. In a read operation charge is transferred via Q_1 from capacitance C to the *BIT LINE*, where it is detected by a *sense amplifier* that is shared by a column of cells. Note that a read operation is *destructive* since it depletes the

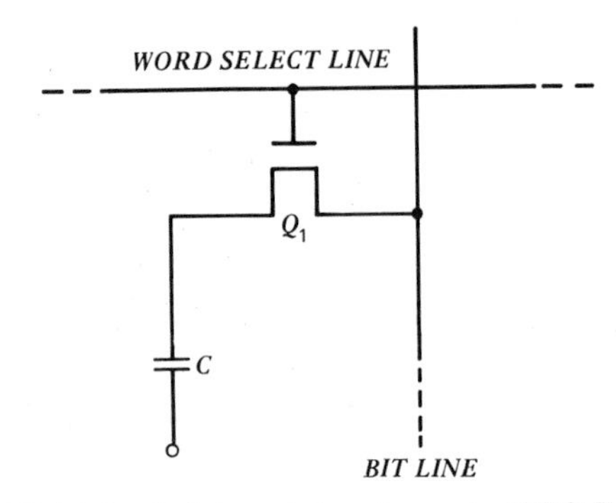

FIGURE 11-6 A 1-transistor dynamic MOS RAM cell.

charge stored on capacitance C; hence, it has to be followed by a rewrite operation.

Because of their simplicity and small size, dynamic MOS RAM cells are especially suited for use in large RAM IC chips: 64 Kbit dynamic LSI RAMs have attained widespread use and 256 Kbit dynamic VLSI RAMs are also available. Note that while the abbreviation k stands for 1000, the abbreviation K stands for 1024; thus 64 Kbits = 65,536 bits and 256 Kbits = 262,144 bits.

11-5.2 Fully Static MOS RAMs

The most widely used fully static MOS RAM cells consist of either 6 or 8 MOS devices. Figure 11-7 shows a 6-transistor fully static MOS RAM cell. It consists of two cross-coupled fully static logic inverters, Q_1-Q_3 and Q_2-Q_4, and of two transfer gates, Q_5 and Q_6.

The cells are laid out in a rectangular array with each horizontal *ROW SELECT LINE* shared by a row of cells and each vertical *BIT LINE* and $\overline{BIT\ LINE}$

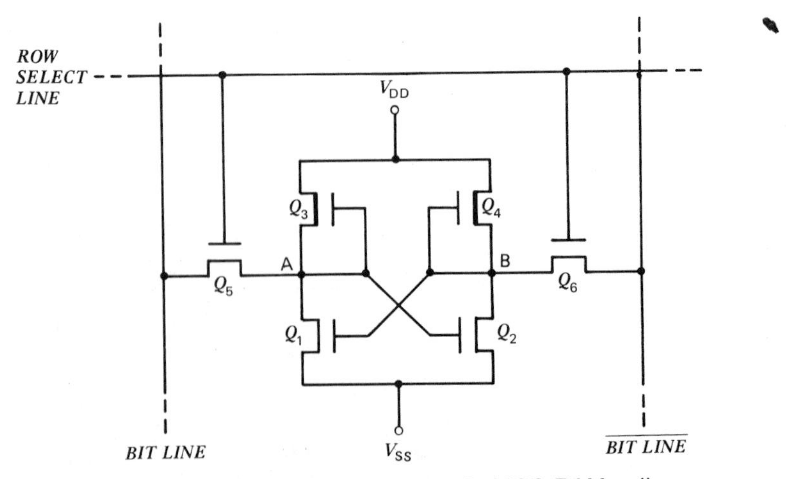

FIGURE 11-7 A 6-transistor fully static MOS RAM cell.

shared by a column of cells. When the *ROW SELECT LINE* of a cell is LOW, transfer gates Q_5 and Q_6 are cut off and the cell is inactive. A cell is selected for a read or a write operation by raising its *ROW SELECT LINE* to HIGH, thus turning on transfer gates Q_5 and Q_6. In a write operation either the *BIT LINE* is LOW and the $\overline{BIT\ LINE}$ is HIGH, or the *BIT LINE* is HIGH and the $\overline{BIT\ LINE}$ is LOW. In the latter case the $\overline{BIT\ LINE}$ forces a LOW state at point B via transfer gate Q_6; in the former case the *BIT LINE* forces a LOW state at point A via transfer gate Q_5. In a read operation the states of the *BIT LINE* and $\overline{BIT\ LINE}$ are sensed by a differential sense amplifier that is shared by a column of cells. Note that, unlike in a dynamic MOS RAM cell, a read operation in a fully static MOS RAM cell is *nondestructive*.

Because their read operations are nondestructive, fully static RAMs do not require rewrite operations; neither do they require refreshing. For these reasons, the peripheral circuitry of fully static RAMs is simpler than those of dynamic RAMs. However, as compared to dynamic RAM cells, fully static RAM cells are more complex and are larger in size, and hence fewer of them can be included on an IC chip.

11-5.3 Bipolar RAMs

The most widely used bipolar RAM cells consist of two dual-emitter or triple-emitter transistors. Figure 11-8 shows a triple-emitter bipolar RAM cell. It consists of two cross-coupled triple-emitter bipolar transistors, Q_1 and Q_2, and of two resistors *R*.

The cells are laid out in a rectangular array with each horizontal *ROW SELECT LINE* shared by a row of cells, and each vertical *COLUMN SELECT LINE*, as well as each vertical *BIT LINE* and $\overline{BIT\ LINE}$, shared by a column of cells. A cell is selected when both its *ROW SELECT LINE* and its *COLUMN*

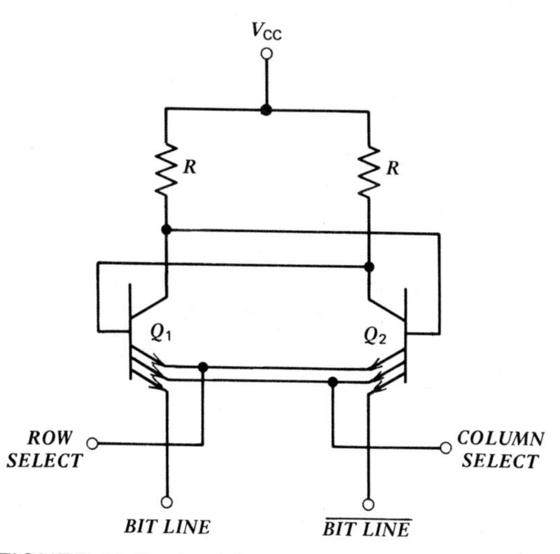

FIGURE 11-8 A triple-emitter bipolar RAM cell.

SELECT LINE are HIGH. In a write operation either the *BIT LINE* is LOW and the $\overline{BIT\ LINE}$ is HIGH, or the *BIT LINE* is HIGH and the $\overline{BIT\ LINE}$ is LOW; the latter turns Q_2 on, the former turns Q_1 on. In a read operation, the states of the *BIT LINE* and the $\overline{BIT\ LINE}$ are sensed by a differential sense amplifier that is shared by a column of cells. Note that, as was the case in a fully static MOS RAM cell, a read operation in a bipolar RAM cell is nondestructive.

Bipolar RAMs are especially suited for applications where short read and write access times are desired. While access times are typically between 30 ns and 500 ns in MOS RAMs, they are in the vicinity of 5 ns in the fastest bipolar RAMs. However, as compared to MOS RAM cells, bipolar RAM cells require more area and dissipate more power; hence, fewer of them can be included on an IC chip: 16 Kbit bipolar RAM ICs are available.

11-5.4 RAM IC Organization

This section describes the organization of RAM IC chips. (Memory systems are outlined in Sec. 11-5.5.) The two principal organization schemes of RAM IC chips are *bit organization* and *word organization.*

In bit organization, each memory access writes or reads a single bit of the memory. In word organization, each memory access writes or reads one m-bit word, where *word length* m is often 4 bits (a *nibble*) or 8 bits (a *byte*). In both organizations, access is selected via n coded *ADDRESS* lines for an overall memory capacity of $m \cdot 2^n$ bits (in bit organization the word length $m = 1$).

In bit organization, data access is usually via two separate lines for writing and for reading: via a unidirectional (one-way) *DATA IN* line for writing and via a unidirectional *DATA OUT* line for reading, although the two lines are sometimes combined into a single bidirectional (two-way) *DATA* line that is used both for writing and for reading. In word organization, data access is usually via a bidirectional *DATA* bus that consists of m bidirectional *DATA* lines used both for writing and for reading, although ICs with separate *DATA IN* and *DATA OUT* buses are also available.

Thus, in general, for a given overall memory capacity there are fewer *ADDRESS* lines but more *DATA* lines in a word-organized memory than in a bit-organized memory.

EXAMPLE 11-1

Find the number of *DATA* lines and the number of *ADDRESS* lines of a RAM IC with an overall memory capacity of 4 Kbits if

a. It has bit organization,

b. It has word organization with a word length of $m = 8$ bits (''*byte organization*'').

Solution

a. In bit organization we have **1** *DATA* line, or one pair of *DATA IN* and *DATA OUT* lines. Also, $2^{12} = 4096 = 4$ K (see Appendix for a table of powers of 2)—thus we have $n =$ **12** *ADDRESS* lines.

b. In word organization with a word length of $m = 8$ bits we have **8** *DATA* lines, or possibly 8 *DATA IN* lines and 8 *DATA OUT* lines. Also, we have an overall memory capacity of $m \cdot 2^n = 4096$ bits. With $m = 8$ this becomes $8 \cdot 2^n = 4096$, that is, $2^n = 4096/8 = 512$. Since $512 = 2^9$, we have $n =$ **9** *ADDRESS* lines.

In a bit-organized memory, we require n *ADDRESS* lines for an overall memory capacity of 2^n bits. Thus, for example, we require 14 *ADDRESS* lines for a bit-organized 16-Kbit RAM since 16 K = 16,384 = 2^{14}. However, the connection of this many *ADDRESS* lines is often impractical in an IC package, because of limitations on the number of available pins. In such cases, we may want to reduce the number of *ADDRESS* line connections. This can be done by using only 7 *ADDRESS* lines in the above example, through which lines the two 7-bit halves of the 14-bit address are transmitted sequentially. This scheme requires two separate *ADDRESS STROBE* lines to store (latch) the two halves of the address in two separate registers in the RAM; however, it reduces the number of address line connections of the IC by a factor of 2.

EXAMPLE 11-2

A bit-organized RAM has an overall memory capacity of 64 Kbits. Find the required number of *ADDRESS* lines and *ADDRESS STROBE* lines if

a. Each *ADDRESS* line carries 1 address bit,

b. Each *ADDRESS* line carries 2 address bits.

Solution

Since 64 K = 65,536 = 2^{16}, we have $n = 16$ address bits.

a. Each *ADDRESS* line carries 1 address bit; hence, there are **16** *ADDRESS* lines; also, **1** *ADDRESS STROBE* line is needed. Thus, a total of 17 IC pins are required for the connection of the *ADDRESS* and *ADDRESS STROBE* lines.

b. Each *ADDRESS* line carries 2 address bits, hence there are only $16/2 =$ **8** *ADDRESS* lines; also, **2** *ADDRESS STROBE* lines are needed. Thus, a total of 10 IC pins are required for the connection of the *ADDRESS* and *ADDRESS STROBE* lines—a saving of 7 pins as compared to (**a**).

The operation of a RAM IC chip may be controlled by several control lines such as $\overline{CE}$ ($\overline{CHIP\ ENABLE}$), $READ/\overline{WRITE}$, $\overline{WE}$ ($\overline{WRITE\ ENABLE}$). In dynamic RAMs we may also have an $\overline{ADDRESS\ STROBE}$, or a $\overline{RAS}$ ($\overline{ROW\ ADDRESS\ STROBE}$) and a $\overline{CAS}$ ($\overline{COLUMN\ ADDRESS\ STROBE}$) when *ADDRESS* lines carry 2 address bits.

When $\overline{CE}$ is HIGH, it completely deactivates the chip irrespective of all other controls. It does this by inhibiting addressing and write operations, and by forcing the 3-state output drivers of the *DATA* or *DATA OUT* lines to their high-impedance output states. When $\overline{CE}$ is LOW, it permits the operation of the other control lines.

When $READ/\overline{WRITE}$ is HIGH, it permits read operations and inhibits write operations. When it is LOW, it permits write operations and forces the 3-state output drivers of the *DATA* or *DATA OUT* lines to their high-impedance output states. When $\overline{WE}$ is HIGH, it inhibits write operations; when it is LOW, it permits write operations.

An $\overline{ADDRESS\ STROBE}$ is provided in dynamic RAMs if each *ADDRESS* line carries 1 address bit. When $\overline{ADDRESS\ STROBE}$ is LOW, it sets the address stored on the chip to the one present on the *ADDRESS* lines. A $\overline{RAS}$ and a $\overline{CAS}$ are provided in dynamic RAMs if each *ADDRESS* line carries 2 address bits. When $\overline{RAS}$ is LOW, it sets the row address half of the address stored on the chip to the row address half of the address present on the *ADDRESS* lines; when $\overline{CAS}$ is LOW, it sets the column address half of the address stored on the chip to the column address half of the address present on the *ADDRESS* lines.

From the above discussion we can see that there may be as many as five control lines in a RAM chip. This may result in a number of pin connections that is undesirably high.

EXAMPLE 11-3

A bit-organized RAM chip with separate *DATA IN* and *DATA OUT* lines has an overall memory capacity of 64 Kbits. Each *ADDRESS* line carries 2 address bits.

a. List all connections of the RAM chip.
b. How many connections are there?
c. Can all connections of the chip be accommodated in a 16-pin DIP package?

Solution

a. In Ex. 11-2 we found that for such a RAM there are 8 *ADDRESS* lines. We also have a *DATA IN* and a *DATA OUT* line, as well as $\overline{CE}$, $READ/\overline{WRITE}$, $\overline{WE}$, $\overline{RAS}$, $\overline{CAS}$, +5 V, and ground.
b. There are altogether **17** connections to the chip.
c. **No**. The 17 connections cannot be accommodated in a 16-pin DIP package.

We should note that not all the control lines discussed above are necessary for the operation of a RAM IC chip, although they are often convenient for the memory system that uses the RAM. Thus, a RAM chip can be operated without a $\overline{CE}$ line; also, depending on the details of the RAM, $READ/\overline{WRITE}$ or $\overline{WE}$ may be omitted.

11-5.5 Memory Systems

The choice of a RAM IC is largely governed by the memory system in which the IC is used. A choice that is dominated by stringent speed requirements is illustrated in Ex. 11-4, which follows.

EXAMPLE 11-4

A high-speed *data acquisition system* operates with a word length of 8 bits. It requires a memory that has a capacity of 1024 words and read and write access times of 15 ns. What RAMs should we use?

Solution

Bipolar bit-organized RAM ICs with a capacity of 1024 bits and read and write access times of 15 ns are readily available. We use 8 such ICs, one for each bit of the 8-bit word.

When speed requirements of the memory system are not too stringent, MOS RAMs may be used. In comparing dynamic and fully static MOS RAMs, we should note that while the memory capacity of fully static MOS RAM ICs is smaller, for dynamic MOS RAMs we have the overhead of the refresh circuitry. Thus, in general, fully static MOS RAMs are preferable for smaller memory systems and dynamic MOS RAMs for larger memory systems. As a rough guide, with present technology the crossover is in the vicinity of 10 Kbytes to 50 Kbytes, that is, 80 Kbits to 400 Kbits. This is illustrated in two examples that follow: in Ex. 11-5 for a small memory system and in Ex. 11-6 for a larger memory system.

EXAMPLE 11-5

A medium-speed data acquisition system operates with a word length of 8 bits. It requires a memory that has a capacity of 4096 words = 4 Kwords and read and write access times of 150 ns. What RAMs should we use?

Solution

The required overall memory capacity is 4 Kwords · 8 bits/word = 32 Kbits = 4 Kbytes. For such a small memory we use fully static MOS RAMs. Bit-organized fully static MOS RAM ICs with a capacity of 4 Kbits (4096 bits) and read and write access times of 150 ns are readily available. We use 8 such ICs, one for each bit of the 8-bit word.

EXAMPLE 11-6

A small computer operates with a word length of 8 bits. It requires a memory that has a capacity of 128 Kwords and read and write access times of 500 ns. What RAMs should we use?

Solution

The required overall memory capacity is 128 Kwords · 8 bits/word = 1024 Kbits = 128 Kbytes. For such a memory size we use dynamic MOS RAMs. Bit-organized dynamic MOS RAM ICs with a capacity of 64 Kbits and read and write access times of 500 ns are readily available. We use 16 such ICs, two for each bit of the 8-bit word.

The choice between the use of a RAM IC that has bidirectional *DATA* lines or one that has *unidirectional DATA IN* and *DATA OUT* lines is governed by the memory system in which the IC is used. If the memory system uses separate *DATA IN* and *DATA OUT* lines, we should use ICs that likewise have separate *DATA IN* and *DATA OUT* lines; such usage eliminates the need for additional routing circuitry. If the memory system uses bidirectional *DATA* lines we may, of course, use ICs that likewise have bidirectional *DATA* lines; however, we may also use ICs that have separate *DATA IN* and *DATA OUT* lines as illustrated in Ex. 11-7, which follows.

EXAMPLE 11-7

The bit-organized 65 Kbit RAM IC used in Ex. 11-6 has a unidirectional *DATA IN* line and a unidirectional *DATA OUT* line. Describe the connection of these lines if

a. The memory system has separate unidirectional *DATA IN* and *DATA OUT* lines.

b. The memory system has bidirectional *DATA* lines.

Solution

a. We connect the two *DATA IN* lines of the ICs to the *DATA IN* line of the memory system, and the two *DATA OUT* lines of the ICs to the *DATA OUT* line of the memory system. Also, at all times we keep one of the two IC *DATA OUT* lines in its high-impedance state by holding the $\overline{CE}$ ($\overline{CHIP\ ENABLE}$) of the IC HIGH—thus also deactivating the *DATA IN* line of the IC.

b. We connect the two *DATA IN* lines *and* the two *DATA OUT* lines of the ICs to the *DATA* line of the memory system. Also, we operate the $\overline{CE}$ lines of the ICs the same way as described in (**a**).

11-6 READ-ONLY MEMORIES (ROMs)

Read-only memories (*ROMs*) are random-access memories into which information has been written permanently. In what follows here, we outline ROM cells, ROM IC organization, programmable and reprogrammable ROMs, and applications and availability.

11-6.1 ROM Cells

The binary state of a ROM cell may be programmed by the connection of a diode, a bipolar transistor, or a MOS device.

EXAMPLE 11-8

Figure 11-9 shows a small ROM using 16 diode cells; resistors R_1 and R_2 are chosen such that $R_2 \gg R_1$. The COLUMN DECODER and the ROW DECODER each decode two input variables onto four lines, of which only one is *selected* (HIGH) at any given time. The state of the *OUTPUT* is determined by the diode at the intersection of the selected column and the selected row. If this diode is not connected, then the *OUTPUT* is pulled LOW via Q_1-D_1, Q_2-D_2, Q_3-D_3, or Q_4-D_4, whichever is selected by the row decoder; if the diode is connected, it does not allow the *OUTPUT* to be pulled LOW and the *OUTPUT* is HIGH.

Demonstrate that the circuit of Fig. 11-9 realizes the logic function $OUTPUT = AB + C\bar{D} + AC + \bar{A}\bar{B}\bar{C}D$.

Solution

The *OUTPUT* is HIGH when the selected diode is connected. In the leftmost column of Fig. 11-9, $A = 0$ and $B = 0$, that is, we have the combination $\bar{A}\bar{B}$. Also, two diodes are connected in this column that are selected when $C = 0$ (LOW) and $D = 1$ (HIGH), or when $C = 1$ and $D = 0$, that is, we have the combinations $\bar{C}D$ and $C\bar{D}$. Thus the leftmost column of the ROM realizes the product terms $\bar{A}\bar{B}\bar{C}D$ and $\bar{A}\bar{B}C\bar{D}$. Similarly, the $AB = 01$ column realizes the product term $\bar{A}BC\bar{D}$, the $AB = 11$ column the product terms $AB\bar{C}\bar{D}$, $AB\bar{C}D$, $ABCD$, and $ABC\bar{D}$, and the $AB = 10$ column the product terms $A\bar{B}CD$ and $A\bar{B}C\bar{D}$. Thus,

$$OUTPUT = \bar{A}\bar{B}\bar{C}D + \bar{A}\bar{B}C\bar{D} + \bar{A}BC\bar{D} + AB\bar{C}\bar{D} + AB\bar{C}D + ABCD + ABC\bar{D} + A\bar{B}CD + A\bar{B}C\bar{D}$$

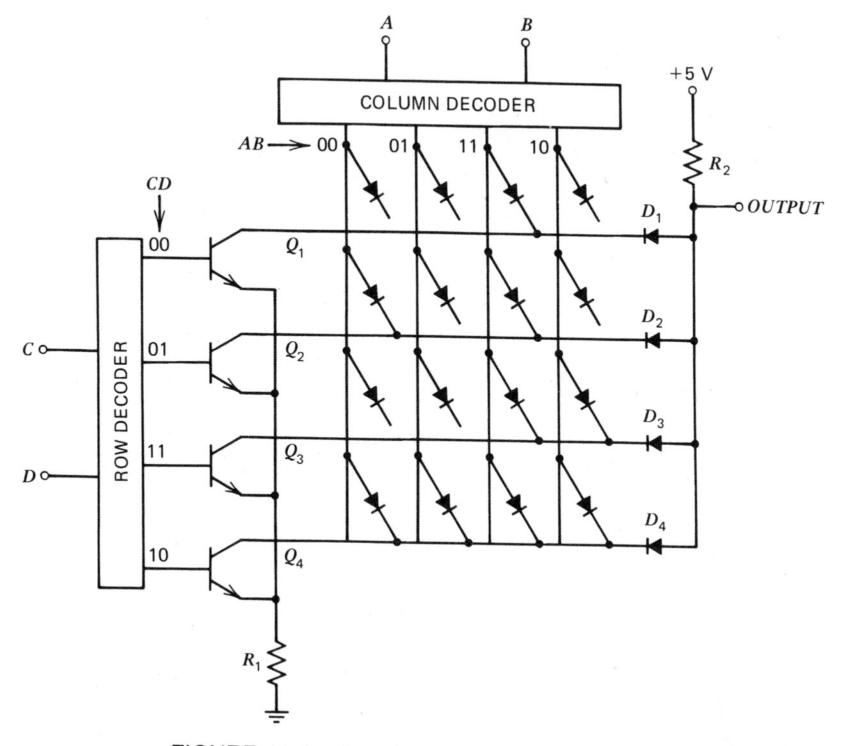

FIGURE 11-9 A 16-bit ROM using diode cells.

which can be simplified to

$$OUTPUT = \boldsymbol{AB} + \boldsymbol{C\overline{D}} + \boldsymbol{AC} + \boldsymbol{\overline{A}\overline{B}\overline{C}D}$$

11-6.2 ROM IC Organization

As was the case in RAM ICs, a ROM IC may be bit-organized—as in Ex. 11-8 and in Fig. 11-9. It may also be word-organized, in which case it is often byte-organized (8-bit word length) or nibble-organized (4-bit word length).

11-6.3 Programmable ROMs (PROMs)

The read-only memories (ROMs) described thus far consisted of an array of cells into which a *fixed pattern* of logic, 0s or 1s, has been entered in the manufacturing process. The basic array together with row and column decoders, output drivers, and any optional control circuitry is made first on the semiconductor wafer. The particular pattern of 0s and 1s is then obtained by providing a metal layer in the *last* manufacturing step: This metal layer carries the desired binary information giving the ROM its "personality."

In the process of developing a digital system, it is often desirable that the users themselves be able to program ROMs. Such ROMs are referred to as *programmable ROMs*, or *PROMs*. In a PROM a complete array of cells is made first, each cell incorporating a fusable link at one of its terminals. During programming such a link may be broken (fused) by the application of a high-current pulse with well-defined shape and duration. A broken link in a cell represents one logic state, an unbroken link the other logic state.

11-6.4 Reprogrammable ROMs, EPROMs, and EAROMs

Information written into a PROM cell is irreversible. Thus, although a user can program a PROM quickly to the desired pattern, there is no recourse if for some reason even one bit of that pattern has to be altered; in such an instance a new PROM would be required—unless the change involves only the breaking (fusing) of additional links. This presents a severe restriction in the development of new systems that require modifications.

Two types of *reprogrammable ROMs* have been developed to solve this problem. The *erasable PROM* (*EPROM*) consists of MOS cells in which no electrical contact has been made to the polysilicon gate terminal. During programming, charge is injected into the gate region, where it remains trapped for years since the gate is well insulated from the remainder of the circuit. However, it is possible to *erase* the information by application of an ultraviolet light that allows the trapped charge under the gate to be neutralized. To this end EPROMs have a quartz window incorporated to allow this erasure to take place when needed.

The *electrically alterable ROM*, *EAROM* (or *EEROM*), uses the same charge-storage mechanism and achieves the same flexibility as an EPROM. In an EAROM, however, information is easier to write: to change the content of an EAROM, new information is written over the old, the latter being erased in the process—no ultraviolet light is needed. Thus, the function of an EAROM is identical to the function of a RAM. However, because of the charge-storage mechanism involved, write times of an EAROM are much longer than read times, and it is for this reason that EAROMs are not considered RAMs. The principal advantage of EAROMs over RAMs is that information stored in an EAROM is not volatile and is preserved even when power is removed, while information stored in a RAM is lost upon removal of power.

EPROMs and EAROMs may be programmed many times without affecting their performance. Once the final bit pattern has been established, conventional ROMs may be manufactured where large quantities are required.

11-6.5 ROM Applications and Availability

ROMs are used primarily when permanent storage with fast access is desired (typical ROM access times are 5 ns to 100 ns). One area of applications uses ROMs for realizing combinational logic as illustrated in Ex. 11-8; such applications include code conversion, display logic, as well as counters (Ch. 9), controllers, and other sequential circuits. Another area of ROM application is in computers; these include storage of initialization sequences (''*bootstrap loaders*''), as well as *function generation* and other *look-up tables*.

With present technology a ROM is often part of a larger IC. However, ROMs are also available as separate ICs: 256-Kbit VLSI ROMs have attained widespread use.

11-7 PROGRAMMABLE LOGIC ARRAYS (PLAs)

A programmable logic array (PLA) is a combinational logic circuit that is customized to realize a given logic function. A PLA becomes more efficient when the logic function can be simplified and when several logic functions share product terms (e.g., $A\bar{B}$ or CD)—which is often the case, for example, in control circuits. This is in contrast with the realization of logic functions by ROMs, where neither simplification nor the simultaneous realization of several logic functions results in a smaller ROM.

The circuit diagram of a small MOS PLA is shown in Fig. 11-10. It consists of an *AND plane* and an *OR plane*, and it includes logic inverters that are often combined with the surrounding circuitry. The AND plane consists of logic inverters (Fig. 7-15, page 158) followed by NOR gates (Fig. 7-17, page 159): These realize AND functions since $\overline{\bar{X} + \bar{Y}} = XY$. The OR plane consists of NOR gates followed by logic inverters: These realize OR functions since $\overline{\overline{X + Y}} = X + Y$.

In the particular PLA of Fig. 11-10, three product terms are generated in the AND plane: $P_1 = A\bar{B}$, $P_2 = \bar{A}B$, and $P_3 = CD$. These product terms are used in the OR plane to generate four outputs as $OUT_1 = P_1 + P_2 = A\bar{B} + \bar{A}B$, $OUT_2 = P_1 + P_3 = A\bar{B} + CD$, $OUT_3 = P_2 + P_3 = \bar{A}B + CD$, and $OUT_4 = P_1 + P_2 + P_3 = A\bar{B} + \bar{A}B + CD$.

For comparison note that realization by ROMs would require four separate 4-variable ROMs, each with the complexity of Fig. 11-9.

As was the case for ROMs, PLAs are used for implementing combinational logic in applications such as code conversion, displays, counters, and controllers; PLAs are also available as fusable *field programmable logic arrays* (*FPLAs*). In addition to their use as separately packaged circuitry, PLAs are also used as part of larger LSI and VLSI circuits; such applications are discussed later in the book.

*11-8 ASSOCIATIVE MEMORIES

The *associative memory*, also known as *content-addressable memory* (*CAM*), includes logic at each word of the memory. It has three operating modes: READ, WRITE, and MATCH. The first two modes are identical to the READ and

*Optional material.

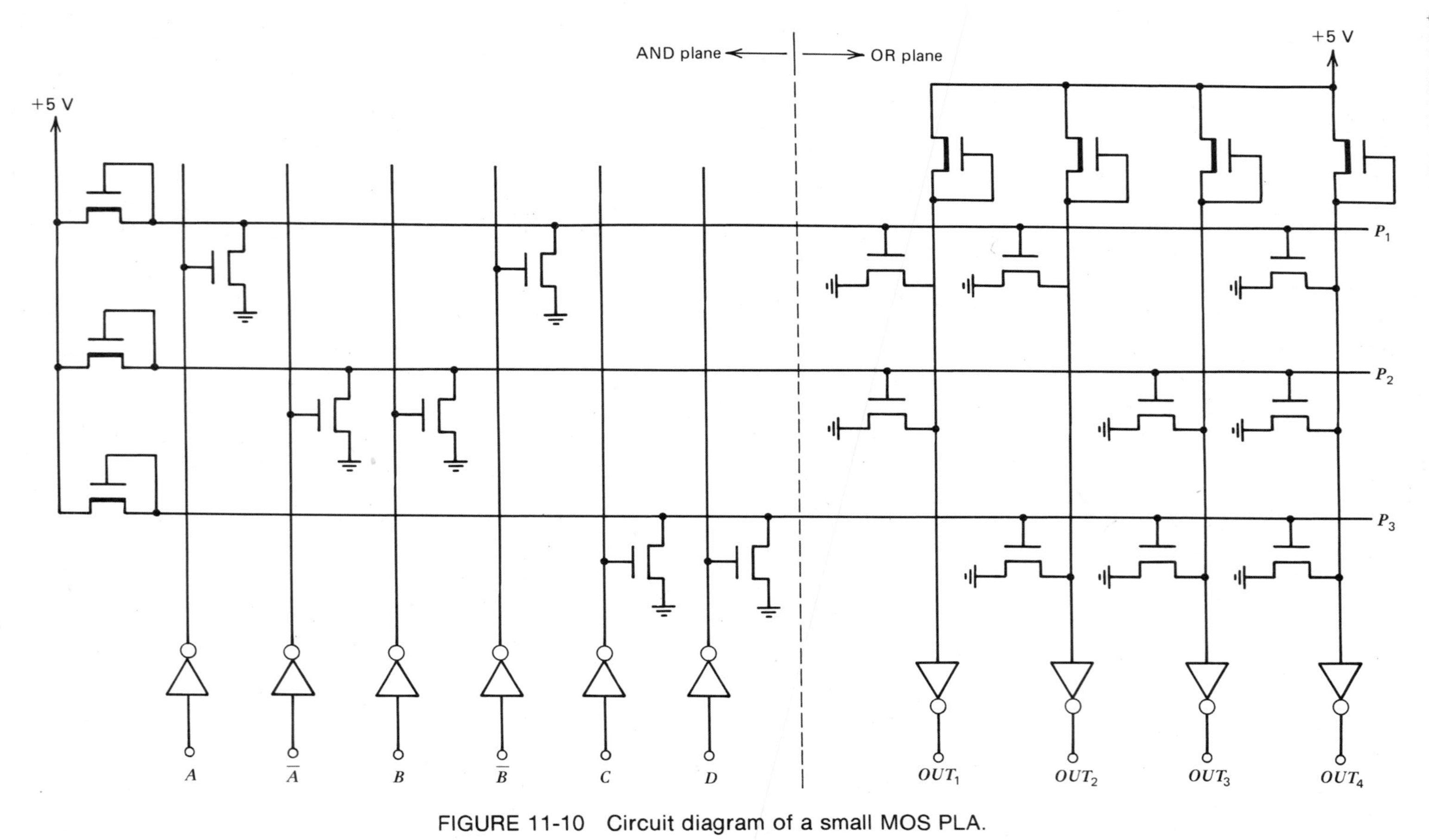

FIGURE 11-10 Circuit diagram of a small MOS PLA.

WRITE operations of a RAM. In the MATCH mode, *DATA* are compared with *selected* bits of words stored in the memory, and such comparisons may be performed simultaneously on several memory words. The combination of logic with memory provides a powerful tool for such operations as search through data files and comparison of numbers.

Figure 11-11 shows the block diagram of a 6-word × 12-bit associative memory. In addition to the memory array consisting of word 1 through word 6,

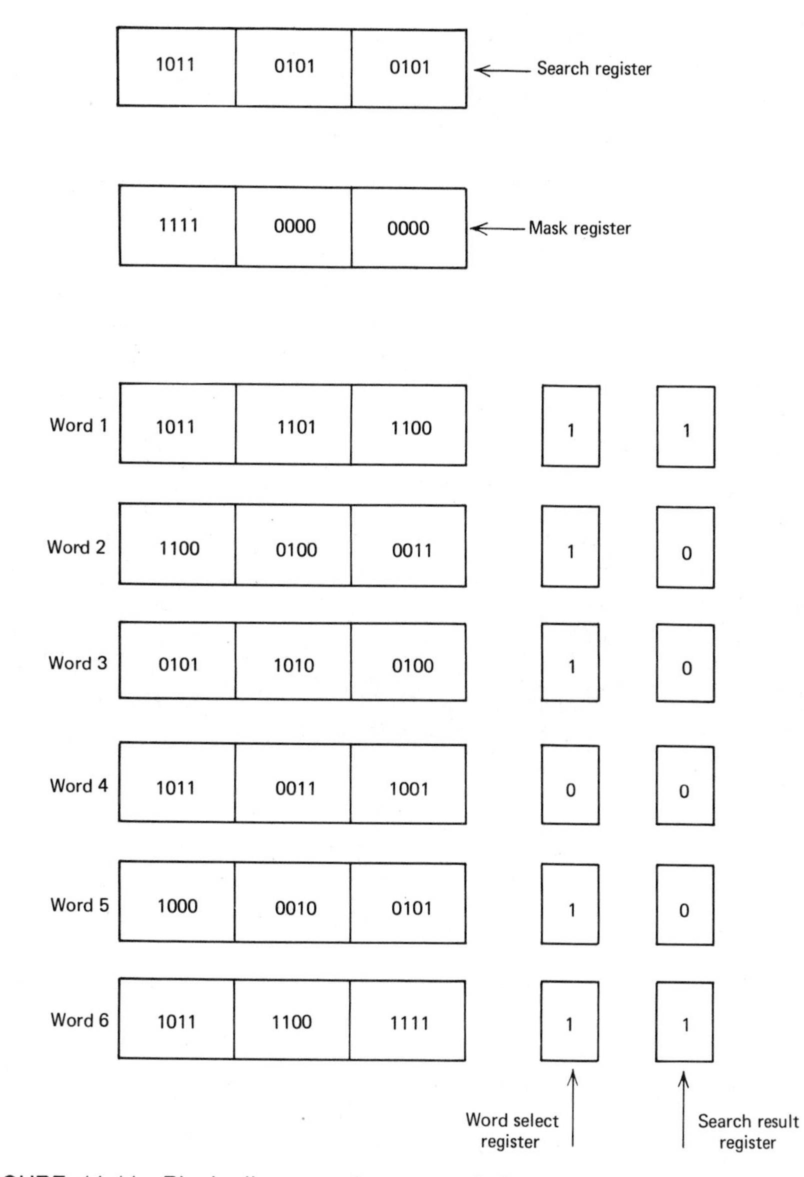

FIGURE 11-11 Block diagram of an associative memory with a capacity of 6 words × 12 bits.

Fig. 11-11 also shows four registers: (1) a *search register* (or *argument register*) that stores the *DATA* word to be compared with words in the memory array; (2) a *mask register* that designates which bits of the search register are to be included in the search—only those bits of the mask register that contain a logic 1 take part in the search operation; (3) a *word select register* that selects those words of the memory that participate in the search; (4) a *search result register* that stores the result of the search for each interrogated memory word: 1 if a match is found, 0 otherwise.

EXAMPLE 11-9

The contents of words 1 through 6 of the memory array are shown in the associative memory of Fig. 11-11. It is desired to find those words that start with 1011—except for word 4, the contents of which are of no interest. Find the contents of the search register, the mask register, the word select register, and the search result register.

Solution

The contents of the four registers are shown in Fig. 11-11.

The search register is loaded with 1011 xxxx xxxx, where each x represents a don't care condition. The mask register has to be instrumental in comparing only the 4 most significant bits, and hence it is loaded with 1111 0000 0000. The word select register has to exclude word 4 from the search, and thus it is loaded (reading from the top to the bottom) with 111011.

A match between the first 4 bits of the search register and the first 4 bits of a memory word is found in memory words 1, 4, and 6. However, the word select register contains logic 0 for word 4; thus the search result register shows a match (logic 1) only for words 1 and 6.

11-9 GATE ARRAYS

A *gate array*, which is also known as a *masterslice*, a *logic array*, or a *cell array*, consists of a fixed array of logic gates that can be interconnected as required by appropriate *customization* of one or more levels of metalization. Also, the layout is repetitive, thus permitting the use of *automated routing* programs that can translate a logic diagram into a chip layout.

Figure 11-12 shows part of a gate array with a layout that is repetitive in both directions. The array extends over most of the IC chip, with interfacing circuits and connecting pads included near the edges of the chip. Interconnections are routed in two *channels*: the first-metal level in the horizontal channel and the second-metal level in the vertical channel. Power connections are either by vertical second-metal lines or by an additional third level of metalization.

Each cell in Fig. 11-12 may include one or more logic gates, or the components required for them. The latter approach results in more flexibility, the former in simpler interconnecting metalization.

Interconnections between a line in a horizontal channel and a line in the vertical channel takes place within an *intersection* by means of a feedthrough called a *via*. This is shown in Fig. 11-13, where the two levels of metalization are

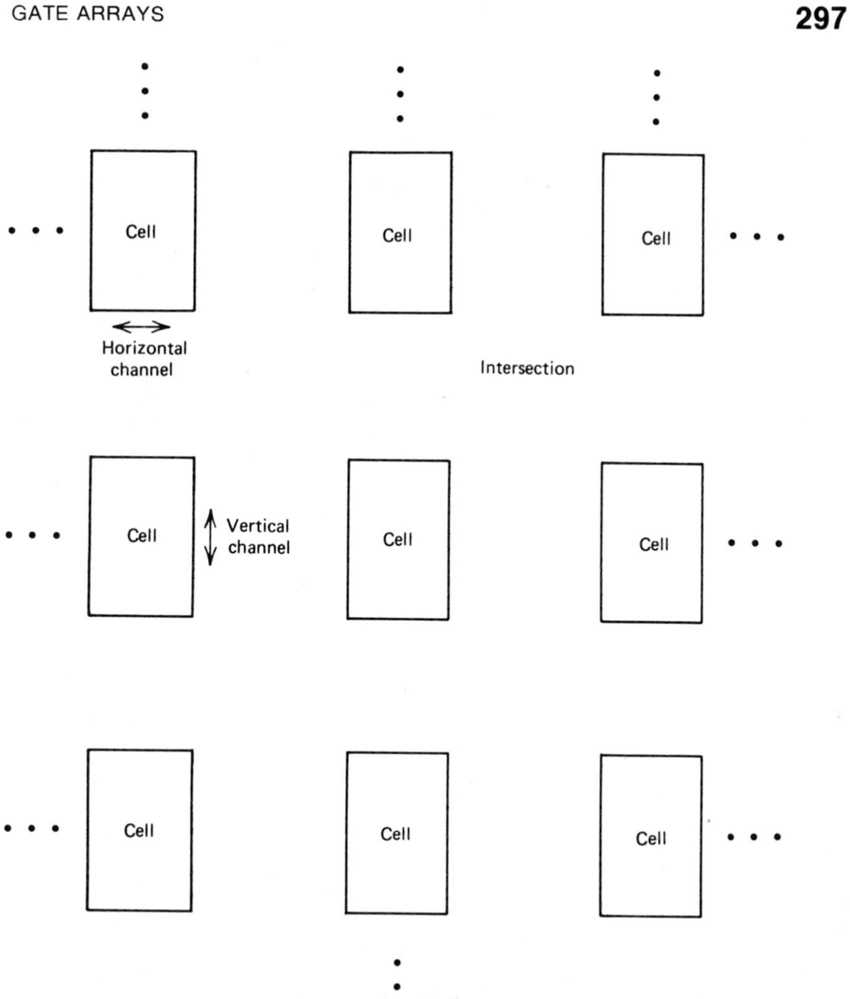

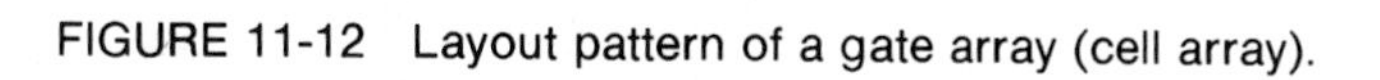
FIGURE 11-12 Layout pattern of a gate array (cell array).

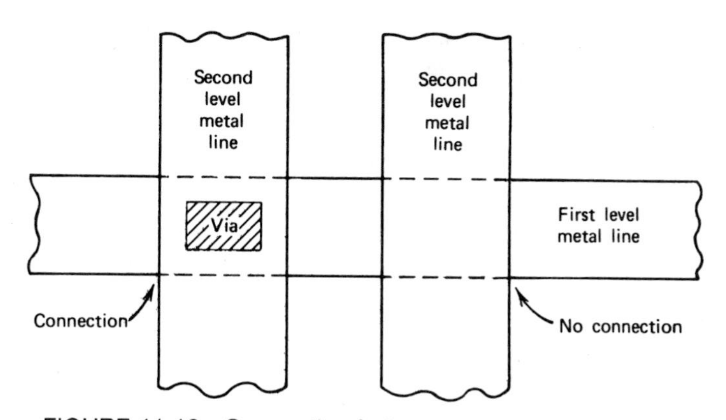

FIGURE 11-13 Connection between two metal levels by a via.

separated by an isolating layer of silicon dioxide (SiO_2), which is assumed to be transparent.

The interconnections between the cells of a gate array are completely uncommitted. As a result, gate arrays are well suited for the realization of complex combinational logic (''*random logic*''), as well as of latches and flip-flops (FFs), which are built up by the interconnection of logic gates as discussed in Chapter 8.

The principal technologies used in gate arrays are ECL, CMOS, and I^2L. These can provide operating speeds that are comparable to those of the SSI and MSI circuits of Chapter 7, even with the large capacitive loads of the interconnections between the gate-array cells. Several thousand logic gates on a chip (LSI) are practical in ECL and I^2L. However, the most promising technology for VLSI gate arrays is CMOS, where over ten thousand logic gates on a chip are now available.

SUMMARY

Dynamic MOS logic inverters and gates were introduced, followed by a description of dynamic MOS shift registers. Fully static and quasi-static MOS circuits were discussed and compared with dynamic MOS circuits. Various designs of random-access memory (RAM) cells were shown and their relative merits discussed. Read-only memories (ROMs), programmable ROMs (PROMs), and reprogrammable ROMs (EPROMs and EAROMs) were discussed.

Programmable logic arrays (PLAs) were introduced and the organization of associative memories was discussed. The chapter was concluded by a description of gate arrays.

SELF-EVALUATION QUESTIONS

1. Discuss the operation of dynamic MOS logic circuits.
2. Describe the operation of dynamic MOS shift registers.
3. Describe the operation of a static MOS clocked D-latch.
4. Sketch a 1-transistor dynamic MOS RAM cell.
5. Describe a fully static MOS RAM cell.
6. Describe a bipolar RAM cell.
7. Discuss the differences between a RAM and a ROM.
8. Discuss the merits of programmable and erasable ROMs.
9. Describe a programmable logic array (PLA).

*10. Explain the functions of the following registers in an associative memory: search register, mask register, word select register, and search result register.

11. Describe the salient features of gate arrays.

*Optional material.

PROBLEMS

11-1. Implement an R-S FF using fully static MOS NOR gates. Show the terminals R, S, Q, and $\overline{Q}$.

11-2. Implement an R-S FF using fully static MOS NAND gates. Show the terminals $\overline{R}$, $\overline{S}$, Q, and $\overline{Q}$.

11-3. A 4096-word × 1-bit memory uses coincidence addressing.

a. How many rows and columns are there in the square cell array?

b. Find the number of address bits applied to the row and to the column decoders.

11-4. A 16,384-bit memory uses coincidence addressing.

a. How many rows and columns are there in the square cell array?

b. Find the number of address bits applied to the row and to the column decoders.

11-5. Draw a circuit of a 4-bit × 4-bit diode ROM, such as shown in Fig. 11-9, and show the connections of the diodes for the Boolean function

$$f = \overline{A}\,\overline{B}\,\overline{C}\,\overline{D} + \overline{A}\,\overline{B}CD + AB\overline{C}D + A\overline{B}\,\overline{C}D + \overline{A}BC\overline{D}$$

11-6. Draw a circuit of a 4-bit × 4-bit diode ROM, such as shown in Fig. 11-9, and show the connections of the diodes for the Boolean function

$$f = \overline{A}\,\overline{B}\,\overline{C}\,\overline{D} + \overline{A}B\overline{C}D + ABCD + A\overline{B}C\overline{D}$$

11-7. The simplified function $f = AB + C\overline{D} + AC + \overline{A}\,\overline{B}\,\overline{C}D$ is to be realized using a ROM of the type shown in Fig. 11-9.

a. Expand the function to obtain it in a form suitable for programming the ROM.

b. Show the connections of the diodes in the ROM implementing the function.

11-8. The simplified function $f = AB + \overline{A}\,\overline{C} + AD$ is to be realized using a ROM of the type shown in Fig. 11-9.

a. Expand the function to obtain it in a form suitable for programming the ROM.

b. Show the connections of the diodes in the ROM implementing the function.

11-9. What change is required in the PLA of Fig. 11-10 to result in $OUT_1 = \overline{A}B$?

11-10. What change is required in the PLA of Fig. 11-10 to result in $OUT_1 = A\overline{B}$?

***11-11.** In a 6-word × 12-bit associative memory, the contents of the registers are as follows:

Search register: 1010 0101 1100

Mask register: 1111 0000 0000

Word select register (reading from the top to the bottom): 101111

The contents of the memory are:

Word 1: 0111 1001 1010

Word 2: 1010 0100 1100

Word 3: 0101 0010 1100

Word 4: 1010 1001 0110

Word 5: 0101 0100 1100

Word 6: 1010 1010 1010

Find the contents of the search result register as read from the top (Word 1) to the bottom.

***11-12.** In a 6-word × 12-bit associative memory, the contents of the search register, the word select register, and the memory array are the same as in Prob. 11-11. However, the contents of the mask register are 0000 0000 1111. Find the contents of the search result register as read from the top (Word 1) to the bottom.

11-13. In the gate array of Fig. 11-12, the horizontal dimension of each cell is 50 μm and the vertical dimension of each cell is 75 μm. Also, the horizontal channel is 50 μm wide and the vertical channel is 75 μm wide. What is the approximate size of the gate array if it includes 50 × 50 = 2500 cells?

11-14. In the gate array of Fig. 11-12, the horizontal dimension of each cell is 25 μm and the vertical dimension of each cell is 50 μm. Also, the horizontal channel is 50 μm wide and the vertical channel is 75 μm wide. What is the approximate size of the gate array if it includes 50 × 50 = 2500 cells?

11-15. Each cell in the gate array of Prob. 11-13 consists of a CMOS logic gate with a dc power dissipation of 50 μW. What is the dc power dissipation of the gate array?

11-16. Each cell in the gate array of Prob. 11-13 consists of a CMOS logic gate with a dc power dissipation of 50 μW and an ac power dissipation of 250 μW. What is the total power dissipation of the gate array?

*Optional problem.

11-17. Each cell in the gate array of Prob. 11-14 consists of the I^2L gate described in Ex. 7-4 (page 156) operated at a current of 0.5 mA. What is the power dissipation of the gate array?

11-18. Each cell in the gate array of Prob. 11-14 consists of the I^2L gate described in Ex. 7-4 (page 156) operated at a current of 0.5 μA. What is the power dissipation of the gate array?

CHAPTER

12 Arithmetic Circuits

Instructional Objectives

This chapter describes the structure and operation of arithmetic circuits. After you read it you should be able to:

1. Analyze the operation of various digital comparators.
2. Compare the properties of ripple comparators, look-ahead carry comparators, and 2-stage comparators.
3. Analyze the operation of the half adder and the full adder.
4. Assemble multibit adders using available ICs.
5. Compare the properties of ripple adders, look-ahead carry adders, and 2-stage adders.
6. Describe the operation of a subtractor.
7. Assemble a subtractor using an adder and logic gates.
8. Describe functions performed by an arithmetic logic unit (ALU).
9. Analyze the operation of a multiplier.

12-1 OVERVIEW

Arithmetic circuits are of great importance in digital computers and in other digital systems. While other circuits such as memories and input–output devices are used for storing and transferring data, operations on the data are performed by *arithmetic circuits*.

This chapter emphasizes basic arithmetic circuits that receive widespread implementation with ICs. It describes the structure and operation of digital comparators, adders, subtractors, arithmetic logic units (ALUs), and multipliers.

12-2 DIGITAL COMPARATORS

The simplest arithmetic circuit operating on two numbers is the *digital comparator*, also known as *magnitude comparator* or simply *comparator*. The comparators described here can be applied for the comparison of two numbers in any binary code; however, for simplicity it is assumed that the numbers are in natural binary code and also that they are not negative.

When two numbers A and B are compared, the *magnitude relations* of interest are $A > B$ (A greater than B), $A < B$ (A less than B), $A = B$ (A equal B), $A \geq B$ (A greater than or equal to B), and $A \leq B$ (A less than or equal to B). It can be shown that the following logic statements are valid:

$$(A > B) = \overline{(A \leq B)} \tag{12-1}$$

$$(A < B) = \overline{(A \geq B)} \tag{12-2}$$

$$(A = B) = \overline{(A > B)} \cdot \overline{(A < B)} \tag{12-3}$$

$$(A \geq B) = \overline{(A < B)} \tag{12-4}$$

$$(A \leq B) = \overline{(A > B)} \tag{12-5}$$

Thus, for example, from eq. 12-1, if $A > B$ is TRUE, then $A \leq B$ is FALSE, and vice versa.

12-2.1 One-Bit Comparators

When A and B are 1-bit numbers, they can take on only the values of 0 or 1, and the logic statements $A > B$, $A < B$, $A = B$, $A \geq B$, and $A \leq B$ can be translated into Boolean expressions as shown in Fig. 12-1.

The logic statement $A > B$ is TRUE only when $A = 1$ and $B = 0$. Hence, the Karnaugh map shows a logic 1 only for this combination of variables. The resulting Boolean function is $A\overline{B}$, which can be also written as $\overline{\overline{A} + B}$. Thus, the logic statement $A > B$ can be realized by an AND gate or by a NOR gate as shown. The situation is similar for the logic statement $A < B$, except that the roles of the two variables A and B are interchanged.

In order for the $A = B$ logic statement to be TRUE, we must have either $A = 0$ and $B = 0$, or $A = 1$ and $B = 1$. Hence, the Karnaugh map shows logic 1s for these combinations of variables. The resulting Boolean function is $\overline{A}\overline{B} + AB$, which can be also written as $\overline{A \oplus B}$. Thus, the logic statement $A = B$ can be realized by an EQUALITY gate, also known as an EXCLUSIVE-NOR gate.

The logic statement $A \geq B$ is TRUE when $A = 0$ and $B = 0$, or when $A = 1$ and $B = 0$, or when $A = 1$ and $B = 1$. Note that this is complementary to the $A < B$ logic statement, which is in accordance with eq. 12-2. The situation is similar for the logic statement $A \leq B$, except that the roles of the two variables A and B are interchanged.

12-2.2 Two-Bit Comparators

This section describes the magnitude comparison of two 2-bit numbers A_1A_0 and B_1B_0, where A_1 and B_1 are the most significant bits (msb) and A_0 and B_0 are the least significant bits (lsb).

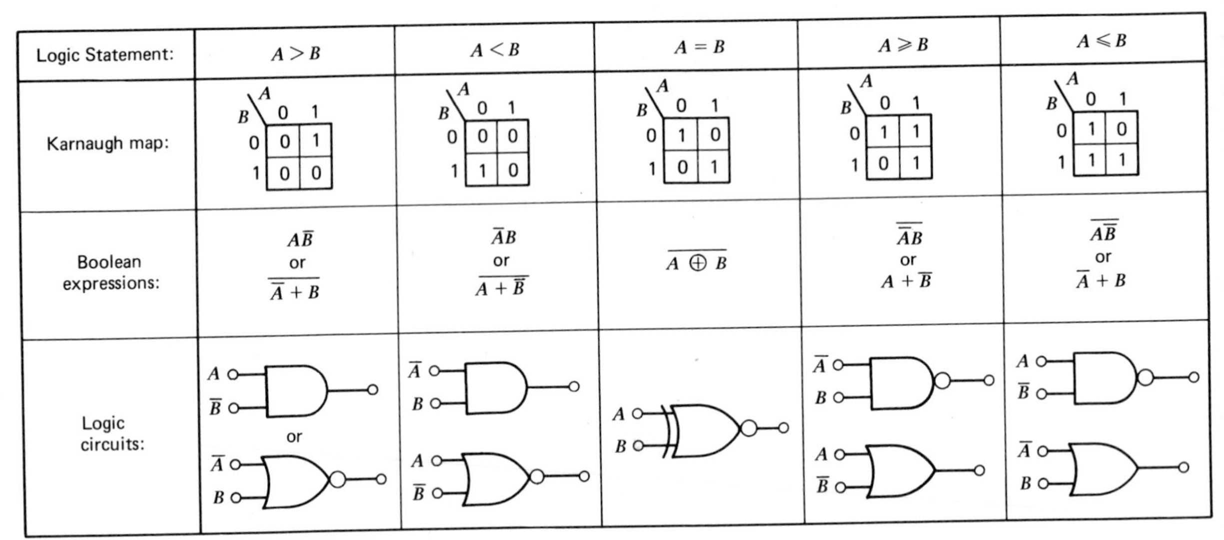

FIGURE 12-1 One-bit digital comparators.

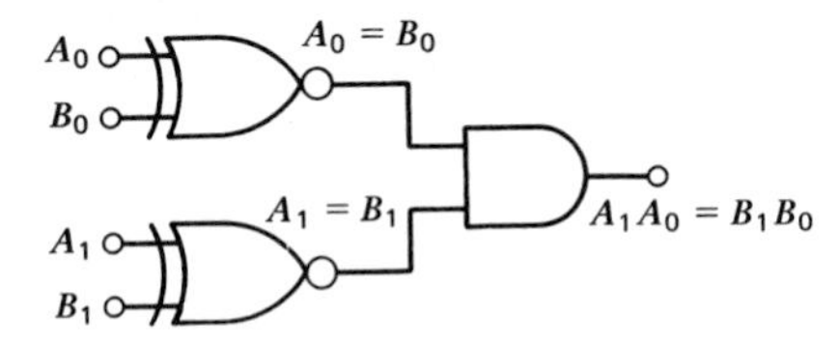

FIGURE 12-2 Two-bit comparator realizing $A_1A_0 = B_1B_0$.

12-2.2.1 Two-Bit Comparator Realizing $A_1A_0 = B_1B_0$

In order for the logic statement $A_1A_0 = B_1B_0$ to be TRUE, we must have $A_1 = B_1$ and $A_0 = B_0$, that is,

$$(A_1A_0 = B_1B_0) = (A_1 = B_1) \cdot (A_0 = B_0) \tag{12-6}$$

By the application of the Boolean expression for $A = B$ from Fig. 12-1, eq. 12-6 becomes

$$(A_1A_0 = B_1B_0) = \overline{(A_1 \oplus B_1)} \cdot \overline{(A_0 \oplus B_0)} \tag{12-7}$$

The realization of eq. 12-7 is shown in Fig. 12-2.

12-2.2.2 Two-Bit Comparator Realizing $A_1A_0 > B_1B_0$

In describing the realization of the logic statement $A_1A_0 > B_1B_0$, we first follow an inspection procedure, then translate it into a Boolean equation, and then into a logic circuit. The inspection procedure is as follows:

Step 1. Inspect the most significant bits A_1 and B_1. If $A_1 = 1$ and $B_1 = 0$, then

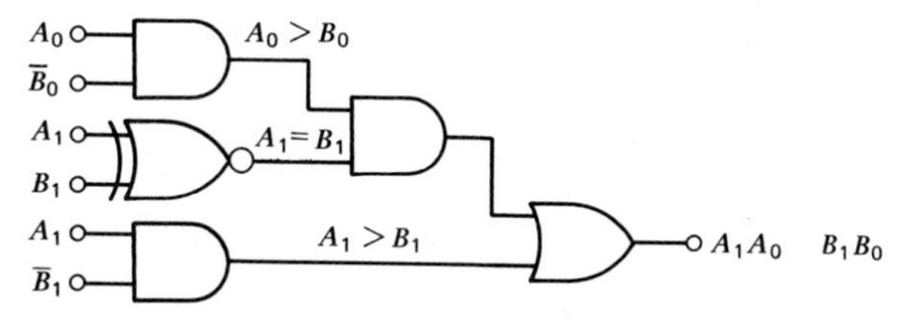

FIGURE 12-3 Two-bit comparator realizing $A_1A_0 > B_1B_0$.

the logic statement $A_1A_0 > B_1B_0$ is TRUE and the procedure is terminated. If $A_1 = 0$ and $B_1 = 1$, then the logic statement $A_1A_0 > B_1B_0$ is FALSE and the procedure is again terminated. However, if $A_1 = B_1$, then we proceed to Step 2.

Step 2. If $A_0 = 1$ and $B_0 = 0$, then the logic statement $A_1A_0 > B_1B_0$ is TRUE; otherwise it is FALSE.

EXAMPLE 12-1

Evaluate the logic statement $A_1A_0 > B_1B_0$ if

a. $A_1A_0 = 11$ and $B_1B_0 = 10$

b. $A_1A_0 = 11$ and $B_1B_0 = 01$

Solution

We follow the inspection procedure described above.

a. Step 1. Since $A_1 = 1$ and $B_1 = 1$, $A_1 = B_1$, and we proceed to Step 2.

Step 2. We have $A_0 = 1$ and $B_0 = 0$; hence, the logic statement $A_1A_0 > B_1B_0$ is TRUE. (Indeed, $A_1A_0 = 11 = 3_{10}$, $B_1B_0 = 10 = 2_{10}$, and $3_{10} > 2_{10}$.)

b. We have $A_1 = 1$ and $B_1 = 0$; thus the logic statement $A_1A_0 > B_1B_0$ is TRUE and the inspection procedure is terminated.

The inspection procedure can be translated into the following Boolean equation:

$$(A_1A_0 > B_1B_0) = (A_1 > B_1) + (A_1 = B_1) \cdot (A_0 > B_0) \tag{12-8}$$

A realization of eq. 12-8 is shown in Fig. 12-3. Also, it can be shown that a realization resulting in a shorter propagation delay can be obtained by writing eq. 12-8 in a sum-of-products form as

$$(A_1A_0 > B_1B_0) = A_1\overline{B}_1 + A_1A_0\overline{B}_0 + A_0\overline{B}_1\overline{B}_0 \tag{12-9}$$

leading to a 2-stage logic circuit.

12-2.2.3 *Two-Bit Comparators Realizing $A_1A_0 < B_1B_0$, $A_1A_0 \geq B_1B_0$, and $A_1A_0 \leq B_1B_0$*

The logic statement $A_1A_0 < B_1B_0$ can be realized by Fig. 12-3 with the roles of the variables A_1 and B_1 interchanged and with the roles of the variables A_0 and B_0 interchanged. When the resulting circuit is followed by a logic inverter, it realizes the logic statement $A_1A_0 \geq B_1B_0$.

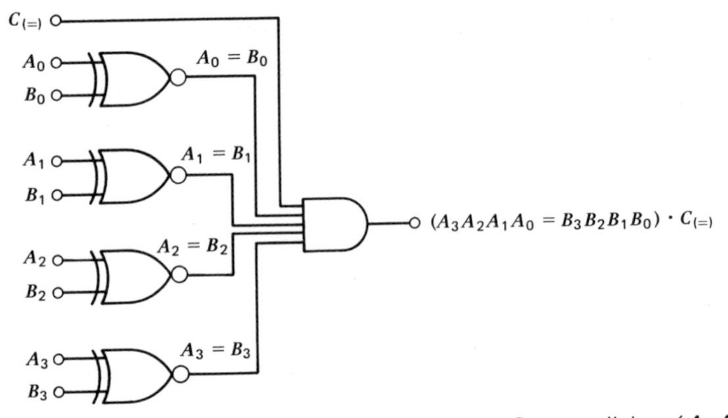

FIGURE 12-4 Four-bit comparator with carry input $C_{(=)}$ realizing $(A_3A_2A_1A_0 = B_3B_2B_1B_0)C_{(=)}$.

The logic statement $A_1A_0 \leq B_1B_0$ can be realized by following the circuit of Fig. 12-3 with a logic inverter.

12-2.3 Four-Bit Comparators

The 2-bit comparators of the preceding section can also be extended to more bits. In what follows, the structure and operation of 4-bit comparators are described.

12-2.3.1 Four-Bit Comparators Realizing $A_3A_2A_1A_0 = B_3B_2B_1B_0$

A 4-bit comparator for the realization of the logic statement $A_3A_2A_1A_0 = B_3B_2B_1B_0$ was introduced in Sec. 6-2.5, Fig. 6-12*b* (page 114). It can be further extended by the addition of a carry input $C_{(=)}$ as shown in Fig. 12-4, realizing the logic statement

$$(A_3A_2A_1A_0 = B_3B_2B_1B_0) \cdot C_{(=)} = (A_3 = B_3) \cdot (A_2 = B_2) \cdot (A_1 = B_1) \cdot (A_0 = B_0) \cdot C_{(=)} \quad (12\text{-}10)$$

Carry input $C_{(=)}$ can be connected to the output of a preceding identical comparator, thus permitting the comparison of more than 4 bits.

12-2.3.2 Four-Bit Comparators Realizing $A_3A_2A_1A_0 > B_3B_2B_1B_0$

The 2-bit comparator of Fig. 12-3 can be extended to 4 bits by following the $A_1A_0 > B_1B_0$ output by additional circuitry. Such a circuit is called a *ripple comparator* because the inputs have to "ripple" through many gates to reach the output. The propagation delay could be reduced by the use of a 2-stage logic circuit, but the resulting circuit would be quite large. For this reason the

Comparison Inputs				Outputs		
A_3, B_3	A_2, B_2	A_1, B_1	A_0, B_0	$A = B$	$A > B$	$A < B$
$A_3 > B_3$	x	x	x	0	1	0
$A_3 < B_3$	x	x	x	0	0	1
$A_3 = B_3$	$A_2 > B_2$	x	x	0	1	0
$A_3 = B_3$	$A_2 < B_2$	x	x	0	0	1
$A_3 = B_3$	$A_2 = B_2$	$A_1 > B_1$	x	0	1	0
$A_3 = B_3$	$A_2 = B_2$	$A_1 < B_1$	x	0	0	1
$A_3 = B_3$	$A_2 = B_2$	$A_1 = B_1$	$A_0 > B_0$	0	1	0
$A_3 = B_3$	$A_2 = B_2$	$A_1 = B_1$	$A_0 < B_0$	0	0	1
$A_3 = B_3$	$A_2 = B_2$	$A_1 = B_1$	$A_0 = B_0$	$C_{(=)}$	$C_{(>)}$	$C_{(<)}$

FIGURE 12-5 Function table for the TTL Type 74L85 4-bit magnitude comparator.

most commonly used realization is the *look-ahead carry comparator* (not shown), which uses a 3-stage logic circuit. Also, often the $=$, $>$, and $<$ operations are combined into a single circuit and three carry inputs $C_{(=)}$, $C_{(>)}$, and $C_{(<)}$ are provided to facilitate cascading for more than 4 bits.

For example, the TTL Type 74L85 is a 4-bit look-ahead carry comparator that has 4 pairs of comparison inputs: A_3, B_3; A_2, B_2; A_1, B_1; and A_0, B_0. It also has 3 carry inputs $C_{(=)}$, $C_{(>)}$, and $C_{(<)}$, as well as 3 outputs $A = B$, $A > B$, and $A < B$.

A function table is shown in Fig. 12-5. When used as a 4-bit comparator, the carry inputs are permanently connected as $C_{(=)} = 1$, $C_{(>)} = 0$, and $C_{(<)} = 0$. In this case output $A = B$ realizes the logic statement $A_3A_2A_1A_0 = B_3B_2B_1B_0$, since the $A = B$ output is at logic 1 only if $A_3 = B_3$, $A_2 = B_2$, $A_1 = B_1$, and $A_0 = B_0$. Also, the $A > B$ output realizes the logic statement $A_3A_2A_1A_0 > B_3B_2B_1B_0$, since it is at logic 1 when $A_3 > B_3$; or when $A_3 = B_3$ and $A_2 > B_2$; or when $A_3 = B_3$, $A_2 = B_2$, and $A_1 > B_1$; or when $A_3 = B_3$, $A_2 = B_2$, $A_1 = B_1$, and $A_0 > B_0$. The $A < B$ output realizes the logic statement $A_3A_2A_1A_0 < B_3B_2B_1B_0$ in a similar way.

When comparing two numbers with more than 4 bits each, carry inputs $C_{(=)}$, $C_{(>)}$, and $C_{(<)}$ of the comparator handling the more significant bits can be connected, respectively, to the $A = B$, $A > B$, and $A < B$ outputs of the comparator handling the next less significant bits. Also, in such applications the carry inputs of the comparator handling the least significant bits are connected as $C_{(=)} = 1$, $C_{(>)} = 0$, and $C_{(<)} = 0$.

12-3 ADDERS

This section describes circuits that perform the addition of numbers by use of digital circuits. For simplicity, the discussion focuses on the addition of two numbers; the addition of more than two numbers can be performed by successive additions of two numbers at a time. Also, the discussion is limited to

Inputs		Outputs	
A	B	S	C_{OUT}
0	0	0	0
0	1	1	0
1	0	1	0
1	1	0	1

(a) Truth table

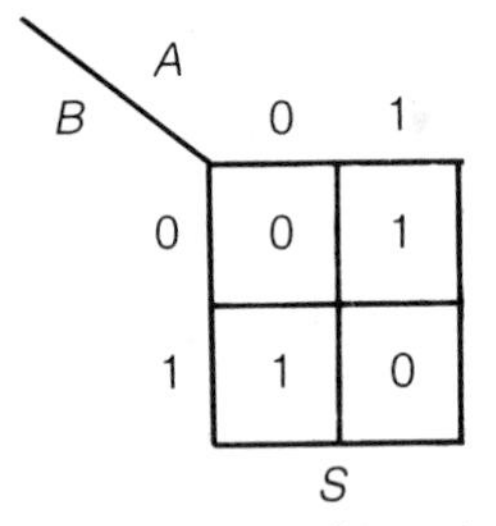

(b) Karnaugh map for output S

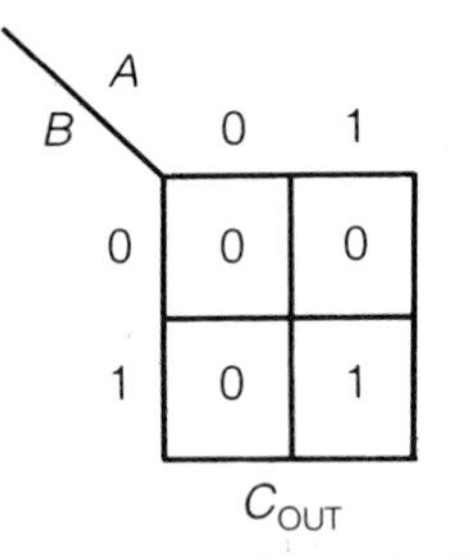

(c) Karnaugh map for output C_{OUT}

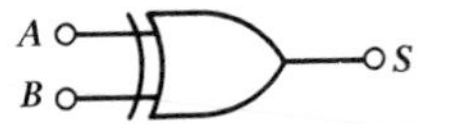

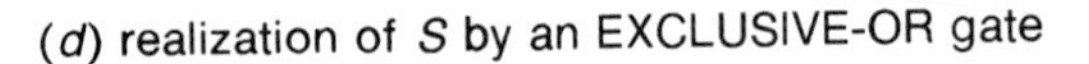

(d) realization of S by an EXCLUSIVE-OR gate

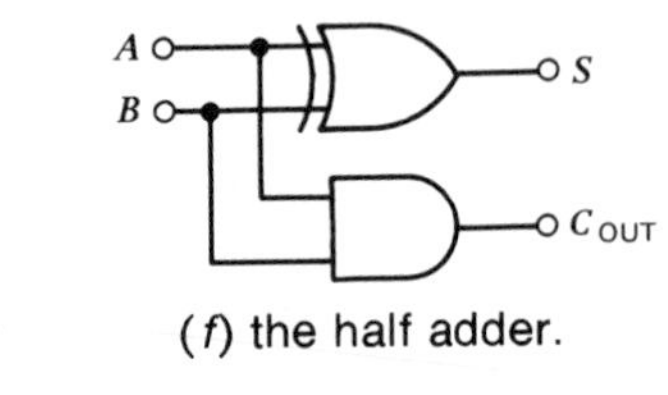

(f) the half adder.

(e) realization of C_{OUT} by an AND gate

FIGURE 12-6 One-bit adder with carry-out.

numbers that are in natural binary code, even though BCD numbers are also in widespread use.

12-3.1 One-Bit Adders with Carry-Out

When two 1-bit numbers are added, there can be four possible results: 0 plus 0 = 0, 0 plus 1 = 1, 1 plus 0 = 1, and 1 plus 1 = 1 carry 1. From the truth table of Fig. 12-6a showing sum S and carry C_{OUT} we can derive the Karnaugh map of Fig. 12-6b for output S and the Karnaugh map of Fig. 12-6c for output C_{OUT}. Note that $S = A \oplus B$ and $C_{OUT} = AB$.

The Karnaugh maps of Figs. 12-6b and 12-6c can be realized by logic gates as shown in Figs. 12-6d and 12-6e, respectively. The realizations of S and C_{OUT} are shown as a single circuit in Fig. 12-6f: this circuit is called a *half adder*.

12-3.2 One-Bit Adder with Carry-In and Carry-Out

The addition of two multibit binary numbers was described in Sec. 3-3.1. We saw that, except for the least significant bits, the addition of two number bits and of a carry bit from the next less significant bits was required. An adder that performs this operation is called a *full adder*.

One way of realizing a full adder is by a combination of two half adders and an OR gate, as shown in Fig. 12-7a. A truth table for the circuit of Fig. 12-7a is shown in Fig. 12-7b. Since the full adder adds 3 bits (A, B, and C_{IN}), it can

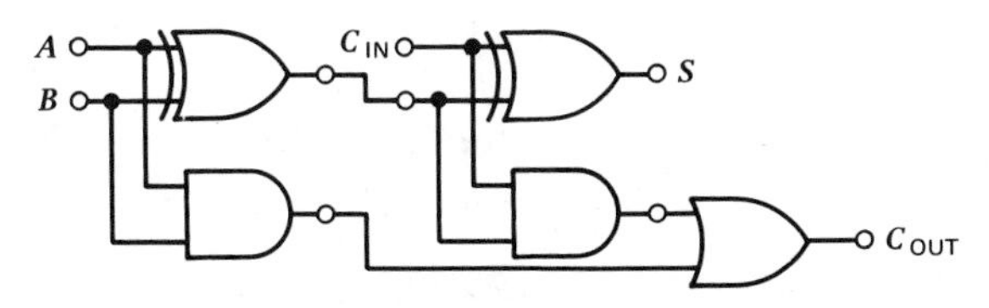

(a) Realization using two half adders and an OR gate

Inputs			Outputs	
A	B	C_{IN}	S	C_{OUT}
0	0	0	0	0
0	0	1	1	0
0	1	0	1	0
0	1	1	0	1
1	0	0	1	0
1	0	1	0	1
1	1	0	0	1
1	1	1	1	1

(b) truth table

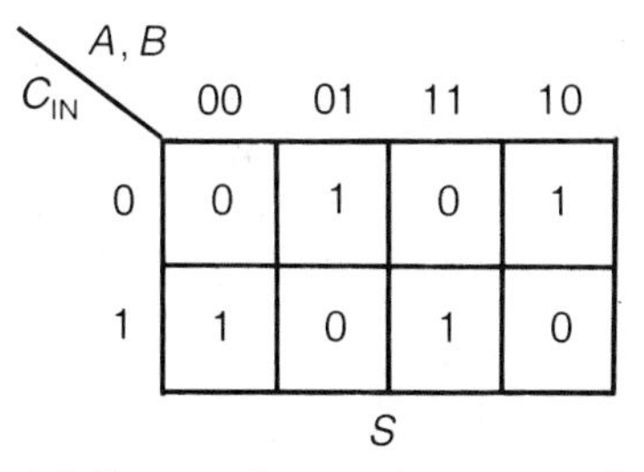

(c) Karnaugh map for output S

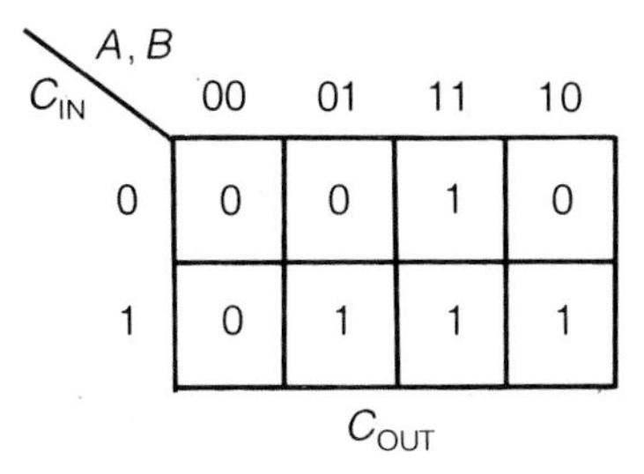

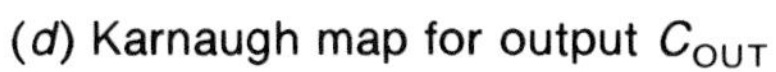

(d) Karnaugh map for output C_{OUT}

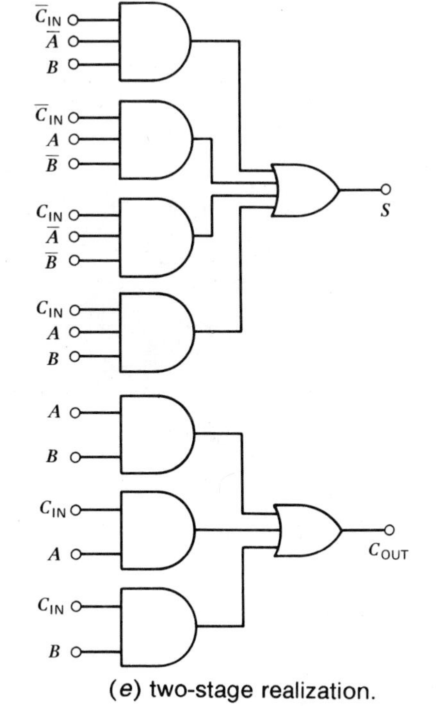

(e) two-stage realization.

FIGURE 12-7 One-bit adder with carry-in and carry-out (full adder).

have an output of 0, 1, 2, or 3. An output of 0 is represented as $S = 0$ and $C_{OUT} = 0$ (first row in Fig. 12-7*b*), an output of 1 is represented as $S = 1$ and $C = 0$ (second, third, and fifth rows), an output of 2 is represented as $S = 0$ and $C_{OUT} = 1$ (fourth, sixth, and seventh rows), and an output of 3 is represented as $S = 1$ and $C_{OUT} = 1$ (last row).

From the truth table of Fig. 12-7*b* we can derive the Karnaugh maps of Fig. 12-7*c* and 12-7*d* for outputs S and C_{OUT}, respectively. Also, from these Karnaugh maps we can write

$$S = \overline{C}_{IN}\overline{A}B + \overline{C}_{IN}A\overline{B} + C_{IN}\overline{A}\,\overline{B} + C_{IN}AB = A \oplus B \oplus C_{IN} \qquad (12\text{-}11)$$

$$C_{OUT} = AB + C_{IN}A + C_{IN}B \qquad (12\text{-}12)$$

The resulting realization of the full adder by a 2-stage logic circuit is shown in Fig. 12-7*e*. It consists of 7 AND gates and 2 OR gates. Thus, this realization is more complex than Fig. 12-7*a*, but it has the advantage of shorter propagation delays.

12-3.3 Two-Bit Adders

The simplest adder for the addition of two 2-bit numbers A_1A_0 and B_1B_0 can be assembled from Fig. 12-6*f* and 12-7*a*. The resulting 2-bit *ripple adder* is shown in Fig. 12-8, where A_1A_0 plus $B_1B_0 = S_1S_0$ carry C_{OUT}. The preparation of a truth table for the circuit of Fig. 12-8 is left to the reader as an exercise (see Probs. 12-3 and 12-4).

The propagation delays from A_0 and B_0 to C_{OUT} in Fig. 12-8 each equal the sum of 3 gate propagation delays. These can be reduced by use of 2-stage adders that are not described here.

12-3.4 Multibit Adders

A 4-bit adder can be built using four of the 1-bit adders of Fig. 12-7*a*, as shown in Fig. 12-9. However, the resulting propagation delays are long: for example,

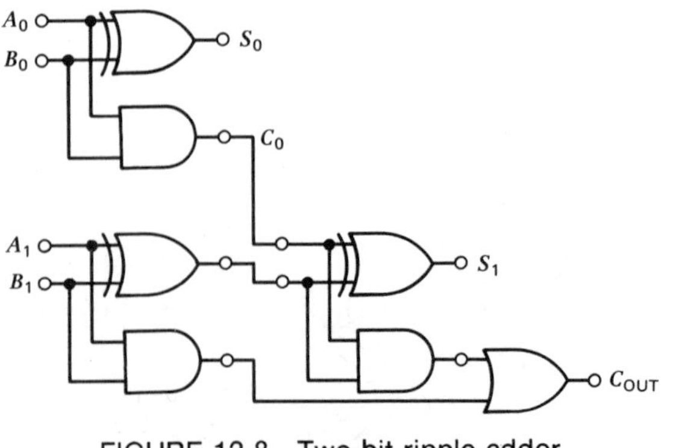

FIGURE 12-8 Two-bit ripple adder.

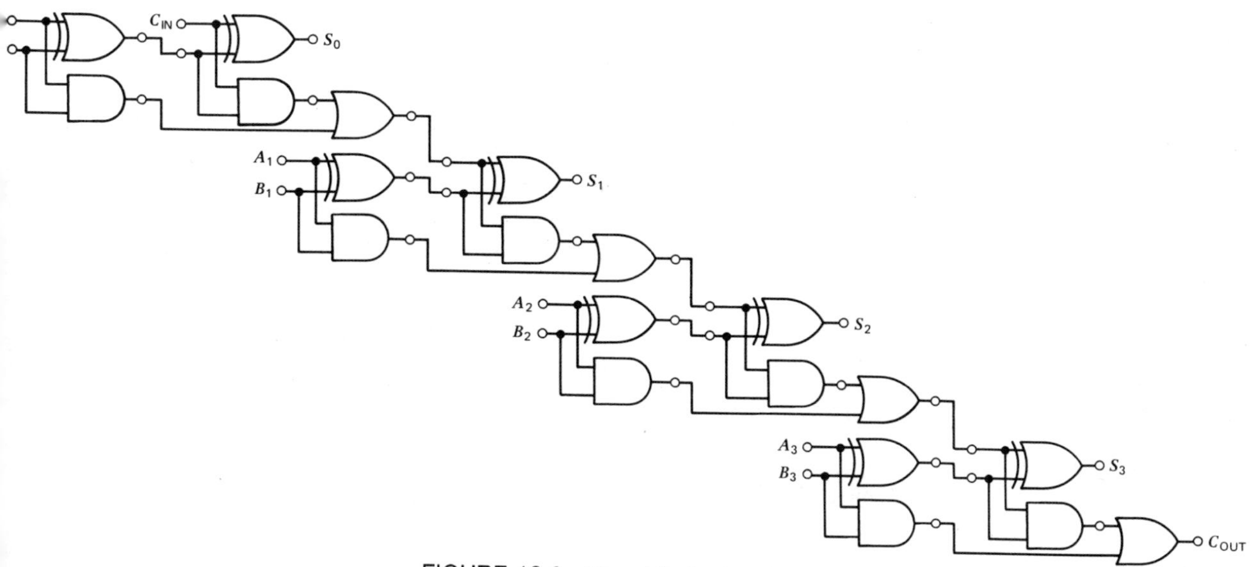

FIGURE 12-9 Four-bit ripple adder.

input A_0 has to ripple through 8 logic gates to reach output S_3. For this reason, present technology usually implements 4-bit adders by use of 3-stage logic circuits that are called *look-ahead carry adders*. A carry-in is also provided to facilitate cascading adders for more than 4 bits. This is illustrated in Ex. 12-2, which follows.

EXAMPLE 12-2

The TTL Type 74283 4-bit look-ahead carry adder has 4 pairs of inputs for addition: A_3, B_3; A_2, B_2; A_1, B_1; and A_0, B_0. It also has 4 sum outputs S_3, S_2, S_1, S_0, and a carry output C_{OUT}. Sketch the interconnection of two adders for adding the two 8-bit numbers $A_7A_6A_5A_4A_3A_2A_1A_0$ and $B_7B_6B_5B_4B_3B_2B_1B_0$.

Solution

The interconnection of the two adders is shown in Fig. 12-10. The sum is given by $S_7S_6S_5S_4S_3S_2S_1S_0$ and the carry-out is available as C_{OUT}.

12-4 SUBTRACTORS

We saw in Sec. 3-6 that negative numbers can be expressed in sign-and-magnitude representation, 1's complement representation, or 2's complement representation. Each of these representations requires a different circuit for subtraction. In what follows, a 2's complement subtractor is described.

According to Sec. 3-6.3, to subtract a subtrahend from a minuend, we first take the 2's complement of the subtrahend and then add it to the minuend. Also, the 2's complement of a number is obtained by first taking the 1's complement of the number and then adding 1 to the least significant bit. These operations can be performed by an adder and logic inverters as illustrated in Ex. 12-3 for 4-bit numbers.

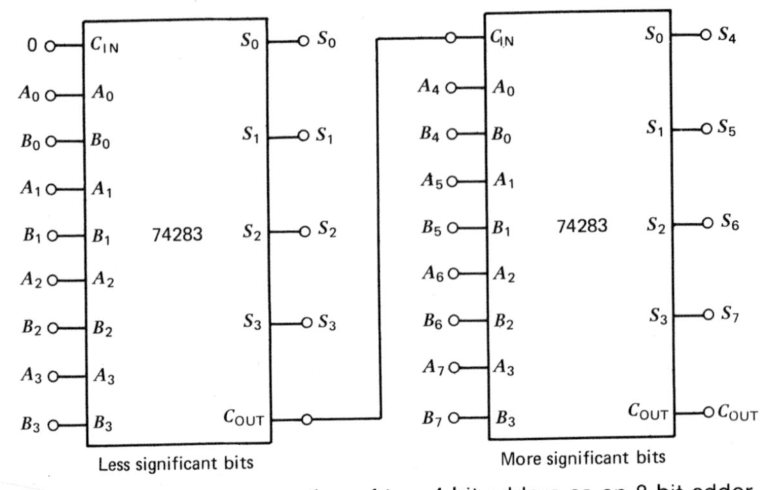

FIGURE 12-10 Interconnection of two 4-bit adders as an 8-bit adder.

EXAMPLE 12-3

Use a TTL type 74283 4-bit adder described in Ex. 12-2 and 4 logic inverters for building a 2's complement subtractor that subtracts the subtrahend $B_3B_2B_1B_0$ from the minuend $A_3A_2A_1A_0$, where B_3 and A_3 are the sign bits.

Solution

The circuit is shown in Fig. 12-11. The 2's complement of $B_3B_2B_1B_0$ is obtained by first taking its 1's complement by means of the four logic inverters and then adding 1 to the least significant bit via C_{IN}. The result of the subtraction is available in 2's complement representation as $S_3S_2S_1S_0$; carry-out C_{OUT} is ignored.

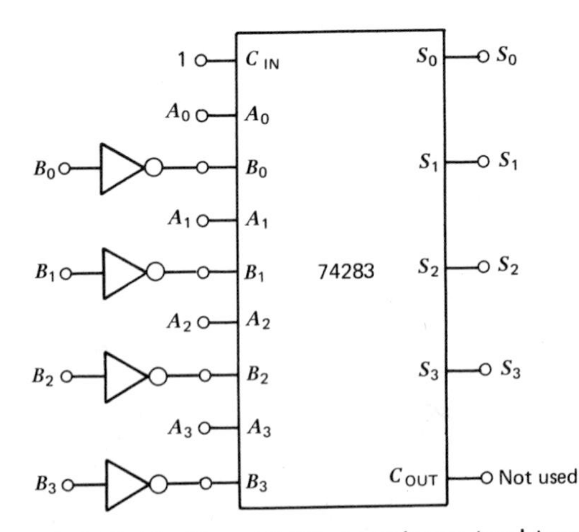

FIGURE 12-11 Four-bit 2's complement subtractor.

TABLE 12-1
Function table for the TTL Type 74181 arithmetic logic unit. Selection between logic operations and arithmetic operations is by means of an additional control line.

Control Lines				Logic Operation	Arithmetic Operation
S_3	S_2	S_1	S_0		
L	L	L	L	$\overline{A}$	A plus $CARRY\text{-}IN$
L	L	L	H	$\overline{A+B}$	$(A+B)$ plus $CARRY\text{-}IN$
L	L	H	L	$\overline{A}B$	$(A+\overline{B})$ plus $CARRY\text{-}IN$
L	L	H	H	0	minus 1 plus $CARRY\text{-}IN$
L	H	L	L	$\overline{AB}$	A plus $A\overline{B}$ plus $CARRY\text{-}IN$
L	H	L	H	$\overline{B}$	$(A+B)$ plus $A\overline{B}$ plus $CARRY\text{-}IN$
L	H	H	L	$A \oplus B$	A minus B minus 1 plus $CARRY\text{-}IN$
L	H	H	H	$A\overline{B}$	$A\overline{B}$ minus 1 plus $CARRY\text{-}IN$
H	L	L	L	$\overline{A}+B$	A plus AB plus $CARRY\text{-}IN$
H	L	L	H	$\overline{A \oplus B}$	A plus B plus $CARRY\text{-}IN$
H	L	H	L	B	$(A+\overline{B})$ plus AB plus $CARRY\text{-}IN$
H	L	H	H	AB	AB minus 1 plus $CARRY\text{-}IN$
H	H	L	L	1	A plus 2 times A plus $CARRY\text{-}IN$
H	H	L	H	$A+\overline{B}$	$(A+B)$ plus A plus $CARRY\text{-}IN$
H	H	H	L	$A+B$	$(A+\overline{B})$ plus A plus $CARRY\text{-}IN$
H	H	H	H	A	A minus 1 plus $CARRY\text{-}IN$

12-5 ARITHMETIC LOGIC UNITS (ALUs)

Present technology permits the combination of a comparator, an adder, and a subtractor into a single IC chip called an *arithmetic logic unit* (*ALU*). In addition to the above operations, a typical ALU can also perform shifting, AND, NAND, OR, NOR, EXCLUSIVE-OR, EXCLUSIVE-NOR, and combinations of logic and arithmetic operations.

EXAMPLE 12-4

The TTL Type 74181 4-bit arithmetic logic unit is capable of performing 16 arithmetic and 16 logic operations as shown in Table 12-1. These include the operation $(A \cdot B)$ plus A, where $A \cdot B$ stands for A AND B, A for $A_3A_2A_1A_0$, and B for $B_3B_2B_1B_0$. Find the result of this operation if $A = 1010$ and $B = 0111$.

Solution

The bit-by-bit AND of A and B is $A \cdot B = 0010$; hence the result of the operation is 0010 plus 1010 = **1100**.

12-6 MULTIPLIERS

Multiplication of two binary numbers can be performed similarly to the pencil-and-paper multiplication algorithm of two decimal numbers: the multiplicand is

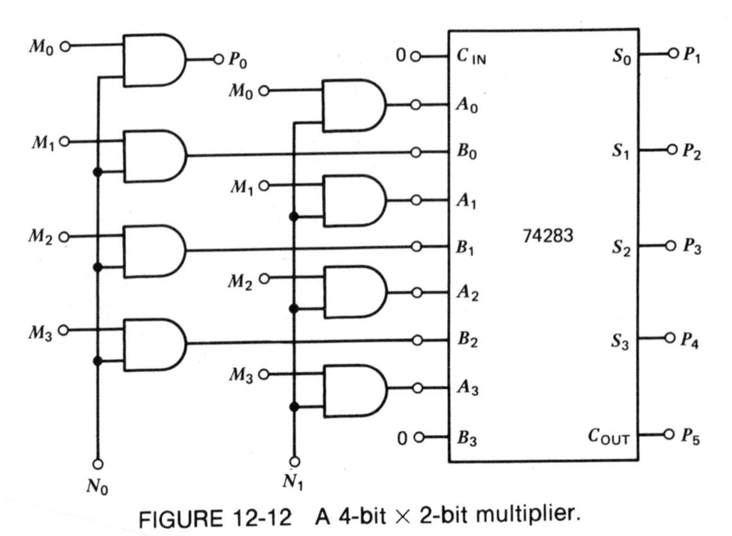

FIGURE 12-12 A 4-bit × 2-bit multiplier.

multiplied by successive digits of the multiplier and the appropriately positioned partial products are added. In binary multiplication each binary digit (bit) can be either 1 or 0; hence, each partial product is either equal to the multiplicand or it is zero.

EXAMPLE 12-5

Perform the binary multiplication $1101_2 \times 10_2$.

Solution

$$\begin{array}{r} 1101 \times 10 \\ \hline 0000 \\ 1101 \\ \hline 11010 \end{array}$$

Thus $1101_2 \times 10_2 = \mathbf{11010_2}$.

Figure 12-12 shows a 4-bit × 2-bit multiplier built by use of 8 AND gates and a TTL Type 74283 adder that was described in Ex. 12-2. It multiplies the 4-bit multiplicand $M_3M_2M_1M_0$ and the 2-bit multiplier N_1N_0 and generates the 6-bit product $P_5P_4P_3P_2P_1P_0$ by forming the partial products $(M_3M_2M_1M_0) \times N_0$ and $(M_3M_2M_1M_0) \times N_1$, and appropriately positioning and adding them.

EXAMPLE 12-6

Write in the logic values (0 or 1) in Fig. 12-12 if $M_3M_2M_1M_0 = 1101$ and $N_1N_0 = 10$.

Solution

The logic values are shown in Fig. 12-13. These show that the result of the multiplication is 11010, which is in agreement with the result of Ex. 12-5.

The multiplier of Fig. 12-12 can be also extended to more bits. Present technology is capable of including on a single IC chip a 16-bit × 16-bit multiplier with a propagation delay in the vicinity of 50 ns.

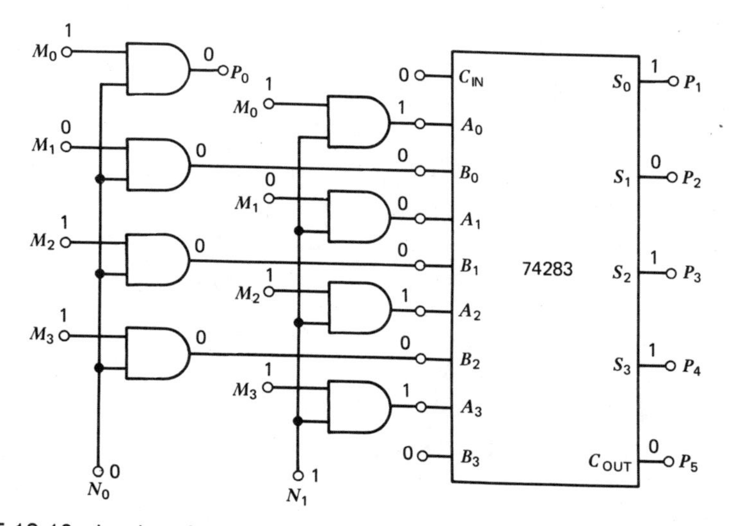

FIGURE 12-13 Logic values in Fig. 12-12 with $M_3M_2M_1M_0 = 1101$ and $N_1N_0 = 10$.

SUMMARY

Digital comparators were introduced realizing the magnitude relations of greater than, less than, equal, greater than or equal, and less than or equal. This was followed by a description of adders, subtractors, arithmetic logic units, and multipliers.

SELF-EVALUATION QUESTIONS

1. List the five magnitude relations realized by digital comparators.
2. Sketch the logic diagram of a 1-bit comparator realizing $A > B$.
3. Sketch the logic diagram of a 2-bit comparator realizing $A_1A_0 = B_1B_0$.
4. What is a half adder?
5. Sketch the logic diagram of a half adder.
6. What is a full adder?
7. What is a ripple adder?
8. Sketch the logic diagram of a 2's complement subtractor.
9. What is an arithmetic logic unit?
10. Sketch the logic diagram of a 4-bit $\times$ 2-bit multiplier.

PROBLEMS

12-1. Realize the logic statement $A > B$, where A and B are 2-bit numbers. Use 3 AND gates and 1 OR gate.

12-2. Realize the logic statement $A > B$, where A and B are 2-bit numbers. Use 5 OR gates and 1 AND gate.

12-3. Write Boolean equations for S_0, S_1, and C_{OUT} in Fig. 12-8 as functions of A_0, B_0, A_1, and B_1.

12-4. Draw a truth table for the circuit of Fig. 12-8. Include all 16 states and show the states of all three outputs S_0, S_1, and C_{OUT}.

12-5. Write in all logic values (0 or 1) in Fig. 12-9 if $A_3A_2A_1A_0 = 1010$ and $B_3B_2B_1B_0 = 0011$. Check whether the result of the addition is correct.

12-6. Write in all logic values (0 or 1) in Fig. 12-9 if $A_3A_2A_1A_0 = 1010$ and $B_3B_2B_1B_0 = 1010$. Check whether the result of the addition is correct.

12-7. The minuend in Fig. 12-11 is $A_3A_2A_1A_0 = 0101$ and the subtrahend is $B_3B_2B_1B_0 = 0011$. Write in logic values (0 or 1) at the outputs of the four logic inverters, and at S_3, S_2, S_1, S_0, and C_{OUT}. Check whether the result of the subtraction is correct.

12-8. The minuend in Fig. 12-11 is $A_3A_2A_1A_0 = 0011$ and the subtrahend is $B_3B_2B_1B_0 = 1110$. Write in logic values (0 or 1) at the outputs of the four logic inverters, and at S_3, S_2, S_1, S_0, and C_{OUT}. Check whether the result of the subtraction is correct.

12-9. Perform the multiplication $1010_2 \times 11_2$. Write into Fig. 12-12 the logic values (0 or 1) that are generated during the multiplication. Check the result.

12-10. Perform the multiplication $1111_2 \times 11_2$. Write into Fig. 12-12 the logic values (0 or 1) that are generated during the multiplication. Check the result.

12-11. The TTL Type 74283 4-bit adder in the multiplier circuit of Fig. 12-12 has a propagation delay of 24 ns. What is the propagation delay of the multiplier if each AND gate has a propagation delay of 20 ns?

12-12. The TTL Type 74283 4-bit adder in the multiplier circuit of Fig. 12-12 is replaced by a Schottky-TTL Type 74S283 4-bit adder with a propagation delay of 18 ns. What is the propagation delay of the multiplier if each AND gate has a propagation delay of 7 ns?

CHAPTER

13 Code Converters and Displays

Instructional Objectives

This chapter introduces combinational and sequential code converters. Display decoders, drivers, and display devices are also discussed. After you read it you should be able to:

1. Convert any 4-bit code to any other 4-bit code, using combinational code converters.
2. Convert any length of binary code to Gray code, and vice versa.
3. Convert any length of binary code to BCD using sequential and combinational techniques including ROMs.
4. Convert any length of BCD code to binary using sequential and combinational techniques including ROMs.
5. Describe decimal, 7-segment, dot-matrix, plasma, and CRT displays.
6. Outline the functions performed by a vector generator and a character generator.

13-1 OVERVIEW

Coding is a means of expressing in binary form a given quantity, a character of the alphabet, or a symbol. Various codes were discussed in Ch. 4; in this chapter we introduce combinational and sequential *code converters*. Such code converters are necessary at the interface of man–computer communication, as well as for transmission of data between various digital subsystems that do not employ the same code.

For example, internal operations of a computer may use the natural binary code, while various *peripheral devices* use, for good technical reasons, a variety of codes. These devices may include *input devices* such as a digital voltmeter requiring BCD to binary code conversion, or a shaft-encoder requiring Gray code to binary code conversion. They may also include *output devices* such as numerical displays requiring binary to BCD, BCD to decimal, and BCD to 7-segment code conversions, as well as cathode-ray tube (CRT) displays that require vector generators and character generators. There are also important *input–output devices*, such as magnetic tape units (not discussed here) that may use the ASCII code (Sec. 4-5.3) or the EBCDIC code (Sec. 4-5.4).

In addition to their use in computer systems, code converters are also used in many other applications such as shaft encoders, counters, clocks, and digital data transmission systems. In what follows here, we first illustrate a general code converter design procedure on an excess-3 Gray code to BCD code converter, then describe binary to Gray, Gray to binary, binary to BCD, and BCD to binary code converters. This is followed by an introduction of various displays such as decimal displays, 7-segment displays, dot-matrix displays, plasma displays, and CRT displays, as well as various code converters required by these displays.

13-2 EXCESS-3 GRAY CODE TO 8-4-2-1 BCD CODE CONVERTER

The excess-3 (XS-3) Gray code is a unit-distance code (see Sec. 4.3-2) that is employed in linear or angular shaft encoders; its truth table is given in Table 13-1.

In Ex. 13-1 below we show a step-by-step procedure for converting from the excess-3 Gray code to the 8-4-2-1 BCD (natural BCD) code. The procedure, with appropriate modifications, is applicable for conversion between any two 4-bit codes.

TABLE 13-1
Truth Table for the Excess-3 Gray Code and the 8-4-2-1 BCD Code

	Excess-3 Gray				8-4-2-1 BCD			
Decimal	G_3	G_2	G_1	G_0	B_3	B_2	B_1	B_0
0	0	0	1	0	0	0	0	0
1	0	1	1	0	0	0	0	1
2	0	1	1	1	0	0	1	0
3	0	1	0	1	0	0	1	1
4	0	1	0	0	0	1	0	0
5	1	1	0	0	0	1	0	1
6	1	1	0	1	0	1	1	0
7	1	1	1	1	0	1	1	1
8	1	1	1	0	1	0	0	0
9	1	0	1	0	1	0	0	1

EXAMPLE 13-1 Convert the excess-3 Gray code to the 8-4-2-1 BCD code.

Solution

Step 1. Draw a truth table for both codes as shown in Table 13-1. Mark the variables of the 8-4-2-1 BCD code B_3, B_2, B_1, and B_0 in descending order of binary weights. Mark the corresponding variables of the excess-3 Gray code G_3, G_2, G_1, and G_0.

Step 2. Draw a 4-variable Karnaugh map (Fig. 13-1*a*) for the variable B_0 as follows: Look up in the table those B_0 terms that have a binary value 1; these terms are for decimals 1, 3, 5, 7, and 9. Enter in the map the corresponding standard product terms from the excess-3 Gray code table. For example $B_0 = 1$ for decimal 1 and its corresponding

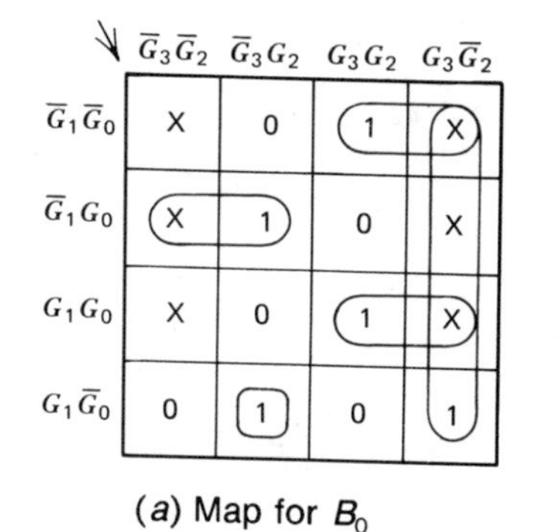

(*a*) Map for B_0

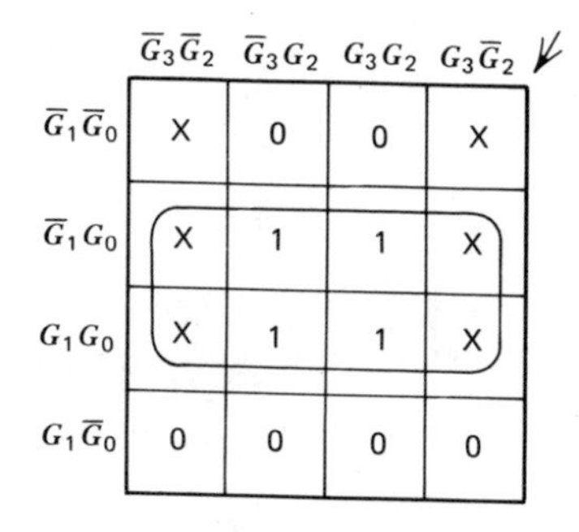

(*b*) Map for B_1

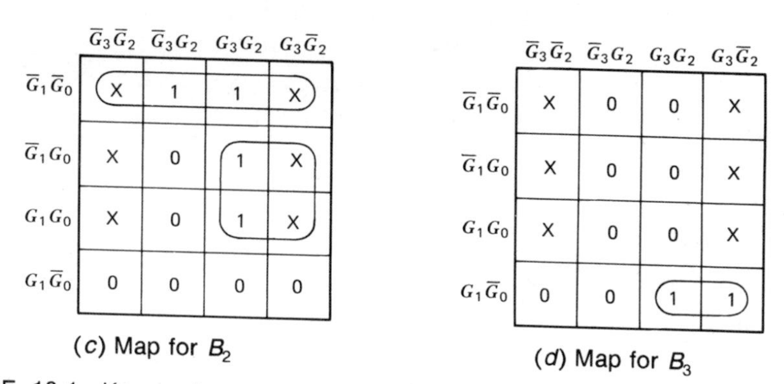

(*c*) Map for B_2

(*d*) Map for B_3

FIGURE 13-1 Karnaugh maps for excess-3 Gray code to 8-4-2-1 BCD code conversion.

standard product term in the excess-3 Gray code table is $\overline{G}_3G_2G_1\overline{G}_0$; enter a 1 for this term in the Karnaugh map of B_0. In another example, $B_0 = 1$ for decimal 3 and its corresponding standard product term in the excess-3 Gray code table is $\overline{G}_3G_2\overline{G}_1G_0$; enter a 1 for this term in the Karnaugh map of B_0. Complete entering 1s in the Karnaugh map of B_0 for the remainder terms where $B_0 = 1$.

Step 3. The excess-3 Gray code has six redundant states, which are: 0000, 0001, 0011, 1000, 1001, and 1011. Mark these terms with x's in the Karnaugh map because they represent don't care conditions.

Step 4. Obtain the minimized expression for B_0 from the adjacencies in the Karnaugh map:

$$B_0 = G_3\overline{G}_2 + G_3\overline{G}_1\overline{G}_0 + \overline{G}_3\overline{G}_1G_0 + G_3G_1G_0 + \overline{G}_3G_2G_1\overline{G}_0$$

Step 5 and subsequent steps. Repeat steps 2 through 4 to obtain the simplified expressions for B_1, B_2, and B_3. The respective Karnaugh maps are shown in Figs. 13-1*b*, 13-1*c*, and 13-1*d*, from which we obtain:

$$B_1 = G_0 \qquad B_2 = G_3G_0 + \overline{G}_1\overline{G}_0 \qquad B_3 = G_3G_1\overline{G}_0$$

13-3 BINARY CODE TO GRAY CODE CONVERTER

The two codes are shown in Table 13-2. By inspection we can see that $G_3 = B_3$. For G_2 we draw a Karnaugh map as shown in Fig. 13-2*a*—there are no don't care conditions since all 16 states of the codes are utilized. From Fig. 13-2*a* we derive the expression $G_2 = \overline{B}_3B_2 + B_3\overline{B}_2 = B_3 \oplus B_2$. Also, from Fig. 13-2*b* we get $G_1 = B_2\overline{B}_1 + \overline{B}_2B_1 = B_2 \oplus B_1$. The expression for G_0 can be obtained by a similar technique as $G_0 = B_1 \oplus B_0$.

The EXCLUSIVE-OR type of solution is applicable to any bit, and the general expression is $G_i = B_{i+1} \oplus B_i$. For example, in a code involving many bits the 12th Gray code bit is $G_{11} = B_{12} \oplus B_{11}$.

Figure 13-2*c* shows an implementation of a 4-bit code converter. The circuit is expandable for the conversion of as many bits as desired.

13-4 GRAY CODE TO BINARY CODE CONVERTER

The two codes are shown in Table 13-2. By inspection we can see that $B_3 = G_3$. For B_2 we draw a Karnaugh map as shown in Fig. 13-3*a*—there are no don't care conditions since all 16 states of the code are utilized. From the map we derive the expression $B_2 = \overline{G}_3G_2 + G_3\overline{G}_2 = G_3 \oplus G_2$. However, since $G_3 = B_3$, we can make the substitution $B_2 = B_3 \oplus G_2$.

TABLE 13-2
Truth Table for the 4-Bit Binary and Gray Codes

Decimal	Binary				Gray			
	B_3	B_2	B_1	B_0	G_3	G_2	G_1	G_0
0	0	0	0	0	0	0	0	0
1	0	0	0	1	0	0	0	1
2	0	0	1	0	0	0	1	1
3	0	0	1	1	0	0	1	0
4	0	1	0	0	0	1	1	0
5	0	1	0	1	0	1	1	1
6	0	1	1	0	0	1	0	1
7	0	1	1	1	0	1	0	0
8	1	0	0	0	1	1	0	0
9	1	0	0	1	1	1	0	1
10	1	0	1	0	1	1	1	1
11	1	0	1	1	1	1	1	0
12	1	1	0	0	1	0	1	0
13	1	1	0	1	1	0	1	1
14	1	1	1	0	1	0	0	1
15	1	1	1	1	1	0	0	0

← Axis of symmetry (between rows 7 and 8)

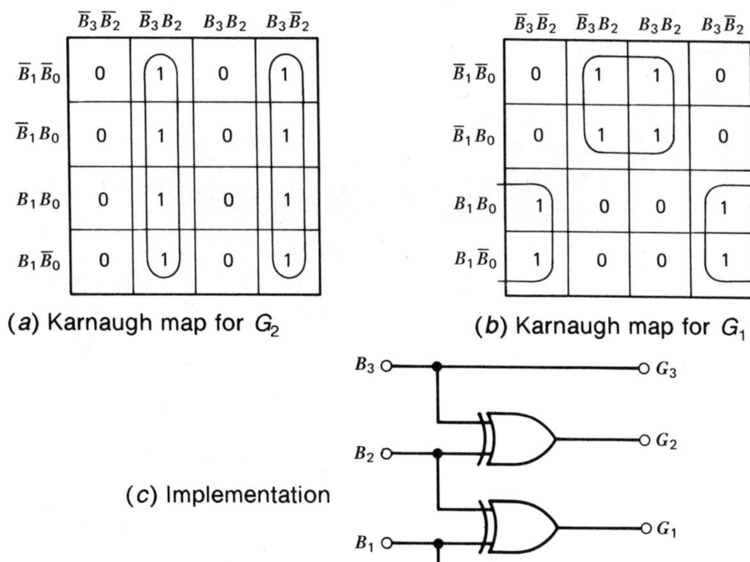

(*a*) Karnaugh map for G_2

(*b*) Karnaugh map for G_1

(*c*) Implementation

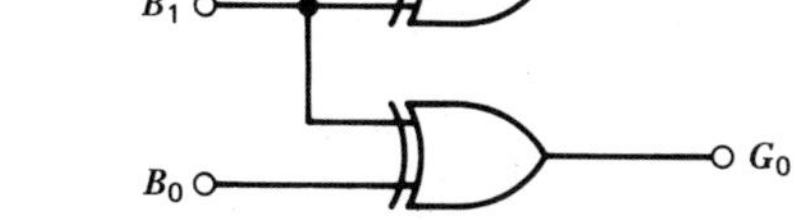

FIGURE 13-2 A 4-bit binary code to Gray code converter.

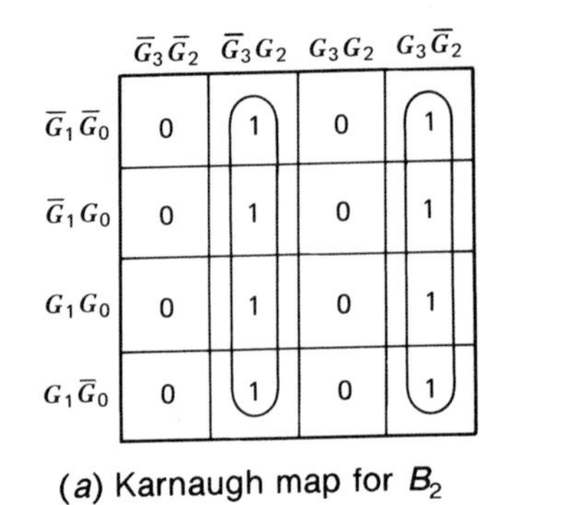

(*a*) Karnaugh map for B_2

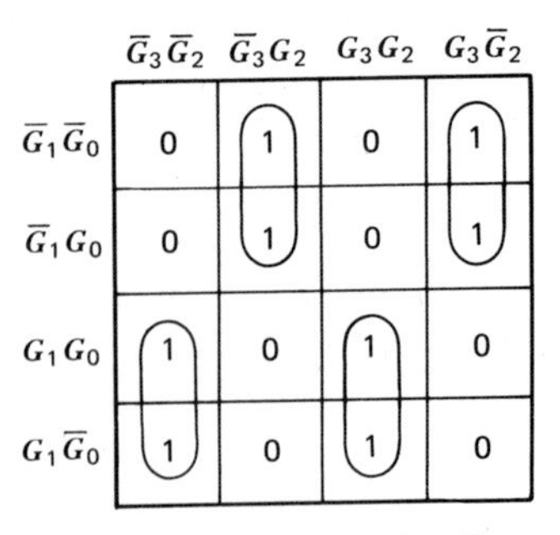

(*b*) Karnaugh map for B_1

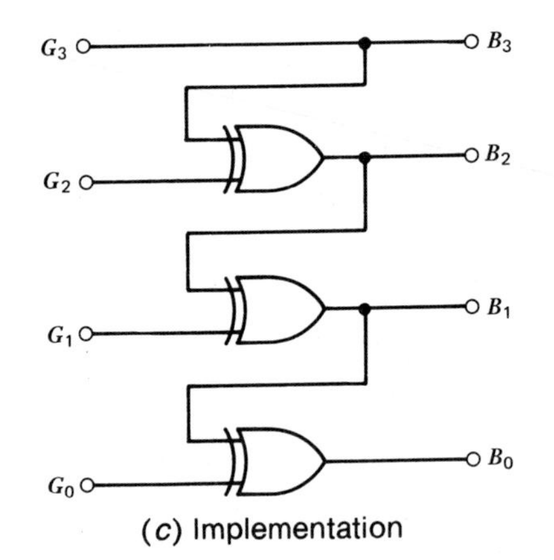

(*c*) Implementation

FIGURE 13-3 A 4-bit Gray code to binary code converter.

From the map of Fig. 13-3*b* we derive the expression for B_1 as $B_1 = \overline{G}_3\overline{G}_2G_1 + \overline{G}_3G_2\overline{G}_1 + G_3G_2G_1 + G_3\overline{G}_2\overline{G}_1 = \overline{G}_3(\overline{G}_2G_1 + G_2\overline{G}_1) + G_3(G_2G_1 + \overline{G}_2\overline{G}_1) = \overline{G}_3(G_2 \oplus G_1) + G_3(\overline{G_2 \oplus G_1})$. This can be further simplified by substituting X for $G_2 \oplus G_1$. Hence, $B_1 = \overline{G}_3X + G_3\overline{X} = G_3 \oplus X = G_3 \oplus G_2 \oplus G_1$. Also, since $G_3 = B_3$ and $B_3 \oplus G_2 = B_2$, it follows that $B_1 = B_2 \oplus G_1$. Using a similar procedure we obtain an expression for B_0 as $B_0 = G_3 \oplus G_2 \oplus G_1 \oplus G_0 = B_1 \oplus G_0$.

From the expressions for B_3 through B_0 we can draw the circuit of a 4-bit Gray code to binary code converter, as shown in Fig. 13-3*c*. The circuit is expandable for the conversion of as many bits as desired. Note, however, that the less significant bits of the converter undergo many propagation delays. In the 4-bit converter of Fig. 13-3*c*, B_0 is obtained only after 3 propagation delays of EXCLUSIVE-OR gates. For a 16-bit word, the least significant bit is obtained only after 15 propagation delays. Compare this with the binary to Gray code converter of Fig. 13-2*c*, in which all converted bits are available after one propagation delay.

13-5 BINARY TO BCD CODE CONVERTERS

The extensive use of BCD and binary numbers in digital systems calls for efficient conversion methods between the two number representations. The Karnaugh map method discussed in the previous three sections is practical for a relatively small number of bits. In the following sections we discuss conversion methods that utilize MSI and LSI components.

13.5.1 Serial Binary to BCD Code Converter

This method is applicable when a relatively long time is available to execute the conversion. The circuit, shown in Fig. 13-4, functions as follows: the binary number to be converted is loaded into a binary down-counter via parallel inputs and a *LOAD* signal. A down-counter decrements by one at each application of a clock pulse. Figure 13-4 also shows a BCD up-counter that must be cleared to zero at the beginning of each conversion. A *START* pulse starts a gated clock that is applied simultaneously to the clock inputs of the binary and the BCD counters. At each clock pulse the binary counter is *decremented* by 1, while the BCD counter is *incremented* by 1. When the down-counter has reached an all-zero state, the output of the active-low NAND gate (all-zero detector) stops the clock. The contents of the BCD up-counter at this time are the BCD equivalent of the binary number that was initially loaded into the binary down-counter.

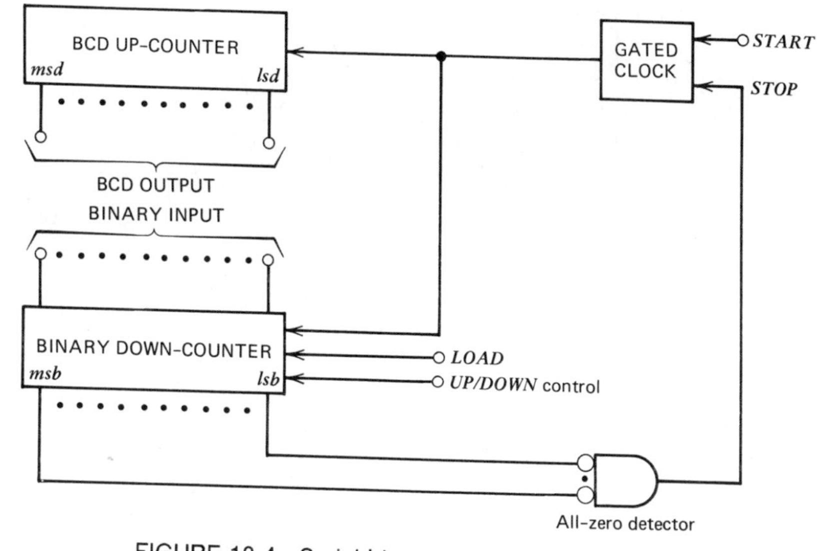

FIGURE 13-4 Serial binary to BCD code converter.

When a binary down-counter is not available we can use a binary up-counter in Fig. 13-4. However, instead of loading into the binary counter the number to be converted we load into it the *2's complement* of the number to be converted.

EXAMPLE 13-2

An 8-bit binary up-counter is utilized in the code converter of Fig. 13-4. What number should be parallel loaded for the conversion of (**a**) 255_{10} and (**b**) 16_{10}?

Solution

a.

$$255_{10} = 1\ 1\ 1\ 1\ 1\ 1\ 1\ 1$$

$$\text{1's complement of } 255_{10} = 0\ 0\ 0\ 0\ 0\ 0\ 0\ 0$$

$$\text{2's complement of } 255_{10} = 0\ 0\ 0\ 0\ 0\ 0\ 0\ 1$$

The number to be parallel loaded for the conversion of 255_{10} is **00000001**. After 254 counts the state of the binary up-counter will be all 1s; on the 255th count it will be all 0s and the negative NAND gate (all-zero detector) will inhibit the gated clock from further incrementing the counters.

b.

$$16_{10} = 0\ 0\ 0\ 1\ 0\ 0\ 0\ 0$$

$$\text{1's complement of } 16_{10} = 1\ 1\ 1\ 0\ 1\ 1\ 1\ 1$$

$$\text{2's complement of } 16_{10} = 1\ 1\ 1\ 1\ 0\ 0\ 0\ 0$$

The number to be parallel loaded for the conversion of 16_{10} is **11110000**.

Serial code converters are applicable to small numbers, and also to large numbers if the time required for the conversion is not a limiting factor.

EXAMPLE 13-3

Calculate the time, t, required to convert a 16-bit binary number to BCD using counters with a clock frequency of 20 MHz.

Solution

The largest number that can be expressed by 16 bits is $2^{16} - 1 = 65{,}535$. Each clock period requires $1/20$ MHz seconds = 50 ns. Hence the total conversion time is 65,535 counts $\times$ 50 ns = 3276.75 μs = **3.27675 ms**.

13-5.2 Combinational Binary to BCD Code Converters Using ROMs

A read-only memory (ROM) may be used as a look-up table (Sec. 11-6). For each combination of input variables there is an output that has been programmed in the ROM. For binary to BCD conversion the input combinations are binary numbers while the output represents the corresponding BCD numbers.

EXAMPLE 13-4

a. Determine the binary input combinations and their corresponding BCD outputs for the numbers 12_{10}, 31_{10}, 48_{10}, and 80_{10}.
b. How many bits are required to express the input combinations?
c. How many bits are required to express the BCD outputs?

TABLE 13-3
Solution to Ex. 13-4a

Decimal Number	Binary Input	BCD Output	
		Tens	Units
12	0 1 1 0 0	0 0 0 1	0 0 1 0
31	1 1 1 1 1	0 0 1 1	0 0 0 1
48	1 1 0 0 0 0	0 1 0 0	1 0 0 0
80	1 0 1 0 0 0 0	1 0 0 0	0 0 0 0

Solution

a. See Table 13-3.
b. The largest binary input is 80_{10} = 1010000, requiring **7** input bits.
c. The largest BCD output is 80_{10} = 1000 0000, requiring **8** output bits.

A ROM is usually specified as having a given number of *words*, each word consisting of a number of *output bits*. The total number of words in a ROM is related to the number of *input bits*: For example, for 8 input bits the ROM has $2^8 = 256$ words. Thus a ROM is well suited for binary to BCD conversion and, as shown later, also for BCD to binary conversion.

EXAMPLE 13-5

It is desired to convert a 10-bit binary number to BCD. What size ROM (words × bits) is required to accomplish the conversion?

Solution

A 10-bit binary number has $2^{10} = 1024$ input combinations; thus the number of *words* in the ROM is 1024. The number of required *output bits* in each word is calculated as follows: The maximum BCD number to be expressed is 1024. This represents 3 BCD decades with 4 bits each, plus 1 additional bit for the BCD digit of one thousand. Thus a total of 13 output bits is required and a ROM capacity of **1024 words** × **13 bits** is required to accomplish the conversion.

Binary to BCD conversion with large ROMs as illustrated in Ex. 13-5 is limited in practice to about 13 bits. For larger numbers the size of the ROM becomes impractical.

EXAMPLE 13-6

Calculate the ROM capacity required for a conversion to BCD of a 16-bit binary number.

Solution

A 16-bit binary number has 2^{16} combinations; thus the number of words in the ROM is $2^{16} = 65{,}536$. The number of bits in this 5-digit BCD word is the number of BCD digits × 4 bits/BCD digit = 20 bits. Thus a ROM of 65,536 words × 20 bits is required, making a total of **1,310,720 bits**, at present an impractical proposition.

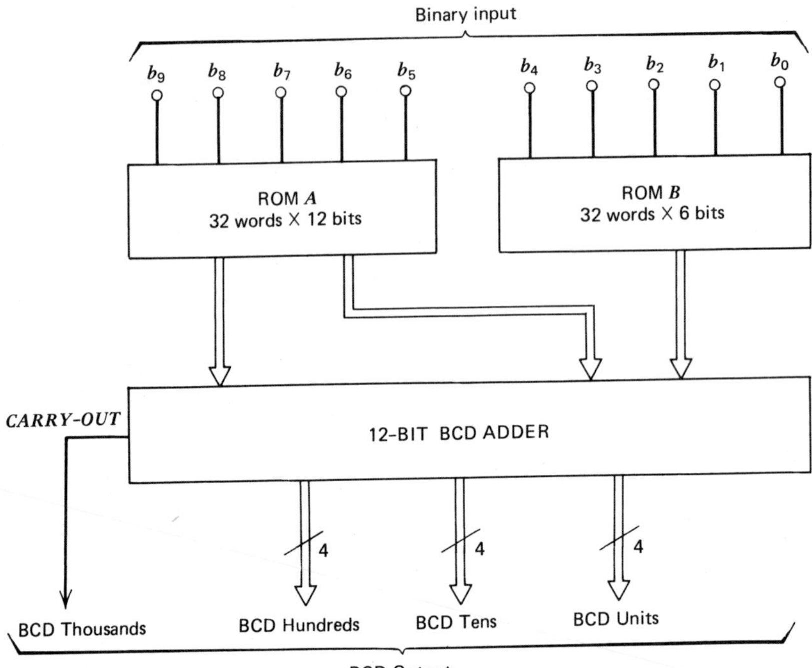

FIGURE 13-5 A 10-bit binary to BCD code converter using two ROMs and a BCD adder.

*13-5.2.1 Conversion of Large Numbers

The problem of conversion of large numbers can be solved by dividing the number into two parts: the most significant bits are converted in one ROM, and the least significant bits are converted in another ROM. The results of these two conversions are then combined in a BCD adder to yield the desired BCD number. The method is, for convenience, illustrated with a 10-bit number; this is followed by an example that shows the circuits required for a 16-bit conversion.

The binary input number, N, is divided into two parts:

$$N = A + B$$

where A represents the 5 most significant bits followed by five 0s,

$$A = a_9 a_8 a_7 a_6 a_5 00000$$

while B represents the 5 least significant bits preceded by five 0s,

$$B = 00000 b_4 b_3 b_2 b_1 b_0$$

The 5 most significant bits of A are converted to BCD in ROM A, which has a capacity of $2^5 = 32$ words, each word having 12 bits (the largest number to be expressed by this ROM is 1111100000, which is 992_{10} and requires 3 BCD digits, i.e., 12 bits). Similarly, the 5 bits of B are converted to BCD in the

*Optional material.

32-word × 6-bit ROM B. The results of these conversions are combined in a BCD adder. The BCD adder must add up to 1024; therefore, it must have a capacity of 3 full BCD digits plus 1 additional bit to represent the digit of one thousand. The block diagram of the code converter is shown in Fig. 13-5. We have thus accomplished the binary to BCD conversion of a 10-bit number with a ROM capacity of 32 words × 12 bits plus 32 words × 6 bits, for a total of $32 \times (12 + 6) = 576$ bits. Compare this with the solution using one ROM, as in Ex. 13-5, which requires a ROM of 1024 words × 13 bits for a total of 13,312 bits.

EXAMPLE 13-7

Using two ROMs calculate the ROM capacity required for the conversion of a 16-bit binary number to BCD.

Solution

We divide the binary number into A representing the 8 most significant bits followed by 8 zeros, and B representing the 8 least significant bits. The ROM for A has $2^8 = 256$ words and the largest number at its output is $1111111100000000 = 65{,}280$, which requires 20 output bits for 5 BCD digits. The ROM for B has $2^8 = 256$ words and the largest number at its output is 255, which requires 10 output bits.

Thus the ROM requirement is 256 words × 20 bits plus 256 words × 10 bits, for a total of $256 \times (20 + 10) =$ **7680 bits**; compare this with the 1,310,720 bits of Ex. 13-6 where a single ROM was used. To complete the code converter using two ROMs, we also require a 5-digit BCD adder.

13-6 BCD TO BINARY CODE CONVERTERS

Thus far we have discussed binary to BCD code converters. Some of their applications are at the output of a binary computer where the number has to be displayed in a convenient form for a human operator, that is, in decimal form. BCD to binary conversion is the reverse process, namely, applying to a computer numbers that for convenience are decimal and are thus BCD coded. Since the internal operation of many computers is binary, a conversion from BCD to binary is desired. Though such a conversion may be executed by the computer itself, much valuable time of the computer is saved by using auxiliary hardware which, incidentally, can do the conversion in a much shorter time than the computer.

This section describes a sequential code converter, followed by a combinational code converter using adders. The section concludes with a discussion of BCD to binary code converters using ROMs.

13-6.1 Serial BCD to Binary Code Converter

This code converter is similar to the serial binary to BCD code converter that was discussed in Sec. 13-5.1. It is applicable when a relatively long time is available to execute the conversion. The circuit shown in Fig. 13-6 functions as

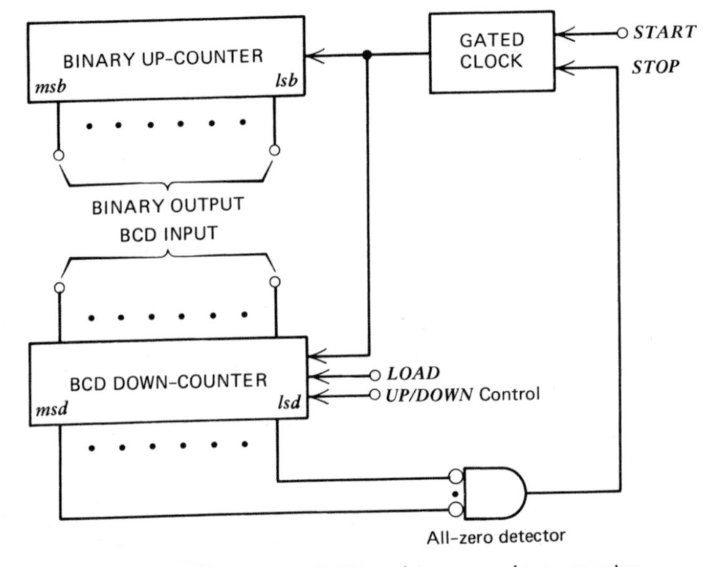

FIGURE 13-6 Serial BCD to binary code converter.

follows: the BCD number to be converted is loaded into a BCD down-counter via parallel inputs and a *LOAD* signal. Figure 13-6 also shows a binary up-counter that must be cleared to zero at the beginning of each conversion. A *START* pulse starts a gated clock that is applied simultaneously to the clock inputs of the BCD and the binary counters. At each clock pulse the BCD counter is *decremented* by 1 while the binary counter is *incremented* by 1. When the down-counter has reached zero, the output of the negative NAND gate (all-zero detector) stops the gated clock. The contents of the binary up-counter at this time are the binary equivalent of the BCD number that was initially loaded into the BCD counter.

When a BCD down-counter is not available, we can use a BCD up-counter in Fig. 13-6 preloaded with the *10's complement* of the number to be converted. The 10's complement of a BCD digit is obtained by subtracting the digit from 1010_2. In the 8-4-2-1 BCD code, for example,

the 10's complement of 0010 is 1000
the 10's complement of 0100 is 0110
the 10's complement of 0111 is 0011

EXAMPLE 13-8

A 2-decade BCD up-counter is utilized in the BCD to binary code converter shown in Fig. 13-6. What number should be parallel loaded for the conversion of 33_{10} to binary?

Solution

The number 33_{10} in BCD is 0011 0011; thus the 10's complement of BCD 33 is **0111 0111**.

This number is parallel loaded into the BCD up-counter. After 32 counts the state of the counter will be all 1s and on the 33rd count it will become all 0s,

TABLE 13-4

BCD	10's Complement
$DCBA$	$WXYZ$
0 0 0 0	1 0 1 0
0 0 0 1	1 0 0 1
0 0 1 0	1 0 0 0
0 0 1 1	0 1 1 1
0 1 0 0	0 1 1 0
0 1 0 1	0 1 0 1
0 1 1 0	0 1 0 0
0 1 1 1	0 0 1 1
1 0 0 0	0 0 1 0
1 0 0 1	0 0 0 1

causing the active-low NAND gate (all-zero detector) to inhibit the gated clock from further incrementing the counters.

A 10's complementing combinational logic circuit may be obtained using the conversion techniques discussed in Sec. 13-2.

EXAMPLE 13-9

Obtain a minimum realization of a 10's complementing combinational logic circuit.

Solution

We first establish a truth table, Table 13-4, showing in the left columns the 10 BCD combinations of variables D, C, B, A, and in the right columns the corresponding 10's complement combinations of variables W, X, Y, Z.

By inspection we can see that the least significant bits are identical: $\boldsymbol{Z = A}$.

Next we establish bit Y in terms of D, C, B, and A. We look up in the table those BCD standard product terms for which the column for Y shows $Y = 1$. Thus

$$Y = \overline{D}\,\overline{C}\,\overline{B}\,\overline{A} + \overline{D}\,\overline{C}BA + \overline{D}C\overline{B}\,\overline{A} + \overline{D}CBA + D\overline{C}\,\overline{B}\,\overline{A}.$$

Taking six don't care terms into consideration, we obtain from a Karnaugh map a simplified expression for Y as

$$\boldsymbol{Y = \overline{A}\,\overline{B} + AB}$$

We proceed similarly to obtain simplified expressions for X and W as

$$\boldsymbol{X = \overline{B}C + \overline{A}C + AB\overline{C}}$$
$$\boldsymbol{W = \overline{D}\,\overline{C}\,\overline{A} + \overline{D}\,\overline{C}\,\overline{B} = \overline{D}\,\overline{C}(\overline{B} + \overline{A})}$$

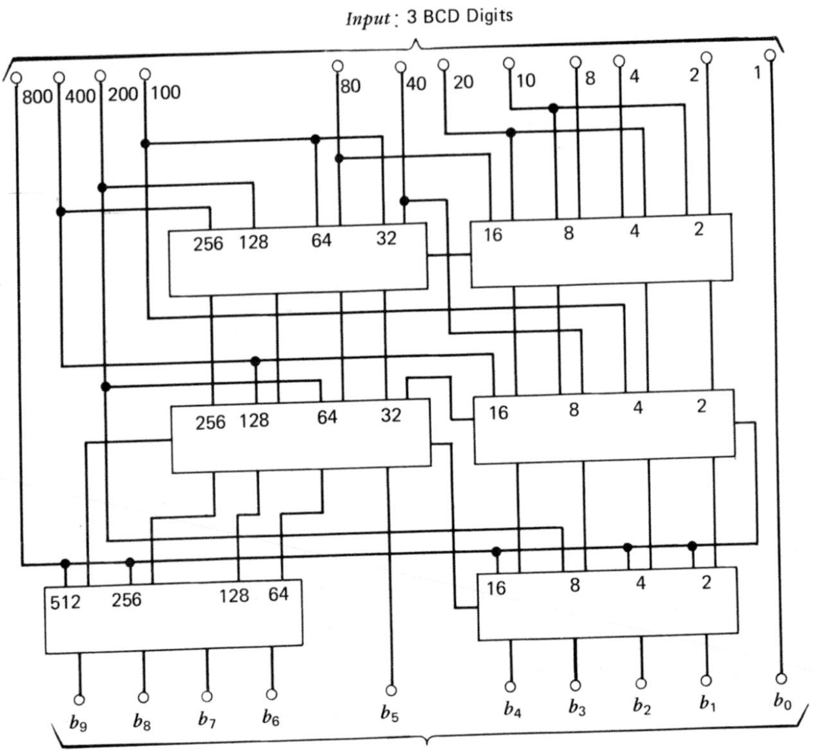

FIGURE 13-7 Conversion of 3 BCD digits to binary using adders. Each block represents a TTL Type 7483 4-bit binary full adder.

*13-6.2 Combinational BCD to Binary Code Converter Using Adders

This conversion method is based on the principle that any BCD number may be expressed as a sum of binary powers. Thus every number in the units decade may be made up from sums of $1(=2^0)$, $2(=2^1)$, $4(=2^2)$, and $8(=2^3)$. For the tens decade we have

$$\begin{aligned}10 &= 8+2 = 2^3+2^1\\ 20 &= 16+4 = 2^4+2^2\\ 40 &= 32+8 = 2^5+2^3\\ 80 &= 64+16 = 2^6+2^4\end{aligned}$$

Similarly, for the hundreds decade we have

$$\begin{aligned}100 &= 64+32+4 = 2^6+2^5+2^2\\ 200 &= 128+64+8 = 2^7+2^6+2^3\\ 400 &= 256+128+16 = 2^8+2^7+2^4\\ 800 &= 512+256+32 = 2^9+2^8+2^5\end{aligned}$$

*Optional material.

From the above discussion we can see that we require only binary adders to carry out a BCD to binary conversion. Figure 13-7 shows six 4-bit adders and their interconnections for the conversion of 3 BCD digits to binary. The method is fast since the conversion process is completed after three adder propagation delays.

*13-6.3 Combinational BCD to Binary Code Converters Using ROMs

ROMs may be used for BCD to binary conversion following a procedure that is similar to that shown for binary to BCD conversion in Sec. 13-5.2.

EXAMPLE 13-10

Convert a 3-digit BCD number to binary. How many ROM bits are required?

Solution

A 3-digit BCD number has $4 \times 3 = 12$ bits; hence we require 12 input bits and these generate $2^{12} = 4096$ different combinations (words). Not all of these combinations are utilized since the largest 3-digit BCD number is only 999. Thus BCD to binary conversion is not as efficient in ROM utilization as binary to BCD conversion. The number of output bits is obtained from the number of bits required to express in binary the largest equivalent BCD number. That number is 999 and requires 10 bits at the output. Thus the total number of bits in the ROM is 2^{12} words $\times$ 10 bits = **40,960 bits**.

*13-6.3.1 *Conversion of Large Numbers*

We saw in Ex. 13-10 that the number of bits in a ROM that are required to convert a 3-digit BCD number to binary is rather large and would become impractical for larger numbers. In binary to BCD conversion we solved a similar problem by segmenting the ROM into two ROMs and incorporating a *BCD adder* to combine the outputs of the two ROMs. A similar strategy used in BCD to binary code converters utilizes two ROMs—for higher order and lower order BCD to binary conversion—and the two quantities are combined in a *binary adder*.

EXAMPLE 13-11

Use two ROMs and a binary adder for the BCD to binary conversion of 3 BCD digits.

Solution

Figure 13-8 shows a schematic diagram of the code converter. The 6 bits of the higher order BCD number require a ROM of $2^6 = 64$ words. The number of ROM bits may be calculated considering that the highest BCD number represented by the six leftmost BCD bits shown in Fig. 13-8 is 980. This requires 10 bits and thus a ROM of 64 words $\times$ 10 bits. The 6 bits of the lower order BCD number require a ROM of $2^6 = 64$ words $\times$ 6 bits. Thus total ROM requirements reduce to 64 words $\times$ (10 + 6) bits = 1024 bits, which compares favorably with the one-ROM method of Ex. 13-10 in which a 40,960-bit ROM was required.

*Optional material.

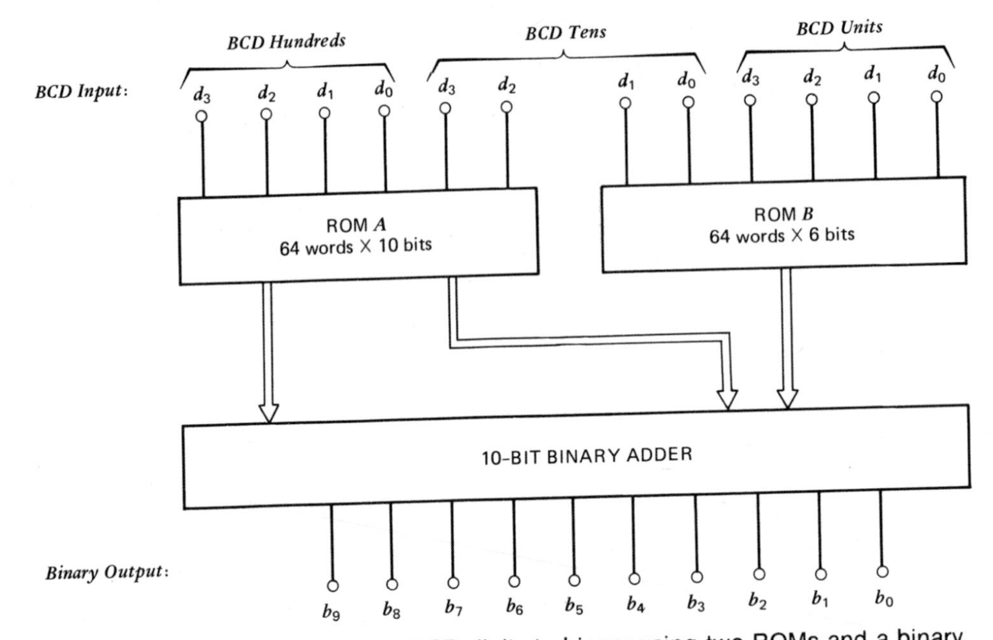

FIGURE 13-8 Conversion of 3 BCD digits to binary using two ROMs and a binary adder.

	BCD input				Decimal Output	Minimized Terms			
	X_3	X_2	X_1	X_0		X_3	X_2	X_1	X_0
	0	0	0	0	0	0	0	0	0
	0	0	0	1	1	0	0	0	1
	0	0	1	0	2		0	1	0
	0	0	1	1	3		0	1	1
	0	1	0	0	4		1	0	0
	0	1	0	1	5		1	0	1
	0	1	1	0	6		1	1	0
	0	1	1	1	7		1	1	1
	1	0	0	0	8	1			0
	1	0	0	1	9	1			1
Nonvalid inputs	1	0	1	0	2	False Outputs			
	1	0	1	1	3				
	1	1	0	0	4, 8				
	1	1	0	1	5, 9				
	1	1	1	0	6, 8				
	1	1	1	1	7, 9				

(*a*) Truth table

FIGURE 13-9 8-4-2-1 BCD to decimal decoder for a NIXIE driver.

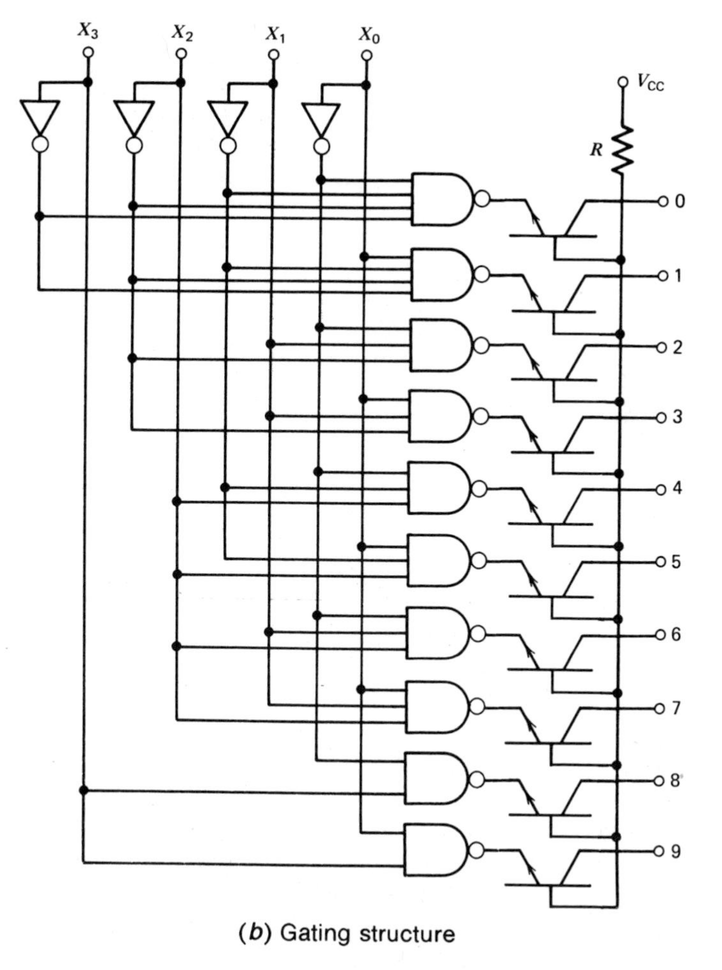

(*b*) Gating structure

FIGURE 13-9 (*Continued*)

13-7 DECIMAL DISPLAYS

In a decimal display, one of the 10 decimal digits 0 through 9 is activated at any given time; also, decimal points are often provided. Decimal displays in use include those lit by *incandescent lights*, *vacuum fluorescent displays* operating at voltages of ≈ 50 V and currents of up to 25 mA, *liquid-crystal displays* (*LCDs*) operating at voltages of ≈ 1 V, and gas-filled *NIXIE displays* operating at voltages of ≈ 80 V and currents of 0.1 mA to 5 mA (NIXIE is a trademark of the Burroughs Corporation).

Figure 13-9 shows a drive circuit used for NIXIE displays: Fig. 13-9*a* shows a truth table of the 8-4-2-1 BCD to decimal decoder, Fig. 13-9*b* the resulting gating structure. One of the 10 output transistors is turned on at any given time, pulling to near ground one of the 10 output lines 0 through 9. These lines are connected to the negative terminals (cathodes) of the 10 display digits (not

TABLE 13-5
Truth Table for a BCD to 7-Segment Decoder

Decimal Number	BCD Number	Segment
	D C B A	*a b c d e f g*
0	0 0 0 0	1 1 1 1 1 1 0
1	0 0 0 1	0 1 1 0 0 0 0
2	0 0 1 0	1 1 0 1 1 0 1
3	0 0 1 1	1 1 1 1 0 0 1
4	0 1 0 0	0 1 1 0 0 1 1
5	0 1 0 1	1 0 1 1 0 1 1
6	0 1 1 0	1 0 1 1 1 1 1
7	0 1 1 1	1 1 1 0 0 0 0
8	1 0 0 0	1 1 1 1 1 1 1
9	1 0 0 1	1 1 1 1 0 1 1

shown in the figure), while the positive terminals (anodes) of the digits are all connected to $\approx +100$ V via a resistor. Note that the circuit is not immune to nonvalid input combinations and provides false outputs for BCD inputs of 10_{10} through 15_{10}.

13-8 SEVEN-SEGMENT DISPLAYS

A 7-segment display is built around a pattern of 7 straight segments of which several are activated at a time for display of the desired numeral or character of the alphabet. The basic 7-segment pattern is shown in Fig. 13-10*a*. Figure

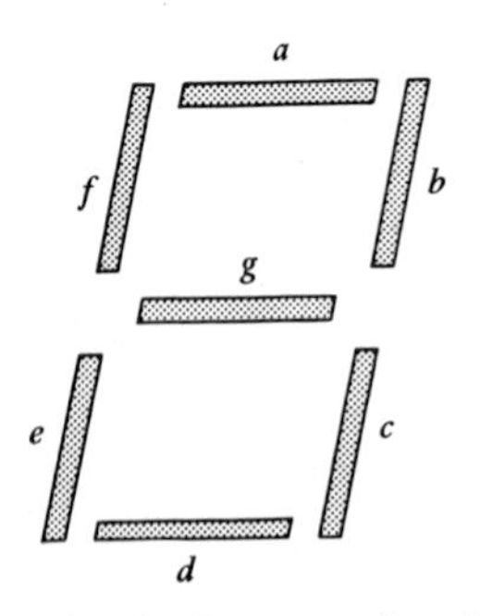

(*a*) The basic 7-segment pattern

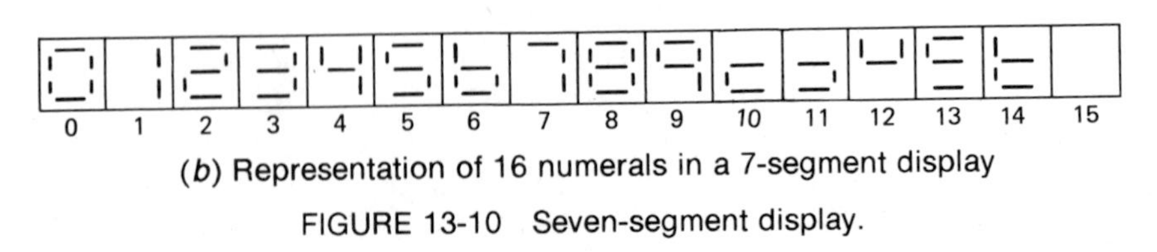

(*b*) Representation of 16 numerals in a 7-segment display

FIGURE 13-10 Seven-segment display.

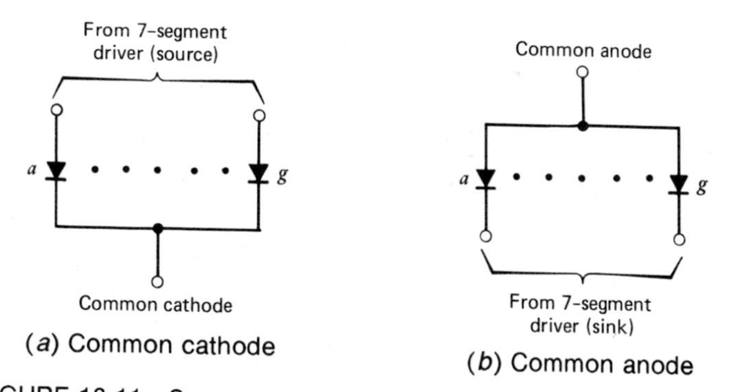

(*a*) Common cathode

(*b*) Common anode

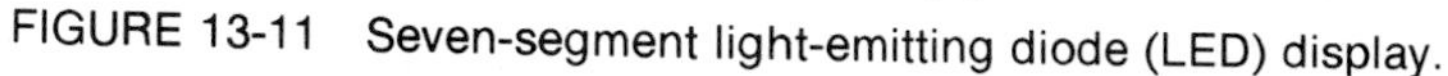

FIGURE 13-11 Seven-segment light-emitting diode (LED) display.

13-10*b* shows the representation of 16 numerals through the selective activation of the individual segments. Table 13-5 shows a truth table.

When using a programmable logic array (PLA) for the realization of the truth table, it is not necessary to simplify the Boolean functions, since simplification would not result in a reduction of the size of the PLA.

Seven-segment displays include those using *incandescent lights, gas-filled displays* operating at voltages of $\approx$ 80 V and currents of 0.1 mA to 5 mA, those using *light-emitting diodes* (*LEDs*) operating at voltages of 2 V to 5 V and currents of up to 100 mA, as well as liquid-crystal displays (LCDs). In what follows here, we outline a few applications of LED displays.

Seven-segment LED displays are available with either common cathodes as shown in Fig. 13-11*a*, or with common anodes as in Fig. 13-11*b*. The common cathode configuration requires a driver circuit that *sources* the requisite current while the common anode configuration requires a driver circuit that *sinks* the current.

Two types of displays are in use: single-digit and multidigit. In a single-digit display the various segments are activated in accordance with the desired numeral to be displayed. In a multidigit display each *digit* has a separate

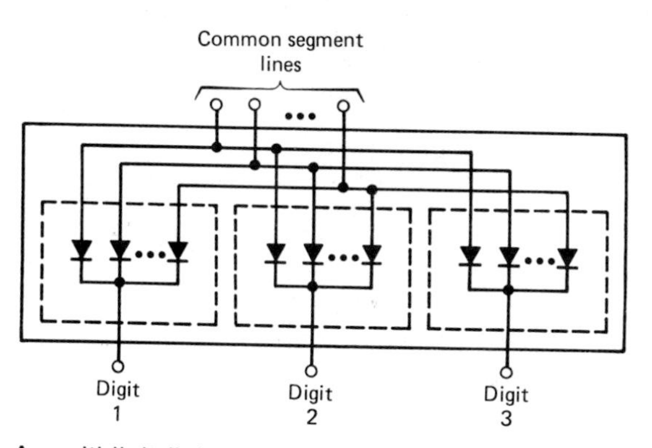

FIGURE 13-12 A multidigit light-emitting diode (LED) assembly for a 7-segment display.

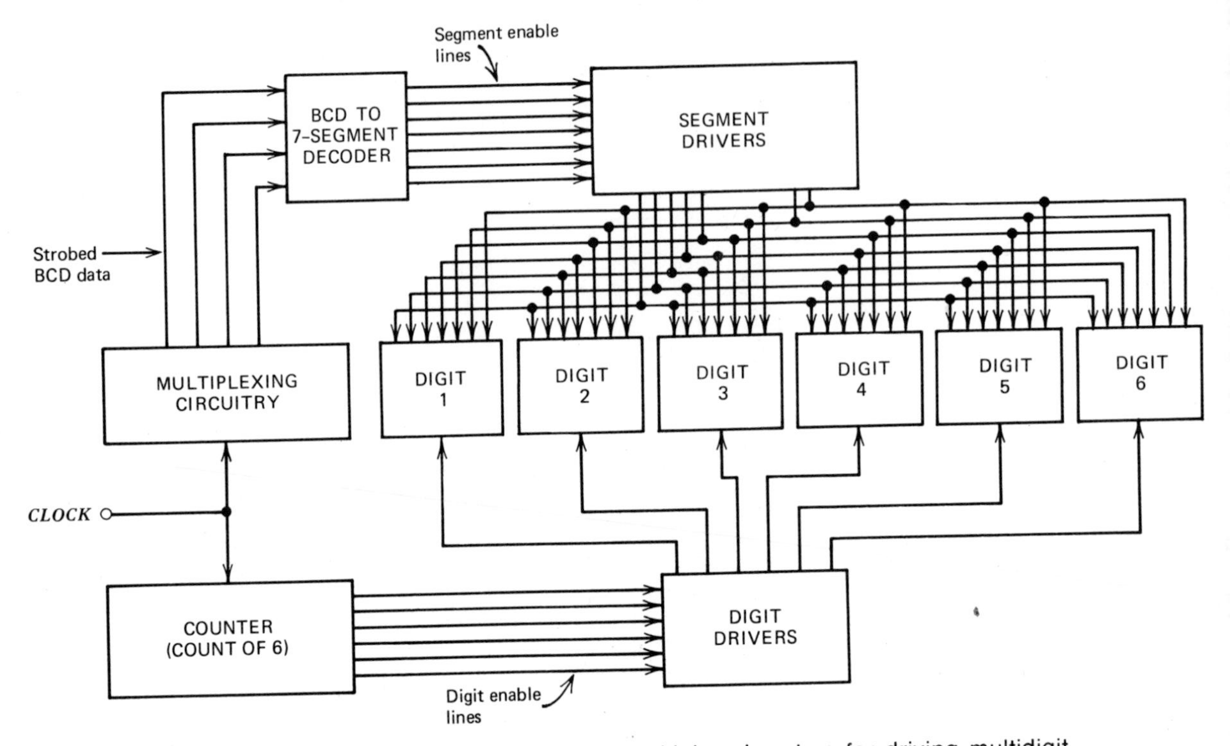

FIGURE 13-13 Block diagram of a time-multiplexed system for driving multidigit LED displays.

cathode (or anode) while the 7 segment lines are common to all digits. As shown in Fig. 13-12, current flows through an LED if both its *segment line and* its *digit line* have been activated by appropriate signals.

Multidigit displays are driven in a *time-multiplexed* mode by a system as shown in Fig. 13-13. A single BCD to 7-segment decoder and 7 segment drivers serve all the digits of the display. A counter operated at a frequency of 250 to 1000 Hz selects sequentially each digit by providing a sink current through the respective digit driver.

13-9 DOT-MATRIX DISPLAYS

In a dot-matrix display a rectangular array of 5 × 7 or 7 × 9 LEDs is used together with a *character generator* that selects which of the LEDs are activated. This type of display has become widely used because a great variety of *fonts* can be displayed and because integrated circuits are available that facilitate the assembly of a display system.

An alphanumeric character is obtained by a process called *scanning*. In a 5 × 7 LED display the character is thought of being made of 7 rows and 5 columns. Two signals are applied simultaneously to activate a row and a column of the dot matrix.

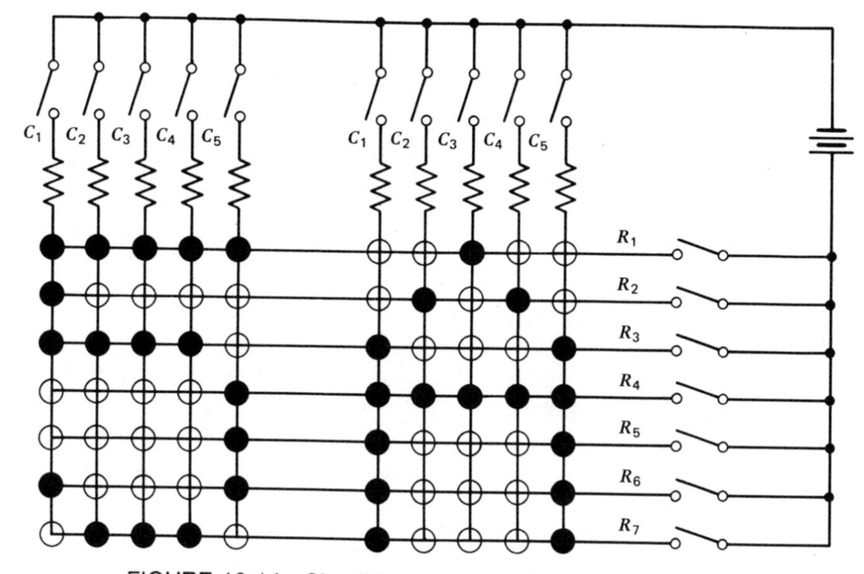

FIGURE 13-14 Simplified diagram of a dot-matrix display.

Figure 13-14 shows a simplified diagram of a dot-matrix display—in a real display the switches are replaced by transistor drivers. The display of Fig. 13-14 uses *scanning* that selects one row at a time, and turns on those LEDs in the selected row that are part of the desired character. The process is completed when all 7 rows have been scanned.

Figure 13-14 also shows how a numeral **5** and a letter **A** are displayed. Scanning of the display is controlled by a *character generator ROM*, which is not shown in Fig. 13-14. A dot-matrix display also requires input storage buffers to store the characters, output storage buffers and drivers for the rows and the columns, and timing circuitry.

13-10 PLASMA DISPLAYS

The gas-filled *plasma displays* operate at voltages of ≈ 80 V and at currents up to 20 mA. They are scanned in a manner that is similar to the dot-matrix display of Fig. 13-14. However, since a plasma display may display hundreds of characters, character generation is usually performed by a computer, not by a ROM.

13-11 CATHODE-RAY TUBE (CRT) DISPLAYS

Cathode-ray tubes (*CRTs*) are widely used as display devices in computer terminals and in television receivers. A beam of electrons is accelerated and is deflected by horizontal and vertical deflection systems. In computer display applications, the horizontal deflection is proportional to a digital control word and the vertical deflection is proportional to another digital control word. Often these control words are referenced to their preceding values by means of additional storage registers and adders ("*vector generator*"). CRT displays often include a *character generator* that is similar to that of a dot-matrix display, but is more complex.

SUMMARY

Code conversion between two 4-bit codes was illustrated using the excess-3 Gray code to BCD conversion. The regular structures of Gray code to binary code converters and binary code to Gray code converters resulted in code converters using only EXCLUSIVE-OR gates. Several types of binary to BCD and BCD to binary converters were also discussed.

The chapter also included descriptions of decimal displays, 7-segment displays, dot-matrix displays, plasma displays, and CRT displays.

SELF-EVALUATION QUESTIONS

1. What is a code converter? Show examples of its application.

2. Describe the combinational method of converting from one 4-bit code to another. When is minimization applicable?

3. After how many propagation delays is the Gray code available in a 16-bit combinational binary code to Gray code converter?

4. After how many propagation delays is the binary code available in a 16-bit combinational Gray code to binary code converter?

5. Draw a block diagram of the binary to BCD converter using the counter method and discuss its operation.

6. Assume that a binary down-counter is not available in the converter of the preceding question; how can you implement a conversion using the counter method?

***7.** Discuss the relative advantages and disadvantages of the combinational and the counter methods for the conversion of a 12-bit binary code to a BCD code.

8. Discuss the practical limits on the word size for binary to BCD conversion using ROMs. Illustrate your discussion with an example of a 16-bit code converter.

***9.** Discuss the method of binary to BCD conversion using two ROMs. What are the advantages?

10. What is the 9's complement of an 8-4-2-1 BCD number?

11. What is the 10's complement of an 8-4-2-1 BCD number?

*Optional material.

*12. What is the basis for the BCD to binary converter using adders? Show how BCD numbers can be expressed as sums of powers of 2.
13. Discuss the criteria for choosing a decoder for a display driver.
14. Describe the following displays: NIXIE, 7-segment, and dot-matrix.
15. Discuss vertical and horizontal scan in a dot-matrix character generator.
16. What is a vector generator?

PROBLEMS

13-1. You are given the 8-4-2-1 BCD code.

a. Convert it to the excess-3 Gray code, obtaining minimized expressions for G_3, G_2, G_1, and G_0.

b. What are the redundant states of the 8-4-2-1 BCD code and how do you use them for minimization? The excess-3 Gray code is given in Table 13-1.

TABLE P13-1

Decimal	2-4-2-1 BCD
0	0 0 0 0
1	0 0 0 1
2	0 0 1 0
3	0 0 1 1
4	0 1 0 0
5	1 0 1 1
6	1 1 0 0
7	1 1 0 1
8	1 1 1 0
9	1 1 1 1

13-2. Design a simplified combinational logic circuit for the conversion of the 8-4-2-1 BCD code to the 2-4-2-1 BCD code given in Table P13-1.

13-3. **a.** Design a simplified combinational logic circuit for the conversion of the excess-3 BCD code to the 8-4-2-1 BCD code. The excess-3 BCD code is given in Table 13-1.

b. What are the don't care conditions of the excess-3 BCD code?

13-4. Verify that the expression for G_0 in the binary code to Gray code converter (Sec. 13-3) is $G_0 = B_1 \oplus B_0$.

*Optional material.

***13-5.** ROMs are used for the binary to BCD conversion of a 12-bit binary number. What ROM capacity (in words $\times$ bits) is required?

***13-6.** ROMs are used for the binary to BCD conversion of a 9-bit binary number. What ROM capacity (in words $\times$ bits) is required?

13-7. Figure P13-1 shows a serial Gray code to binary code converter. The Gray code input is applied, with the most significant bit first, to the control input of a T FF (see Sec. 8-11). Assuming the Gray code input is 1100101001, what will be the binary output at the *Q* terminal of the flip-flop? Show the waveforms of the clock input and the binary output and discuss the operation of the circuit.

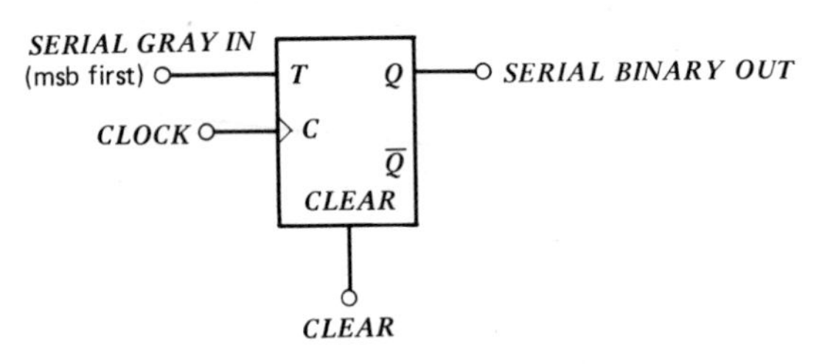

FIGURE P13-1.

13-8. Figure P13-2 shows a serial binary code to Gray code converter. The binary input is applied, with the most significant bit first, to the control input of a D FF (see Sec. 8-9). Assume that the binary input is 1100101001. What will be the Gray code output at the *SERIAL GRAY OUT* terminal? Show the waveforms of the clock input and the Gray code output, and discuss the operation of the circuit.

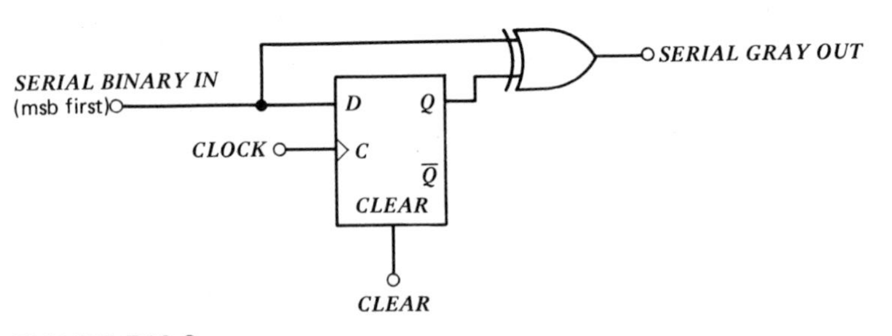

FIGURE P13-2.

13-9. Verify the expressions for *W*, *X*, and *Y* obtained in Ex. 13-9.

***13-10.** A 9's complementing circuit is sometimes required for subtraction in BCD arithmetic. The circuit accepts one 8-4-2-1 BCD digit and delivers its 9's complement. The 9's complement of an 8-4-2-1 BCD

*Optional problem

digit is obtained by subtracting the digit from 1001_2. Show an implementation of the 9's complementer using combinational logic circuits.

***13-11.** In Fig. 13-5 we show the ROM for the *B*-word as being of 32-word × 6-bit capacity. Show how the bit capacity of this ROM was arrived at.

***13-12.** What is the bit capacity of the ROMs and the BCD adder used in a 12-bit binary to BCD converter? The method divides the 12-bit number into two 6-bit numbers and uses two ROMs, similar to the scheme shown in Fig. 13-5.

13-13. Establish a truth table and a simplified gating structure for an excess-3 BCD to decimal code converter.

13-14. Establish a truth table and a gating structure for a decimal to 8-4-2-1 BCD code converter.

13-15. The circuit of Fig. P13-1 can also be used for serial parity checking. Explain the operation of such a use. (Even and odd parities are discussed in Sec. 4-4.1.)

*Optional problem.

CHAPTER

14 Computers and Microcomputers

Instructional Objectives

This chapter presents an overview of digital computers with an emphasis on microcomputers. After you read it you should be able to:

1. Describe basic properties of digital computers.
2. Describe the basic structure of a digital computer.
3. Outline the structure of microprocessors.
4. Use basic programming techniques.
5. Use flow charts.
6. Describe the input–output (I/O) section of a computer.
7. Describe the operation of the main memory.
8. Outline the operation of the control unit in a microcomputer.

14-1 OVERVIEW

The most commonly used digital system of our age is the *digital computer* and in particular the *microcomputer*, which has attained widespread use in calculators, appliances, personal computers, and many other applications. This chapter describes the structure, operation, and use of digital computers. Although the principles are general, we concentrate on microcomputers, including both the *hardware* and *programming* required for their efficient use.

The subject matter is extensive and is rapidly growing. In what follows here, only the basic features are outlined.

14-2 BASIC CHARACTERISTICS

One of the most significant steps in the development of digital computers was the introduction of the *stored-program computer*. Unlike an abacus or a manually operated desk-calculator, the operating sequence in a stored-program computer is controlled by an internally stored *program*.

EXAMPLE 14-1

Vehicular traffic at the intersection of a principal and a secondary road is regulated by a traffic controller that has a 60-second timing cycle. Traffic lights for the principal road are green for 30 second, followed by a 5-second amber and a 25-second red. Is this traffic controller a stored-program computer?

Solution

The sequencing of the lights is stored in the controller. This makes the traffic controller a stored-program computer.

In current interpretation, however, a stored-program computer has an additional feature: it is capable of *branching* between various segments of its program. Such branching, or *decision making*, can be controlled by a result of previous computations; it can also be controlled by information received from an *input device* of the computer.

EXAMPLE 14-2

The traffic controller of Ex. 14-1 is expanded to include two vehicle sensors that are connected as input devices to the controller. The sensors, located on the secondary road, indicate when a vehicle is waiting for the traffic light to change. At the completion of the 30-second green light for the principal road, the controller interrogates the sensors and changes the lights only if a vehicle is waiting on the secondary road.

Identify branching in this controller.

Solution

Branching takes place at the completion of the 30-second green light for the principal road. The lights remain unchanged *if* there is no vehicle waiting on the secondary road; otherwise they change.

Stored-program digital computers have come into widespread use during the last three decades. This has been due primarily to technological developments such as the introduction of transistors that now permeate all parts of the computer, improvements in the storage elements used in the *memory*, increased reliability of the electromechanical *input* and *output devices*, and the increasing use of integrated circuits. Present-day digital computers include *special-purpose computers* tailored to a single use and *general-purpose computers* utilized in many diverse areas such as control, data processing, and scientific calculations.

Parallel to the improved reliability, computational capability, and ease of use of general-purpose computers came general-purpose *minicomputers*, which, though limited in computational capability, were smaller and less costly. Principally because of their lower cost, minicomputers have penetrated into many applications that were previously in the exclusive domain of small special-purpose computers. The remaining gap separating general-purpose computers from special-purpose computers and controllers is being filled by the latest, and smallest, general-purpose computer, the *microcomputer*.

The simplicity and reduced cost that make microcomputers widely applicable also result in programming techniques that are often difficult and clumsy compared to those required by minicomputers. Furthermore, the circuit, or *hardware*, aspects are often more enmeshed with the programming, or *software*, aspects in a microcomputer than in a minicomputer. Thus, although the work may sometimes be divided between "hardware experts" and "software experts," the development of a system that uses a microcomputer frequently requires basic knowledge in both fields. For this reason, hardware and software are interwoven in most of this chapter, aiming to provide an overall understanding of both.

14-3 BASIC STRUCTURE

A simplified block diagram of a computer is shown in Fig. 14-1. It consists of three functional blocks: the *input–output (I/O) section*, the *central processor unit (CPU)*, and the *main memory*. (A more detailed block diagram appears in Sec. 14-7.5.)

14-3.1 The Input–Output (I/O) Section

The lines at the left of the I/O section shown in Fig. 14-1 connect the computer to the *input* and *output (I/O) devices*, also known as *peripheral devices*.

EXAMPLE 14-3

A handheld calculator has 10 numerical keys labeled 0 through 9, five functions keys +, −, ×, ÷, and =, and 6 digits of decimal display, and incorporates a microcomputer that processes and stores the data. Identify the input and output devices.

Solution

The keys are the input devices and the displayed digits are the output devices.

The operation of the I/O section is described in detail in Sec. 14-5. A simplified block diagram of an I/O section is shown in Fig. 14-2. Selection of the I/O devices is performed by *input* and *output (I/O) multiplexers*, often abbreviated as *MPX* or *MUX* (Sec. 6-7). Output information is usually stored in the *output storage buffers*. The *I/O register* provides temporary storage during the transmission of information between the CPU and the I/O section.

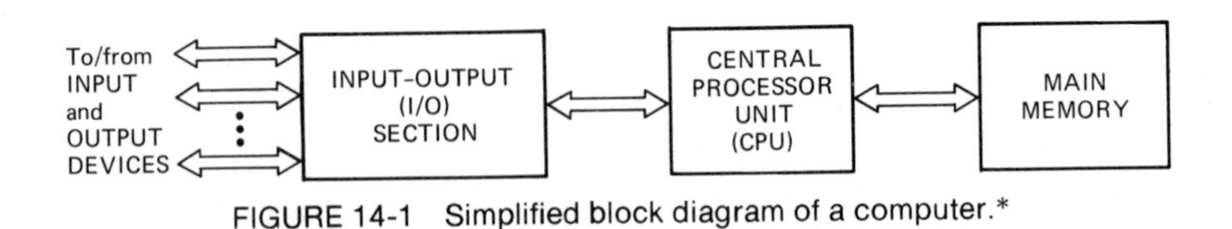

FIGURE 14-1 Simplified block diagram of a computer.*

*Wide interconnections in block diagrams represent multiple lines.

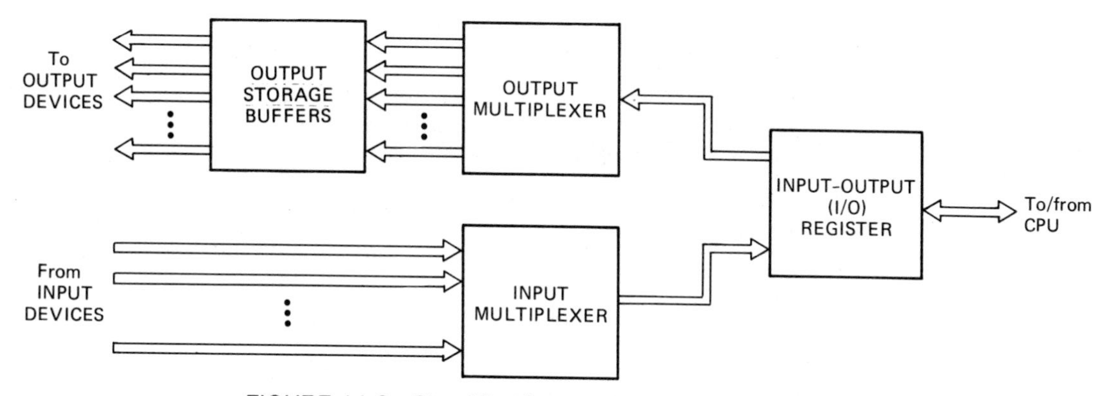

FIGURE 14-2 Simplified block diagram of an I/O section.

14-3.2 The Central Processor Unit (CPU)

The internal structure of the CPU varies widely among various computers. Here we outline a simple CPU of a microcomputer. It consists of an *arithmetic-logic unit* (*ALU*), several *registers*, and a *control unit*, as shown in the block diagram of Fig. 14-3.

The number of lines interconnecting the ALU, the *accumulator* (register A), and register B is determined by the *word length*, which is the maximum number of bits the arithmetic-logic unit can process in parallel. The word length in

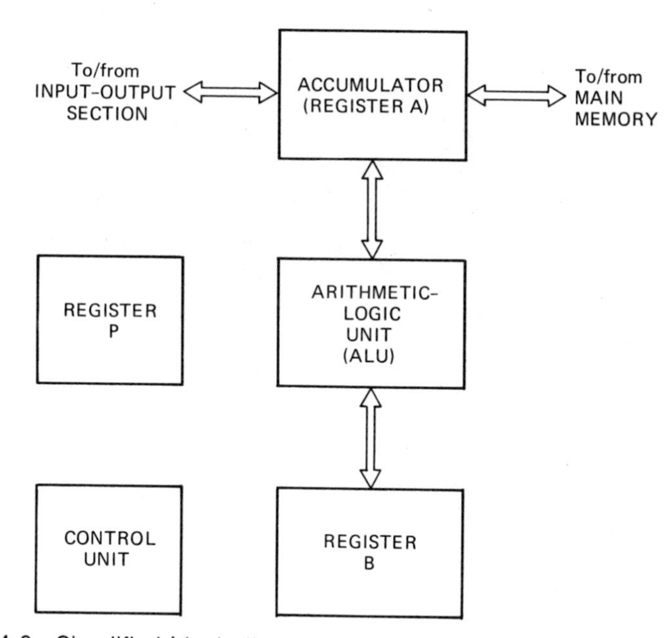

FIGURE 14-3 Simplified block diagram of the CPU of a microcomputer. Connections of the control unit and register P are not shown.

present-day microcomputers varies between 4 and 32 bits while in minicomputers and full-size computers the range is commonly between 12 and 64 bits.

Figure 14-3 also shows a *P register*, also known as a *program counter* or *PC*. The operation of this register is discussed in Sec. 14-6.1. The operation of the control unit is described in Sec. 14-7.

14-3.2.1 The Arithmetic-Logic Unit (ALU)

The ALU operates on one or two numbers. It performs arithmetic and logic operations such as addition, subtraction, and the detection of equality. The structure and operation of an ALU were discussed in Sec. 12-5.

14-3.2.2 Registers

The CPU includes several registers that are ofen called *data registers*, *working registers*, or *scratch-pad memory*. Register A (accumulator) and register B provide storage for data operated on by the ALU. Thus, for example, addition in the CPU of Fig. 14-3 is performed by adding the contents of register B to the contents of the accumulator and placing the result of the addition in the accumulator.

The bit capacity of the accumulator as well as that of register B are determined by the word length. In some arithmetic operations the two registers can be used together as a single register of *double word length* when required, for example, for the result of a multiplication.

EXAMPLE 14-4

Multiplication in the CPU of a microcomputer with an 8-bit word length is performed by multiplying the contents of the accumulator by the contents of register B. The result of the multiplication is a 16-bit number that is stored by placing the most significant 8 bits in the accumulator and the least significant 8 bits in register B.

Find the contents of register A (accumulator) and register B after the multiplication, if before the multiplication the contents of the A register were $0010\ 0000 = 32_{10}$ and the contents of the B register were $0001\ 0100 = 20_{10}$.

Solution

The result of the multiplication is $0000\ 0010\ 1000\ 0000 = 640_{10}$. Thus after multiplication the contents of the A register become **0000 0010** and the contents of the B register become **1000 0000**.

14-3.3 The Main Memory

The CPU and the I/O section of a computer contain several registers, or *buffers*, that store changing or temporary digital information. The principal data storage in a computer, however, is in its *main memory*. Compared to the registers of the CPU and the output buffers in the I/O section, the storage elements of the main memory are more numerous and usually slower.

The main memory may include several types of storage elements. The most prevalent types in microcomputers are *random-access memories*, RAMs (Sec. 11-5), and *read-only memories*, ROMs (Sec. 11-6). Actually both can be accessed randomly and *access times* to all locations are substantially identical. Data can be read from any location of a ROM; in a RAM, data can be read from *or* written into any location. RAM data are, however, usually destroyed upon the removal of power, while in a ROM data are preserved when the power is turned off. ROMs are therefore preferred for the storage of *permanent* information such as a program. The operation of the main memory is discussed in Sec. 14-6.

14-3.4 Microprocessors

Modern technology often permits the inclusion of the entire microcomputer of Fig. 14-1 on a single semiconductor chip; other approaches put most or all of the I/O section, the memory, and the CPU on a chip or on a few chips. Such chips are called *microprocessor chips*, *microprocessor chip sets*, or *microprocessors*.

14-4 BASIC PROGRAMMING TECHNIQUES

The operating sequence of a computer is determined by a *program*. A program consists of a sequence of *instructions* and of *data*. Each instruction is stored in the main memory as one or more *instruction words*; data are stored as *data words*. Each instruction consists of an *operator* and some number of *operands*. The operator describes the *operation* to be performed on the data specified by the operand or operands.

EXAMPLE 14-5

The instruction "ADD 4" is an abbreviation for "Add to the contents of the accumulator the contents of memory location 4." Identify the operator and the operand.

Solution

The word "ADD" is the operator, and the number "4" is the operand.

The program controlling the operating sequence is stored in the main memory of the computer. Each instruction of the program is first *fetched* from the main memory and then *executed*. In the course of the execution, the control unit sets up appropriate *data paths* and *operations*.

EXAMPLE 14-6

Identify the data paths and operations required for the execution of the multiplication described in Ex. 14-4.

Solution

One data path routes the contents of the accumulator to the ALU and another data path routes the contents of register B to the ALU. Next the ALU is set to perform a multiplication operation. Finally, data paths are set up for storing the result of the multiplication in registers A and B.

14-4.1 Machine-Language Instructions

Each instruction in the main memory of the computer is stored in the form of a *machine-language instruction code*, which includes the operator and also the operand (or operands).

EXAMPLE 14-7

In a microcomputer with an 8-bit word length the instruction abbreviated as "ADD 4" (see Ex. 14-5) is stored as a machine-language instruction code consisting of two 8-bit words. The first word stores the operator; the second word stores the operand.

What are the contents of the two words if the operator "ADD" is represented by the machine-language instruction code 0011 1000?

Solution

The first word contains the operator **0011 1000** and the second word contains the operand **0000 0100** ($= 4_{10}$).

The *instruction set* of a computer may include loading, storing, arithmetic, jump, logic, and I/O instructions. Some examples of these instructions are given below; additional instructions are discussed in later sections.

14-4.1.1 Loading Instructions

Load the contents of a specified memory location into the accumulator ("*Load accumulator*"). If the operator of this instruction is followed by the operand 0000 0110 ($= 6_{10}$), then the *contents* of memory location 6 are loaded into the accumulator. The contents of memory location 6 remain undisturbed.

Load a number into the accumulator ("*Load immediate*"). If the operator of this instruction is followed by the operand 0000 1111 ($= 15_{10}$), then the *number* 15 is loaded into the accumulator.

14-4.1.2 Storing Instructions

Store the contents of the accumulator in a specified memory location. If the operator of this instruction is followed by the operand 0000 1101 ($= 13_{10}$), then the contents of the accumulator are stored in memory location 13. The contents of the accumulator remain undisturbed.

14-4.1.3 Arithmetic Instructions

Add to the contents of the accumulator the contents of a specified memory location. The result is stored in the accumulator. The contents of the specified memory location remain undisturbed.

Subtract from the contents of the accumulator the contents of a specified memory location. The result is stored in the accumulator. The contents of the specified memory location remain undisturbed.

14-4.1.4 Jump Instructions

Jump to a specified memory location ("*Unconditional jump*"). The contents of the program counter (register P) are changed to the specified location and the next instruction is fetched from that location.

Jump to a specified memory location if the content of a specified flag is a binary 1 (''*Conditional jump*''). The *flag*, which is a 1-bit register, may be an arithmetic *carry*, *borrow*, or *overflow flag*, or a *sign* or a *zero flag*, or some other *condition flag*.

14-4.1.5 Logic Instructions

Comparison instruction. Compare the contents of the accumulator with the contents of register B and set a flag if the former is greater.

Logical AND instruction. A logical AND operation is performed bit by bit on the accumulator and a specified memory location. The result is stored in the accumulator. The contents of the specified memory location remain undisturbed.

EXAMPLE 14-8

An AND instruction is performed bit by bit on the accumulator and a specified memory location. Before the execution of the instruction, the contents of the accumulator are 1111 0101. The contents of the specified memory location are 0011 0011. What are the contents of the accumulator after the execution of the instruction?

Solution

The bit-by-bit logical AND of 1111 0101 and 0011 0011 is 0011 0001; thus the contents of the accumulator become **0011 0001**.

14-4.1.6 Input–Output Instructions

Receive data from a specified input device. Data from the specified input device are loaded into the I/O register (or often the accumulator) via the input multiplexer.

Send data to a specified output device. Data from the I/O register (or often from the accumulator) are loaded into the output storage buffer of the specified output device via the output multiplexer.

14-4.2 Assembly-Language Instructions

The preceding section described machine-language instructions. A program may consist of hundreds of such instructions and must be prepared with meticulous care. A limitation arising in the use of machine-language instructions is the inability of humans to handle long binary numbers without making errors. For example, the machine-language instruction code 0011 1000 0000 0100 may stand for ''Add to the contents of the accumulator the contents of memory location 4'' (see Exs.14-5 and 14-7). The number of mistakes in copying such an instruction code can be reduced considerably by use of a *mnemonic* abbreviation such as ''ADD 4,'' as was done in Exs.14-5 and 14-7. A programming language in which each machine-language instruction code is replaced by a mnemonic abbreviation is called an *assembly language* or *symbolic language*.

To obtain a program in machine language as required by the computer, the program that was written in an assembly language must be translated; this *translation* is performed by an *assembler* program. The assembler may be *resident* in the main memory of the computer; alternatively it may be located in a different computer, in which case it is called a *cross-assembler*.

For example, consider the simple case of a microcomputer program written in an assembly language by using a *paper-tape punch* controlled by a typewriter-like keyboard. The resulting roll of tape is read into a large computer by use of a *paper-tape reader* as an input device. A cross-assembler program that is written in a language suitable for the large computer is also read in. The large computer translates the assembly-language program into a machine-language program for the microcomputer as directed by the cross-assembler program. The resulting machine-language program is punched out by the large computer on paper tape, which is then read into the microcomputer.

14-4.2.1 Macroinstructions

Once an assembler is available that translates assembly-language instructions into machine-language instructions, we can use it for other tasks too. For example, a large group of instructions that is repeated many times in the program can be defined as a *macroinstruction*, and then be used instead of the group of instructions. Each time the macroinstruction appears in the program, the assembler substitutes the group of instructions for each macroinstruction as required.

As an example, consider a group of 5 instructions occurring 20 times in an assembly-language program. To save writing, we define a macroinstruction named MAC1 as follows:

```
DEFINE MACRO MAC1
LOAD ACCUMULATOR 6
ADD 5
ADD IMMEDIATE 12
SUBTRACT 7
STORE 8
END
```

Subsequently we write MAC1 in 20 different places in the program. The assembler substitutes the 5 instructions for each of the 20 MAC1s as it encounters them.

Thus the use of macroinstructions does not alter the resulting machine-language program, but it saves repeated writing of a long sequence of instructions in the assembly-language program. An unfortunate side effect in using macroinstructions is that the assembly-language program may look deceptively short. Hence care must be exercised in estimating the program size (memory space allocation) whenever macroinstructions are used.

EXAMPLE 14-9

An assembly-language program consists of 20 instructions, of which 12 are MAC1 macroinstructions defined above (without the lines of DEFINE and END).

a. Find the number of instructions in the resulting machine-language program.
b. Will the resulting machine-language program fit into a memory space allocation of 64 machine-language instructions?

Solution

a. We have 12 macroinstructions and $20 - 12 = 8$ other machine-language instructions. Thus the total number of machine-language instructions is $(12 \times 5) + 8 = 68$.
b. No, since $68 > 64$.

14-4.3 Higher-Level Programming Languages

The use of an assembly language represents an improvement over a machine language for the programmer who must write or understand a program. Nevertheless, a significant amount of writing still remains, and even simple programs may require many instructions.

The work of the programmer has been further alleviated by the introduction of *higher-level programming languages*. These programming languages combine several instructions into a *statement*, and a sequence of statements constitutes a program. For example, in ALGOL, a higher-level programming language, the statement $Y \leftarrow Z$ means: "The value of Z is assigned to Y." A more complex statement is the following: IF $V < W$ THEN $X \leftarrow Y - 1$ ELSE $X \leftarrow Y + 1$.

A higher-level programming language is translated into a machine language by a *compiler* or an *interpreter* program. This program inspects each statement, assigns memory locations for the storage of variables and constants, and generates the machine-language instructions.

An important question that arises in connection with the operation of a compiler is its efficiency. Since it is only a program that operates in a routine manner, the machine-language program it generates usually consists of considerably more instructions and may use more memory locations than does the machine-language program written by a skilled programmer. This inefficiency may be more critical in a small microcomputer, although it may also be significant in larger computers. For this reason the programmer of a computer is often given the option of combining assembly language and higher-level programming language in a single program.

14-4.4 Subroutines

We have already seen that the use of macroinstructions simplifies the writing of an assembly-language program but that the resulting machine-language program is not shortened. When a sequence of instructions or statements occurs many times in a program, it may preferably be part of the machine-language program only once and should be referred to, or *called*, whenever it is needed. Such a sequence of instructions is designated a *subroutine*, also known as a *procedure*. It can be written in a machine language, an assembly language, or a higher-level programming language. When used many times in a program, a subroutine provides significant savings in memory space. For example, one of the solutions for a root of a quadratic equation, $ROOT = (-B + \sqrt{B^2 - 4AC})/2A$, is used 100 times in a program, each time with different values of A, B, and C. A subroutine computing *ROOT* may be written in ALGOL as follows:

```
PROCEDURE ROOT(A, B, C);
ROOT ← 0;
D ← B × B − 4 × A × C;
IF D ≥ 0 THEN ROOT ← ( − B + SQRT(D))/(2 × A);
PRINT (A, B, C, D, ROOT);
END;
```

Note that SQRT(D) stands for $\sqrt{D}$ when $D \geq 0$. When $D < 0$, the printout will show a $ROOT = 0$ and the negative value of D, thus indicating that the correct (complex) *ROOT* cannot be computed by this simple procedure.

14-4.5 Flow Charts

An important tool in the development of a program is the *flow chart*. Unless the programmer is very experienced and the problem at hand is very simple, it is desirable to first draw a flow chart based on the original presentation of the problem and then proceed to write the program. Selected *elements* of flow charts are shown in Fig. 14-4. A flow chart consists of a combination of these elements.

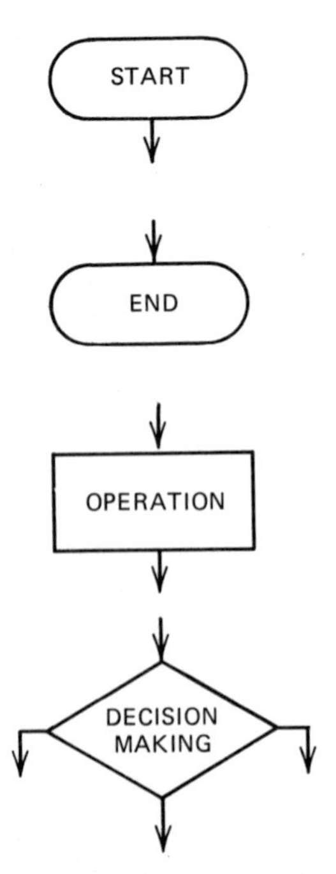

FIGURE 14-4 Selected elements of a flow chart.

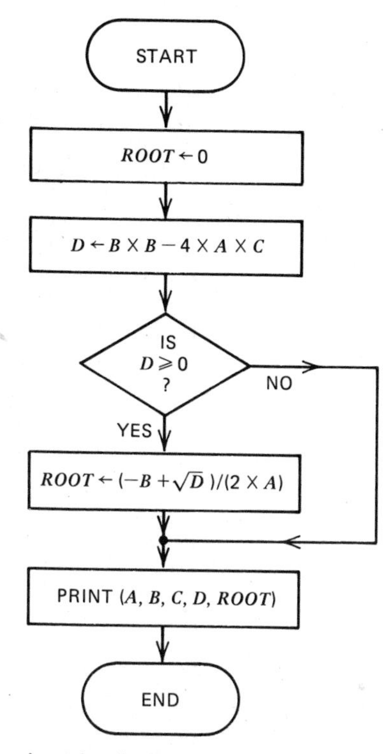

FIGURE 14-5 Flow chart for finding a root of a quadratic equation.

Figure 14-5 shows a flow chart describing the subroutine that was given above for finding a root of a quadratic equation. It consists of several blocks: one START, one END, four OPERATION, and one DECISION MAKING (branching).

14-4.6 The DO-Loop

Each element in a flow chart may be passed through any finite number of times during the execution of the program except that a START or an END may be passed through only once. An important structure for the multiple execution of a program segment is the *loop*, also known as *DO loop*. For example, Fig. 14-6 shows a flow chart of a program that types out the contents of a block of memory locations. After selecting the first memory location to be typed, it types the contents of each subsequent location by passing through the loop in the flow chart until all data are typed out.

14-5 INPUT AND OUTPUT (I/O)

Communication between the computer and the outside world is through *I/O devices*, also known as *peripheral devices*. This section discusses the structure of instructions used for the operation of I/O devices, the structure and oper-

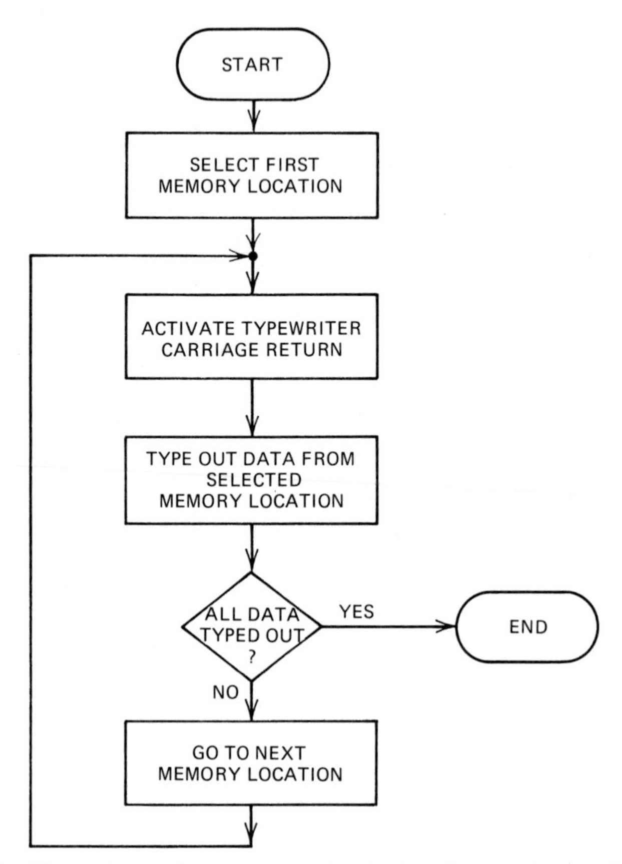

FIGURE 14-6 Flow chart of a program for typing the contents of a block of memory locations.

ation of the I/O section, interconnection of I/O devices, interrupts, and direct memory access.

14-5.1 I/O Instructions

The I/O instructions of a computer were described in Sec. 14-4.1.6 as "Data from the specified input device are loaded into the I/O register via the input multiplexer" and "Data from the I/O register are loaded into the output storage buffer of the specified output device via the output multiplexer." In addition to data, these instructions may also transfer *control information*. Control information may include a "ready" signal from a printer, an "end of card" signal from a punched-card reader, a "ready for next card" signal from a card punch, a "rewind" command to a magnetic-tape unit, a "rewind completed," or an "end of tape" signal from a magnetic-tape unit, and others.

14-5.2 The I/O Section

The basic components of the I/O section, which is also known as *I/O controller*, were given in Fig. 14-2 as the I/O register, the I/O multiplexers, and

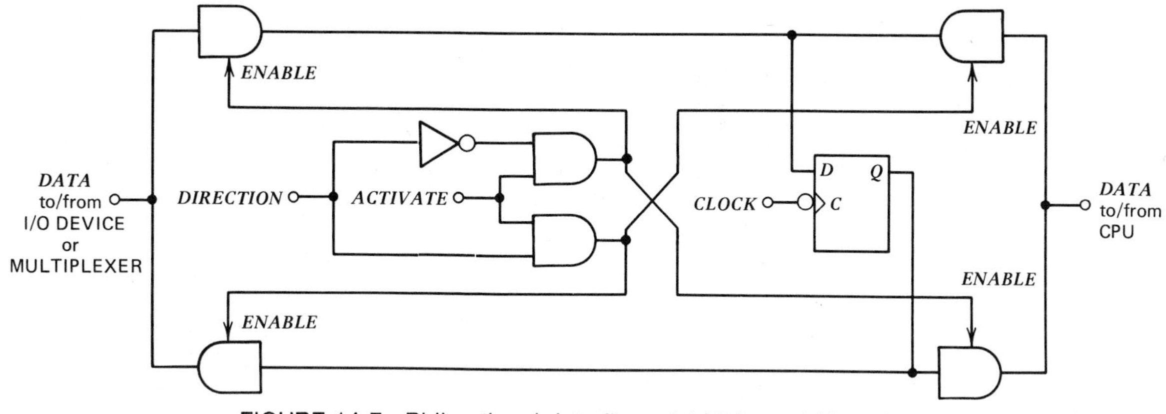

FIGURE 14-7 Bidirectional data flow of 1 bit in an I/O register.

the output storage buffers. This section presents more details on the operation of these circuits.

14-5.2.1 The I/O Register

The I/O register handles the transmission of information between the I/O section and the CPU of the computer. When the bit capacity of the I/O register is inadequate for the simultaneous transmission of data, control information, and *device identification*, transmission takes place in sequential groups.

For example, consider a microcomputer that has an 8-bit word length and 16 I/O devices, some requiring 4 bits of control information. Thus the I/O section, hence the I/O register, has to handle 4 bits of device identification ($2^4 = 16$) and 4 bits of control information in addition to the 8 bits of data. These are handled in two subsequent groups: the first group of 8 bits consists of the 4 bits of device identification and the 4 bits of control information, and the second group consists of 8 bits of data.

A significant feature of the I/O register is that it provides for *bidirectional* data flow; this is illustrated for a single bit in the circuit of Fig. 14-7. The storage element in the circuit is a D FF whose next state is set by clock C to the present state of input D. Also used are logic circuits with 3-*state outputs* that present an open circuit when the *ENABLE* input (frequently also designated as $\overline{INHIBIT}$) is logic 0 (Sec. 7-2.7). The *DIRECTION* input determines whether data flow is to the right or to the left, provided that the *ACTIVATE* input permits data flow. (See also Probs. 14-3 and 14-4.)

14-5.2.2 Multiplexers and Buffers

As mentioned in Sec. 14-3.1, a computer may include several *multiplexers*. The structure of a multiplexer is governed by whether its data flow is in only one direction or in both directions. In the simple case of the I/O section of Fig. 14-2, the input and output multiplexers are both *unidirectional*, but their structures are different.

The circuit diagram of a *unidirectional input multiplexer* is shown in Fig. 14-8. It handles four input devices with 4-bit word lengths. Merging of data is

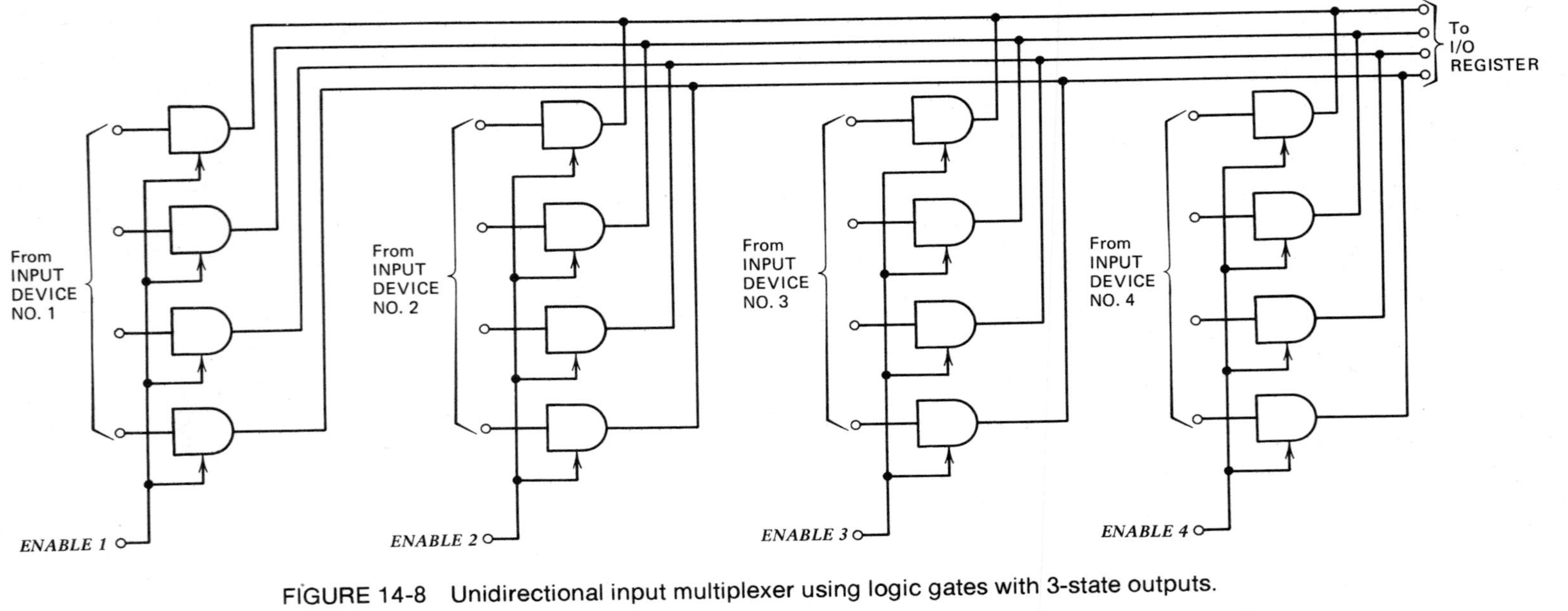

FIGURE 14-8 Unidirectional input multiplexer using logic gates with 3-state outputs.

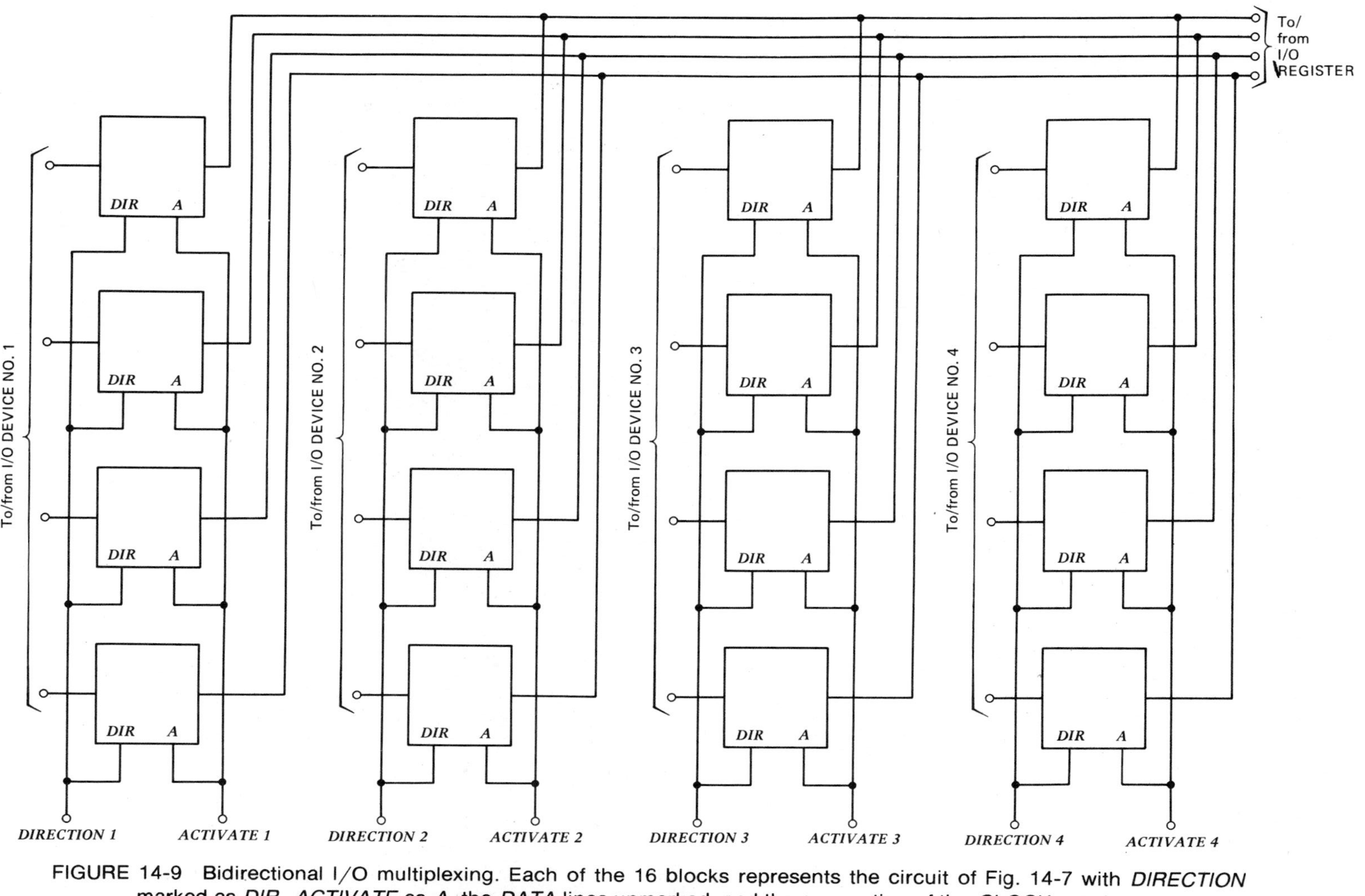

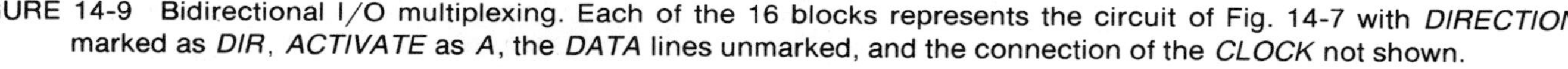
FIGURE 14-9 Bidirectional I/O multiplexing. Each of the 16 blocks represents the circuit of Fig. 14-7 with *DIRECTION* marked as *DIR*, *ACTIVATE* as *A*, the *DATA* lines unmarked, and the connection of the *CLOCK* not shown.

accomplished by logic gates with 3-state outputs that are activated by one of four *ENABLE* inputs; only one *ENABLE* input is allowed to be a logic 1 at any given time.

Unlike a unidirectional input multiplexer, a *unidirectional output multiplexer* does not require the use of 3-state outputs. Such a multiplexer is the subject of Probs. 14-5 and 14-6.

When the lines connecting the input and output devices are not separated, a *bidirectional multiplexer* must be utilized. The basic circuit of a bidirectional multiplexer is identical to Fig. 14-7, and its use is illustrated in Fig. 14-9. All *DIRECTION* (*DIR*) inputs for a given device are connected in parallel, as well as all *ACTIVATE* (*A*) inputs. A device is selected by making its *ACTIVATE* a logic 1; only one of the *ACTIVATE* lines is allowed to be a logic 1 at any given time.

Control information can be transferred together with data when both flow in the same direction. Otherwise data and control information of a device can be treated as if they belonged to two separate devices with opposite directions of data flow.

14-5.2.3 Overall Structure of an I/O Section

The overall structure of an I/O section is outlined in Fig. 14-10. I/O device identification is transferred from the I/O register to the *I/O device ID register*. Bidirectional transfer of data and control information takes place between the *I/O data register* and the I/O register. The contents of the I/O device ID register are decoded by the *I/O device ID decoder*.

Often, especially when many I/O devices are involved, it is more economical to locate the I/O device ID decoders, as well as the *I/O data storage buffers*, at the I/O devices rather than in the computer. In such cases the I/O multiplexer is realized as a *data bus* that is shared by the I/O data storage buffers.

14-5.3 Interrupts

Thus far computing and I/O operations have been interleaved by operating the I/O devices under program control. In another mode of operation, an I/O device may request servicing by sending an *interrupt request* to the computer.

For example, consider the operation of a keyboard as an input device. Every time a key is depressed, the keyboard requests the computer to accept information.

Interrupt requests may also be issued by output devices. Some output devices, such as a printer, can wait indefinitely for the computer to supply the next character and thus can operate either under program control or via interrupt. However, other output devices, such as a magnetic tape unit, may require servicing within a certain time, and hence must use interrupt for prompt attention.

For example, consider the case of a low-cost cassette tape recorder that is started by a start command from the computer. After attaining its normal operating speed of 7.5 ft/s, it records 1000 characters/s. Every time it is ready to record a character, the cassette recorder sends an interrupt request signal to

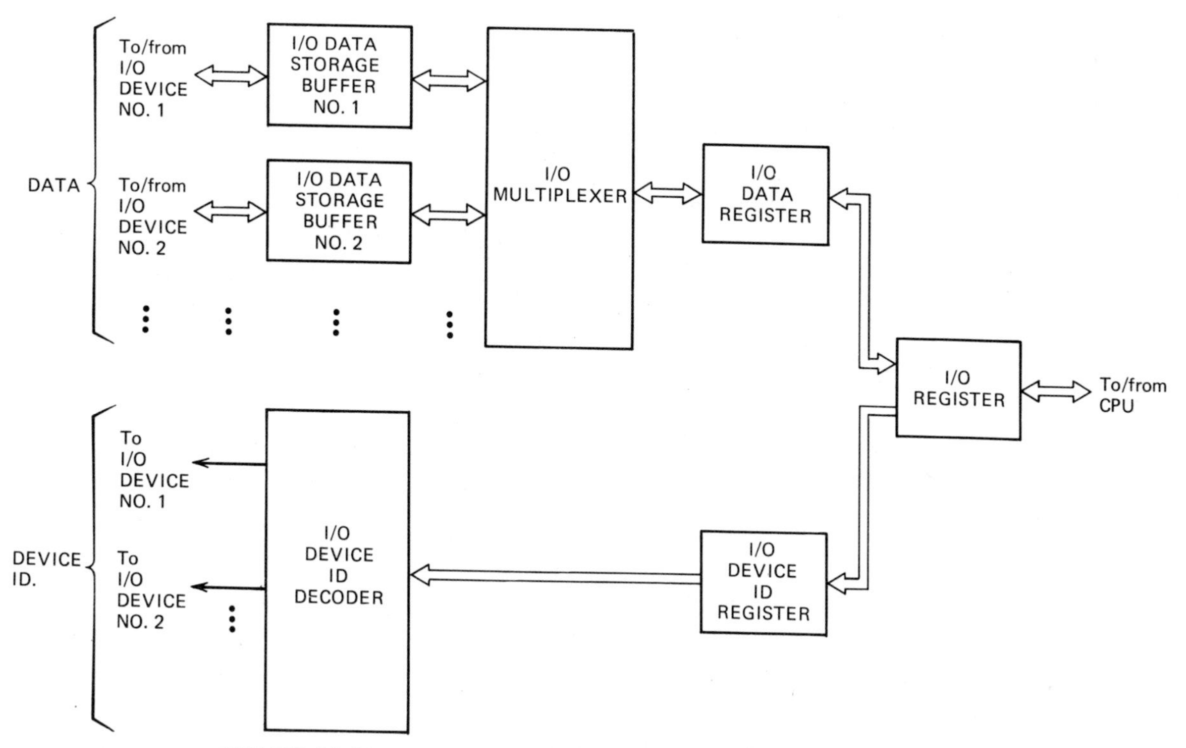

FIGURE 14-10 Simplified overall block diagram of an I/O section.

the computer requesting the next character. For correct operation of the tape unit, the computer must service this interrupt within a time period of about 1 ms.

Many computers provide a *multiple interrupt* system, to which several I/O devices may be connected. The devices may be assigned equal or different *priorities* in obtaining access by interrupt (see Sec. 6-9 for circuitry).

14-5.4 Direct Memory Access (DMA)

Direct memory access (DMA) bypasses the I/O section and provides high-speed direct data transfers in either direction between the main memory and an I/O device. Operation of the I/O section and of the CPU usually halt during DMA, and memory control is performed by an external DMA logic circuit. The necessity for this external circuitry, however, offsets somewhat the advantage gained in increased speed; as a result, DMA is installed primarily when high-speed transfer of large blocks of data is required.

14-5.5 Serial Input and Output

In the preceding, connections to and from the computer were via several parallel data lines. However, *serial I/O* transmitting data on a single data line is also in use. For example, the *RS232* Interface Standard specifies a *bit-serial* data transmission rate of 0 to 20,000 bits per second, as well as a voltage of +3 V to +25 V for logic 0 and a voltage of −3 V to −25 V for logic 1. Additional interconnections are provided for control signals such as Request to Send, Clear to Send, and so on.

14-5.6 I/O Processors

LSI and VLSI technology often permits the construction of a complete self-contained I/O section on a single IC chip—such I/O sections are called *I/O processors.* An I/O processor has a structure similar to that of a computer; that is, in addition to I/O multiplexers and buffers, it also includes a simple ALU, registers, RAM and ROM, and a control unit.

14-6 THE MAIN MEMORY

This section describes the organization and operation of the *main memory* of the computer. Although in principle the main memory could include any storage medium that can be organized to perform a memory function, the discussion here focuses on the use of *semiconductor memories* that were described in Ch. 11.

14-6.1 The Memory Registers

In addition to memory elements, orderly flow of information to and from the main memory also requires several auxiliary registers as well as data and address buses. A block diagram that includes several auxiliary registers is shown in Fig. 14-11. The *memory data register* (*MDR*) is instrumental in transferring the data bidirectionally between the memory and the CPU; the *memory address register* (*MAR*) has the capacity to address uniquely each word in the main memory. (These registers are often part of the CPU.) The information of the MAR is derived from the *instruction register* in the CPU, or from the *program counter* (*PC*), which contains the address of the next instruction to be fetched and usually is incremented after this address has been transferred to the MAR.

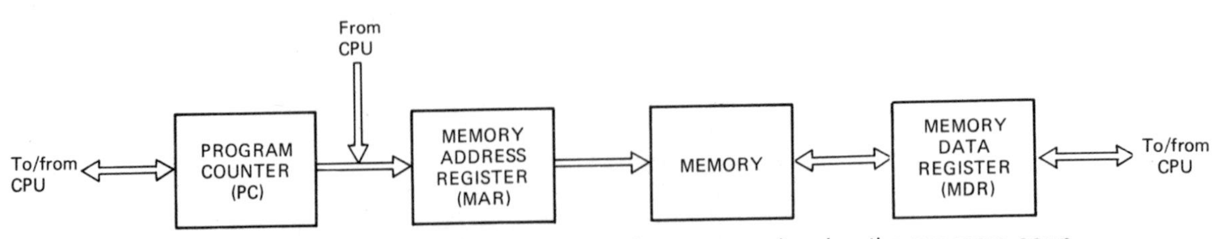

FIGURE 14-11 Simplified block diagram of a memory showing the program counter, the memory address register, and the memory data register.

15	14	13	12	11	10	9	8	7	6	5	4	3	2	1	0
OPERATOR (6 BITS)						ADDRESSING MODE (2 BITS)		DISPLACEMENT FIELD (8 BITS)							

FIGURE 14-12 Sixteen-bit instruction format of a computer with four different addressing modes.

14-6.2 Addressing Modes

The length of an instruction in a computer is in many cases adequate to include the full address of the specified memory location. Some computers, however, have a limited number of bits in their memory reference instruction formats, which, therefore, do not permit the addressing of the whole memory. This limitation is alleviated by the use of various *addressing modes*. In most of these, an *effective address* is calculated by combining the instruction word address bits with another word that was previously entered into a register or into a specified memory location; hence two words are required to completely specify one memory location.

In what follows here, four of these addressing modes are illustrated with a 16-bit instruction word shown in Fig. 14-12. The 6 most significant bits are reserved for the operator while the 8 least significant bits represent the *address field* (also called the *displacement field*) that allows 256 memory words to be addressed. Two additional bits, namely bits 8 and 9, designated *addressing mode bits*, select one of the four addressing modes described below.

14-6.2.1 Base Page Addressing

Base page addressing is the simplest addressing mode; it is also known as *addressing on page 0*. In this addressing mode, bits 8 and 9 of Fig. 14-12 are 00. The range of memory words that may be addressed by this scheme lies between 0 and 255, since the displacement field in Fig. 14-12 consists of 8 bits and $2^8 = 256$.

14-6.2.2 Addressing with a Page Register

The base page addressing scheme described above would suffice only for the smallest memory systems. To increase the addressing capacity as required by a larger memory, we set up a *page register*. As in base page addressing, each page consists of only 256 words. However, the page register allows the program to recognize which set of 256 words, that is, which *page*, is to be used in memory referencing. In effect, the page register adds higher-order bits to the address, thus permitting the addressing of additional memory locations as shown in Fig. 14-13.

Although Fig. 14-13 requires an additional instruction for page register setting, the scheme may be attractive if care is exercised by the programmer to effect the majority of memory referencing *within* the selected page; thus resetting of the page register will be infrequent compared to the frequency of its use.

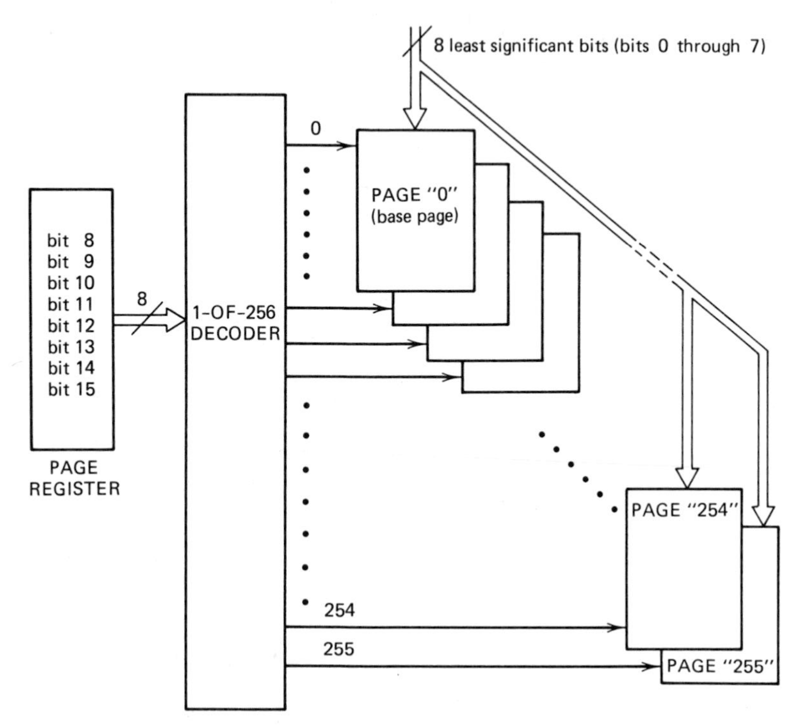

FIGURE 14-13 Addressing with a page register. Each of the 256 pages is selected via the auxiliary page register, and each page has a capacity of 256 words.

EXAMPLE 14-10

Assume that bits 8 and 9 of Fig. 14-12 are 01, which designates addressing with a page register. The page register of Fig. 14-13 is loaded with 1000 0001 ($= 129_{10}$), which are the higher-order bits of an effective 16-bit address; hence page 129_{10} is selected by the decoder. What addresses can be accessed?

Solution

By including the 8 lower-order bits, the addresses that can be accessed are in the range of (129×2^8) to $(129 \times 2^8) + 255$, that is, $\mathbf{33{,}024_{10}}$ to $\mathbf{33{,}279_{10}}$.

14-6.2.3 Addressing Relative to the Program Counter

In this addressing mode the 8 least significant bits of the instruction word are treated as a signed binary number in which bit 7 takes on the function of the sign. The effective address is obtained by adding the contents of the program counter (PC) to the signed number thus formed. This permits relative addressing of 256 addresses that are centered (within one address) around the address contained in the program counter.

Base page addressing and addressing relative to the program counter are economical, since they do not require additional instructions for page setting as does addressing with a page register. Although their addressing ranges are modest, careful programming may effect many memory reference instructions within their ranges.

14-6.2.4 *Addressing Relative to an Index Register*

In this addressing mode the effective address is generated by the sum of the contents of an *index register* and the contents of the displacement field. The advantage of using index registers lies in their capability of being easily modified (i.e., cleared, loaded, incremented, or decremented).

14-6.3 Indirect Addressing

Indirect addressing means that the address found by any one of the addressing modes discussed thus far does not itself contain the desired operand (data) but only its address. Thus the memory reference instruction addresses a memory location whose contents serve only as a *pointer* to the desired address. Usually the instruction contains one bit that determines whether the addressing mode is direct or indirect.

Indirect addressing is illustrated in Fig. 14-14. Unlike in Fig. 14-12, the instruction word of Fig. 14-14*a* consists of a 7-bit operator, a D/I bit designating direct or indirect addressing mode, and an 8-bit displacement field. Figure 14-14*b* shows portions of the main memory. We assume that the D/I bit

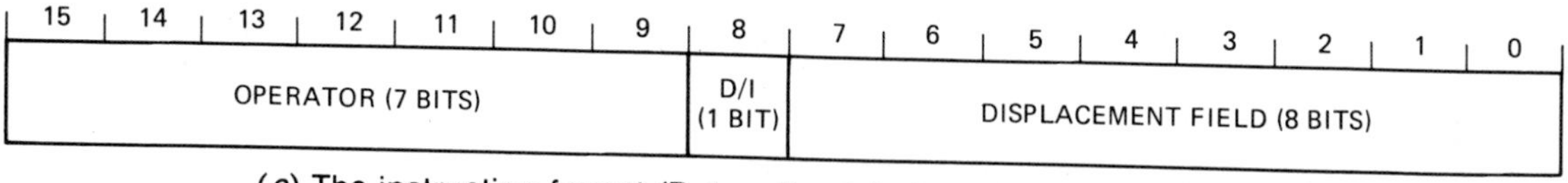

15	14	13	12	11	10	9	8	7	6	5	4	3	2	1	0
OPERATOR (7 BITS)							D/I (1 BIT)	DISPLACEMENT FIELD (8 BITS)							

(*a*) The instruction format (D/I = direct/indirect addressing mode bit)

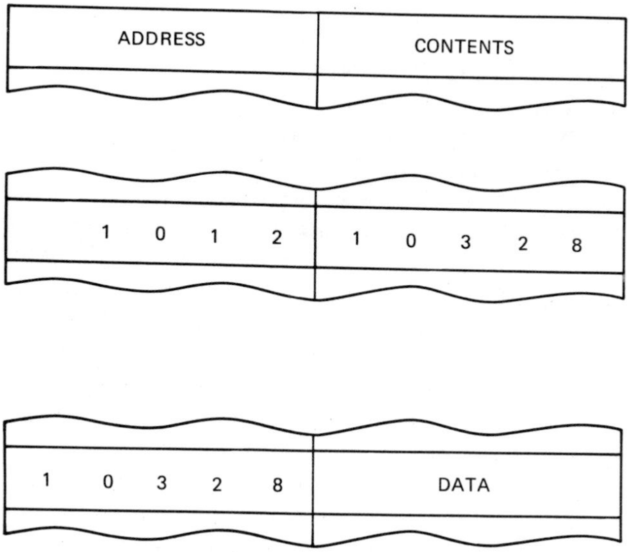

ADDRESS	CONTENTS
1 0 1 2	1 0 3 2 8
1 0 3 2 8	DATA

(*b*) Portions of the main memory

FIGURE 14-14 Illustration of indirect addressing.

of the instruction register indicates indirect addressing mode. Thus when address 1012 is referenced by the MAR, the contents of location 1012 are interpreted as the *address*, that is, location 10,328, where the next operand (data) can be found.

14-7 THE CONTROL UNIT

The control unit is the "nerve center" of the computer. It provides sequencing and timing for the processing of instructions and controls the data paths between the various parts of the computer. In what follows, the principal properties of a microcomputer control unit are outlined.

14-7.1 Sequencing

An instruction in a microcomputer is performed in one or more *processor cycles*, also designated *machine cycles*, or *basic instruction cycles*. A processor cycle consists of two principal *phases*: the *fetch* (or *addressing*) *phase* and the *execute phase*. Each fetch phase is utilized for addressing a memory location via the memory address register (MAR) and for transferring data between the memory data register (MDR) and the addressed location. In the case of some simple instructions, a complete instruction is fetched in a single phase and is executed in a single execute phase.

Thus, for example, in Sec. 14-3.2.2 addition was performed by adding the contents of register B to the contents of the accumulator and placing the result in the accumulator. The instruction for this addition occupies only one memory location; thus it can be fetched in a single fetch phase. Furthermore, since it involves only the CPU and does not require additional memory referencing, it can be executed in a single execute phase. As a result, the instruction is completed in a single fetch phase and a single execute phase, hence in a single processor cycle.

Other instructions, especially in microcomputers that have word lengths of less than 16 bits, may occupy more than one memory location. For example, consider a microprocessor with a word length of 8 bits and with a main memory that consists of $2^{16} = 65{,}536$ words. The storage of a jump instruction requires three memory locations: the first location contains the jump instruction code in machine language; the second and third locations contain the 16-bit address that specifies the memory location to which the program should jump.

An instruction that occupies more than one memory location requires more than one fetch phase, hence more than one processor cycle. The execute phase of an instruction, however, usually cannot begin until the entire instruction has been transferred into the *instruction register*. For this reason, execute phases are idle in (or are omitted from) the processor cycles until all required fetch phases have been completed.

14-7.2 Timing

A processor cycle is divided into *time states* of identical durations. This duration is determined by the system clock, hence is fixed in a given microcomputer; it is typically between 0.1 and 3 μs. The number of time states is usually fixed for a fetch phase but may vary for an execute phase.

14-7.3 Data Paths and Bus Structure

Data flow between the various parts of a computer takes place via interconnecting data paths. Data flow, which includes the flow of instructions as well, may be unidirectional such as the transmission of an address from the MAR to the main memory, or bidirectional such as transmission of data between the MDR and the main memory. When the data path is unidirectional, it can be activated (enabled) by an AND gate in each bit of the path. When the data path is bidirectional, a direction control signal must also be supplied to a circuit similar to that of Fig. 14-7.

The *bus structure* of a computer is comprised of data paths that are accessible at more than two connections. The operation of a bus is illustrated in Fig. 14-15. It shows data communication among three registers in a microcomputer with a word length of 4 bits.

The *D-E FFs* in Fig. 14-15 are characterized by a next state set by clock *C* to the present state of input *D* if *ENABLE* input *E* is at logic 1; the state remains unchanged irrespective of *D* if *E* is at logic 0. The *ENABLE* (*E*) inputs of all bits belonging to the same output device are connected in parallel. The outputs of the D-E FFs are connected to the data bus via logic gates with 3-state outputs that are activated by the *SEND* controls. The *D* inputs of the D-E FFs are also connected to the data bus and are activated by the *RECEIVE* controls. A maximum of one *SEND* control and any number of *RECEIVE* controls may be activated at a given time. Figure 14-15 illustrates the operation of the data bus with three 4-bit registers; additional registers may be connected in parallel to the data bus.

During the processing of an instruction, the control unit activates the data paths that are required to fetch the instruction from the main memory in the instruction register. The instruction is decoded by the *instruction decoder*, which determines the additional data paths that are required for the execution of the instruction.

14-7.4 Microprogramming

The data paths and their sequence of activation are governed by the instruction set of the computer. Thus once an instruction set is selected and the required data paths are built into the *hardware* of the computer, it is not possible to alter or expand the instruction set. A more flexible arrangement is provided by realizing the instruction decoder and the data path controls by *firmware* such as read-only memory (ROM): the result is a *microprogrammed* computer. It should

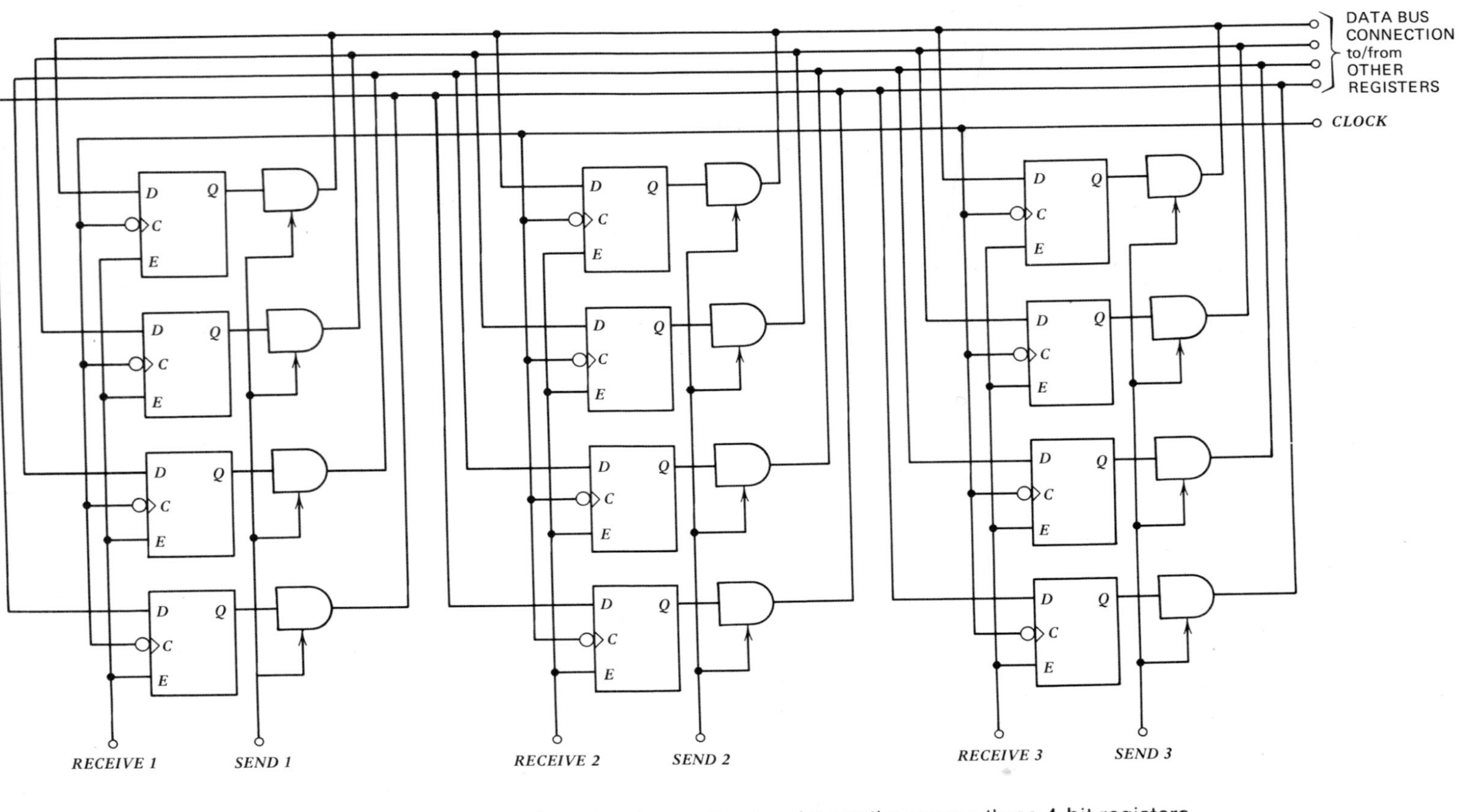

FIGURE 14-15 Data bus structure activating data paths among three 4-bit registers.

be noted that the word *microprogramming* is *not* an abbreviation of microcomputer programming, but describes a method for the activation of data paths.

In microprogramming, the data paths of a machine-language instruction are activated by a *microprogram*, the data paths of a processor cycle are activated by a *microinstruction*, and the data paths of a time state are activated by a *microoperation*. Thus a microprogram consists of one or more microinstructions, and a microinstruction consists of one or more microoperations.

The circuitry required by the microprogramming includes a ROM that is programmed for a particular instruction set. The ROM provides *data path control*, which activates the required data paths, and *next address control*, which determines the location of the next microoperation. Logic circuits external to the ROM include a *ROM address control* and a *ROM address register*.

14-7.5 Block Diagram of a Microcomputer

A simplified block diagram that shows details of the CPU is given in Fig. 14-16 for a hypothetical microcomputer. Note that the structure is somewhat arbitrary and varies among the various microcomputer types.

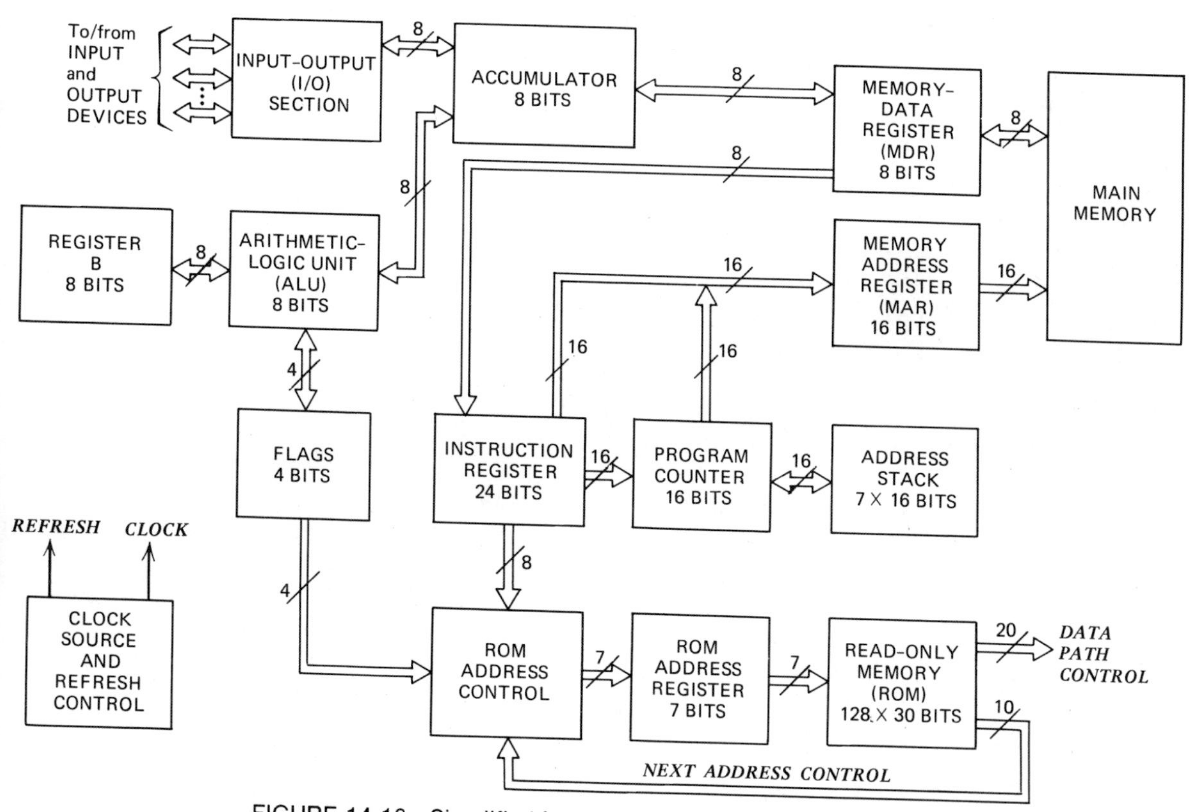

FIGURE 14-16 Simplified block diagram of a microcomputer.

The word length is 8 bits; thus the ALU, the accumulator, register B, and the MDR all have 8-bit storage capacities. The 16-bit MAR can address $2^{16} = 65{,}536$ memory locations. The program counter has a capacity of 16 bits. The *address stack* is instrumental in handling subroutines; its operation is not described here. The instruction register can store 24 bits: 8 bits for the operator and 16 bits for the operand. Data path control is performed by a microprogrammed ROM consisting of $2^7 = 128$ words of 30 bits each; auxiliary circuitry includes a ROM address control and a 7-bit ROM address register.

Communication with the outside world is via the I/O section that was described in detail in Sec. 14-5. Arithmetic and logic instructions are performed by the ALU, the accumulator, register B, and the 4 flags.

The main memory communicates data with the CPU via the bidirectional MDR and is addressed by the MAR, which may be loaded either from the program counter or from the operand field of the instruction register. (In some cases the MAR is loaded directly from the ALU; data paths for such loading are not shown in the figure.) Instructions are transferred from the main memory to the instruction register via the MDR in three 8-bit *bytes*.

Instructions are executed by the ROM, governed by the ROM address control, which combines information from the flags, the operator field of the instruction register, and the next address control of the ROM. Data paths and next address, as well as internal operations of the ALU and the I/O section, are controlled by the data path control of the ROM.

SUMMARY

After a short introduction, basic properties of digital computers were outlined. This was followed by an overview of the basic structure of computers: the I/O section, the CPU, and the main memory. Basic programming techniques were described, and details of input and output, the main memory, and the control unit were discussed.

SELF-EVALUATION QUESTIONS

1. What is a stored-program computer?
2. Describe branching in a digital computer.
3. What are the three principal parts of a computer?
4. What is a microprocessor?
5. Describe the terms machine language, assembly language, and higher-level programming language.
6. What is a macroinstruction?
7. Sketch a flow chart of a program for typing out the contents of a block of memory locations.
8. Describe the structure of an input-output section.
9. What are the principal features of the RS232 bit-serial interface?

10. What are the functions of the program counter, the memory address register, the memory data register, and the instruction register?

11. Describe addressing with a page register.

12. Describe the terms microprogram, microinstruction, and microoperation.

PROBLEMS

14-1. The machine-language instruction code for the operator of the "load accumulator" instruction is 0000 1000, and for the "load immediate" it is 0000 1001. The contents of memory location 11_{10} are 5_{10}. What are the contents of the accumulator after the execution of each of the following instructions: 0000 1000 0000 1011, 0000 1001 0000 1011, and 0000 1001 1111 1111?

14-2. The machine-language instruction code for the operator of the "load accumulator" instruction is 0000 1000, and for the "load immediate" it is 0000 1001. The contents of memory location 5_{10} are 11_{10}. What are the contents of the accumulator after the execution of each of the following instructions: 0000 1000 0000 0101, 0000 1001 0000 0101, and 0000 1001 1111 1111?

14-3. Determine the direction of data flow in Fig. 14-7 when (**a**) the *DIRECTION* input is at logic 0 and the *ACTIVATE* input is at logic 0, (**b**) the *DIRECTION* input is at logic 0 and the *ACTIVATE* input is at logic 1.

14-4. Determine the direction of data flow in Fig. 14-7 when (**a**) the *DIRECTION* input is at logic 1 and the *ACTIVATE* input is at logic 0, (**b**) the *DIRECTION* input is at logic 1 and the *ACTIVATE* input is at logic 1.

14-5. A D-E FF is characterized by a next state that is set by clock input C to the present state of input D if input E is at logic 1; the state remains unchanged irrespective of D if E is at logic 0. Prepare a state table and a function table for the D-E FF.

14-6. Sketch the logic diagram of a unidirectional output multiplexer for four output devices, each with 4 bits of data. Use D-E FFs described in Prob. 14-5. Connect the E inputs of the FFs belonging to the same output device in parallel.

14-7. How many 64-word $\times$ 1-bit RAM circuits are required for a memory system of 256 words $\times$ 4 bits?

14-8. How many 256-word $\times$ 1-bit RAM circuits are required for a memory system of 1024 words $\times$ 8 bits?

14-9. In the addressing with the page register shown in Fig. 14-9, we set the page register to 1010 0100; the lower-order bits are 0101 1011.

Which page will be selected and what memory location will be addressed?

14-10. In the addressing with the page register shown in Fig. 14-13, we set the page register to 1100 1100; the lower-order bits are 1010 1111. Which page will be selected and what memory location will be addressed?

14-11. Establish the logic values (0 or 1) of the *RECEIVE* and *SEND* signals that are required in Fig. 14-15 to load the contents of the leftmost 4-bit register into both other registers. Assume that a logic 1 is required for activation.

14-12. Establish the logic values (0 or 1) of the *RECEIVE* and *SEND* signals that are required in Fig. 14-15 to load the contents of the rightmost 4-bit register into both other registers. Assume that a logic 1 is required for activation.

14-13. Is the data bus in Fig. 14-8 unidirectional or bidirectional?

14-14. Is the data bus in Fig. 14-9 unidirectional or bidirectional?

CHAPTER

15 Digital-to-Analog and Analog-to-Digital Converters

Instructional Objectives

This chapter deals with the interface between analog and digital quantities. After you study it you should be able to:

1. Describe digital-to-analog converters for binary and BCD quantities.
2. Use four different digital input codes.
3. Describe digital-to-analog converters using binary weighted resistances or *R*-2*R* ladder networks.
*4. Specify operational amplifier parameters such as open-loop gain, input bias current, input offset current, input offset voltage, input impedance, slew rate, and settling time.
5. Describe a multiplying digital-to-analog converter.
6. Measure parameters of digital-to-analog converters.
7. Specify analog voltage comparators.
8. Describe sample-and-hold circuits.
9. Sketch block diagrams of the following analog-to-digital converter types: counter ramp, successive approximation, and flash.

15-1 OVERVIEW

This chapter describes digital-to-analog converters (DACs) in Sec. 15-2 through Sec. 15-8, and analog-to-digital converters (ACDs) in Sec. 15-9 through Sec. 15-14.

*Optional material.

Digital-to-analog conversion translates digital information into a continuous quantity, such as voltage, current, or mechanical motion. Applications of digital-to-analog converters are found, for example, in computer-controlled processes for actuation of various mechanisms, in positioning an electron beam on the face of a cathode-ray tube (CRT) display, and in waveform generation.

Analog-to-digital conversion translates a continuous quantity, such as a voltage or current, into digital form. Applications of analog-to-digital converters are, for example, in data-acquisition systems that measure physical quantities, such as temperature or pressure. The measuring elements (transducers) convert the physical quantities into an analog voltage or current. A conversion from analog to digital is then required to obtain this voltage or current in a digital form that is suitable as an input to a data-acquisition system, such as a computer, where it can be stored and/or mathematically manipulated.

In this chapter we present a variety of digital-to-analog (D/A) and analog-to-digital (A/D) conversion methods, consider limits of their accuracies, and discuss properties of circuit components utilized in various D/A and A/D converters.

15-2 DIGITAL-TO-ANALOG CONVERTERS (DACs)

We introduce the *digital-to-analog* converter (*DAC*) first because it is also used as a component in a variety of analog-to-digital converters. A DAC accepts a digital word and delivers a voltage or current that is proportional to the numerical value represented by the word.

Following this introductory Sec. 15-2, we discuss circuit components used in DACs: operational amplifiers in Sec. 15-3 and current switches for DACs in Sec. 15-4. Then we describe DACs using ladder networks in Sec. 15-5, DACs using current sources in Sec. 15-6, multiplying DACs in Sec. 15-7, and DAC parameter measurements in Sec. 15-8.

15-2.1 The Basic DAC

Figure 15-1 shows a simplified diagram of a *basic DAC*. It converts each bit of its *DIGITAL DATA IN* input that has a binary value of 1 to a current, and sums these currents into I_{SUM} at its *ANALOG OUT* output. Each switch, *S*, controls a current source, *I*. When a switch is ON, its respective current source contributes its current to output current I_{SUM}, which becomes proportional to the parallel input presented as *DIGITAL DATA IN*.

15-2.2 Block Diagram of a DAC System

A simplified block diagram of a complete DAC system is shown in Fig. 15-2. The input register (buffer) ensures that the data bits controlling the switches are

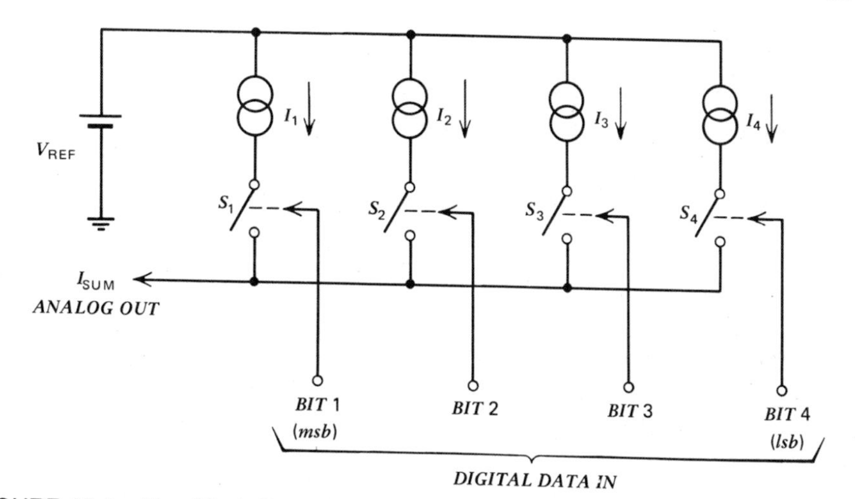

FIGURE 15-1 Simplified diagram of a basic digital-to-analog converter with 4 bits. Digital data inputs bit 1 through bit 4 control the states of the switches S_1 through S_4.

presented to the DAC simultaneously and that they are held for the duration of the conversion. At the *STROBE* signal, the register inputs (*DIGITAL DATA IN*) are simultaneously transferred to the outputs of the input register, which are connected to the inputs of the digital-to-analog converter. The reference voltage, V_{REF}, is a stable voltage source that provides the currents to the current sources. The operational amplifier converts the current output of the digital-to-analog converter into a proportional output voltage, V_{OUT}.

15-2.3 The Binary Weighted Resistance DAC

A *binary weighted resistance DAC* consists of the following four major elements, as shown in Fig. 15-3: (1) A weighted resistance network in which the values of the resistors increase by a factor of 2 for each bit; the value of the resistor in the msb position is $R \times 2^0$, in the next to the msb it is $R \times 2^1$, in the next bit it is $R \times 2^2$, and so on; the value of the resistor in the lsb position of an n-bit DAC is $R \times 2^{n-1}$. (2) n switches, one for each bit applied to the DAC.

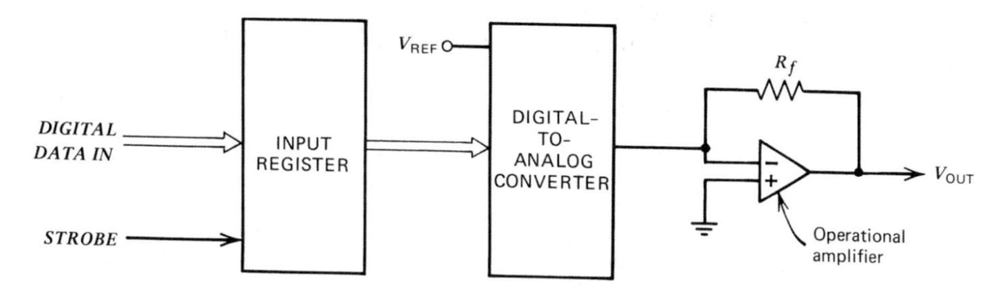

FIGURE 15-2 Simplified block diagram of a complete DAC system.

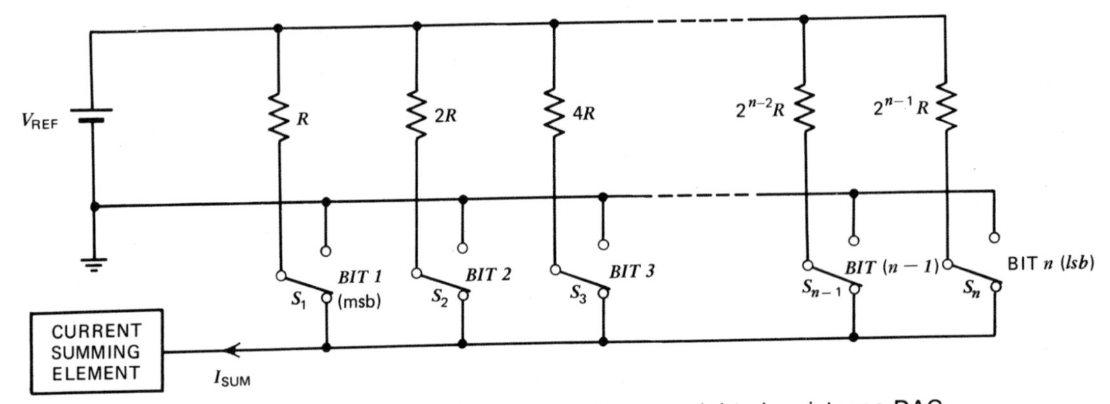

FIGURE 15-3 Simplified diagram of a binary weighted resistance DAC.

(3) A stable reference voltage, V_{REF}. (4) A summing element that adds the currents flowing from the resistive network.

A register, which is assumed to be external to the DAC, delivers a binary number to the input of the DAC and controls the states of the switches. As described above, the analog output current, I_{SUM}, is proportional to the number obtained from the bit positions for which the input is logic 1. The current due to bit 1 (msb) at logic 1 is V_{REF}/R; the current due to bit 2 at logic 1 is $V_{REF}/2R$, and so on; the smallest current is due to the lsb at logic 1 and equals $V_{REF}/(2^{n-1} \times R)$.

EXAMPLE 15-1

In an 8-bit DAC ($n = 8$) the value of the resistor in the msb position is 10 kΩ and $V_{REF} = +10$ V. What is the current for:

a. The msb input at logic 1.
b. The lsb input at logic 1.
c. Both the msb and the lsb inputs at logic 1.

Solution

a. The current for the msb input at logic 1 is 10 V/10 kΩ = **1 mA**.
b. To calculate the current for the lsb input at logic 1, we first find the value of the resistance in the lsb position: $R_{lsb} = R_{msb} \times 2^{n-1} = 10$ kΩ × 128 = 1.28 MΩ. From this value we find $I_{lsb} = 10$ V/1.28 MΩ = **7.8125 μA**.
c. When both the msb and the lsb inputs are at logic 1, the output is the sum of the respective currents. From solutions (**a**) and (**b**) this is I_{SUM} = 1 mA + 7.8125 μA = **1.0078125 mA**.

The maximum value, I_{max}, of current I_{SUM} in the summing element of a binary weighted resistance DAC occurs when the digital input consists entirely of 1s. It can be shown that

$$I_{max} = \frac{V_{REF}}{R} \times \frac{2^n - 1}{2^{n-1}} \tag{15-1}$$

where n is the number of bits in the DAC.

EXAMPLE 15-2

Calculate I_{max} for the DAC described in Ex. 15-1.

Solution

For an 8-bit DAC, $n = 8$, $2^n - 1 = 255$ and $2^{n-1} = 128$. Thus from eq. 15-1,

$$I_{max} = \frac{10\text{ V}}{10\text{ k}\Omega} \times \frac{255}{128} = \mathbf{1.992\ mA}$$

The *accuracy* of a DAC depends on the precision and the stability of its components. Resistors change with time because of aging and with variations in the ambient temperature. Since the value of the weighted resistor increases by a factor of 2 for each lower bit, the contribution from the lower-order bits is smaller and hence the tolerances in precision and stability of the resistors of these bits are not as exacting as those of the higher-order bits.

EXAMPLE 15-3

Given a 10-bit DAC with full-scale output of +10 V, $R_{msb} = 10\text{ k}\Omega \pm 0.05\%$, and R_{lsb} having a tolerance of ±5%. Calculate the errors introduced in:

a. The lsb position.
b. The msb position.

Solution

a. For a 10-bit DAC the value of the resistor in the lsb position is $R_{msb} \times 2^{10-1} = 10\text{ k}\Omega \times 512 = 5.12\text{ M}\Omega$. The current I_{lsb} due to the lsb only being ON is thus

$$I_{lsb} = +10\text{ V}/5.12\text{ M}\Omega = \mathbf{+1.953\ \mu A}$$

The error in the lsb position is **±5% of the lsb**, consistent with the tolerance of the lsb resistor.

b. The current I_{msb} due to only the msb being ON is

$$I_{msb} = +10\text{ V}/10\text{ k}\Omega = 1\text{ mA}$$

A ±0.05% tolerance of the msb resistor causes an error of $\pm 0.05\% \times 1\text{ mA} = \pm(0.05/100) \times 1\text{ mA} = \pm 0.5\ \mu\text{A}$. Since $I_{lsb} = 1.953\ \mu\text{A}$, the error in the msb position expressed as a fraction of I_{lsb} is $\pm 0.5\ \mu\text{A}/1.953\ \mu\text{A} \approx$ **±25% of the lsb**.

In practice we are concerned principally with tolerances of the *ratios* of the weighted resistors. In monolithic fabrication it is easier to achieve a close tolerance of the resistance ratios rather than a high precision in their absolute values. Also *tracking* of the resistance ratios with changes of temperature is very good when all resistors are on one monolithic substrate. Resistance ratio tracking of 1 part per million per one degree centigrade (1 ppm/°C) has been achieved. The effects of tracking error are illustrated in Ex. 15-4 below.

EXAMPLE 15-4

In a DAC the resistance ratios are specified to track within 5 ppm/°C. What is the ratio mismatch, in percent, due to a 30°C change in the ambient temperature of the DAC?

Solution

For a 30°C change in temperature, tracking is within (5 ppm/°C) × 30°C = 150 ppm = 150×10^{-6}. In percentage this equals $150 \times 10^{-6} \times 100 =$ **0.015%**.

TABLE 15-1
Output Voltage for Selected 8-Bit Binary Inputs in an 8-Bit Unipolar Binary DAC

Binary Input								Fraction of Full Scale	V_{OUT} for 10 V Full Scale, V
1	1	1	1	1	1	1	1	255/256	9.961
1	0	0	0	0	0	0	0	1/2	5.0
0	1	0	0	0	0	0	0	1/4	2.5
0	0	1	0	0	0	0	0	1/8	1.25
0	0	0	1	0	0	0	0	1/16	0.625
0	0	0	0	1	0	0	0	1/32	0.312
0	0	0	0	0	1	0	0	1/64	0.156
0	0	0	0	0	0	1	0	1/128	0.078
0	0	0	0	0	0	0	1	1/256	0.039
0	0	0	0	0	0	0	0	0	0

15-2.4 Digital Input Codes

A *unipolar DAC* is a digital-to-analog converter that generates analog signals of only one polarity. Table 15-1 shows the relationship between the binary input and the analog output for an 8-bit unipolar binary DAC. Notice that the all-1 code represents 255/256 of full scale. With a full-scale voltage of 10 V this corresponds to an analog output voltage of 9.961 V.* Half scale is indicated by only the msb being at 1, 1/4 scale by only the next bit being at logic 1, and so forth. The lsb represents 1/256 of full scale; with full-scale voltage of 10 V this corresponds to 0.039 V.

Table 15-2 shows the relationship between the BCD input and the analog output for an 8-bit unipolar BCD DAC. Note that the *resolution* of the BCD input code is not as good as that of the corresponding binary code, since resolution is defined as the value of the lsb relative to the full scale.

EXAMPLE 15-5

Calculate the resolution of:

a. A 12-bit binary DAC.
b. A 12-bit BCD DAC.

Solution

a. The resolution of a 12-bit binary DAC is $1/2^{12} = 1/4096 = 0.0244\% =$ **244 ppm**.
b. The resolution of a 12-bit BCD (3 digits) DAC is $1/1000 = 0.1\% =$ **1000 ppm**, which is inferior to the resolution of the binary DAC by a factor of about 4.

*It is assumed that the DAC output current is transformed to a proportional output voltage. Such transformations are mostly performed by use of operational amplifiers that are discussed in Sec. 15-3.

TABLE 15-2
Output Voltage for Selected 2-Digit BCD Inputs in an 8-Bit Unipolar BCD DAC

BCD Input								Percent of Full Scale	V_{OUT} for 10 V Full Scale, V
1	0	0	1	1	0	0	1	99	9.90
0	1	0	1	0	0	0	0	50	5.0
0	1	0	0	0	0	0	0	40	4.0
0	0	1	0	0	0	0	0	20	2.0
0	0	0	1	0	0	0	0	10	1.0
0	0	0	0	1	0	0	1	9	0.9
0	0	0	0	0	1	0	0	4	0.4
0	0	0	0	0	0	1	0	2	0.2
0	0	0	0	0	0	0	1	1	0.1
0	0	0	0	0	0	0	0	0	0

A *bipolar DAC* is a digital-to-analog converter that generates positive *and* negative analog signals. The bipolar output is obtained in response to one of the following commonly used binary input codes: *sign-and-magnitude, offset binary, 2's complement,* and *1's complement*. These codes, except the offset binary, were discussed in Sec. 3-6. They are shown for comparison in Table 15-3, which gives codes for 4-bit numbers.

TABLE 15-3
Binary Bipolar Codes for 4-Bit Numbers

Decimal Number	Sign and Magnitude	Offset Binary	2's Complement	1's Complement
+7	0 1 1 1	1 1 1 1	0 1 1 1	0 1 1 1
+6	0 1 1 0	1 1 1 0	0 1 1 0	0 1 1 0
+5	0 1 0 1	1 1 0 1	0 1 0 1	0 1 0 1
+4	0 1 0 0	1 1 0 0	0 1 0 0	0 1 0 0
+3	0 0 1 1	1 0 1 1	0 0 1 1	0 0 1 1
+2	0 0 1 0	1 0 1 0	0 0 1 0	0 0 1 0
+1	0 0 0 1	1 0 0 1	0 0 0 1	0 0 0 1
+0	0 0 0 0	1 0 0 0	0 0 0 0	0 0 0 0
−0	1 0 0 0	1 0 0 0	0 0 0 0	1 1 1 1
−1	1 0 0 1	0 1 1 1	1 1 1 1	1 1 1 0
−2	1 0 1 0	0 1 1 0	1 1 1 0	1 1 0 1
−3	1 0 1 1	0 1 0 1	1 1 0 1	1 1 0 0
−4	1 1 0 0	0 1 0 0	1 1 0 0	1 0 1 1
−5	1 1 0 1	0 0 1 1	1 0 1 1	1 0 1 0
−6	1 1 1 0	0 0 1 0	1 0 1 0	1 0 0 1
−7	1 1 1 1	0 0 0 1	1 0 0 1	1 0 0 0
−8		0 0 0 0	1 0 0 0	

Digital-to-analog conversion using the sign-and-magnitude code is easily implemented by employing two reference voltages of opposite polarities. One of these two is connected electronically to the DAC circuit for the duration of the conversion. The sign bit is used to control which reference voltage is to be switched in for a given input code.

EXAMPLE 15-6

A 10-bit bipolar DAC has a full-scale output of ± 10 V and uses a sign-and-magnitude code. What is the output for a digital input of 11100 00000?

Solution

The leading 1 in the digital input signifies that the reference voltage of -10 V has to be connected for the duration of the conversion. The input *magnitude* bits are 1100 00000 with a corresponding analog output of $-(\frac{1}{2} \times 10 \text{ V}) - (\frac{1}{4} \times 10 \text{ V}) = \mathbf{-7.5\ V}$.

The offset binary code is also easily implemented. Note in Table 15-3 that zero is represented by the half-scale value 1000. Thus when a current is added to the summing point in Fig. 15-1 with a value of $-\frac{1}{2}$ of full scale, the output is negative full scale for a DAC input of 0000, it is zero for a DAC input of 1000, and so on.

The 2's complement code is identical to the offset binary code except for the msb where in one code it is the complement of the msb in the other code. Thus a 2's-complement code DAC uses the same circuitry as the offset binary DAC, but with the msb complemented.

The 1's-complement DAC is awkward to implement and is infrequently used.

15-3 OPERATIONAL AMPLIFIERS

Thus far we assumed an ideal summing element in the DAC that generates an output signal proportional to the sum of the currents flowing in the closed switches. The function of summing is performed by an *operational amplifier* that delivers a voltage proportional to its current input. The name "operational amplifier" is derived from the early use of this amplifier in analog computers. Today the operational amplifier is widely used in a variety of applications.

An operational amplifier with its input resistor R_1 and its feedback resistor R_f is shown in Fig. 15-4. Without the feedback resistor R_f, its amplification equals the *open-loop gain* A, which is very high, typically between 10,000 and 1,000,000. We show below that an operational amplifier with *feedback* has a well-defined *feedback gain*, V_{OUT}/V_{IN}, which is mainly dependent on the precision and stability of the external components and has negligibly small dependence on the variations of the open-loop gain A of the operational amplifier.

An *ideal operational amplifier* has infinite open-loop gain A, infinite *input impedance* Z_s, zero *output impedance* Z_{OUT}, and zero *input bias currents* $I_{s1} = I_{s2} = 0$. We now find the gain of an ideal operational amplifier with feedback, as shown in Fig. 15-4.

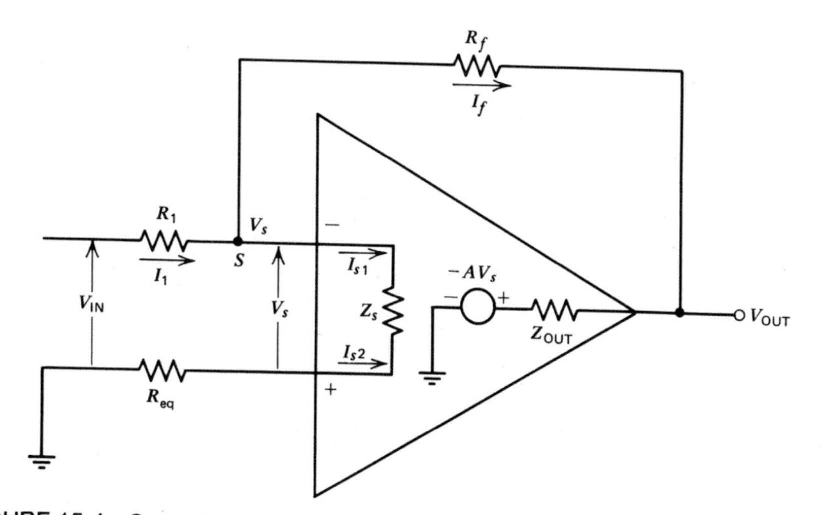

FIGURE 15-4 Operational amplifier with feedback. A = open-loop gain, Z_s = input impedance, Z_{OUT} = output impedance, R_1 = input resistance, R_f = feedback resistance.

We made the assumption that input bias current $I_{s2} = 0$; hence, the voltage across resistor R_{eq} is also zero. From Fig. 15-4 for $Z_{OUT} = 0$ we also have $V_{OUT} = -AV_s$, hence

$$V_s = -\frac{V_{OUT}}{A} \tag{15-2}$$

Since we also made the assumption that A approaches infinity, it follows from eq. 15-2 that V_s must approach 0 V for finite V_{OUT}. Also, since the voltage across resistor R_{eq} is zero, the voltage between *summing junction S* and ground is zero too. Thus, the current flowing through input resistor R_1 is

$$I_1 = \frac{V_{IN}}{R_1} \tag{15-3}$$

The voltage drop across feedback resistor R_f is $I_f R_f$. Also, since the voltage between summing junction S and ground is zero, with the polarities as shown in Fig. 15-4 we have

$$V_{OUT} = -I_f R_f \tag{15-4}$$

However, input bias current $I_{s1} = 0$, thus $I_f = I_1$, and by use of eqs. 15-3 and 15-4 we get

$$V_{OUT} = -\frac{V_{IN}}{R_1} R_f \tag{15-5}$$

The *feedback gain* A_f is *defined* in Fig. 15-4 as the ratio of output voltage V_{OUT} to input voltage V_{IN}:

$$A_f \equiv \frac{V_{OUT}}{V_{IN}} \tag{15-6}$$

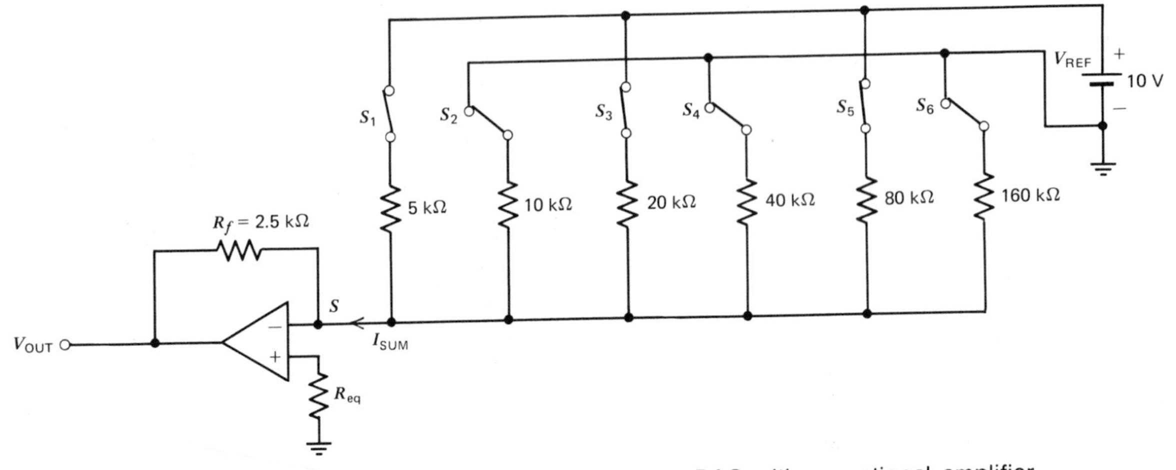

FIGURE 15-5 A 6-bit binary weighted resistance DAC with operational amplifier current summing element. Switches show a binary input 101010.

By use of eq. 15-5 this becomes

$$A_f = -\frac{R_f}{R_1} \tag{15-7}$$

An operational amplifier with feedback is also used for the *current summing element* of Fig. 15-3, where the current flowing into the current summing element is I_{SUM}. In such an application the output voltage in Fig. 15-4 becomes

$$V_{OUT} = -I_{SUM}R_f \tag{15-8}$$

with current I_{SUM} flowing into summing junction S through input resistor R_1 in Fig. 15-4. Also note that, for correct operation, $R_1 + R_f/A$ of Fig. 15-4 has to be negligibly small as compared to R of Fig. 15-3.

Figure 15-5 shows a 6-bit DAC using binary weighted resistors and an operational amplifier. Its operation is illustrated in Ex. 15-7, which follows.

EXAMPLE 15-7

The switches in the DAC of Fig. 15-5 are controlled by a digital input of 101010. What is the analog output, V_{OUT}, of the operational amplifier?

Solution

Currents will be flowing to the summing junction, S, only because of those switches that are connected to $V_{REF} = +10$ V. The summing current, I_{SUM}, is thus

$$I_{SUM} = 10\ \text{V}/5\ \text{k}\Omega + 10\ \text{V}/20\ \text{k}\Omega + 10\ \text{V}/80\ \text{k}\Omega$$
$$= 2 + 0.5 + 0.125\ \text{mA} = 2.625\ \text{mA}$$

This current of 2.625 mA also flows through the feedback resistor R_f. The output voltage is thus, from eq. 15-8,

$$V_{OUT} = -I_{SUM}R_f = -2.625\ \text{mA} \times 2.5\ \text{k}\Omega = \mathbf{-6.5625\ V}$$

A practical operational amplifier differs from the ideal operational amplifier assumed above, among others, in the following parameters: open-loop gain, input bias currents, input impedance, input offset voltage, slew rate, and settling time. These parameters and their impact on the DAC's accuracy are discussed below.

*15-3.1 Open-Loop Gain

The *open-loop gain, A,* in modern operational amplifiers varies from 10^3 to 10^7, and in most applications it is in the range of 10^4 to 10^6. Thus the assumption of an infinite open-loop gain A cannot strictly be made in high-resolution DACs.

It can be shown that for a finite value of A, feedback gain A_f of Fig. 15-4 becomes

$$A_f = \frac{-R_f/R_1}{1 + 1/A + R_f/(AR_1)} \tag{15-9}$$

This is smaller than the gain given by eq. 15-7. When $A \gg 1$ and $R_f/(AR_1) \ll 1$, eq. 15-9 reduces to eq. 15-7.

*15-3.2 Input Bias Currents

The *input bias currents* I_{s1} and I_{s2} are the currents flowing into the input terminals of the operational amplifier as shown in Fig. 15-4—they range in values from 100 μA to less than 1 pA ($= 10^{-12}$ A). When $I_{s1} \neq I_{s2}$, a single input bias current may be defined as $(I_{s1} + I_{s2})/2$.

The input bias currents flow through resistors R_1, R_f, and R_{eq} in Fig. 15-4, and they generate a voltage that is amplified by feedback gain A_f, resulting in an error voltage at the output of the operational amplifier. It can be shown that when the circuit of Fig. 15-4 is driven from a voltage source V_{IN}, the resulting output error voltage $(V_{OUT})_{I_s}$ contributed by the input bias current becomes

$$(V_{OUT})_{I_s} \approx A_f \frac{R_1 R_f}{R_1 + R_f}(I_{s2} - I_{s1}) - A_f\left(\frac{R_1 R_f}{R_1 + R_f} - R_{eq}\right)\frac{I_{s1} + I_{s2}}{2} \tag{15-10}$$

It can be seen from eq. 15-10 that the error voltage contributed by the bias current would be zero if I_{s1} and I_{s2} were equal, *provided* R_{eq} is *chosen* such that it is equal to the parallel combination of R_1 and R_f. This explains why R_{eq} has been included in Fig. 15-4 even though it does not seem to provide any function as far as the amplification is concerned.

Once R_{eq} has been chosen to equal the parallel combination of R_1 and R_f, it follows from eq. 15-10 that a small value of $I_{s2} - I_{s1}$ is desired to keep the error output voltage as small as possible. The quantity $|I_{s2} - I_{s1}|$ is called the *input offset current* of the operational amplifier.

*15-3.3 Input Impedance

In an operational amplifier there are two input impedances that have to be taken into consideration. (1) The *differential input impedance* (Z_s in Fig. 15-4) which

*Optional material.

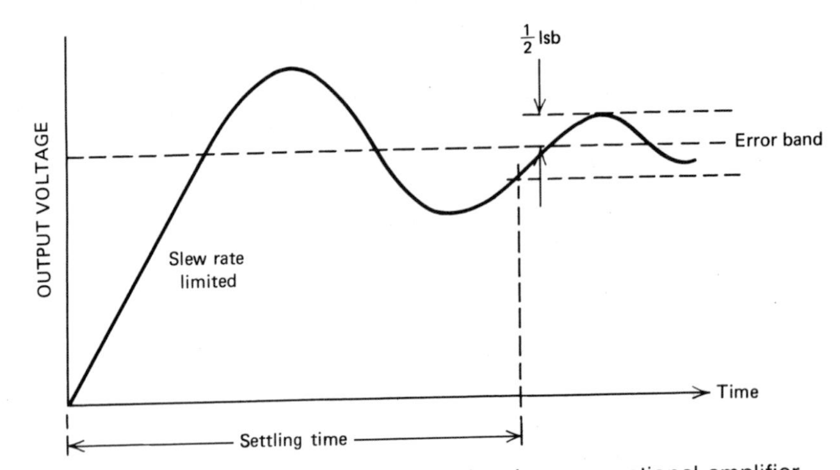

FIGURE 15-6 Slew rate and settling time in an operational amplifier.

is measured as the impedance between the two input terminals. (2) The *common-mode impedance* (not shown in Fig. 15-4) which is measured as the impedance between ground and the two input terminals connected together. Differential input impedances vary typically between 0.5 MΩ and 10^{12} Ω, while common-mode impedances are usually higher.

15-3.4 Input Offset Voltage

The *input offset voltage*, V_{OS}, is the voltage change required at the input to bring the output of an operational amplifier to 0 V; the output voltage is often adjusted to zero by an external control. The error at the output due to nonzero input offset voltage is

$$(V_{OUT})_{offset} = -A_f V_{OS}. \tag{15-11}$$

*15-3.5 Slew Rate and Settling Time

The *slew rate* is the maximum rate of change in output voltage, usually expressed in V/μs, in response to a large input step change. This limitation arises from the finite currents available to charge internal and external capacitances to the new voltage.

Settling time is defined as the time elapsed from the application of a large input step change to the time when the output voltage has entered and remains within a specified error band that is symmetrical about its final value. Settling time in a DAC application is usually specified to $\pm \frac{1}{2}$ lsb.

A typical response to a step function input is shown in Fig. 15-6. It consists of a slew rate limited period followed by an overshoot and several oscillations.

*Optional material.

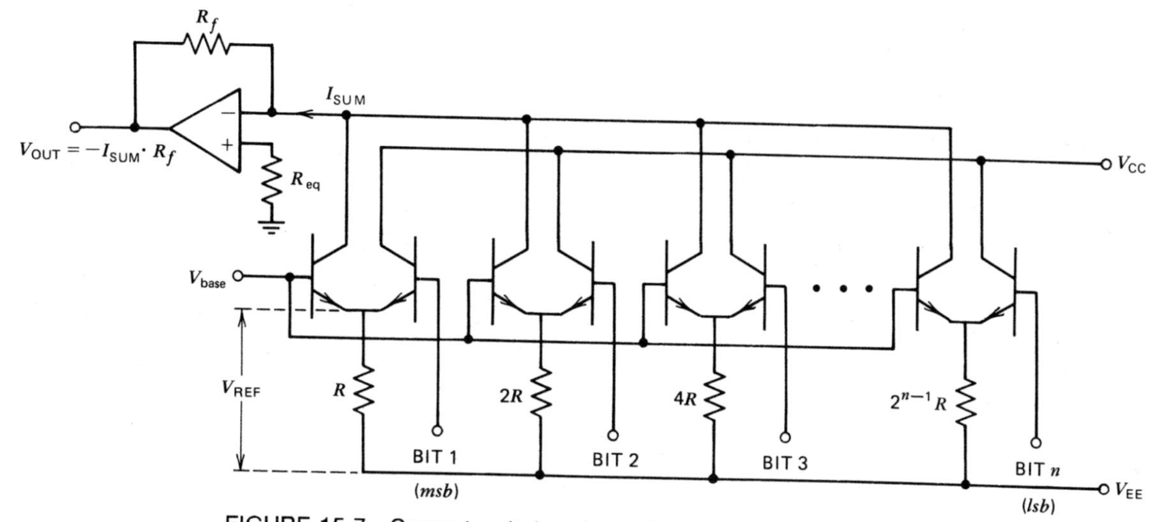

FIGURE 15-7 Current switches for a digital-to-analog converter.

15-4 CURRENT SWITCHES FOR DACs

The preceding figures showed schematically switches that connect the weighted resistors either to ground or to V_{REF}. Three types of switches are used in conjunction with DACs: *relays*, *semiconductor voltage switches*, and *semiconductor current switches*. The latter are most commonly used because of their precision and speed.

A circuit using semiconductor current switches is shown in Fig. 15-7. Each switch consists of a pair of *npn* transistors. The left transistor of each pair is biased at its base by a stabilized voltage V_{base}. (This voltage changes slightly with temperature to compensate for the base-emitter voltage changes in the transistors.) The right transistor of each pair accepts the logic levels corresponding to the digital value that has to be converted to analog. Active-low logic is employed for these signals: $V_{base} + 1$ V represents logic 0, while $V_{base} - 1$ V represents logic 1. When logic 0 is applied to any transistor pair, the right-hand transistor will be on and the collector current of the left-hand transistor will not contribute to I_{SUM}. Conversely, when logic 1 is applied to any transistor pair, the right-hand transistor will be off and the left-hand transistor will conduct. The collector current of this transistor will depend on the value of V_{REF} and the common-emitter resistors $R, 2R, 4R \ldots 2^{n-1}R$. This collector current is summed in a common line and appears as I_{SUM} at the input of the operational amplifier.

15-5 DACs USING LADDER NETWORKS

In Fig. 15-5 we showed a 6-bit binary weighted resistance DAC. The value of the msb resistor was 5 kΩ while the value of the lsb resistor was 160 kΩ. Since the resistor value changes by a factor of 2 for each bit, a high-resolution DAC would have a large ratio between the msb and lsb resistors.

EXAMPLE 15-8

In a 14-bit DAC of the weighted resistance type the msb resistor is 5 kΩ. What is the value of the lsb resistor?

Solution

In the msb bit the resistance value is $5\ k\Omega \times 2^0 = 5\ k\Omega$, in the next bit it is $5\ k\Omega \times 2^1 = 10\ k\Omega$, in the third bit it is $5\ k\Omega \times 2^2 = 20\ k\Omega$ and so on; in the 14th bit the resistance value is $5\ k\Omega \times 2^{14-1} = 5\ k\Omega \times 12^{13} = 5\ k\Omega \times 8192$ $=$ **40.96 MΩ**.

Several difficulties would arise in such a DAC. First, it would be very difficult to manufacture a set of resistors of such a wide resistance range that has accurate ratios and good temperature tracking. Second, the current in the lsb position would be very low due to the high resistance. Such low currents make the DAC sensitive to noise currents that may be of the same magnitude.

EXAMPLE 15-9

Assume that the DAC of Ex. 15-8 has a reference voltage $V_{REF} = +10$ V. What is the current in the lsb position when a logic 1 is applied to it?

Solution

$$I_{lsb} = 10\ V/R_{lsb} = 10\ V/(5\ k\Omega \times 2^{13})$$
$$= 10\ V/(5\ k\Omega \times 8192) = \mathbf{0.244\ \mu A}$$

The difficulties associated with the binary weighted resistance DAC are eliminated in a DAC with an *R-2R ladder network*, where the resistance values span only a range of 2. One such DAC is shown in Fig. 15-8.

In Fig. 15-8, the current flowing through resistor R_T is dependent on which switches are connected to reference voltage V_{REF}. Specifically, if we assign 1 to each switch S_1 through S_n that is connected to V_{REF} and 0 to each switch that is

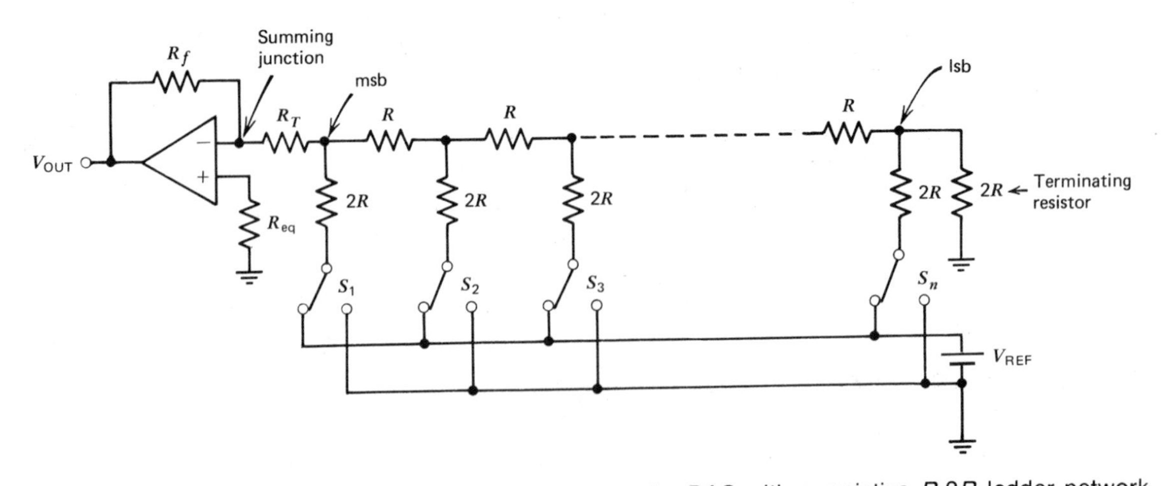

FIGURE 15-8 Simplified diagram of a DAC with a resistive *R*-2*R* ladder network.

connected to ground, then it can be shown that the current I_{R_T} flowing through resistor R_T is proportional to $S_1 2^{-1} + S_2 2^{-2} + \cdots + S_n 2^{-n}$. This holds for any value of R_T. Also note that the output voltage V_{OUT} is proportional to the current flowing through R_T.

For example, assume that the current through R_T is $I_{R_T} = 1$ mA when only switch S_1 is connected to V_{REF} and all other switches are connected to ground, that is, $S_1 = 1$ while $S_2 = S_3 = \cdots = S_n = 0$. Then, if only $S_2 = 1$, the current through R_T is $I_{R_T} = 0.5$ mA; if only $S_3 = 1$, $I_{R_T} = 0.25$ mA, and so on. Further, if $S_1 = 1$, $S_2 = 1$, and $S_3 = 1$, while $S_4 = S_5 = \cdots = S_n = 0$, then $I_{R_T} = 1$ mA + 0.5 mA + 0.25 mA = 1.75 mA. Thus the summing current in an R-$2R$ ladder network is the same as in a binary weighted resistance network without the disadvantages of the latter. An R-$2R$ ladder network also permits higher precision of conversion because all the switches carry comparable currents, whereas in a DAC such as in Fig. 15-5 the currents change by a factor of 2 for each binary step; such disparities in current cause slightly different base-to-emitter voltages in the current switches (see Fig. 15-7), deteriorating the performance.

*15-6 DACs USING QUAD CURRENT SOURCES

Two methods of digital-to-analog conversion were discussed thus far: The binary weighted resistance DAC and the ladder-network DAC. However, both of these approaches have their disadvantages.

The binary weighted resistance DAC requires a wide range of resistance values with high precision and good temperature tracking, and the currents in the less significant bits are low, which makes the DAC sensitive to noise. Also, the switches have stringent requirements since their on-resistance has to be much lower than the *lowest* resistance of the binary weighted chain (typically a 12-bit DAC would demand an R_{ON} of at most 10 kΩ and an R_{OFF} of at least 10 MΩ).

The ladder-network DAC has the disadvantage that three precision-matched resistors are used in each bit position. This may be costly in discrete or hybrid technology.

A third alternative is provided by using *quad current sources*. This approach is outlined below.

Figure 15-9 shows a DAC using three quad current sources. As the word *quad* (four) implies, each quad current source consists of four current sources; these are binary weighted as 8 : 4 : 2 : 1 by the ratios of the emitter resistors. The operational amplifier presents a near-zero impedance at its negative (−) input terminal; hence, the entire current of the leftmost quad current source flows into feedback resistor R_f. Also, it can be shown that 1/16 of the current from the center quad current source flows into R_f, and that 1/256 of the current from

*Optional material.

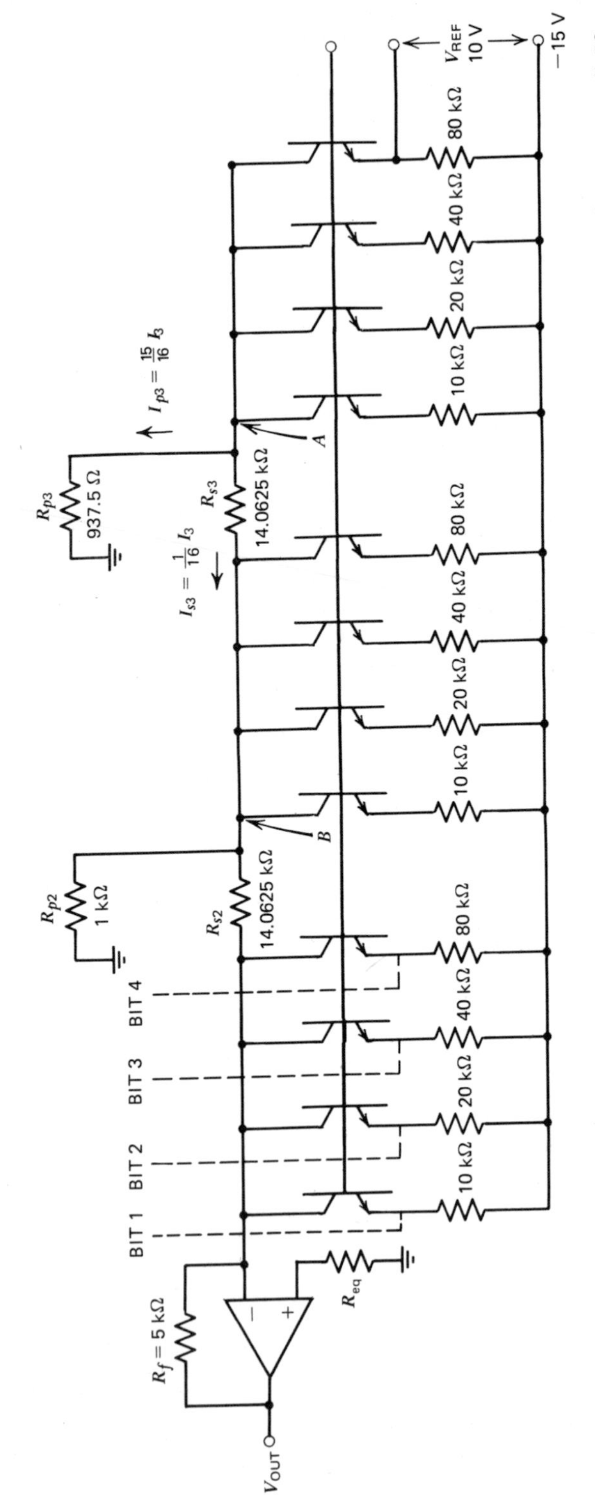

FIGURE 15-9 A DAC employing quad current sources. (Courtesy Analog Devices, Inc., *A / D Conversion Handbook*, page II-58, Copyright 1972.)

the rightmost quad current source flows into R_f. Thus Fig. 15-9 operates as a 12-bit binary coded DAC.

EXAMPLE 15-10

Verify that 1/16 of the current from the center quad current source flows into resistor R_f in Fig. 15-9.

Solution

The current division ratio can be written as $R_{shunt}/(R_{s2} + R_{shunt}) = 1/(1 + R_{s2}/R_{shunt})$, where $R_{shunt} = 1/[1/R_{p2} + 1/(R_{s3} + R_{p3})] = 1/[1/1\ \text{k}\Omega + 1/(14.0625\ \text{k}\Omega + 0.9375\ \text{k}\Omega)] = 0.9375\ \text{k}\Omega$. Hence, the current division ratio is

$$\frac{1}{1 + \dfrac{R_{s2}}{R_{shunt}}} = \frac{1}{1 + \dfrac{14.0625\ \text{k}\Omega}{0.9375\ \text{k}\Omega}} = \mathbf{\frac{1}{16}}$$

Binary weighted quad current sources can be also adapted to DACs with binary-coded decimal, BCD, inputs.

EXAMPLE 15-11

Assume that you are given a 12-bit DAC with binary weighted quad current sources such as shown in Fig. 15-9. The input is a 3-digit BCD quantity. Also given is $R_{s2} = R_{s3} = 9\ \text{k}\Omega$. Calculate

a. R_{p3}.
b. R_{p2}.

Solution

a. The attenuation between decades in a BCD DAC is 10. Refering to Fig. 15-10, at node A, $I_{p3} = 9I_3/10$ has to flow through R_{p3} to ground while $I_{s3} = I_3/10$ has to propagate to the left toward node B (where it is attenuated again by a factor of 10). Thus

$$\frac{R_{p3}}{R_{s3}} = \frac{I_{s3}}{I_{p3}} = \frac{1/10}{9/10} = \frac{1}{9}, \qquad \text{thus } R_{p3} = \frac{R_{s3}}{9} = \mathbf{1.0\ k\Omega}$$

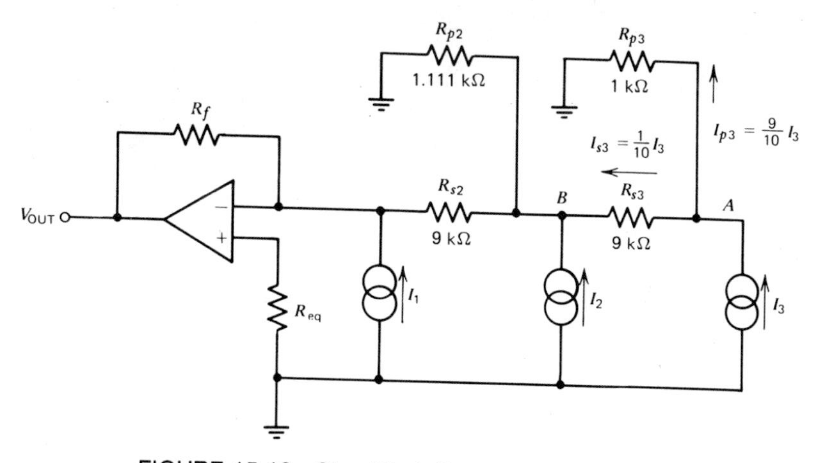

FIGURE 15-10 Simplified diagram for Ex. 15-11.

b. Here we demand that $R'_{p2}/R_{s2} = 1/9$ where $R'_{p2} = 1$ kΩ and is made up of R_{p2} in parallel with (9 kΩ + 1 kΩ). Thus

$$R'_{p2} = 1\ \text{k}\Omega = \frac{R_{p2} \times 10}{R_{p2} + 10}\ \text{k}\Omega; \qquad R_{p2} = \mathbf{1.111\ k\Omega}$$

15-7 MULTIPLYING DACs

We showed earlier that the output of a DAC is proportional to the *product* of the reference voltage, V_{REF}, and the applied binary or BCD input. In order to obtain a precise relationship between V_{OUT} and the digital input, we have made the reference voltage, V_{REF}, fixed and stable. Assume, however, that V_{REF} is made a changing analog voltage source. In such a case the output of the DAC is proportional to the *product* of an analog and a digital input. The DAC can thus be used as a digitally controlled amplifier or attenuator in which the gain is determined by the digital input. Several combinations of multiplying DACs exist: unipolar or bipolar analog inputs in conjunction with unipolar or bipolar digital inputs.

15-8 DAC PARAMETER MEASUREMENTS

The parameters that we are mostly concerned with in a DAC are (1) *zero*, (2) *gain*, (3) *integral nonlinearity*, (4) *differential nonlinearity*, and (5) *monotonicity*. These are discussed below.

15-8.1 Zero and Gain

A *zero* input to a unipolar DAC should, ideally, deliver a 0 V output. However, because of the input offset voltage of the operational amplifier (see Sec. 15-3.4), a slight positive or negative voltage may be present at the output of the DAC with zero input, as shown in Fig. 15-11. This voltage can be zeroed with a

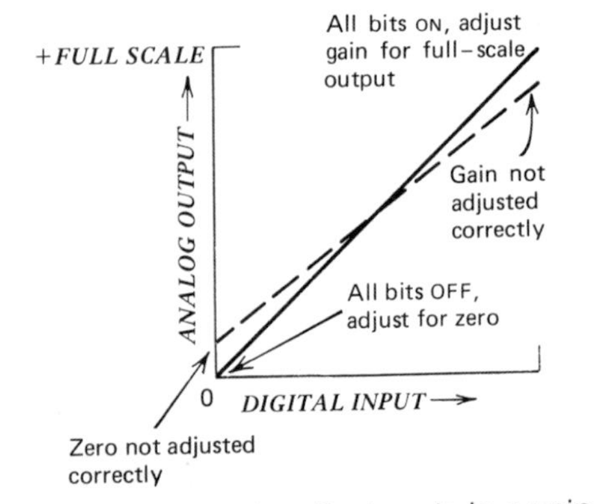

FIGURE 15-11 Zero and gain adjustments in a unipolar DAC.

zero adjustment control on the operational amplifier. In an offset-binary DAC, zero output should result when only the msb is ON and all other input bits are OFF.

Gain is measured at full scale. For a unipolar DAC all bits are set to logic 1 and the gain of the operational amplifier is adjusted for full-scale output. For an offset-binary bipolar DAC, the gain has to be checked first with all bits ON for positive full-scale output and then all bits OFF for negative full-scale output.

15-8.2 Integral and Differential Nonlinearity

The *integral nonlinearity*, or simply *nonlinearity*, is a measure of how linearly the analog output is related to the digital input over the entire operating range. It is the deviation from the expected straight line, as shown in Figs. 15-12*a* and

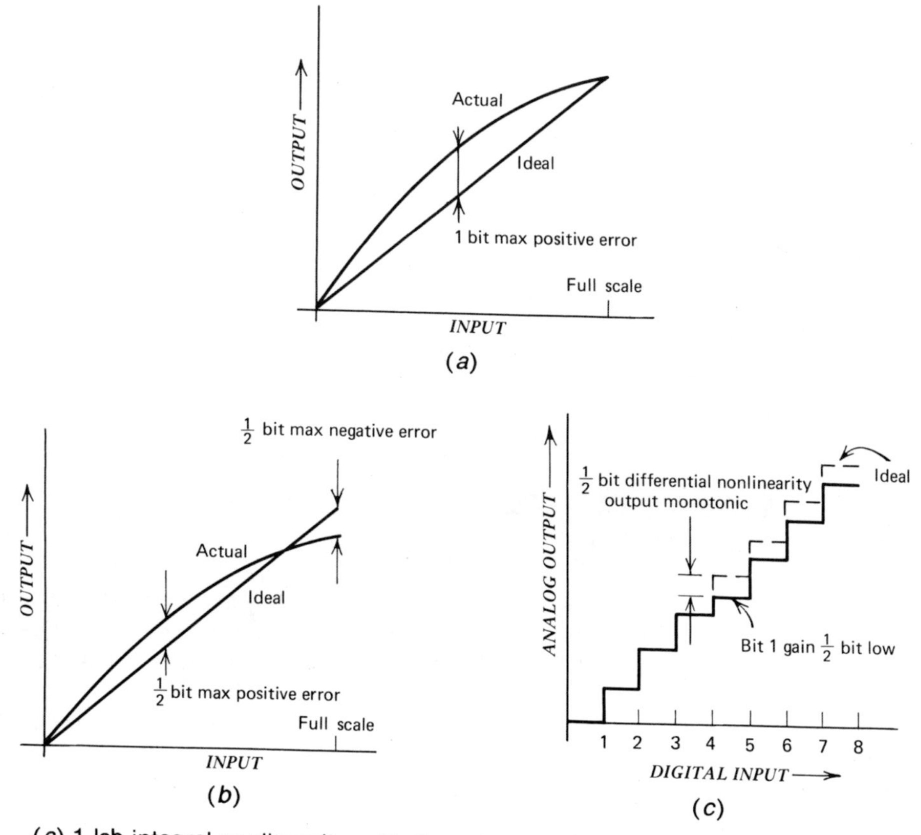

(*a*) 1 lsb integral nonlinearity, with the gain adjusted for zero full-scale error
(*b*) $\pm \frac{1}{2}$ lsb integral nonlinearity, with the gain adjusted for minimum peak error
(*c*) differential nonlinearity

FIGURE 15-12 Nonlinearities in a DAC. (Courtesy Analog Devices, Inc., *A / D Conversion Handbook*, Copyright 1972.)

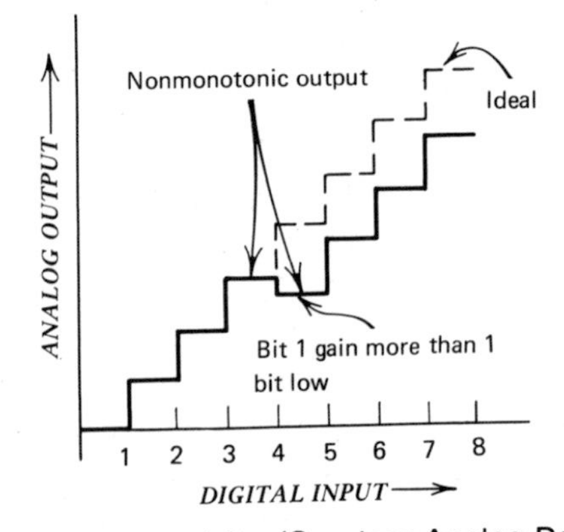

FIGURE 15-13 DAC non-monotonicity. (Courtesy Analog Devices, Inc., *A / D Conversion Handbook*, page II-103, Copyright 1972.)

15-12*b*, and is measured in terms of an lsb. Usually a $\pm \frac{1}{2}$ lsb nonlinearity is accepted in a DAC.

Differential nonlinearity is a measure of the analog output *change* for each one-bit digital input change. Ideally each step should be of the same magnitude over the entire operating range. Because of circuit imperfections, the actual changes in analog voltage with 1-bit digital change are not equal. This is illustrated in Fig. 15-12*c*.

15-8.3 Monotonicity

A *monotonic response* is one in which the output increases (or stays constant) for an increasing input. A non-monotonic DAC is shown in Fig. 15-13, indicating a serious deterioration of the DAC's performance.

15-9 ANALOG-TO-DIGITAL CONVERTERS (ADCs)

An *analog-to-digital converter* (*ADC*) is a circuit that delivers a digital serial or parallel output in response to an analog input that may be unipolar or bipolar. A great variety of ADCs has been developed to satisfy a broad spectrum of requirements.

An important ADC parameter is the *resolution* of the conversion. The resolution of an ADC is the smallest digital step with which the input analog voltage can be approximated.

EXAMPLE 15-12

An 8-bit ADC has a full-scale input voltage of +5 V. Calculate the smallest analog voltage step that can be recognized by the converter.

Solution

Eight bits represent $2^8 = 256$ steps. One step in a 5-V full-scale input voltage represents 5/256 V = **19.53 mV**.

Another important ADC parameter is the *conversion word rate*, which is the number of digital words with the required resolution that are encoded per unit time. In a given technology, there is usually a tradeoff between resolution and conversion word rate; that is, the number of bits can be increased only at the cost of a lower conversion word rate.

In the following, we discuss circuit components for ADCs in Sec. 15-10, counter ramp ADCs in Sec. 15-11, tracking ADCs in Sec. 15-12, successive approximation ADCs in Sec. 15-13, and flash ADCs in Sec. 15-14.

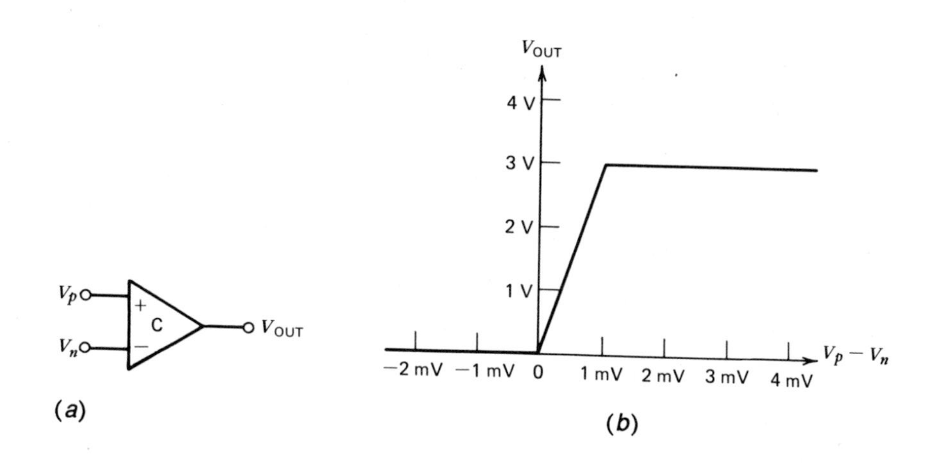

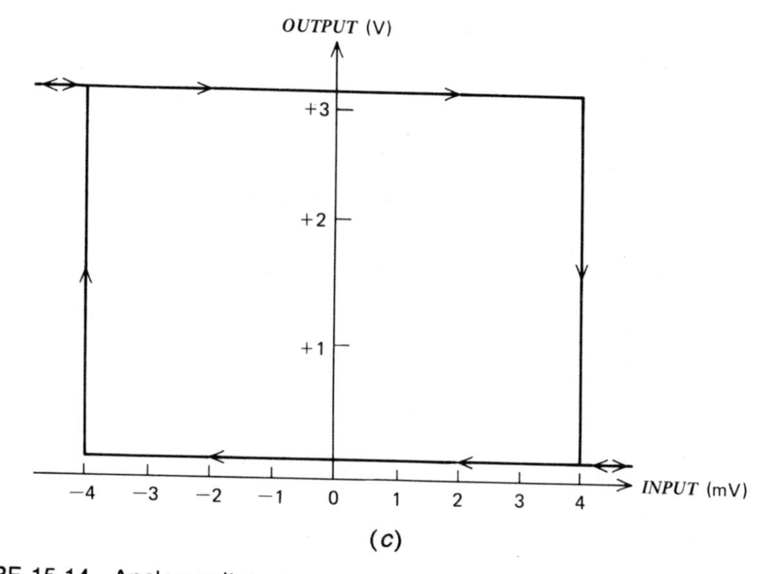

FIGURE 15-14 Analog voltage comparator. (*a*) Symbol (*b*) Output voltage versus input voltage (note different voltage scales of the two axes) (*c*) Output voltage versus input voltage with built-in hysteresis (note different voltage scales of the two axes)

15-10 CIRCUIT COMPONENTS FOR ADCs

In this section we discuss two important analog components used in ADCs: the analog voltage comparator and the sample-and-hold circuit.

15-10.1 Analog Voltage Comparators

Analog voltage comparators are similar to operational amplifiers (Sec. 15-3); however, the output voltage of an analog voltage comparator limits at the logic levels with which it is intended to be used. Figure 15-14*a* shows a symbol of analog voltage comparators; Fig. 15-14*b* shows the dc characteristics of one that limits at 0 V and +3 V and is intended for use with TTL levels. (Recall that the logic threshold voltage of a TTL circuit is about +1.5 V.)

Some analog voltage comparators have a built-in positive feedback that results in a hysteresis (backlash). Figure 15-14*c* shows the dc characteristics of such a circuit. The comparator changes states from its HIGH (+ 3 V) state to its LOW (0 V) state at an input voltage of approximately +4 mV. To change the state from LOW to HIGH requires an input voltage of −4 mV. The 8-mV difference between the "turn-on" voltage of +4 mV and the "turn-off" voltage of −4 mV is the *hysteresis* of the comparator. An analog voltage comparator with hysteresis usually has a better noise immunity, a worse input sensitivity, and a faster response time than an analog voltage comparator without hysteresis.

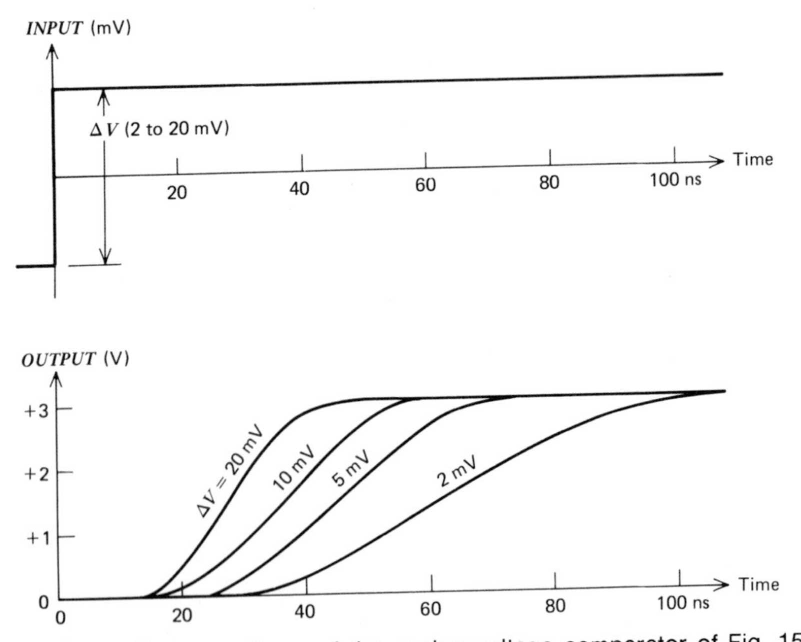

FIGURE 15-15 Response times of the analog voltage comparator of Fig. 15-14*b* for input overdrives of 2 to 20 mV.

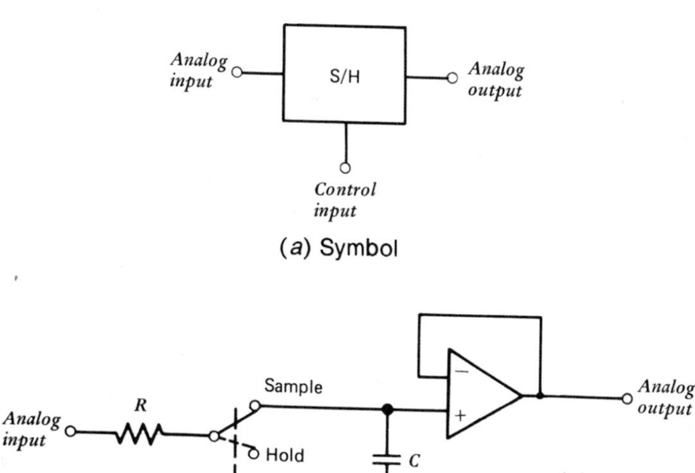

(b) Simplified diagram including a noninverting unity-gain operational amplifier (voltage follower)

FIGURE 15-16 Sample-and-hold circuit.

The *response time* of an analog voltage comparator is the time between the application of an input step voltage and the time when the output voltage crosses the logic threshold voltage. Response time is a function of *overdrive*, as illustrated in Fig. 15-15.

15-10.2 Sample-and-Hold Circuits

Sample-and-hold circuits are used in ADCs to *sample* during a short time interval the time-varying value of an analog input voltage, and then *hold* the sampled value constant for the duration of the conversion. Figure 15-16 shows a symbol and a simplified diagram of a sample-and-hold circuit.

The necessity of sampling an analog signal arises from the nonzero conversion time of an ADC: if the analog input signal of the ADC changes during conversion by more than the resolution of the ADC, the digital output of the ADC may not be representative of its analog input. This is illustrated in Ex. 15-13, which does not use a sample-and-hold circuit.

EXAMPLE 15-13

Determine the maximum allowable conversion time, t_c, for an 8-bit ADC with a sinusoidal input signal $V_{IN} = (V_{p-p}/2)\sin 2\pi ft = (V_{p-p}/2)\sin 100t$, as shown in Fig. 15-17. The error due to input voltage changes *during* the conversion should not exceed $\pm\frac{1}{2}$ lsb. Do not use a sample-and-hold circuit.

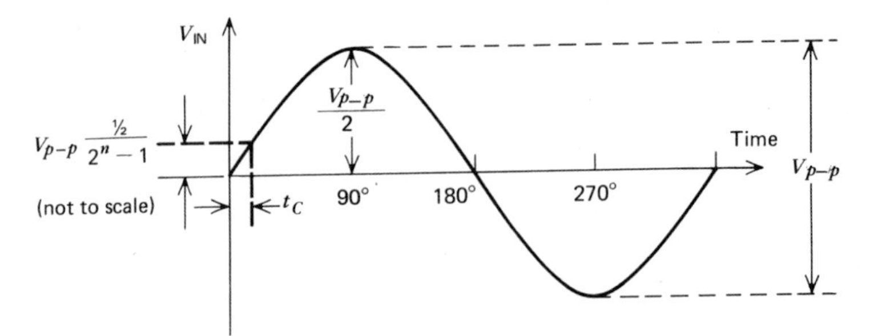

FIGURE 15-17 Input waveform in Ex. 15-13.

Solution

The number of discrete steps in an 8-bit ADC is $2^8 = 256$. Thus 1 lsb is equivalent to $V_{p-p}/256$ V and $\frac{1}{2}$ lsb is equivalent to $\frac{1}{2} \times V_{p-p}/256$. This is the maximum amount of allowed voltage change during the conversion time, t_c.

From the shape of the sine wave it is evident that the amplitude will hardly change while the analog-to-digital conversion is carried out at its peaks, that is, at 90° and at 270° phase angles. The highest rate of change of the sine wave is at 0° and 180° phase angles. For the duration of the conversion, t_c, at small angles, the change in amplitude will be

$$\frac{V_{p-p}}{2} \sin 2\pi f t_c$$

which may be approximated as $(V_{p-p}/2)2\pi f t_c$ since at small angles $\sin 2\pi f t_c \cong 2\pi f t_c$. This change must not exceed the equivalent of $\frac{1}{2}$ lsb, that is,

$$\frac{V_{p-p}}{2} 2\pi f t_c \leq \frac{1}{2} \times \frac{V_{p-p}}{2^n} = \frac{1}{2} \times \frac{V_{p-p}}{256}$$

$$t_c \leq \frac{1}{256} \times \frac{1}{2\pi f} = \frac{1}{256 \times 100} = \mathbf{39.06\ \mu s} \qquad (15\text{-}12)$$

Hence, to obtain an 8-bit digital representation at any given point of a sinusoidal voltage with $2\pi f = 100$, the analog-to-digital conversion time must not exceed 39.06 μs.

Thus, without a sample-and-hold circuit, the conversion time of an ADC must be quite short compared to the time scale of the changes in its analog input signal. However, the situation is different when the ADC is preceded by a sample-and-hold circuit. In this case the output of the sample-and-hold circuit, and of the subsequent ADC, will be representative of the analog input signal as long as the sampling is completed before the analog input signal changes by more than the resolution of the ADC.

A question still remains: how often do we have to sample an analog signal, that is, what is the minimum required *sampling rate*? Theory predicts that a time-varying input signal can be faithfully reproduced if the sampling rate is at least twice as high as the highest frequency component of the signal.

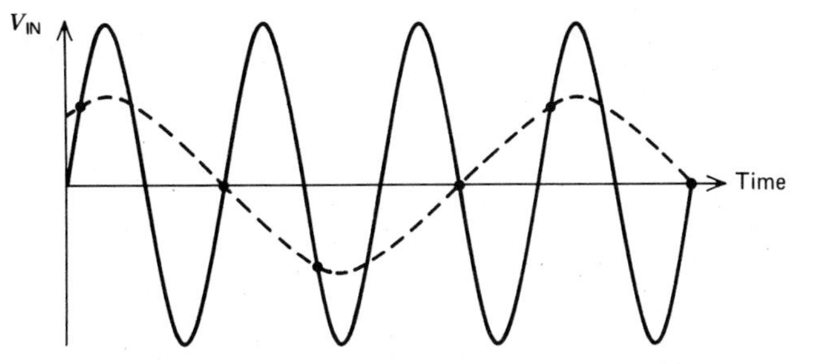

FIGURE 15-18 Aliasing. The apparent frequency (dashed line) derived from a too low sampling rate is lower than the frequency of the analog input (full line). Dots represent the instants of sampling.

EXAMPLE 15-14

The highest frequency component of an analog input signal is 1000 Hz. Determine the minimum required sampling rate.

Solution

The minimum required sampling rate is $2 \times 1000 =$ **2000 samples per second**.

If the sampling rate is too low, *aliasing* will occur, that is, the signal reconstructed from the samples will have different (lower) frequency components than the actual sampled signal. Aliasing is illustrated in Fig. 15-18.

We now return to the sample-and-hold circuit of Fig. 15-16*b*. In practical circuits the switch is realized by a CMOS transmission gate such as shown in Fig. 7-22*a* (page 164); when *p*-channel MOS devices are not available, then we can use an NMOS transmission gate (that is, we omit Q_2 and the logic inverter in Fig. 7-22*a*).

Resistor R in Fig. 15-16*b* represents the resistance of the transmission gate that realizes the switch; R also includes the resistance of the source that drives

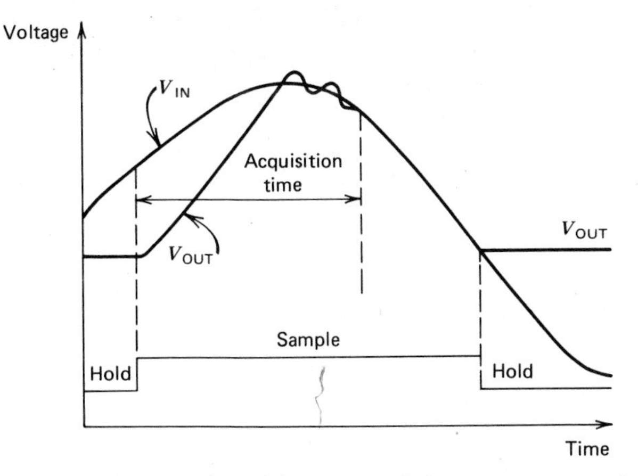

FIGURE 15-19 An error is introduced because of the nonzero acquisition time of the sample-and-hold circuit.

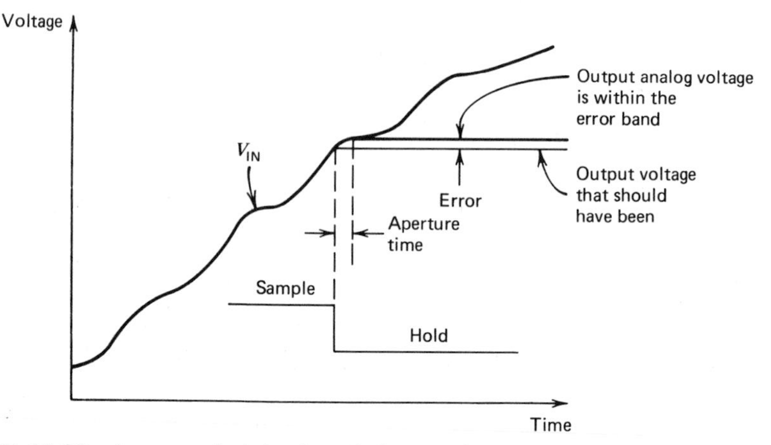

FIGURE 15-20 An error is introduced due to the nonzero aperture time of the sample-and-hold circuit.

the *Analog input*. In the sample mode the switch connects capacitor C to the *Analog input* through resistor R. The capacitor is thus charged with a time constant of RC, and after 2.3 time constants it will have charged to 90% of the input voltage, assuming that its initial voltage was zero. It takes an additional 2.3 time constants to charge the capacitor to 99% of the input voltage, and a total of 6.9 time constants for the capacitor to charge to 99.9% of the input voltage. Thus for an error of 0.1% a charging time of 6.9 RC is required.

The *acquisition time* of an ADC is defined as the length of time that elapses between a sample command and the instant at which the output voltage tracks the input voltage within a specified accuracy. This is illustrated in Fig. 15-19.

When the sampling period is completed, the sample-and-hold circuit receives a *hold* command. However, a semiconductor switch opens only gradually over an interval called the *aperture time*. This introduces an error in the output voltage as shown in Fig. 15-20. The uncertainty in the actual output voltage is dependent on the amplitude of the analog input signal and increases with the rate of change of the signal.

An additional error is introduced by the *aperture time jitter*, which is the uncertainty in the timing of the transition from sample to hold (not shown in Fig. 15-19 or Fig. 15-20).

15-11 COUNTER RAMP ADCs

An analog-to-digital converter of the counter-ramp type is shown in Fig. 15-21; it incorporates a binary counter and a DAC in a feedback loop. The analog input is applied to one input of an analog voltage comparator, while the feedback signal is applied to the other input. The counter is initially reset to zero; at the beginning of a conversion cycle the gated clock is actuated and increments the counter until the output of the DAC (which has the shape of a ramp) has just

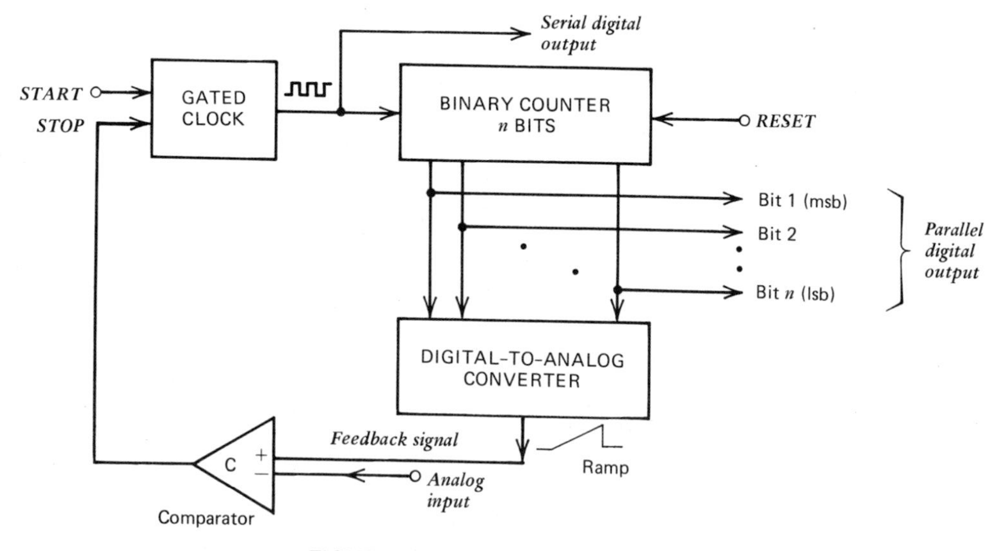

FIGURE 15-21 Counter ramp ADC.

exceeded the analog input signal by the threshold voltage, V_{TH}, of the comparator. The resultant change in the state of the comparator stops the clock. The digital output may be obtained in serial or in parallel form.

The maximum conversion time, t_c, of a counter ramp ADC equals $2^n - 1$ clock periods where n is the number of bits of the counter. The clock period, T, of a counter ramp ADC is limited by the sum of the propagation delays in the feedback loop, which are the propagation delays of the counter, the DAC, and the comparator.

EXAMPLE 15-15

In a 10-bit ($n = 10$) counter ramp ADC, a ripple counter is used with a maximum propagation delay of 200 ns, a DAC with a maximum propagation delay of 350 ns, and a comparator with a maximum propagation delay of 40 ns. Determine:

a. The maximum allowable clock frequency f_{max} of the clock.
b. The conversion time, t_c, for maximum input.
c. The worst case conversion word rate.

Solution

a. The maximum allowable frequency f_{max} is determined by the minimum allowable clock period $T_{min}: f_{max} = 1/T_{min}$. T_{min} is the sum of the propagation delays

$$T_{min} = 200 \text{ ns} + 350 \text{ ns} + 40 \text{ ns} = 590 \text{ ns}$$

Thus $f_{max} = 1/T_{min} = 1/590 \text{ ns} = \mathbf{1.695\ MHz}$.

b. The conversion time for maximum input is $t_c = (2^n - 1) \times T_{min} = 1023 \times 590 \text{ ns} = \mathbf{603.57\ \mu s}$.

c. The worst case conversion word rate is the reciprocal of the conversion time for maximum input:

$$1/t_c = 1/603.57\ \mu\text{s} = \mathbf{1.66\ kHz}$$

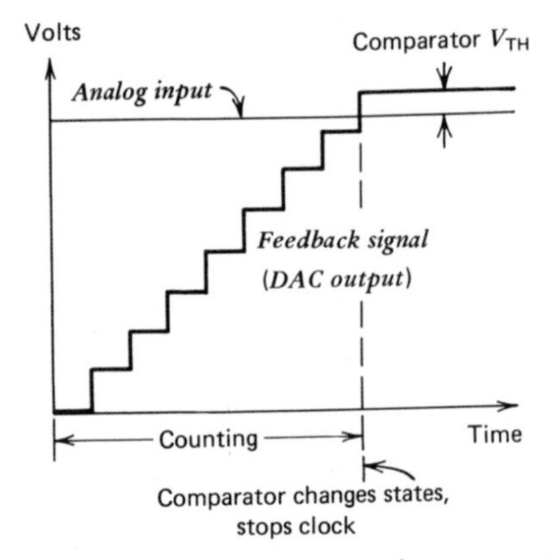

FIGURE 15-22 Input signals to the analog voltage comparator of Fig. 15-21.

Figure 15-22 shows in detail how the comparator changes states when the DAC output has exceeded the analog input. Thus there is an error introduced because of the discrete steps; this error is called the *quantizing error* and in the worst case it could be +1 lsb. In order to divide the errors equally in the positive and negative directions, that is, in order to obtain quantizing errors of $\pm \frac{1}{2}$ lsb, it is customary to offset the analog input by $+\frac{1}{2}$ lsb (not shown in Fig. 15-22).

The counter ramp ADC is used today whenever an ADC with many bits is desired and a long conversion time is acceptable. Available circuits include, for example, a 13-bit ADC with a conversion time of 50 ms built on a single monolithic IC chip.

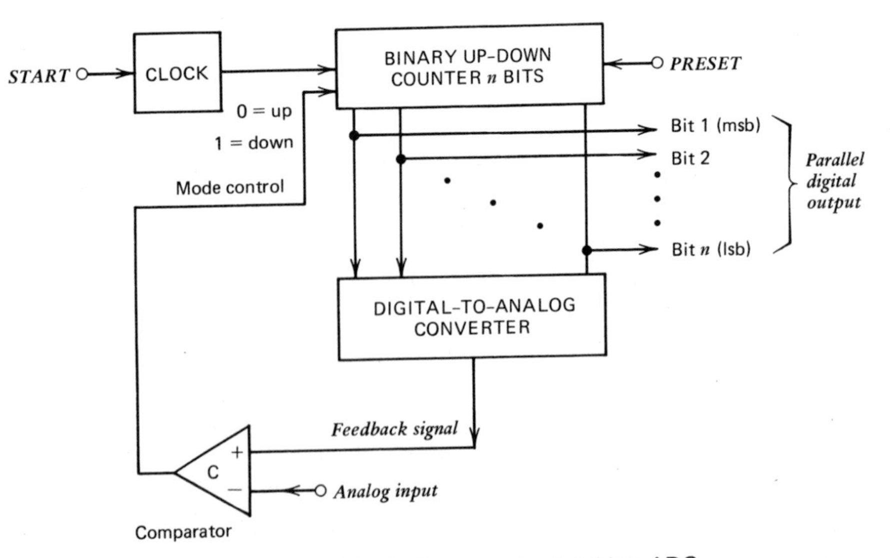

FIGURE 15-23 Block diagram of a tracking ADC.

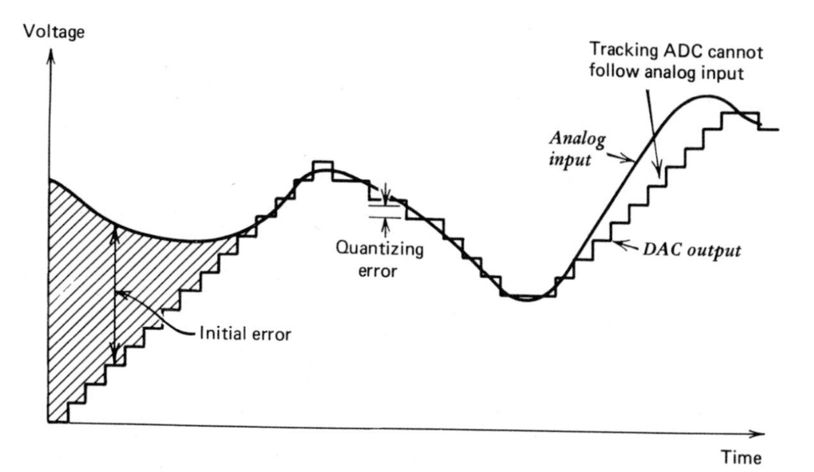

FIGURE 15-24 Tracking ADC. The analog input (smooth curve) and the DAC output (steps) are applied to the comparator that controls the up-down counter.

15-12 TRACKING ADCs

The *tracking ADC* is a modification of the counter ramp ADC in which the binary counter is replaced by an up-down counter. It is used when a continuous digital output is required from a single analog source. A block diagram of the tracking ADC is shown in Fig. 15-23.

The output from the comparator is applied to the mode control of the counter: when LOW (logic 0), it causes the counter to count up and, conversely, when HIGH (logic 1), it causes the counter to count down. Figure 15-24 shows that once the tracking ADC has achieved the proper range, the converter can continuously follow the analog voltage. Tracking speed presents no problem as long as the input variations are relatively slow. To reduce the initial error of the first conversions, the counter is usually set to half scale, that is, to 100 . . . 000.

15-13 SUCCESSIVE APPROXIMATION ADCs

The operation of this most widely used converter is based on successive comparisons between the analog input, V_{IN}, and a feedback voltage V_f. It is similar to the weighing process in which the unknown quantity is compared with a reference quantity. The "weighing" is done in a systematic way: the first comparison determines in which *half* of the range V_{IN} is found, that is, whether V_{IN} is greater or smaller than $\frac{1}{2}V_{max}$, where V_{max} is the maximum possible analog input to the ADC. The next step determines in which *quarter* of the range V_{IN} is found. Each successive step narrows the range of the possible result by a factor of $\frac{1}{2}$.

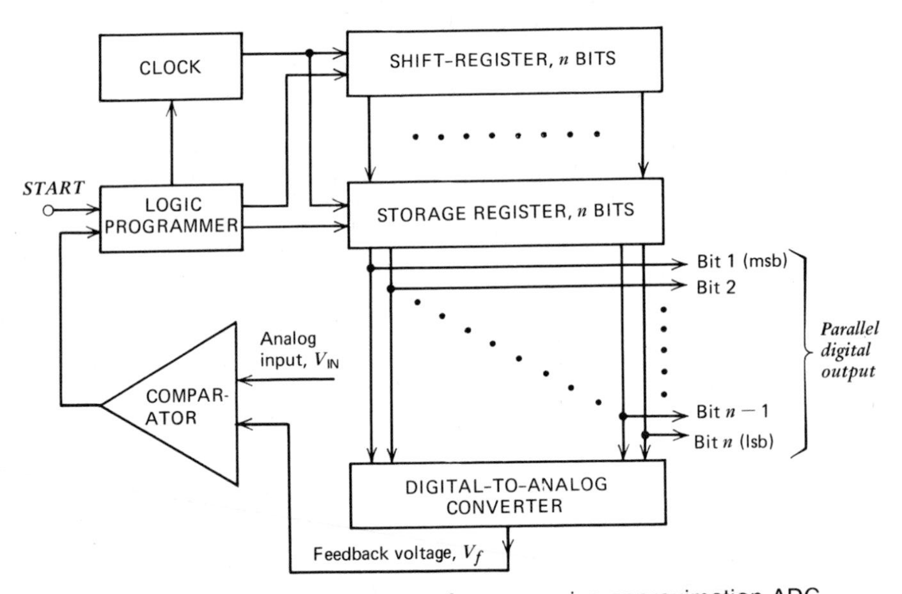

FIGURE 15-25 Block diagram of a successive approximation ADC.

The principal components of a successive approximation ADC are shown in Fig. 15-25. It is essentially a feedback circuit, similar to the counter ramp ADC of Fig. 15-21: the digital output from the storage register is converted in the DAC to an analog quantity, V_f, and is compared in the comparator with the unknown analog input, V_{IN}. The addition of the shift register and the *logic programmer* enable the circuit to decide on the next "weighing" step to be taken. The counter ramp ADC increments in small equal steps until the comparator threshold has just been exceeded, as shown in Fig. 15-22. In contrast, the successive approximation ADC increments or decrements by different-size steps, each step being $\frac{1}{2}$ of the former step. The decision whether to increment or decrement is made by the comparator and executed by the logic programmer after each "weighing."

An n-bit successive approximation register requires n clock periods to complete the conversion. In comparison, the counter ramp ADC of n-bit resolution requires $2^n - 1$ clock periods for conversion of a full-scale input signal.

Each conversion cycle is initiated by a *START* pulse that sets to logic 1 bit 1 (msb) of the two registers in Fig. 15-25. The 100 . . . 00 output of the storage register is converted by the DAC to an analog quantity, V_f, which is thus proportional to one-half of the full-scale input signal. If V_f is smaller than the input voltage, V_{IN}, the first FF in the storage register is left in state 1 and the shift register shifts its 1 to the next bit, bit 2, loading also a 1 in bit 2 of the storage register. If, however, $V_f > V_{IN}$, the msb in the storage register is reset. In either case the next to the most significant bit, bit 2, is examined next; this bit represents $\frac{1}{4}$ of full scale. The conversion process continues along similar lines and is completed after the lsb has been examined and its storage register bit has been set to the appropriate state.

EXAMPLE 15-16

A 4-bit successive approximation ADC with a full-scale input voltage of +16.0 V has an input of +13.3 V applied to it. Show the steps that a successive approximation ADC will undergo in obtaining the corresponding digital output. Ignore offsets required to reduce the quantizing error.

Solution

As the first step, bit 1, the msb, in the registers (Fig. 15-25) is set to 1 and a comparison is made by the comparator. The DAC output due to bit 1 is +8.0 V, which is smaller than the analog input voltage. Thus the msb = 1 is left in the storage register and the shift register shifts the logic 1 to the next bit, bit 2. The output of the DAC is now due to bit 1 (msb) and bit 2, which represent the half and the quarter scale, respectively. Thus the output of the DAC is now $\frac{1}{2} + \frac{1}{4} = \frac{3}{4}$ of full scale corresponding to +12.0 V. The comparator compares this with the +13.3 V of analog input and as a result of the comparison it leaves bit 2 of the storage register in the 1 state. The parallel digital output and the digital input to the DAC is now 1100.

Next, the shift register shifts a 1 into bit 3, resulting in a digital input to the DAC of 1110. Bit 3 carries a weight of $\frac{1}{8}$ of full scale; this added to the weights of bits 1 and 2 results in an output from the DAC of 8 + 4 + 2 = +14 V. The output from the DAC is now *greater* than the analog input voltage, signifying that the contribution from bit 3 was too much. Bit 3 of the storage register is, therefore, reset to zero. Next a logic 1 is shifted into the shift register, bit 4, which represents $\frac{1}{16}$ of full scale, that is, +1 V. The input to the DAC is now 1101, which is converted to an analog voltage of 8 + 4 + 0 + 1 = +13 V. Since this is lower than the +13.3 volts, bit 4 of the storage register is left ON, yielding a result **1101**.

The *quantizing error* is the uncertainty in digitizing of the analog signal because of the nonzero resolution of the ADC.

EXAMPLE 15-17

Determine the quantizing error obtained in the result of Ex. 15-16.

Solution

The digital result obtained in Ex. 15-16 is 1101, which, for a full-scale voltage of +16.0 V, represents an input analog voltage of +13.0 V. The actual input to the ADC in Ex. 15-16 was +13.3 V. Hence the quantizing error is +13.3 V − 13.0 V = **0.3 V**.

The conversion time, t_c, of the successive approximation ADC is constant and equals

$$t_c = n \times T \tag{15-13}$$

In eq. 15-13 n is the number of bits, and $T = 1/f$, where f is the clock frequency of the shifting. In comparison with the counter ramp ADC, the successive approximation ADC has a much higher conversion rate.

EXAMPLE 15-18

Calculate the time required to convert to digital form a full-scale analog input using (**a**) a counter ramp ADC, and (**b**) a successive approximation ADC. The conversion resolution is 12 bits and the clock frequency is 2 MHz.

Solution

a. The counter ramp ADC requires $2^{12} - 1 = 4095$ steps at 0.5 μs per step. Total conversion time is thus $4095 \times 0.5\ \mu s =$ **2.0475 ms**.

b. The successive approximation ADC requires 12 steps at 0.5 μs per step. Total conversion time is thus **6 μs**.

The successive approximation ADC is the most widely used ADC today. It is utilized whenever simplicity and low cost are of primary importance, and when moderate resolutions and conversion times are acceptable. Available circuits include, for example, an 8-bit ADC with a conversion time of 20 μs built on a single monolithic IC chip. The performance can be significantly improved if the blocks of Fig. 15-25 are built as separate ICs that are then assembled on a hybrid substrate. Also, a successive approximation ADC may be combined with other circuitry, such as a digital processor, on a single monolithic IC chip when a lesser performance is acceptable.

15-14 FLASH ADCs

Figure 15-26 shows a block diagram of a *flash ADC*, which is also known as a *parallel ADC*. In this ADC the signal is applied simultaneously to $2^n - 1$

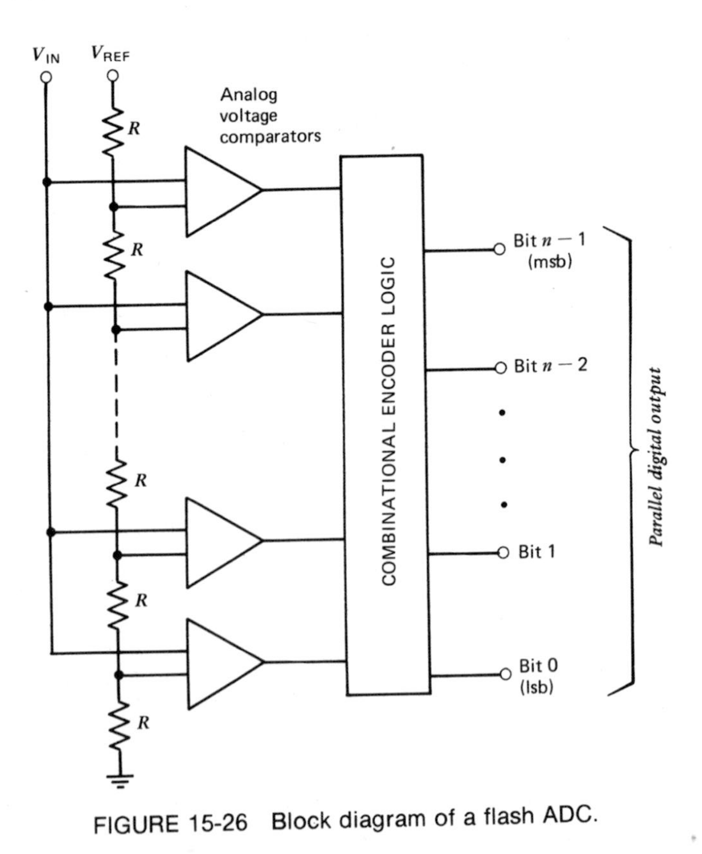

FIGURE 15-26 Block diagram of a flash ADC.

comparators for a resolution of n bits. Each comparator is biased one voltage step away from its neighboring comparators by the resistor string. When the threshold voltage of a comparator has been exceeded by the analog input signal, the comparator will change its logic state. The outputs thus obtained are encoded and latched. Conversion is asynchronous and the conversion time depends on the propagation delays of the comparators and the encoder logic.

Flash ACDs are used primarily when short conversion times are required and limited resolutions are acceptable. Available circuits include, for example, an 8-bit ADC with a conversion time of 20 ns built on a single monolithic IC chip.

SUMMARY

Digital-to-analog converters, DACs, were introduced first because they also serve as a component in many types of analog-to-digital converters. Characteristics of DAC components were discussed, including binary weighted resistance networks, operational amplifiers, and current switches. The treatment on DACs concluded with a discussion of their parameters including zero, gain, integral and differential nonlinearities, and monotonicity.

A description of analog-to-digital converter components preceded the discussion of various types of ADCs: counter ramp ADC, tracking ADC, successive approximation ADC, and flash ADC. The successive approximation ADC is the most widely used converter, while the flash ADC is capable of the highest conversion rates but is limited in its resolution.

SELF-EVALUATION QUESTIONS

1. Discuss applications of a digital-to-analog converter.
2. Discuss applications of an analog-to-digital converter.
3. Draw a schematic diagram of a 4-bit binary weighted DAC and discuss its operation.
4. Define (a) a unipolar DAC and (b) a bipolar DAC.
5. Draw a truth table for a 4-bit offset binary code.
6. Define resolution of a DAC.
***7.** With reference to an operational amplifier, define the following: (a) open-loop gain, (b) input bias current, (c) input offset current, (d) input offset voltage, (e) input impedance, (f) slew rate, (g) and settling time.
8. Describe the operation of current switches as used in DACs.
9. Describe the operation of a multiplying DAC.
10. With reference to a DAC define the following: (a) zero, (b) gain, (c) integral nonlinearity, (d) differential nonlinearity, (e) monotonicity.
11. Discuss the similarities and differences between analog voltage comparators and operational amplifiers.

*Optional material.

12. Define hysteresis as applied to analog voltage comparators.
13. Describe the operation of a sample-and-hold circuit. Define "aperture time" and "acquisition time."
14. Draw a block diagram of a counter ramp ADC.
15. What are the similarities and differences between the counter ramp ADC and the tracking ADC?
16. Describe the operation of a successive approximation ADC.
17. Draw a block diagram of a flash ADC.

PROBLEMS

15-1. Given a 16-bit binary DAC with a full-scale output of 10 V, determine its resolution in percent and the voltage corresponding to 1 lsb.

15-2. Given a 14-bit binary DAC with a full-scale output of 10 V, determine its resolution in percent and the voltage corresponding to 1 lsb.

15-3. The 8-bit binary input to a unipolar DAC with −5-V full-scale output is 11100000. What is the analog output?

15-4. The 10-bit binary input to a unipolar DAC with 10-V full-scale output is 1110000000. What is the analog output?

15-5. In a 10-bit resolution DAC of the binary weighted resistance type the msb resistor is 10 kΩ. What is the value of the resistor in the lsb position?

15-6. In a 12-bit resolution DAC of the binary weighted resistance type the lsb resistor is 10.24 MΩ. What is the value of the resistor in the msb position?

15-7. In a 10-bit DAC the full-scale output is +10 V. Determine the output voltage for an input of:

a. All 1s.

b. Only msb = 1.

c. Only lsb = 1.

15-8. In a 14-bit DAC the output is +610.35 μV when only the lsb = 1.

a. Calculate the full-scale output voltage.

b. Determine the output voltage for an all 1s input.

15-9. State the resolution in percent of:

a. A 14-bit binary DAC.

b. A 16-bit BCD DAC.

c. Which DAC has the better resolution?

15-10. In a 4-bit DAC, employing an offset binary input code, the input is 0011. Assuming a +7-V to −8-V full-scale range, what output voltage will the DAC deliver?

15-11. An 8-bit bipolar DAC has a full-scale output of ±5 V and employs sign-and-magnitude representation. What is the output with a digital input of 10110000?

15-12. In a 6-bit binary weighted resistance type DAC the value of the resistor in the msb position is 20 kΩ. What is the value of the resistor in the lsb position?

15-13. In a 6-bit DAC ($n = 6$) the value of the resistor in the msb position is 10 kΩ and $V_{REF} = +10$ V. What will be the current for:

a. The msb input = 1?

b. All inputs at logic 1?

15-14. What is the resistance tolerance of the msb resistor in an 8-bit DAC with $V_{REF} = 10$ V and $R = 10$ kΩ? Assume a maximum error of $\pm\frac{1}{2}$ lsb.

15-15. In a DAC the resistance ratios are specified to track within 8 ppm/°C. Calculate the maximum ratio mismatch in percent produced because of a rise in ambient temperature of 30°C.

15-16. An 8-bit binary weighted resistance DAC has $V_{REF} = +10$ V and R (msb) = 5 kΩ. It employs an operational amplifier (assumed having ideal properties) with $R_f = 10$ kΩ. Calculate the output voltage for a binary input of 10101010.

15-17. An operational amplifier such as shown in Fig. 15-4 has $I_{s1} = 1\ \mu$A, $I_{s2} = 1.01\ \mu$A, $R_1 = 10$ kΩ, $R_f = 100$ kΩ, and $R_{eq} = 9$ kΩ. Calculate the output voltage V_{OUT} as a result of the nonzero bias currents. Assume an otherwise ideal operational amplifier with an open loop gain $A = \infty$.

15-18. A 12-bit ADC has a full-scale input voltage of +10 V. Calculate the smallest input voltage step that can be recognized by the converter.

15-19. Determine the maximum allowable time, t_c, for a 6-bit ADC with a sinusoidal input signal $V_{IN} = (V_{p-p}/2)\sin 200t$ similar to that shown in Fig. 15-17. The error due to input voltage changes during the conversion should not exceed $\pm\frac{1}{2}$ lsb.

15-20. An ADC of 10-bit resolution ($n = 10$) is utilized for conversion of sinusoidal signals to a digital output. The conversion time of the ADC is $t_c = 10\ \mu$s. Determine the maximum frequency of the input signal.

15-21. A 14-bit successive approximation ADC operates at a clock frequency of 1 MHz.

a. Determine the time required for conversion of an analog voltage to digital form.

b. Does the time depend on the value of the analog voltage?

15-22. The following are the maximum propagation delays of components in a 12-bit ($n = 12$) counter ramp ADC: synchronous counter, 50 ns; DAC, 80 ns; comparator, 30 ns. Determine:

a. The maximum allowable frequency, f_{max}, of the clock.

b. The conversion time for maximum input.

c. The worst case conversion word rate.

15-23. A flash ADC of Fig. 15-26 has 255 comparators. What is its approximate resolution in percent of full scale?

15-24. A flash ADC of Fig. 15-26 has 511 comparators. What is its approximate resolution in percent of full scale?

CHAPTER
16 System Considerations

Instructional Objectives

This chapter describes interactions and constraints that arise in digital systems, several analog-digital circuits, and trouble-shooting instrumentation. After you read it you should be able to:

1. Evaluate tolerances, noise margins, and loading in a digital system.
2. Interface different logic families.
3. Evaluate supply bypass requirements.
4. Isolate grounds in digital systems.
5. Use Schmitt trigger circuits.
6. Use monostable multivibrators.
7. Describe digital diagnostic tools.

16-1 OVERVIEW

This chapter describes interactions and constraints that arise in the design and operation of digital systems, such as tolerances, noise margins, loading rules, interfacing, power, and grounding. It also introduces Schmitt trigger circuits and monostable multivibrators, as well as troubleshooting instrumentation used in diagnosing digital systems.

16-2 TOLERANCES, NOISE MARGINS, LOADING RULES

We assumed throughout this text that the logic 0 and logic 1 levels can always be clearly distinguished. However, in order for this assumption to hold, the design and layout of a digital system must take into account circuit tolerances, provide satisfactory margins against noise and other undesired effects, and consider the behavior of circuits under various load conditions.

16-2.1 Tolerances and Noise Margins

Transfer characteristics of the DTL logic inverter in Fig. 2-19*d* (page 19) show that for predictable operation the LOW input voltage has to be below 1 V and the HIGH input voltage has to be above 2 V. The situation is similar in TTL circuits (Fig. 7-1*d*, page 141) and in CMOS circuits (Fig. 7-18*c*, page 160, and Fig. 7-19*b*, page 161): specifications of these circuits permit a maximum LOW input voltage of 0.8 V and a minimum HIGH input voltage of 2 V. (Specifications of ECL circuits are described in Prob. 16-4.)

In order to ensure reliable operation, specifications are also set on the output voltages of logic circuits. The differences between the maximum permitted LOW input and the maximum possible LOW output, and the difference between the minimum permitted HIGH input and the minimum possible HIGH output are the *noise margins*.

EXAMPLE 16-1

The TTL Type 7404 IC consists of six logic inverters. The LOW output voltage is specified to be between 0 V and 0.4 V when the current *into* the output is between 0 and 16 mA ("*sink current*"). The HIGH output voltage is specified to be between 2.4 V and 5 V for a current of 0 to 0.4 mA flowing *out of* the output ("*source current*"). The maximum permitted LOW input voltage is 0.8 V and the minimum permitted HIGH input voltage is 2 V.

What are the noise margins if the output of one logic inverter is connected to the input of a second logic inverter with negligible input current?

Solution

By neglecting the input current of the second logic inverter, the output current of the first logic inverter becomes 0. Thus its LOW output voltage is between 0 V and 0.4 V. Since the maximum permitted LOW input voltage is 0.8 V, the operating margin in the LOW state is 0.8 V − 0.4 V = **0.4 V**.

The maximum HIGH output voltage is between 2.4 V and 5 V and the minimum HIGH input voltage is 2 V; thus the operating margin in the HIGH state is 2.4 V − 2 V = **0.4 V**.

The noise margins (0.4 V in the example above) are available for covering unavoidable external interference and noise picked up along the interconnections; these are discussed later in this chapter.

The specified input voltages, output voltages, and the resulting noise margins may be displayed in a *band diagram.* A band diagram for the noise margins of TTL circuits is shown in Fig. 16-1.

16-2.2 Loading Rules

The output voltages and propagation delays of logic circuits are specified for given load currents. Thus, if we want to make use of the specifications, we have

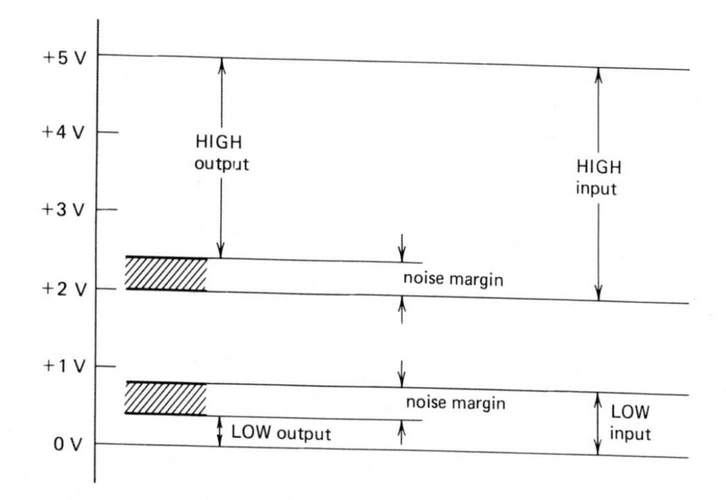

FIGURE 16-1 Band diagram showing the noise margins of TTL circuits.

to make sure that the maximum permissible current loading is not exceeded. A typical loading limitation in standard TTL is illustrated in Ex. 16-2 below.

EXAMPLE 16-2

The TTL Type 7404 IC consists of six logic inverters. Each logic inverter has a maximum input current of 40 μA flowing *into* the input when the input voltage is 2.4 V (HIGH state). Also, at an input voltage of 0.4 V (LOW state) it has a maximum input current of -1.6 mA, that is, a maximum input current of 1.6 mA flowing *out of* the input of the circuit.

Use additional data from Ex. 16-1 and find the maximum number of logic inverter inputs that are permitted to be connected to the output of a preceding logic inverter.

Solution

According to data in Ex. 16-1, at a load current of 0.4 mA the output of a logic inverter can deliver at least 2.4 V (HIGH state). Since at this voltage the input current of a logic inverter is not more than 40 μA, in the HIGH state we can connect 0.4 mA/40 μA = 10 inputs to one output.

When a load current of 16 mA flows into the output, it delivers a LOW output voltage of at most 0.4 V. Since at this voltage the input current of a logic inverter is not more than 1.6 mA, in the LOW input state we can connect 16 mA/1.6 mA = 10 inputs to one output.

Thus we conclude that the number of logic inverter inputs that are permitted to be connected to the output of a preceding logic inverter is **10**.

16-2.3 Interfacing Different Logic Families

We saw that there is a tradeoff between operating speed and power dissipation among the various logic families. Often it is desirable to use several logic families within a single system, requiring the use of *interfacing circuits*, also known as *level shifters*.

EXAMPLE 16-3

A digital system includes a small but fast central processor unit (CPU) and an extensive but relatively slow memory. In order to satisfy operating speed requirements, the CPU is built using ECL circuits. Since the CPU is small, the resulting power dissipation is acceptable. The memory uses CMOS circuits that are comparatively slow but draw little power, which is an important consideration since there are many circuits involved. What interfacing circuits are required in this system?

Solution

Two different types of interfacing circuits are required. Outputs of the ECL circuits are connected to the inputs of the CMOS circuits via interfacing circuits that have an input threshold of -1.2 V and an output voltage swing of approximately 0.4 V to 4 V. Outputs of the CMOS circuits are connected to the inputs of the ECL circuits via interfacing circuits that have an input threshold of approximately 1.5 V and an output voltage swing of approximately -1.6 V to -0.8 V.

Interfacing circuits are available as ICs, usually several circuits in a package. In some cases the use of interfacing circuits is not required and two logic families can be directly connected together. Such a situation occurs when the output of a TTL circuit is connected to the input of *some* (but not all) CMOS circuits.

16-3 POWER DISTRIBUTION

Several limitations are encountered in the distribution of supply voltages to digital circuits. The most significant of these are the stability of the dc voltages, the dc voltage losses due to wire resistances, and the fluctuations in the voltages due to transient load currents.

Supply voltage accuracy requirements of digital ICs are typically $\pm 5\%$: such stabilities are easy to guarantee by use of regulated power supplies. The dc voltage losses can be controlled by use of heavy wiring and by a suitable location of power supplies.

A significant limitation of power distribution in digital systems, especially in larger systems, originates from the transient load currents that occur during switching. Such currents are most prominent in TTL circuits with totem-pole outputs (Sec. 7-2).

EXAMPLE 16-4

During switching from its LOW output state to its HIGH output state, a standard TTL circuit with totem-pole output draws from the $+5$-V supply voltage a current of $I = 50$ mA for a duration of $t = 5$ ns. Find the resulting voltage change V_Δ of the power supply voltage, if it is bypassed to ground by a $C = 0.01\ \mu$F capacitor and is supplied through a wire with an inductance of $L = 0.1\ \mu$H (about 6 inches of wire).

Solution

First we compute the voltage change on capacitance C by *assuming* that the current through inductance L is negligible, then we verify that the assumption is reasonable.

When a current of $I = 50$ mA flows out of a capacitance of $C = 0.01\ \mu$F for a duration of $t = 5$ ns, the resulting voltage change V_Δ can be obtained from circuit theory as

$$V_\Delta = -\frac{It}{C} = -\frac{50 \text{ mA} \times 5 \text{ ns}}{0.01\ \mu\text{F}} = -25 \text{ mV}$$

The current buildup I_L in the inductance can be obtained from circuit theory as

$$I_L = \frac{V_\Delta t}{L} = \frac{-25 \text{ mV} \times 5 \text{ ns}}{0.1\ \mu\text{H}} = -1.25 \text{ mA}$$

Since 1.25 mA is much less than 50 mA, our assumption of neglecting I_L is reasonable; hence a reasonably good approximation is given by the solution $V_\Delta = -$**25 mV**.

The voltage fluctuations of the supply voltages are most detrimental to dynamic random-access memories, RAMs (Sec. 11-5), which typically require that such fluctuations be less than 50 mV. For this reason, power voltages of dynamic RAMs are often bypassed by a capacitor of 0.01 μF to 0.1 μF at each IC. However, bypass requirements are less severe for most other digital ICs, and fewer capacitors suffice.

16-4 GROUNDING

The logic levels of the various logic families are all generated and sensed with respect to 0 V. The 0-V point, which is usually referred to as "ground," may be a metal enclosure, a specified location within the system, or some other reference point. In small digital systems it is often possible to connect all 0-V terminals of the ICs to a single ground point; however, this is rarely possible in larger systems. In such systems voltage drops along ground conductors have to be taken into account.

Voltage drops arise as a result of dc resistance, inductance, and magnetic coupling from other conductors via mutual inductance. A detrimental voltage drop resulting from inductance is the subject of Ex. 16-5 below.

EXAMPLE 16-5

A digital system is separated into two parts that are interconnected by a ground strap with an inductance of $L = 0.1\ \mu$H. Find the voltage drop V_L along the ground strap if the current flowing in it changes by $I_\Delta = 50$ mA in $t = 5$ ns.

Solution

From circuit theory, the voltage drop can be written as

$$V_L = \frac{LI_\Delta}{t} = \frac{0.1\ \mu\text{H} \times 50 \text{ mA}}{5 \text{ ns}} = \mathbf{1\ V}$$

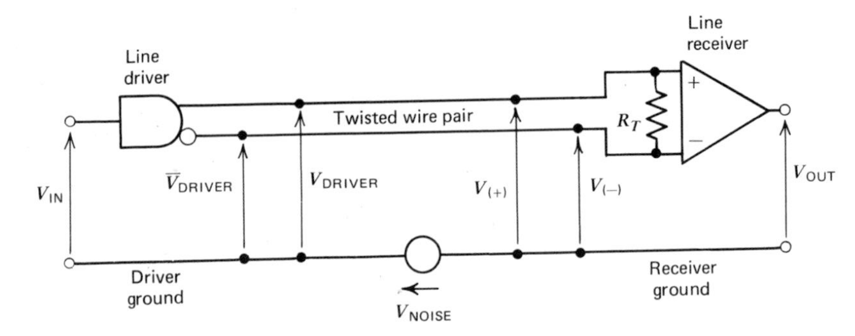

FIGURE 16-2 Use of a line driver and a line receiver.

If the voltage drop along a ground conductor is significant compared to the logic voltage swing, as was the case in Ex. 16-5, it may lead to malfunctioning. In many applications the situation may be alleviated by use of special circuits: line drivers, line receivers, and optical isolators. In what follows, the structure and operation of these circuits are described.

16-4.1 Line Drivers and Line Receivers

Figure 16-2 shows the use of a line driver and a line receiver. The line driver is a logic gate with complementary outputs and the line receiver is a voltage comparator (Sec. 15-10.1). Noise voltage V_{NOISE} represents undesirable but unavoidable noise introduced along the ground connection between the two circuits.

We have several different voltages in Fig. 16-2. Voltage $V_{(+)}$ is the sum of V_{DRIVER} and V_{NOISE}. Voltage $V_{(-)}$ is the sum of $\overline{V}_{DRIVER}$ and V_{NOISE}. Thus, the voltage difference $V_{(+)} - V_{(-)}$ equals $V_{DRIVER} - \overline{V}_{DRIVER}$ and is independent of V_{NOISE}. Since the comparator responds to the voltage difference $V_{(+)} - V_{(-)}$, noise voltage V_{NOISE} has no effect on comparator output voltage V_{OUT}.

The interconnection between the line driver and the line receiver is made by a twisted wire pair. Thus, any voltage that is picked up along *both* wires of the pair appears at both input terminals of the line receiver, and hence has no effect on output voltage V_{OUT}. Terminating resistor R_T is typically between 75 Ω and 200 Ω and is chosen to minimize reflections in the twisted wire pair. The line driver, unlike a usual TTL gate, is capable of delivering a HIGH logic voltage into a load of R_T.

Line drivers are built such that the propagation delay from the input to the true output equals the propagation delay from the input to the complementary output. Line receivers commonly incorporate resistive attenuators at the inputs, thus permitting noise voltages with magnitudes greater than the 5-V supply voltage.

EXAMPLE 16-6

A line receiver is operated with a supply voltage of +5 V. For correct operation the inputs of the line receiver must be within the range of −30 V to +30 V.

The line receiver is used in the circuit of Fig. 16-2. The line driver supplies a voltage of 0 V or +2.5 V at each of its two outputs. Noise voltage V_{NOISE} is a 400-Hz sine wave with an amplitude of 20 V. V_{IN} is 0 V for $t < 1$ ms and $t > 3$ ms and it is +2.5 V for 1 ms $< t <$ 3 ms. Neglect all propagation delays and sketch the waveforms of the following voltages:

1. Input voltage V_{IN}.
2. The voltage between the upper (true) output of the line driver and the driver ground, V_{DRIVER}.
3. The voltage between the lower (complementary) output of the line driver and the driver ground, $\overline{V}_{DRIVER}$.
4. Noise voltage V_{NOISE}.
5. The voltage between the (+) input of the line receiver and the receiver ground, $V_{(+)}$.
6. The voltage between the (−) input of the line receiver and the receiver ground, $V_{(-)}$.
7. The voltage difference $V_{(+)} - V_{(-)}$.
8. The output voltage of the comparator, V_{OUT}.

Solution

The waveforms are shown in Fig. 16-3. The line receiver inputs are always within the range of −30 V to +30 V. Therefore the line receiver operates correctly, and its output voltage, V_{OUT}, is immune to noise voltage V_{NOISE}.

16-4.2 Optical Isolators

Another circuit that facilitates noise isolation is the *optical isolator*, also known as *optical coupler*. The use of optical isolators permits the complete elimination of connections between two digital systems or between two parts of the same digital system, and such isolation can withstand kilovolts of voltage differences. Figure 16-4 shows the principle of the optical isolator. The light-emitting diode (LED) is connected to an output of one of the two digital systems. The light emitted by the LED carries the information (logic 0 or logic 1) to a photodiode that turns on a transistor in the other digital system.

In reality, the current of the photodiode is weak and needs to be amplified. Also, an *ENABLE* input is often provided. For example, Fig. 16-5 shows a simplified schematic diagram of an optical isolator. When the LED is activated by a current of I_{LED} = 5 mA to 10 mA, output *OUT* can sink a current of I_{SINK} = 13 mA, provided that the *ENABLE* input is HIGH (+ 2.4 V to +5 V).

EXAMPLE 16-7

The use of an optical isolator is shown in Fig. 16-6. Supply voltage $(+5\ V)_{IN}$ is provided from the system to the left of the optical isolator (System no. 1); the $(+5\ V)_{OUT}$ is provided from the system to the right of the optical isolator (System no. 2). Also, power to the TTL gate at the input is supplied from the $(+5\ V)_{IN}$ and the $(0\ V)_{IN}$ (connections not shown). Thus there is no connection between the two systems.

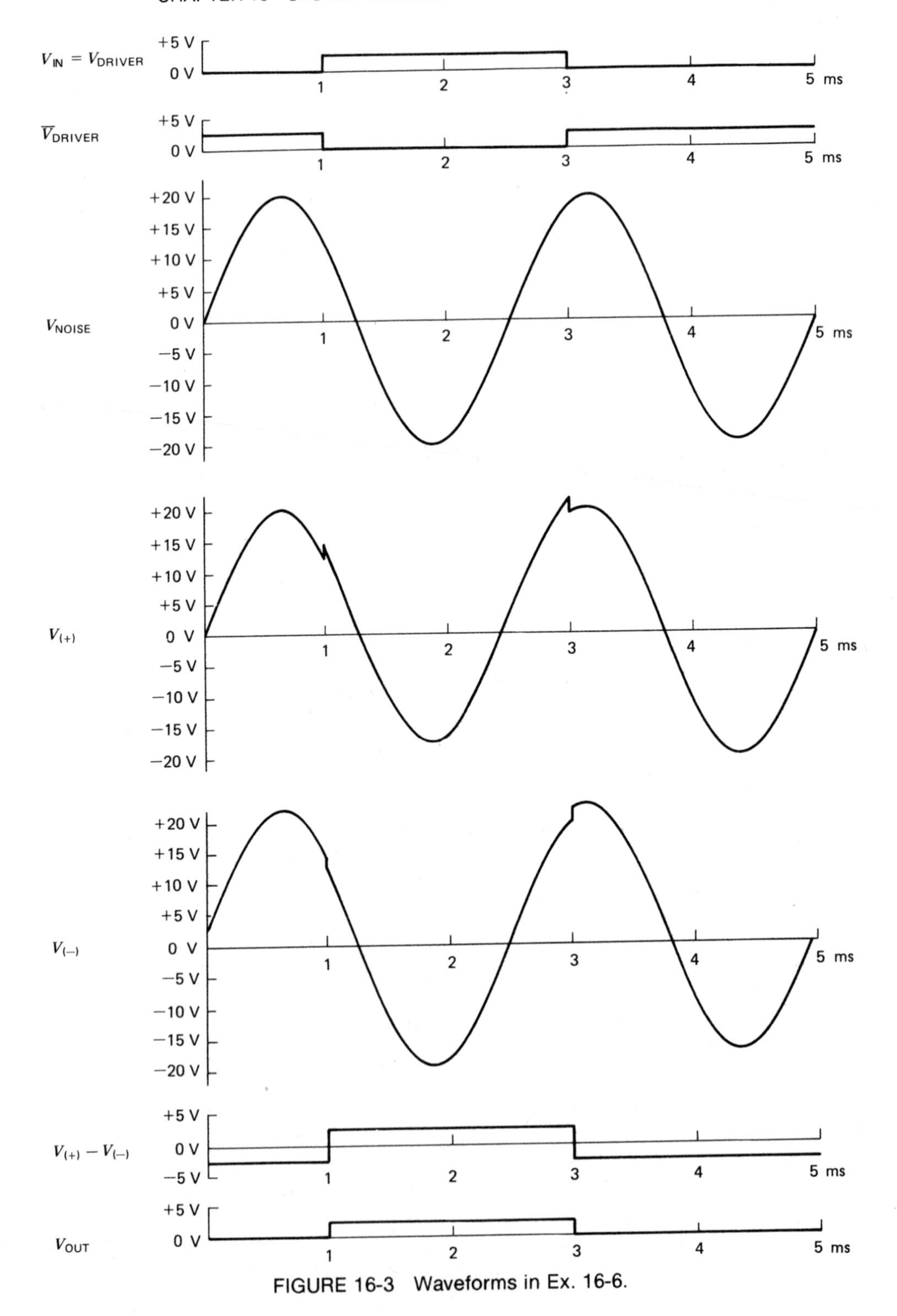

FIGURE 16-3 Waveforms in Ex. 16-6.

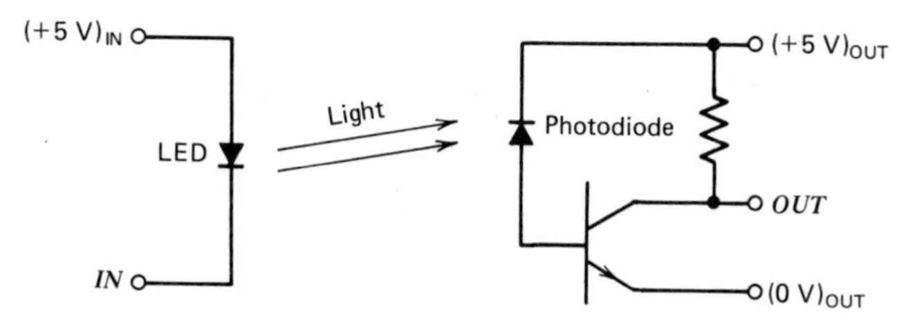

FIGURE 16-4 Principle of the optical isolator.

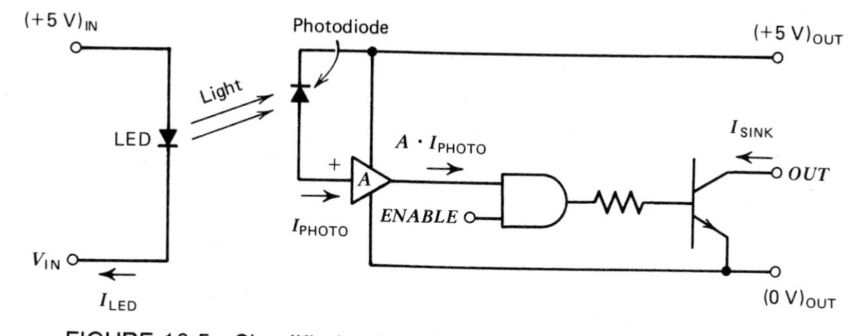

FIGURE 16-5 Simplified schematic diagram of an optical isolator.

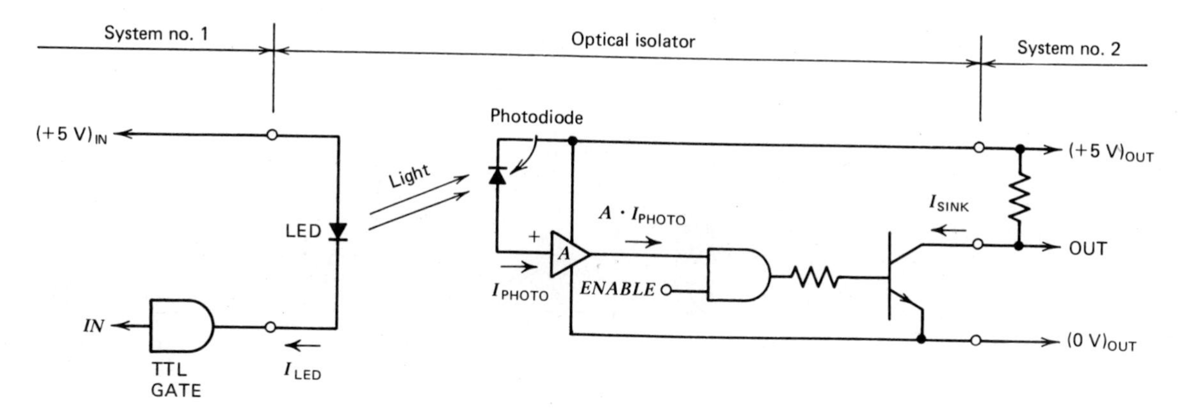

FIGURE 16-6 Use of an optical isolator.

IN	*ENABLE*	*LED*	*OUT*
LOW	LOW	ON	HIGH
LOW	HIGH	ON	LOW
HIGH	LOW	OFF	HIGH
HIGH	HIGH	OFF	HIGH

FIGURE 16-7 Truth table for the circuit of Fig. 16-6.

Prepare a truth table and write Boolean equations for the LED and for output *OUT*.

Solution

The truth table is shown in Fig. 16-7. The LED is ON whenever input *IN* is LOW, thus $LED = \overline{IN}$. Also, output *OUT* is LOW when the LED is ON *and* the *ENABLE* is HIGH, that is, $OUT = \overline{LED \cdot ENABLE} = \overline{\overline{IN} \cdot ENABLE} = IN + \overline{ENABLE}$.

16-5 SCHMITT TRIGGER CIRCUITS

In a digital system a need often arises for reshaping signals that have long transition times. Such signals may originate from a sensor such as a temperature transducer, or may have long transition times as a result of passing through long wires.

We may be tempted to use a logic gate for sharpening up long transition times. However, this may lead to malfunctions if there is an irregularity (e.g., noise) on the waveform when it crosses the threshold voltage of the logic gate.

EXAMPLE 16-8

The upper part of Fig. 16-8 shows a pulse that is applied to the input of a TTL logic inverter with a threshold voltage of +1.5 V. Sketch the output waveform

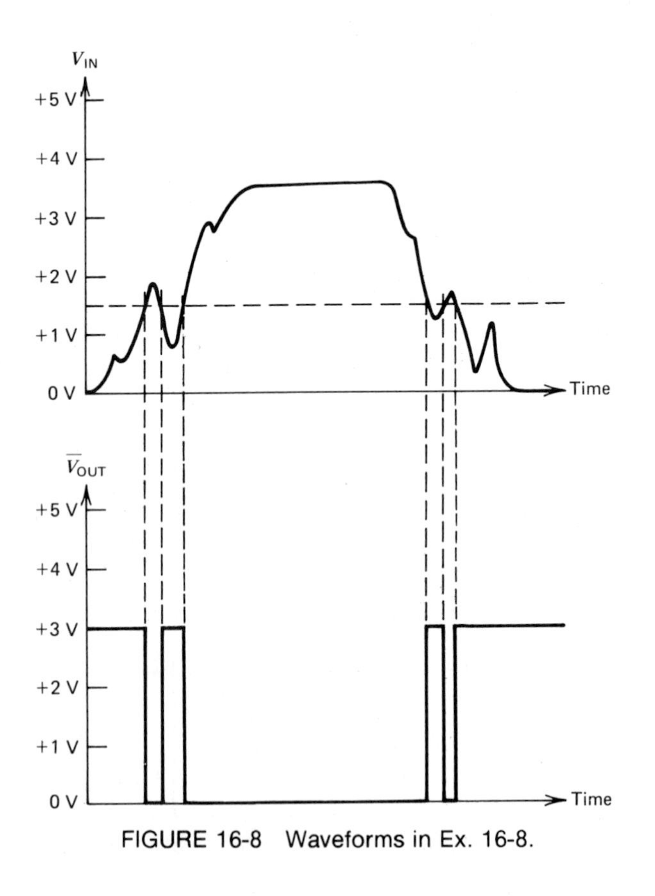

FIGURE 16-8 Waveforms in Ex. 16-8.

of the logic inverter if its propagation delay and transition times are invisibly short.

Solution

The output waveform of the logic inverter is shown in the lower part of Fig. 16-8. It is HIGH when the input signal is below +1.5 V and it is LOW otherwise. When used in a digital system, the output waveform of the logic inverter may lead to malfunction: for example, it could be incorrectly counted as three pulses instead of the correct count of one.

A more reliable way of reshaping slow transition times is by use of a *Schmitt trigger* circuit. In what follows, we discuss the structure and operation of a Schmitt trigger circuit that uses an analog voltage comparator (Sec. 15-10.1) even though simpler implementations are more common.

Consider the circuit shown in Fig. 16-9*a*. It consists of an analog voltage comparator, two resistors in a positive feedback configuration, and a logic inverter. Figure 16-9*b* shows the characteristics of the analog voltage comparator based on Fig. 15-14*b*.

Output voltage $\overline{V}_{OUT} = 0$ V when $V_p - V_{IN} < 0$ V, and it is +3 V when $V_p - V_{IN} > 1$ mV. Also, by inspection of Fig. 16-9*a* we can write

$$V_p = 1.5\ \text{V} + (\overline{V}_{OUT} - 1.5\ \text{V})\frac{R_1}{R_1 + R_F}. \qquad (16\text{-}1)$$

Using the resistor values of $R_1 = 1$ kΩ and $R_F = 2$ kΩ, we get

$$V_p = 1\ \text{V} + \frac{\overline{V}_{OUT}}{3} \qquad (16\text{-}2)$$

From eq. 16-2 we can see that $V_p = +1$ V when $\overline{V}_{out} = 0$ V, it is +2 V when $\overline{V}_{OUT} = +3$ V, and it is between +1 and +2 V when $\overline{V}_{OUT}$ is between 0 and +3 V.

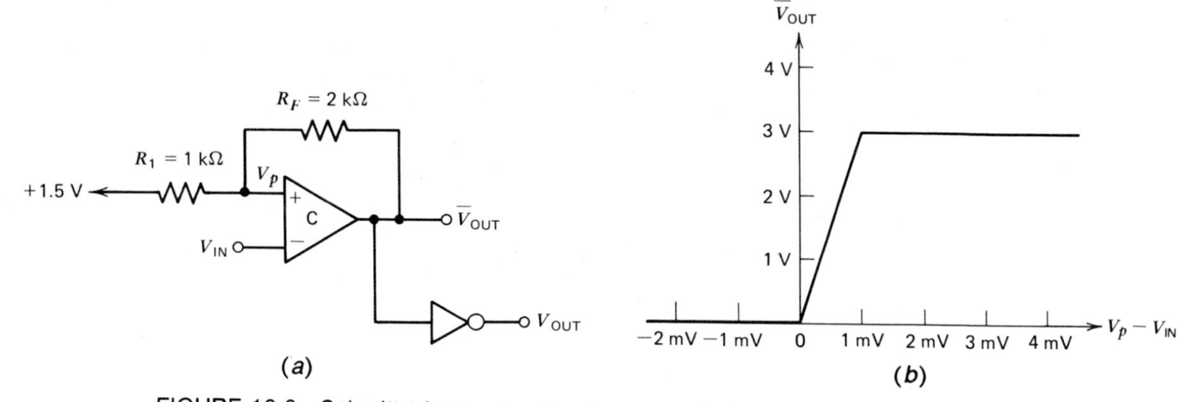

FIGURE 16-9 Schmitt trigger circuit using an analog voltage comparator. (*a*) The circuit, (*b*) output voltage as a function of input voltage for the analog voltage comparator used in (*a*).

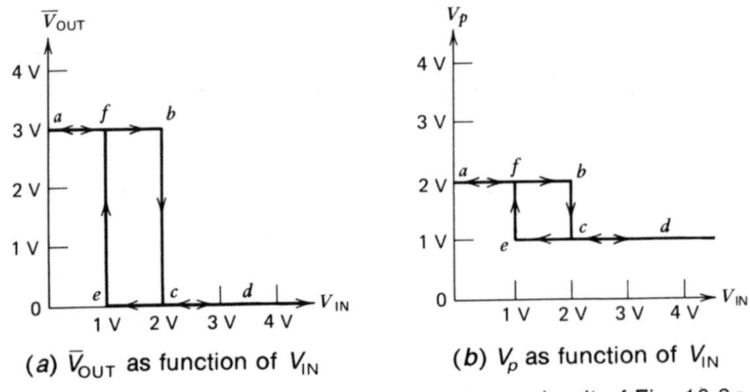

(*a*) $\overline{V}_{OUT}$ as function of V_{IN} (*b*) V_p as function of V_{IN}

FIGURE 16-10 Voltages in the Schmitt trigger circuit of Fig. 16-9*a*.

We now examine the operation of the circuit of Fig. 16-9*a* for various input voltages. As we go along, we also refer to Fig. 16-10, which shows $\overline{V}_{OUT}$ and V_p as functions of input voltage V_{IN}.

When V_{IN} is more negative than +1 V, as is the case for points *a* in Fig. 16-10*a* and Fig. 16-10*b*, $\overline{V}_{OUT}$ becomes +3 V; hence, $V_p = +2$ V. If we now raise input voltage V_{IN}, the state of the circuit does not change until V_{IN} gets within 1 mV of the $V_p = +2$ V voltage (points *b*), since at this point $V_p - V_{IN} =$ 1 mV in Fig. 16-9*b*.

As we further raise V_{IN}, $V_p - V_{IN}$ becomes less than 1 mV, $\overline{V}_{OUT}$ starts decreasing, and V_p starts decreasing according to eq. 16-2. However, a lower V_p results in a lower $V_p - V_{IN}$ in Fig. 16-9*b*; thus the circuit makes a sudden transition to $\overline{V}_{OUT} = 0$, $V_p = +1$ V (points *c* in Fig. 16-10). The circuit remains in this state as V_{IN} is raised further (points *d*). However, as V_{IN} is lowered to +1 V (points *e*), $V_p - V_{IN}$ becomes positive and $\overline{V}_{OUT}$ starts to rise from 0 V (see Fig. 16-9*b*). This starts raising V_p, and a sudden transition to $\overline{V}_{OUT} = +3$ V, $V_p =$ +2 V takes place (points *f*). The circuit now remains in this state as V_{IN} is lowered further (points *a*).

From the above we conclude that when input voltage V_{IN} is between $V_{T-} = +1$ V and $V_{T+} = +1.999$ V, the Schmitt trigger circuit of Fig. 16-9*a* can be in one of two states; thus there is a backlash or *hysteresis* of $V_h = V_{T+} - V_{T-}$, which is approximately 1 V in this circuit. This has two consequences. One of these is that the transitions between the output states of $\overline{V}_{OUT} = 0$ V and $\overline{V}_{OUT} = +3$ V are fast even when input V_{IN} changes slowly. The other consequence is that there is only one output transition on the rising edge of the input signal and one output transition on the falling edge of the input signal, as shown in Fig. 16-11. (Propagation delays and transition times are again assumed to be invisibly short.) However, such favorable behavior is exhibited only as long as the irregularities, such as noise, on input signal V_{IN} are less than hysteresis V_h. Unfortunately, the magnitude of hysteresis voltage V_h may vary widely even among circuits of the same type. This is illustrated in Ex. 16-9, which follows.

EXAMPLE 16-9

The TTL Type 7414 IC consists of six Schmitt trigger circuits. Threshold voltages and hysteresis are specified as follows. Positive-going threshold

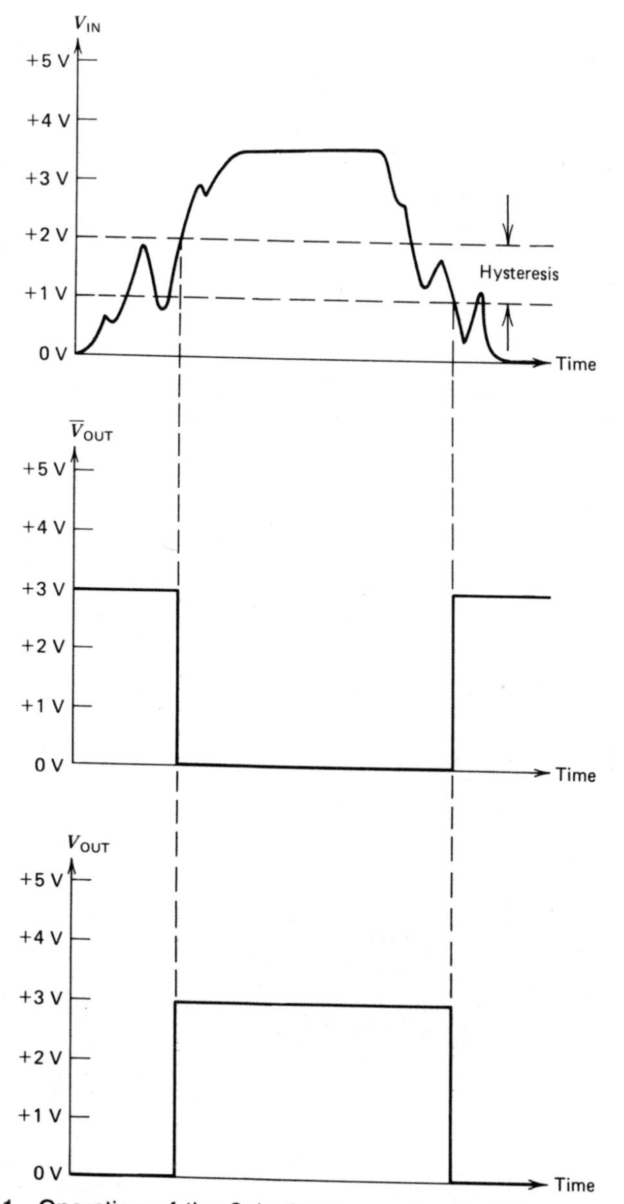

FIGURE 16-11 Operation of the Schmitt trigger circuit of Fig. 16-9*a* for an input signal V_{IN} that is identical to V_{IN} of Fig. 16-8.

voltage, V_{T+}: 1.5 V minimum, 1.7 V typical, 2 V maximum. Negative-going threshold voltage, V_{T-}: 0.6 V minimum, 0.9 V typical, 1.1 V maximum. Hysteresis, $V_h = V_{T+} - V_{T-}$: 0.4 V minimum, 0.8 V typical. No maximum is given for the hysteresis.

Verify the minimum and typical values of the hysteresis, and compute its maximum value.

Solution

The hysteresis is minimum when V_{T+} is minimum (1.5 V) and V_{T-} is maximum (1.1 V); thus the minimum value of the hysteresis is $V_h = V_{T+} - V_{T-} = 1.5\text{ V} - 1.1\text{ V} = \mathbf{0.4\ V}$.

At the typical values of $V_{T+} = 1.7$ V and $V_{T-} = 0.9$ V, $V_h = V_{T+} - V_{T-} = 1.7\text{ V} - 0.9\text{ V} = \mathbf{0.8\ V}$.

The hysteresis is maximum when V_{T+} is maximum (2 V) and V_{T-} is minimum (0.6 V). Thus, the maximum value of the hysteresis is $V_h = V_{T+} - V_{T-} = 2\text{ V} - 0.6\text{ V} = \mathbf{1.4\ V}$.

16-6 MONOSTABLE MULTIVIBRATORS

It is usually straightforward to generate a time interval with a specified duration in a digital system that has a clock. We can generate any time interval that is an integer multiple of the clock period by counting clock pulses in a counter. While this is commonly done in *timing circuits* of computers, it may be unnecessarily complex in many applications.

EXAMPLE 16-10

At the completion of its task, a digital system alerts the operator by ringing a bell for approximately one second. How many FFs are required in a counter generating this time interval if the system has a clock with a frequency of 1 MHz?

Solution

Twenty FFs can divide by $2^{20} \approx 10^6$. Since the clock period is 1/1 MHz = 1 μs, a counter consisting of 20 FFs can generate a time interval with a duration of $1\ \mu\text{s} \times 10^6 = 1$ s. Thus, the solution is **20**.

A simple circuit for generating time intervals in the range of nanoseconds to minutes with 0.1% to 10% accuracies is the *monostable multivibrator*, also known as *single-shot* or *one-shot*, that delivers an output pulse with a fixed duration. Instead of a clock, such a circuit relies for timing on passive elements such as resistors, capacitors, and delay lines. The active circuitry may consist of an analog voltage comparator, a logic gate, or circuitry designed specifically for the purpose.

We distinguish two types of monostable multivibrators. One of these is the monostable multivibrator with *dead time*, which is discussed in Sec. 16-6.1. The other type is the *dead-time-less*, or *retriggerable*, monostable multivibrator, discussed in Sec. 16-6.2.

16-6.1 Monostable Multivibrators with Dead Time

Figure 16-12*a* shows a monostable multivibrator that uses an analog voltage comparator. Its operation is described referring to the timing diagram shown in Fig. 16-12*b*, where invisibly short propagation delays and transition times are assumed.

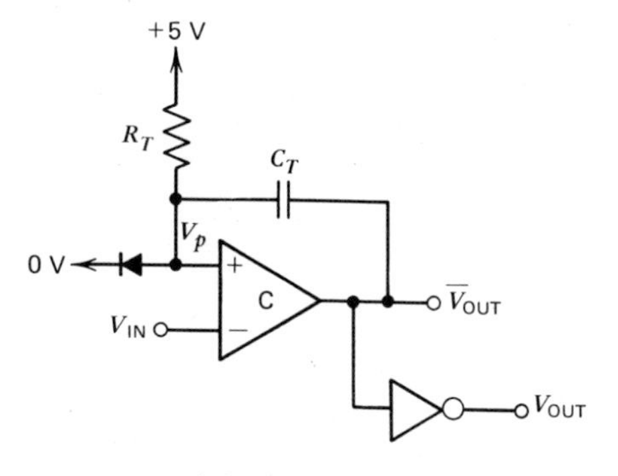

(*a*) The circuit

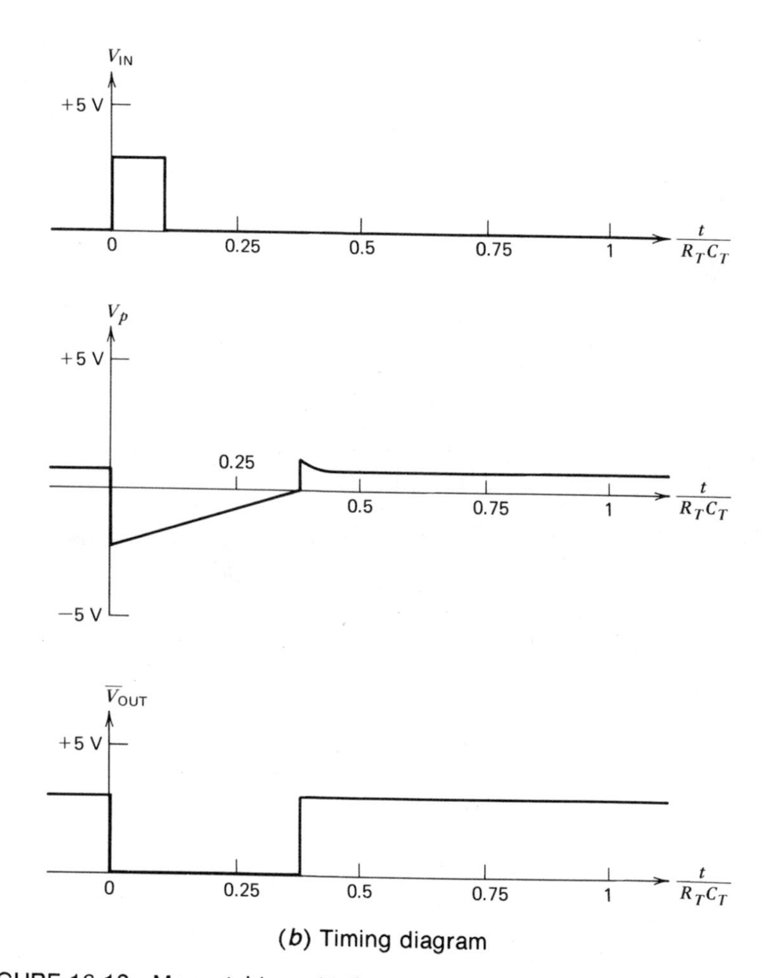

(*b*) Timing diagram

FIGURE 16-12 Monostable multivibrator using an analog voltage comparator.

Initially $V_{IN} = 0$, $\bar{V}_{OUT} = +3$ V, and $V_p = +0.7$ V. For an input pulse V_{IN} with a height of 3 V at time $t = 0$, $\bar{V}_{OUT}$ instantly changes to 0 V. Since the voltage across C_T can not change instantly, V_p is also lowered instantly by 3 V from $V_p = +0.7$ V to $V_p = -2.3$ V. For $t > 0$, V_p can be written as

$$V_p = -2.3\text{ V} + (5\text{ V} + 2.3\text{ V})(1 - e^{-t/R_TC_T}), \tag{16-3}$$

which holds as long as $V_p < +0.7$ V. This waveform (see Fig. 16-12*b*) reaches $V_p = 0$ at a time T given by

$$T = R_TC_T \ln \frac{7.3\text{ V}}{7.3\text{ V} - 2.3\text{ V}} = 0.38\ R_TC_T \tag{16-4}$$

If by this time V_{IN} has returned to 0 V, as is the case in Fig. 16-12*b*, $\bar{V}_{OUT}$ changes instantly to $+3$ V and V_p to $+0.7$ V after a short settling time because of the nonzero forward resistance of the diode. Thus a time interval with a duration of $T = 0.38\ R_TC_T$ has been generated at output $\bar{V}_{OUT}$.

EXAMPLE 16-11

The monostable multivibrator circuit of Fig. 16-12*a* is used for generating a time interval with a duration of $T = 1$ s. Input V_{IN} is $+3$ V for a duration of 0.5 μs and is 0 V otherwise. The value of resistor R_T is 100 kΩ. Find the value of capacitor C_T.

Solution

By use of eq. 16-4, we can write

$$C_T = \frac{T}{0.38\ R_T} = \frac{1\text{ s}}{0.38 \times 100\text{ k}\Omega} \approx \mathbf{26.3\ \mu F}$$

The accuracy of the time interval T given by eq. 16-4 depends principally on the accuracies of R_T, C_T, the supply voltages, and the input impedance of the comparator. While R_T, C_T, and the supply voltages can be held to accuracies of better than 1% (often 0.1%), the input impedance of the comparator limits the maximum permissible R_T.

EXAMPLE 16-12

The TTL Type 74121 IC consists of a monostable multivibrator preceded by a Schmitt trigger. Its circuitry is different from Fig. 16-12*a* and its output pulse width is $T = 0.7\ R_TC_T$. The maximum permitted value of R_T is 40 kΩ and the maximum permitted value of C_T is 1000 μF. The accuracy of T is $\pm 1\%$ when $R_T = 10$ kΩ; no accuracy is specified for other values of R_T.

a. Find duration T of the longest time interval that can be generated.

b. Find values of R_T and C_T for generating a time interval of $T = 1$ s with an accuracy of $\pm 1\%$.

c. Find the values of R_T and C_T for generating a time interval of $T = 1$ s if the use of the smallest possible capacitor value is desired and if the accuracy of T is of no importance.

Solution

a. The longest time interval can be generated by use of $R_T = 40$ kΩ and $C_T = 1000\ \mu$F, whereby

$$T = 0.7\ R_TC_T = 0.7 \times 40\text{ k}\Omega \times 1000\ \mu\text{F} = \mathbf{28\ s}$$

b. For $\pm 1\%$ accuracy we use $R_T = 10\ \text{k}\Omega$. Thus, since $T = 0.7\ R_T C_T$,

$$C_T = \frac{T}{0.7\ R_T} = \frac{1\ \text{s}}{0.7 \times 10\ \text{k}\Omega} = \mathbf{140\ \mu F}$$

c. For a given T, the value of C_T is minimum when R_T is maximum, that is, when

$$R_T = \mathbf{40\ k\Omega}.$$

With this value of R_T we get a

$$C_T = \frac{T}{0.7\ R_T} = \frac{1\ \text{s}}{0.7 \times 40\ \text{k}\Omega} = \mathbf{35\ \mu F}$$

The timing diagram of Fig. 16-12*b* shows two additional properties of the monostable multivibrator of Fig. 16-12*a*. One of these is that a second $V_{IN} = +3$ V input pulse arriving during time interval $T = 0.38\ R_T C_T$ has no effect. The other property is that if a second input pulse arrives shortly after T, it results in a period different from T, since V_p did not yet settle back to $+0.7$ V. Hence, there is a *dead time*, D, following T during which time an input pulse is not allowed if an accurate time interval T is desired. Thus, the *maximum duty cycle*, h_{max}, of the monostable multivibrator is

$$h_{max} = \frac{T}{T + D} \tag{16-5}$$

From eq. 16-5 we can also write

$$D = \frac{1 - h_{max}}{h_{max}} T \tag{16-6}$$

EXAMPLE 16-13

When operated with $R_T = 40\ \text{k}\Omega$, the TTL Type 74121 monostable multivibrator has a maximum duty cycle of $h_{max} = 90\%$. Find dead time D if $T = 1$ s, and sketch a timing diagram showing D.

Solution

The maximum duty cycle is $h_{max} = 90\% = 0.9$. Thus, by use of eq. 16-6,

$$D = \frac{1 - h_{max}}{h_{max}} T = \frac{1 - 0.9}{0.9} 1\ \text{s} \approx \mathbf{0.11\ s}$$

Figure 16-13 shows a timing diagram with two output pulses; each output pulse has a duration of $T = 1$ s. The two pulses are separated by a time that

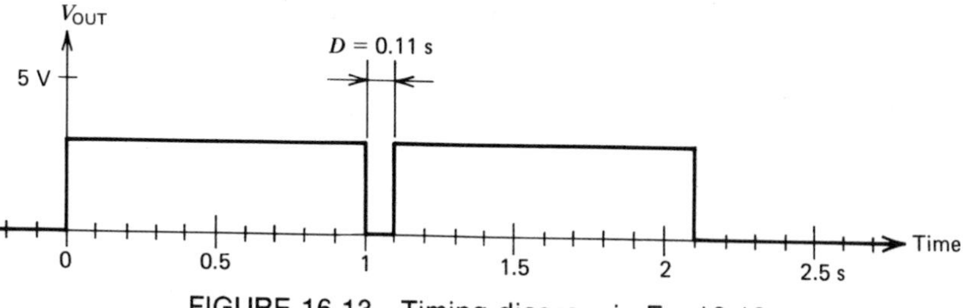

FIGURE 16-13 Timing diagram in Ex. 16-13.

equals the dead time $D = 0.11$ s. Thus the second output pulse cannot occur earlier than shown if an accurate time interval T is desired.

16-6.2 Retriggerable Monostable Multivibrators

Monostable multivibrators with dead time are suitable for many applications, such as the one described in Ex. 16-10. However, they are not suitable for all applications.

EXAMPLE 16-14

A copying machine is capable of operating at a rate of one copy per second. The operation of a cooling fan is required while a copy is made and for an additional period of 10 s. Is a monostable multivibrator with dead time suitable for this application?

Solution

If a second copy is made shortly following the fan operation, the fan is not turned back on because of the nonzero dead time D of the monostable multivibrator. Also, if an additional copy is made within 10 s after the preceding copy, the operation of the fan is not extended for 10 s after this additional copy.

Thus, a monostable multivibrator with dead time is *not* suitable for this application.

A *retriggerable*, or *dead-time-less, monostable multivibrator* is shown in Fig. 16-14. Its operation is illustrated in the timing diagram of Fig. 16-15. Initially input *IN* is at 0 V and the output of the logic inverter with open-collector output, V_C, which is also the input of the Schmitt trigger, is at +2.2 V. An input pulse of +3 V forces V_C to 0 V, and after the input pulse returns to 0 V, V_C rises toward +5 V as

$$V_C = 5\text{ V}(1 - e^{-t/R_T C_T}) \tag{16-7}$$

The positive-going threshold $V_{T+} = +2$ V of the Schmitt trigger is reached at time T, which can be obtained from eq. 16-7 as

$$T = R_T C_T \ln \frac{5\text{ V}}{5\text{ V} - 2\text{ V}} = 0.51\, R_T C_T \tag{16-8}$$

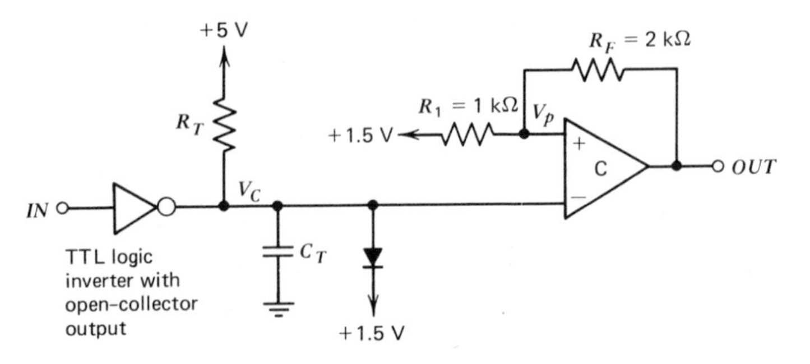

FIGURE 16-14 Retriggerable monostable multivibrator.

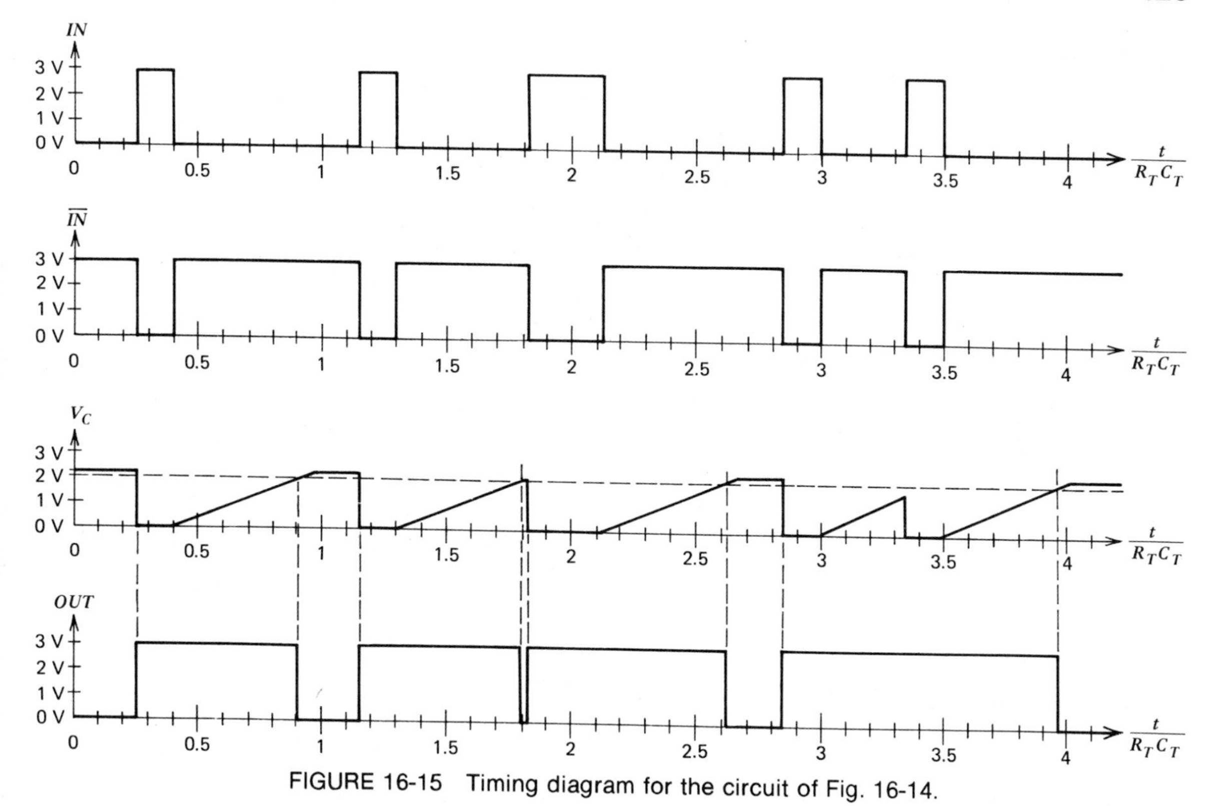

FIGURE 16-15 Timing diagram for the circuit of Fig. 16-14.

If a second input pulse arrives *after* time interval T, then a second identical output pulse is generated as shown in Fig. 16-15. If a second input pulse arrives *during* time interval T, then the interval is *extended* to last until a time T following the trailing edge of the second input pulse (see last output pulse in Fig. 16-16.)

An application of a retriggerable monostable multivibrator is described in Ex. 16-15, which follows.

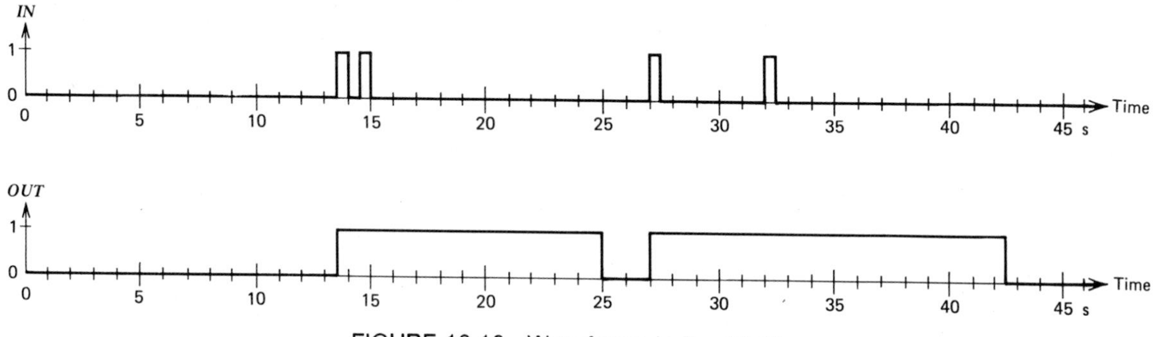

FIGURE 16-16 Waveforms in Ex. 16-15.

EXAMPLE 16-15

The TTL Type 74122 IC is a retriggerable monostable multivibrator. For $C_T \leq 1000\ \mu F$, T is given as

$$T = 0.32(700\ \Omega + R_T)C_T \tag{16-9}$$

where an internal resistance of 700 Ω is in series with the external timing resistor R_T. The maximum permitted value of R_T is 50 kΩ.

a. Find the value of C_T that results in $T = 10$ s if $R_T = 50$ kΩ.

b. The retriggerable monostable multivibrator is used with $T = 10$ s for operating the cooling fan in the copier described in Ex. 16-14. Sketch the output waveform of the retriggerable monostable multivibrator if its input waveform is as shown in the upper part of Fig. 16-16.

Solution

a. From eq. 16-9 we can write

$$C_T = \frac{T}{0.32(700\ \Omega + R_T)} = \frac{10\ \text{s}}{0.32(700\ \Omega + 50{,}000\ \Omega)} = \mathbf{616\ \mu F}$$

b. The output waveform of the retriggerable monostable multivibrator is shown in the lower part of Fig. 16-16. We can see that it operates the cooling fan as required.

16-7 TROUBLESHOOTING INSTRUMENTATION

Initially the only instruments available for digital troubleshooting were general-purpose instruments such as oscilloscopes and signal generators. As digital techniques found more applications, these instruments became more oriented toward digital uses and special-purpose instruments dedicated to digital applications were also developed. In what follows, several of these instruments and their operation are outlined.

16-7.1 Word Triggering

Before the advent of digital circuits, oscilloscopes had free-running sweeps that could be synchronized only to a repetitive waveform. Subsequent developments included internally triggered sweep, externally triggered sweep, delayed sweep, and armed sweep. A further development enhancing the usefulness of oscilloscopes in diagnosing digital systems is *word triggering*, which can be *parallel* or *serial*.

The principle of *parallel word triggering* is shown in Fig. 16-17 for a 4-bit trigger word $A_3A_2A_1A_0$. The *CLOCK IN* is reshaped by a Schmitt trigger, whose output *ST OUT* is delayed and shaped into a narrow *STROBE* pulse by a monostable multivibrator.

Figure 16-18 shows a timing diagram for Fig. 16-17. If input word $A_3A_2A_1A_0$ agrees with the position of switches $S_3S_2S_1S_0$ (= 1011 in Fig. 16-17) at the time of the *STROBE*, then a *TRIGGER* pulse is generated.

The principle of *serial word triggering* is illustrated in Fig. 16-19 for a 4-bit word. A *TRIGGER* pulse is generated when 4 subsequent logic values in the

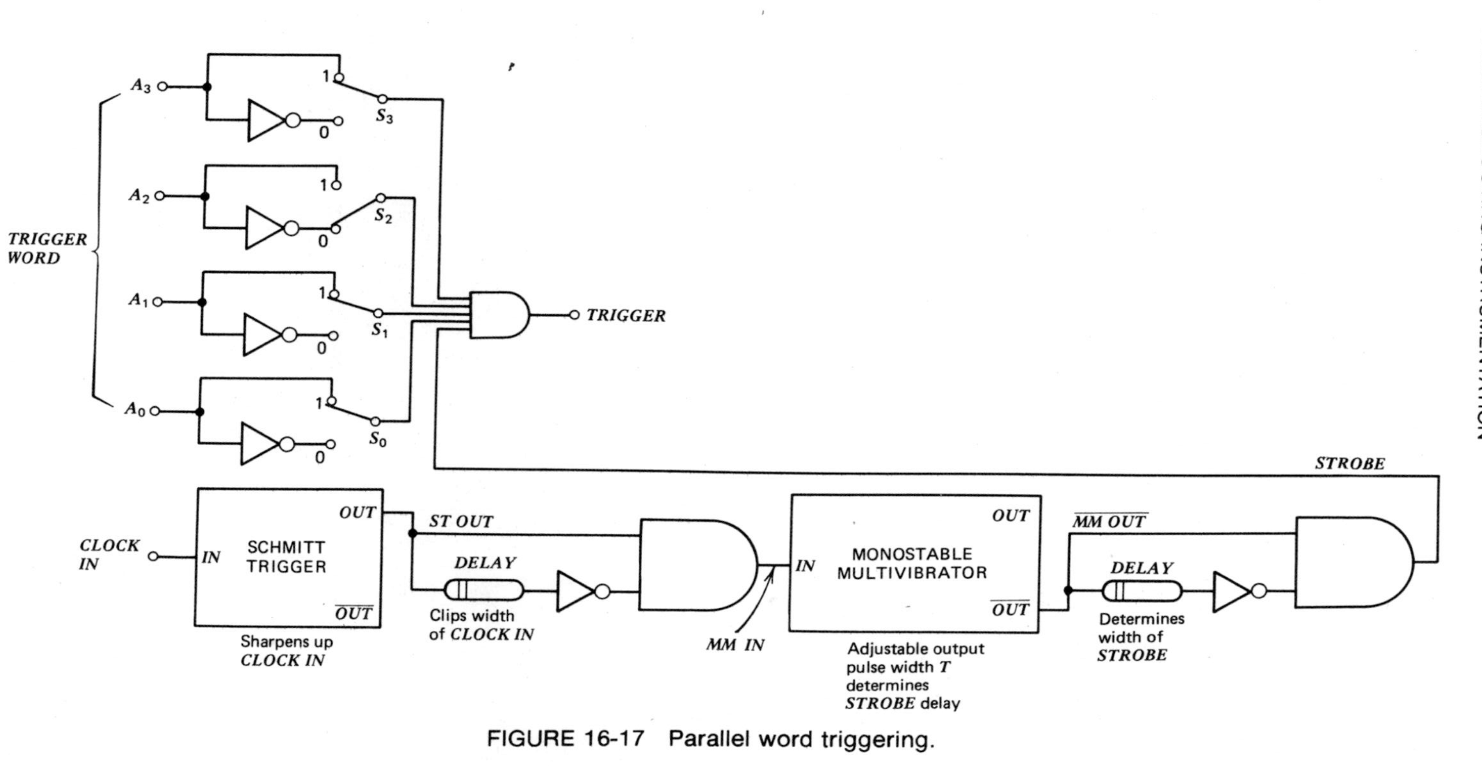

FIGURE 16-17 Parallel word triggering.

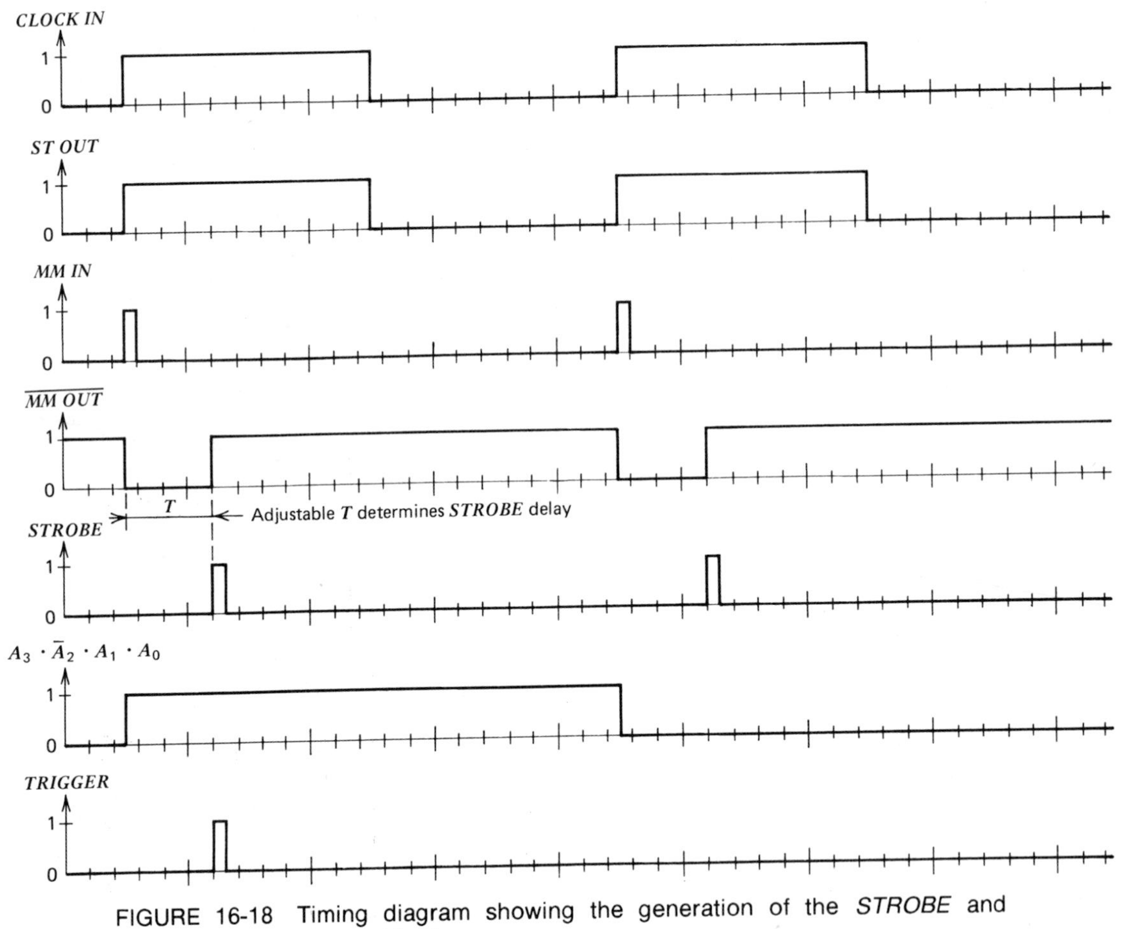

FIGURE 16-18 Timing diagram showing the generation of the *STROBE* and *TRIGGER* pulses in Fig. 16-17.

TRIGGER WORD stream agree with the logic values set by switches S_4, S_3, S_2, and S_1. This is accomplished by storing 4 subsequent *TRIGGER WORD* bits in the 4-stage shift register that is clocked by *STROBE*; the width of the *STROBE* pulse is chosen to exceed the maximum $t_{p_{CLOCK}}$ of the FFs. (The circuitry generating the *STROBE* is identical to that of Fig. 16-17.) The $\overline{TRIGGER}$ pulse is generated by ANDing outputs of the 4 shift register FFs and $\overline{STROBE}$, as shown in the timing diagram of Fig. 16-20.

16-7.2 Digital Delay

An oscilloscope can display events that occur *after* a trigger input signal. However, often we also want to examine signals that happened *before* a trigger. One way to do this is to delay the signal to be examined until after the trigger, by a delay line or by a cable. However, if we are interested only in digital signals

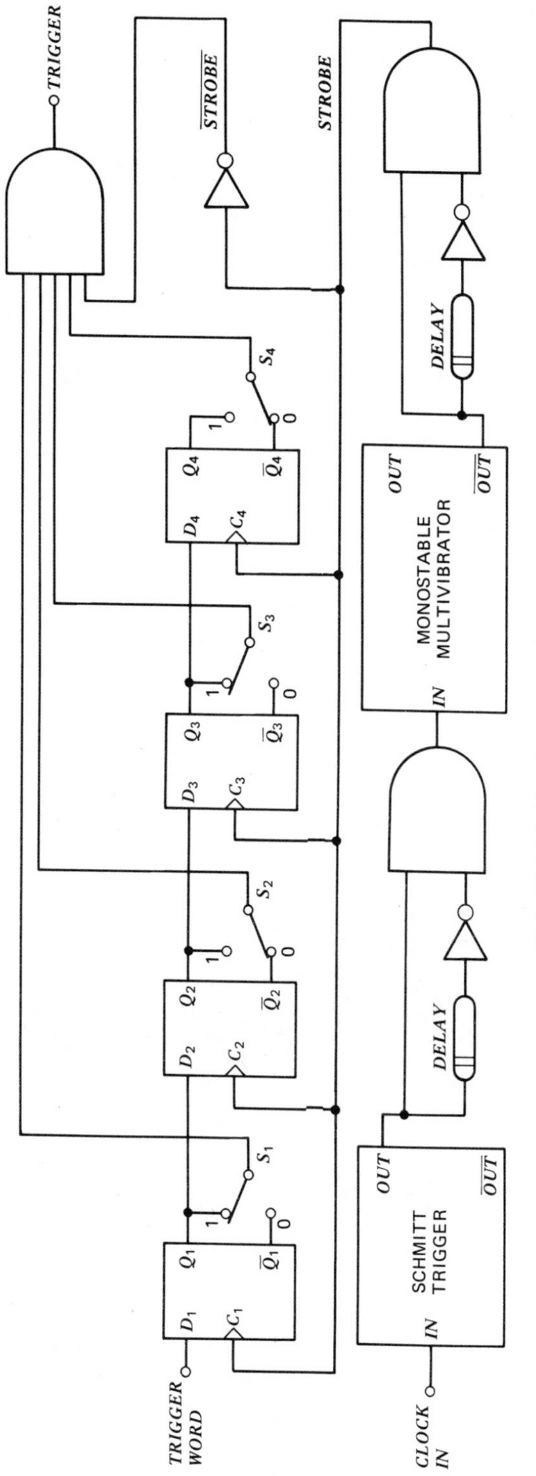

FIGURE 16-19 Serial word triggering.

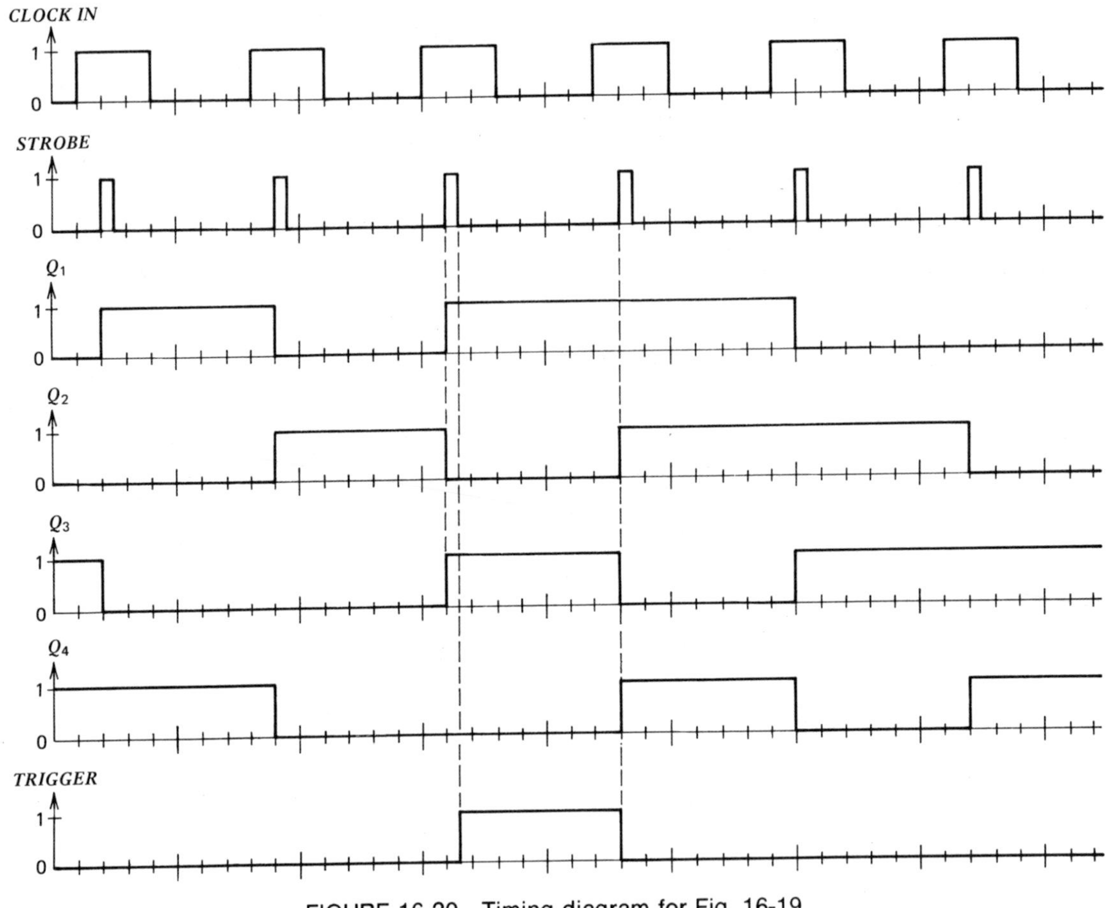

FIGURE 16-20 Timing diagram for Fig. 16-19.

in a clocked digital system, then a shift register may be used for delay. Such *digital delays* may be thousands of bits long, and may be part of an oscilloscope, may be an accessory, or may be a separate instrument.

16-7.3 Logic Analyzers

An instrument that is designed specifically for diagnosing digital systems is the *logic analyzer*. The block diagram of a logic analyzer is outlined in Fig. 16-21; it shows one of the many possible schemes.

Figure 16-21 includes 8 to 32 channels of DIGITAL DELAY (see Sec. 16-7.2) with a delay of 16 to 2048 bits per channel. The WORD TRIGGERING (see Sec. 16-7.1) may be either parallel as shown in Fig. 16-17, or serial as shown in Fig. 16-19. The clock of the digital system under test may be used via the *EXTERNAL CLOCK* input, or it may be substituted by the INTERNAL CLOCK SOURCE.

A wide variety of displays is available such as binary lights, BCD or hexade-

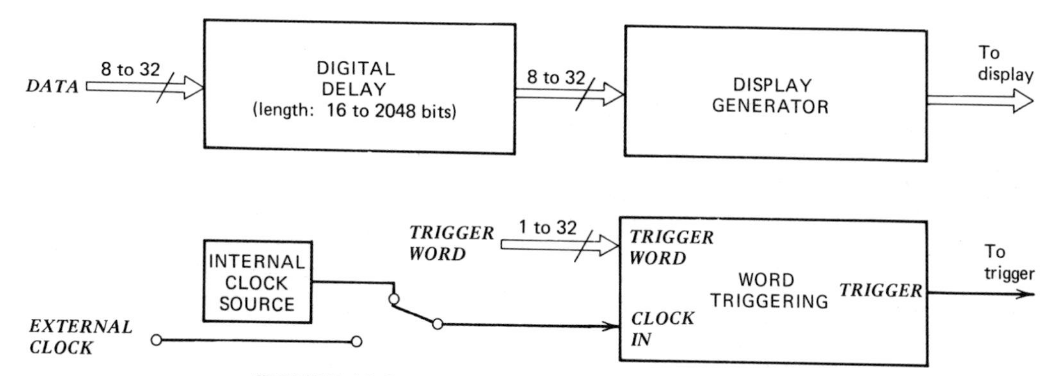

FIGURE 16-21 Basic structure of a logic analyzer.

cimal coded numerals and symbols, 0s and 1s, or square waves displayed on an oscilloscope screen.

16-7.4 Other Diagnostic Tools

Several other tools are in use for diagnosing digital circuits and systems. One of these is the *logic clip*, which displays on individual lights the logic states at all pins of an IC package. A *logic probe* has a single light, which displays the logic state at the tip of the probe, with the light flashing if the level complements even for a very short time. A *current tracer* is similar to a logic probe, but is activated by the magnetic field originating from a current near its tip. Another useful tool is the *logic pulser*, which can override low-impedance outputs of logic gates for a fraction of a microsecond—which is long enough to permit tracing the pulse in subsequent circuits.

SUMMARY

This chapter presented overall system design considerations and system components. After describing tolerances, noise margins, loading rules, and interfacing, it discussed power distribution and grounding. This was followed by a description of Schmitt triggers, monostable multivibrators, and troubleshooting instrumentation.

SELF-EVALUATION QUESTIONS

1. Describe noise margins.
2. What is a band diagram?
3. Discuss loading considerations in TTL circuits.
4. Describe an interface circuit between the output of an ECL circuit and the input of a CMOS circuit.

5. What factors enter into finding the required value of a bypass capacitor in a digital system?
6. What is a line driver?
7. What is a line receiver?
8. What is an optical isolator?
9. Describe a Schmitt trigger circuit.
10. How is the hysteresis of Schmitt trigger circuits utilized for noise rejection?
11. What is a monostable multivibrator?
12. Compare the properties of monostable multivibrators with dead time with the properties of retriggerable monostable multivibrators.
13. Describe word triggering.
14. What is a digital delay?
15. What is a logic analyzer?
16. What is a logic clip?
17. What is a logic probe?
18. What is a current tracer?
19. What is a logic pulser?

PROBLEMS

16-1. The loading specifications of low-power TTL logic inverters are as follows: The LOW output voltage is between 0 V and 0.4 V when the current into the output is between 0 and 3.6 mA. The HIGH output voltage is between 2.4 V and 5 V for a current of 0 to 0.2 mA flowing out of the output. The maximum input current is 10 μA flowing into the circuit when the input voltage is 2.4 V (HIGH state). At an input voltage of 0.4 V (LOW state) the maximum input current is less than −0.18 mA.

Using the above data and data from Exs. 16-1 and 16-2, find the maximum number of low-power TTL inverter inputs that are permitted to load the output of a TTL Type 7404 logic inverter. (Consider only one logic inverter out of the six included in a Type 7404 IC.)

16-2. Using the data given in Ex. 16-1, Ex. 16-2, and Prob. 16-1, find the maximum number of TTL Type 7404 logic inverter inputs that are permitted to load the output of a low-power TTL logic inverter. (Consider only one logic inverter out of the six included in a Type 7404 IC.)

16-3. Using the data given in Prob. 16-1, find the maximum number of low-power TTL inverter inputs that are permitted to load a low-power TTL inverter output.

16-4. Sketch a band diagram for the noise margins of ECL circuits. Use the following data: The minimum LOW output voltage is -2 V, the maximum LOW output voltage is -1.6 V, the minimum HIGH output voltage is -0.8 V, the maximum HIGH output voltage is -0.6 V, the maximum permitted LOW input voltage is -1.5 V, and the minimum permitted HIGH input voltage is -0.9 V.

16-5. A digital system is separated into two parts that are interconnected by a 3-inch long ground strap. The inductance of the ground strap material is specified as 0.1 μH/ft. Find the voltage drop V_L along the ground strap if the current flowing in it changes by $I_\Delta = 20$ mA in $t = 2$ ns.

16-6. A digital system is separated into two parts that are interconnected by a 3-inch long ground strap. The inductance of the ground strap material is specified as 0.05 μH/ft. Find the voltage drop V_L along the ground strap if the current flowing in it changes by $I_\Delta = 40$ mA in $t = 2$ ns.

16-7. Sketch a figure similar to Fig. 16-3, but with a noise voltage V_{NOISE} that is a 50-Hz sine wave with an amplitude of 20 V. Use a V_{IN} that is LOW for $t < 10$ ms and $t > 30$ ms and is HIGH for 10 ms $< t <$ 30 ms.

16-8. Sketch a figure similar to Fig. 16-3, but with a noise voltage V_{NOISE} that is an 800-Hz sine wave with an amplitude of 10 V.

16-9. Sketch $\bar{V}_{OUT}$, V_{OUT}, and V_p in the Schmitt trigger circuit of Fig. 16-9 with V_{IN} as shown in Fig. P16-1.

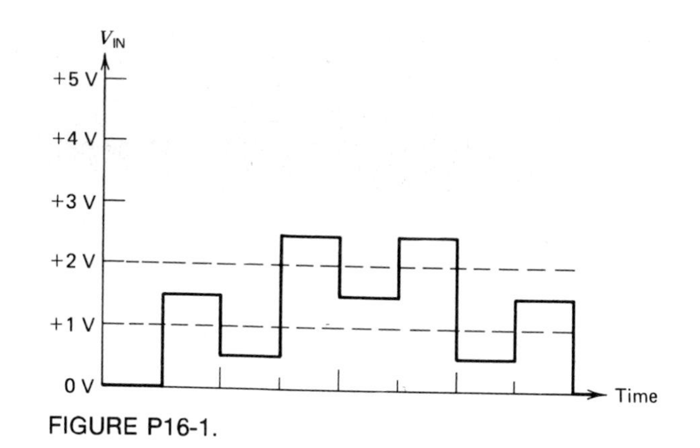

FIGURE P16-1.

16-10. Sketch $\bar{V}_{OUT}$, V_{OUT}, and V_p in the Schmitt trigger circuit of Fig. 16-9 with V_{IN} as shown in Fig. P16-2.

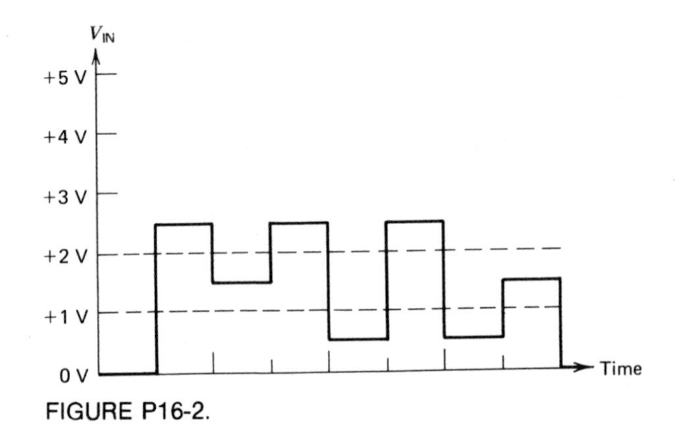

FIGURE P16-2.

16-11. The TTL Type 74121 monostable multivibrator described in Ex. 16-12 is operated with $R_T = 10$ kΩ and $C_T = 10$ μF.

a. Find the value of T.

b. What is the accuracy of T?

16-12. The TTL Type 74121 monostable multivibrator described in Ex. 16-12 is operated with $C_T = 20$ μF. Find the value of R_T that results in $T = 50$ ms.

16-13. When operated with $R_T = 2$ kΩ, the TTL Type 74121 monostable multivibrator has a maximum duty cycle of $h_{max} = 67\%$. Find the dead time D if $T = 0.5$ s.

16-14. When operated with $T = 0.75$ s, a monostable multivibrator has a dead time of $D = 0.8$ s. Find the maximum duty cycle, h_{max}.

16-15. The monostable multivibrator of Prob. 16-13 is operated with the input waveform shown in the upper part of Fig. 16-16. Sketch the resulting output waveform.

16-16. The monostable multivibrator of Prob. 16-14 is operated with the input waveform shown in the upper part of Fig. 16-16. Sketch the resulting waveform.

16-17. Show in a sketch how the *TRIGGER* waveform in the timing diagram of Fig. 16-20 changes if the switches in Fig. 16-19 are set as $S_1S_2S_3S_4 = 1101$.

16-18. Show in a sketch how the *TRIGGER* waveform in the timing diagram of Fig. 16-20 changes if the switches in Fig. 16-19 are set as $S_1S_2S_3S_4 = 0110$.

16-19. List the four principal parts of a logic analyzer.

16-20. What diagnostic tool would you use to indicate the steady state of a digital signal?

16-21. What diagnostic tool would you use to indicate the current flowing between two points in a digital circuit?

16-22. What diagnostic tool would you use for generating pulses to be entered at an internal point of a digital circuit?

Answers to Odd-Numbered Problems

CHAPTER 1

1-1. LSI.

1-3. $10^9 = 1$ billion.

CHAPTER 2

2-1.

Switch *A*	Switch *B*	Switch *C*	Switch *D*	Lamp
OPEN	OPEN	OPEN	OPEN	OFF
OPEN	OPEN	OPEN	CLOSED	OFF
OPEN	OPEN	CLOSED	OPEN	OFF
OPEN	OPEN	CLOSED	CLOSED	OFF
OPEN	CLOSED	OPEN	OPEN	OFF
OPEN	CLOSED	OPEN	CLOSED	OFF
OPEN	CLOSED	CLOSED	OPEN	OFF
OPEN	CLOSED	CLOSED	CLOSED	OFF
CLOSED	OPEN	OPEN	OPEN	OFF
CLOSED	OPEN	OPEN	CLOSED	OFF
CLOSED	OPEN	CLOSED	OPEN	OFF
CLOSED	OPEN	CLOSED	CLOSED	OFF
CLOSED	CLOSED	OPEN	OPEN	OFF
CLOSED	CLOSED	OPEN	CLOSED	OFF
CLOSED	CLOSED	CLOSED	OPEN	OFF
CLOSED	CLOSED	CLOSED	CLOSED	ON

***2-3.**

Switch S_A	Switch S_B	Relay *A*	Relay *B*	Relay *C*
OPEN	OPEN	OFF	OFF	OFF
OPEN	CLOSED	OFF	ON	ON
CLOSED	OPEN	ON	OFF	ON
CLOSED	CLOSED	ON	ON	OFF

Relay *A*	Relay *B*	Relay *C*
OFF	OFF	OFF
OFF	ON	ON
ON	OFF	ON
ON	ON	OFF

*Optional problem.

2-5.

IN_1	IN_2	IN_3	IN_4	OUT
0 V	0 V	0 V	0 V	+0.7 V
0 V	0 V	0 V	+5 V	+0.7 V
0 V	0 V	+5 V	0 V	+0.7 V
0 V	0 V	+5 V	+5 V	+0.7 V
0 V	+5 V	0 V	0 V	+0.7 V
0 V	+5 V	0 V	+5 V	+0.7 V
0 V	+5 V	+5 V	0 V	+0.7 V
0 V	+5 V	+5 V	+5 V	+0.7 V
+5 V	0 V	0 V	0 V	+0.7 V
+5 V	0 V	0 V	+5 V	+0.7 V
+5 V	0 V	+5 V	0 V	+0.7 V
+5 V	0 V	+5 V	+5 V	+0.7 V
+5 V	+5 V	0 V	0 V	+0.7 V
+5 V	+5 V	0 V	+5 V	+0.7 V
+5 V	+5 V	+5 V	0 V	+0.7 V
+5 V	+5 V	+5 V	+5 V	+5 V

2-7. 1.25 V.

2-9.

IN_1	IN_2	IN_3	OUT
0 V	0 V	0 V	+5 V
0 V	0 V	+5 V	+5 V
0 V	+5 V	0 V	+5 V
0 V	+5 V	+5 V	+5 V
+5 V	0 V	0 V	+5 V
+5 V	0 V	+5 V	+5 V
+5 V	+5 V	0 V	+5 V
+5 V	+5 V	+5 V	0 V

2-11. 0 V.

CHAPTER 3

3-1. $4 \times 10^3 + 0 \times 10^2 + 2 \times 10^1 + 6 \times 10^0$, $5 \times 10^{-1} + 2 \times 10^{-2} + 1 \times 10^{-3}$, $-(1 \times 10^1 + 2 \times 10^0)$.

3-3. $1 \times 2^3 + 0 \times 2^2 + 0 \times 2^1 + 1 \times 2^0$, $1 \times 2^{-1} + 0 \times 2^{-2} + 1 \times 2^{-3} + 1 \times 2^{-4}$, $1 \times 2^2 + 0 \times 2^1 + 1 \times 2^0 + 0 \times 2^{-1} + 1 \times 2^{-2}$.

3-5. 5, 13, 29.

3-7. 0.5, 0.25, 0.75.

3-9. 2.5, 1.25, 3.75, 5.125.

3-11. 10101, 11000, 11111, 1111110.

3-13. 101.1, 1001.01, 1101.11, 10011.001.

3-15. **(a)** 110001_2, **(b)** 10_2, **(c)** 10101_2, ***(d)** 100_2.

3-17. 159_{10}.

3-19. 0, 1, 2, 3, 4, 5, 6, 7, 8, 9, A, B, C, D, E, F, 10, 11, 12, 13, 14, 15, 16, 17, 18, 19, 1A, 1B, 1C, 1D, 1E, 1F ($= 31_{10}$).

***3-21.** **(a)** 84.06_{10}, **(b)** 649_{10}.

3-23. 010.01, 0011, 1.011, 00.11.

3-25. Shift the binary point right by 1, 2, 3, or 4 positions, respectively. Add trailing zeros where required.

3-27. Shift the octal point right by 1, 2, or 3 positions, respectively. Add trailing zeros where required.

3-29. 10000, 00110.

CHAPTER 4

4-1. **(a)** 1001 0101 0111; **(b)** 0011 0100 0111 0001; **(c)** 1000 1001 0010.

4-3. **(a)** 58; **(b)** 935; **(c)** 3471.

4-5. **(a)** 0111; **(b)** 0001 0101; **(c)** 0111 0011.

4-7. **(a)** 36; **(b)** 729; **(c)** 653.

4-9. **(a)** 91; **(b)** 472; **(c)** 508.

4-11. **(a)** 1100 1010; **(b)** 1011 1000 0101; **(c)** 1001 0111 0110.

4-13. 1.

4-15. 10001001.

4-17. 1, 0, 1, 1, 0, 0, 1, 1, 0, 1.

***4-19.** COMPUTER.

4-21. end.

CHAPTER 5

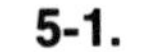

5-1.

A	B	OUT
0	0	0
0	1	1
1	0	1
1	1	0

5-3.

A	B	C	AB	$\bar{B}$	$AB(\bar{B} + C)$
0	0	0	0	1	0
0	0	1	0	1	0
0	1	0	0	0	0
0	1	1	0	0	0
1	0	0	0	1	0
1	0	1	0	1	0
1	1	0	1	0	0
1	1	1	1	0	1

*Optional problem.

5-5. (a)

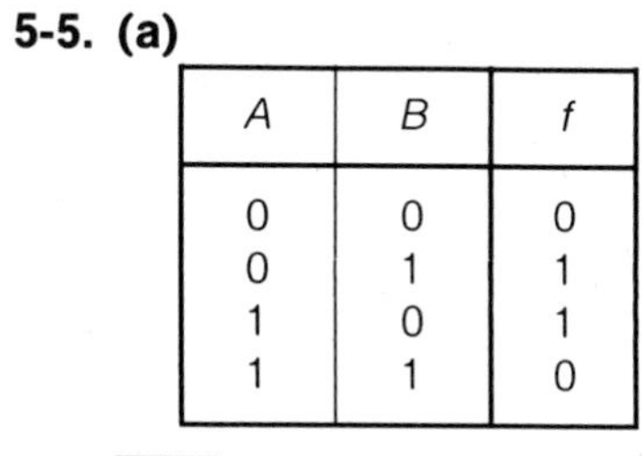

A	B	f
0	0	0
0	1	1
1	0	1
1	1	0

(b)

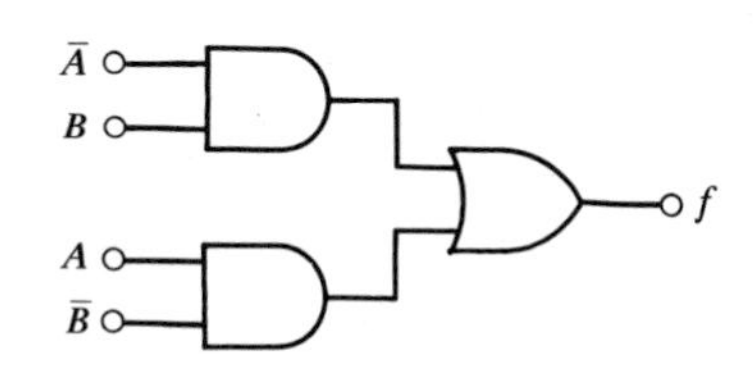

5-7. $\overline{A\bar{B}C} + B\bar{C} = \bar{A} + B + \bar{C} + B\bar{C}.$

5-9.

A	B	AB	$\overline{AB}$	$\bar{A}$	$\bar{B}$	$\bar{A}+\bar{B}$
0	0	0	1	1	1	1
0	1	0	1	1	0	1
1	0	0	1	0	1	1
1	1	1	0	0	0	0

↑ Identity ↑ ($\overline{AB}$ and $\bar{A}+\bar{B}$)

5-11. $\overline{(A + \bar{B})\overline{(\bar{C} + D)} + E\bar{F}} = (\bar{A}B + \bar{C} + D)(\bar{E} + F).$

5-13. $\bar{A}B\bar{C}D + ABC\bar{D} + \bar{A}\bar{B}.$

5-15. $AB\bar{C} + ABC.$

5-17.

X	Y	Z	XYZ	$\overline{XYZ}$	$\bar{X}$	$\bar{Y}$	$\bar{Z}$	$\bar{X}+\bar{Y}+\bar{Z}$
0	0	0	0	1	1	1	1	1
0	0	1	0	1	1	1	0	1
0	1	0	0	1	1	0	1	1
0	1	1	0	1	1	0	0	1
1	0	0	0	1	0	1	1	1
1	0	1	0	1	0	1	0	1
1	1	0	0	1	0	0	1	1
1	1	1	1	0	0	0	0	0

↑ identity ↑ ($\overline{XYZ}$ and $\bar{X}+\bar{Y}+\bar{Z}$)

5-19. (a) $f = BC + AC + AB.$

5-21.

A	B	$\bar{A}$	$\bar{A}B$	$A+\bar{A}B$	$A+B$
0	0	1	0	0	0
0	1	1	1	1	1
1	0	0	0	1	1
1	1	0	0	1	1

↑ Identity ↑ ($A+\bar{A}B$ and $A+B$)

5-23. $\overline{AB + \overline{A}\overline{B}} = (\overline{A} + \overline{B})(A + B) = \overline{A}B + A\overline{B}$.

5-25. $B\overline{C} + AC$.

5-27. $\overline{C}D + AB + B\overline{D}$.

5-29. $X + Y$.

CHAPTER 6

6-1. The wired-OR eliminates one 4-input AND gate at the output.

6-3.

		Positive Logic			Negative Logic		
A	B	A	B	EXCLUSIVE-OR	A	B	EXCLUSIVE-NOR
L	L	0	0	0	1	1	1
L	H	0	1	1	1	0	0
H	L	1	0	1	0	1	0
H	H	1	1	0	0	0	1

6-5. $f = A\overline{C} + B\overline{C} + AB$.

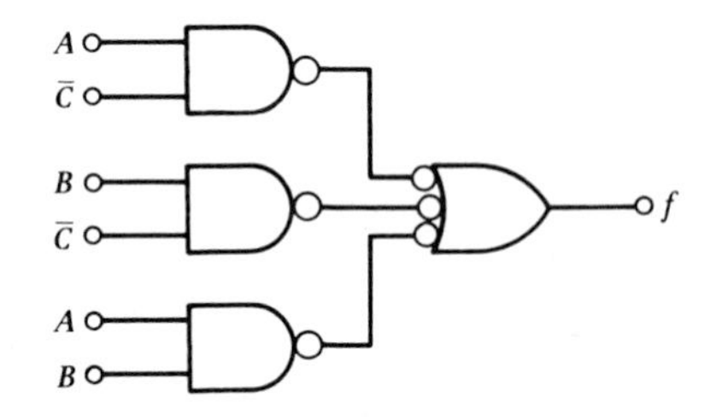

***6-7.** **(a)** 0; **(b)** 1.

6-9. $CD + \overline{A}D + \overline{B}D$.

6-11.

A	B	OUT
1	0	1
1	1	0

6-13. EXCLUSIVE-OR.

6-15. $\overline{C}(A \oplus B)$.

*Optional problem.

6-17.

A	B	C	f
0	0	0	0
0	0	1	1 $\}\ I_0 = C$
0	1	0	0
0	1	1	1 $\}\ I_1 = C$
1	0	0	0
1	0	1	1 $\}\ I_2 = C$
1	1	0	1
1	1	1	0 $\}\ I_3 = \overline{C}$

6-19. $\overline{21}$.

***6-21.** EXCLUSIVE-OR.

CHAPTER 7

7-1. 0.725 mA, out of the collector of Q_1.

7-3. 0.725 mA, out of the collector of Q_1.

***7-5.** 3.7 V.

7-7. 18.5 ns.

***7-9.** 1.4 V, 0 V, +0.7 V, +0.7 V.

7-11. 51.

7-13. 0, 0, 3.6 mA.

7-15.

IN_1	IN_2	IN_3	IN_4	OUT_A	OUT_B
−1.6 V	−1.6 V	−1.6 V	−1.6 V	−0.8 V	−1.6 V
−1.6 V	−1.6 V	−1.6 V	−0.8 V	−0.8 V	−0.8 V
−1.6 V	−1.6 V	−0.8 V	−1.6 V	−0.8 V	−0.8 V
−1.6 V	−1.6 V	−0.8 V	−0.8 V	−0.8 V	−0.8 V
−1.6 V	−0.8 V	−1.6 V	−1.6 V	−0.8 V	−0.8 V
−1.6 V	−0.8 V	−1.6 V	−0.8 V	−1.6 V	−0.8 V
−1.6 V	−0.8 V	−0.8 V	−1.6 V	−1.6 V	−0.8 V
−1.6 V	−0.8 V	−0.8 V	−0.8 V	−1.6 V	−0.8 V
−0.8 V	−1.6 V	−1.6 V	−1.6 V	−0.8 V	−0.8 V
−0.8 V	−1.6 V	−1.6 V	−0.8 V	−1.6 V	−0.8 V
−0.8 V	−1.6 V	−0.8 V	−1.6 V	−1.6 V	−0.8 V
−0.8 V	−1.6 V	−0.8 V	−0.8 V	−1.6 V	−0.8 V
−0.8 V	−0.8 V	−1.6 V	−1.6 V	−0.8 V	−0.8 V
−0.8 V	−0.8 V	−1.6 V	−0.8 V	−1.6 V	−0.8 V
−0.8 V	−0.8 V	−0.8 V	−1.6 V	−1.6 V	−0.8 V
−0.8 V	−0.8 V	−0.8 V	−0.8 V	−1.6 V	−0.8 V

*Optional problem.

***7-17.** 0.625 mA.

***7-19.** S_1.

7-21. 70 ns.

7-23. 11.1 ns.

7-25. 10.05 mW.

CHAPTER 8

8-1.

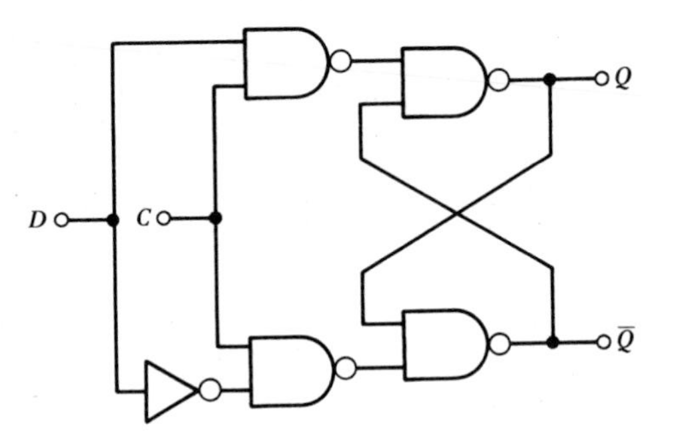

8-3.

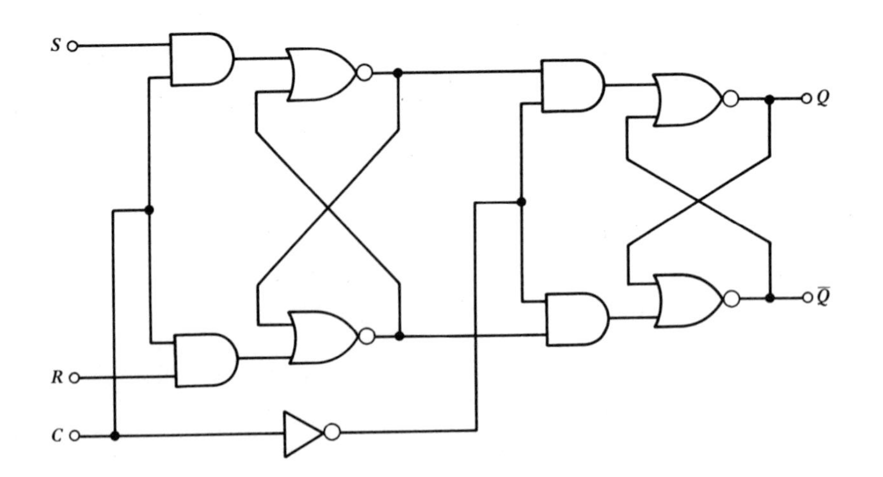

*Optional problem.

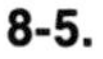

8-5.

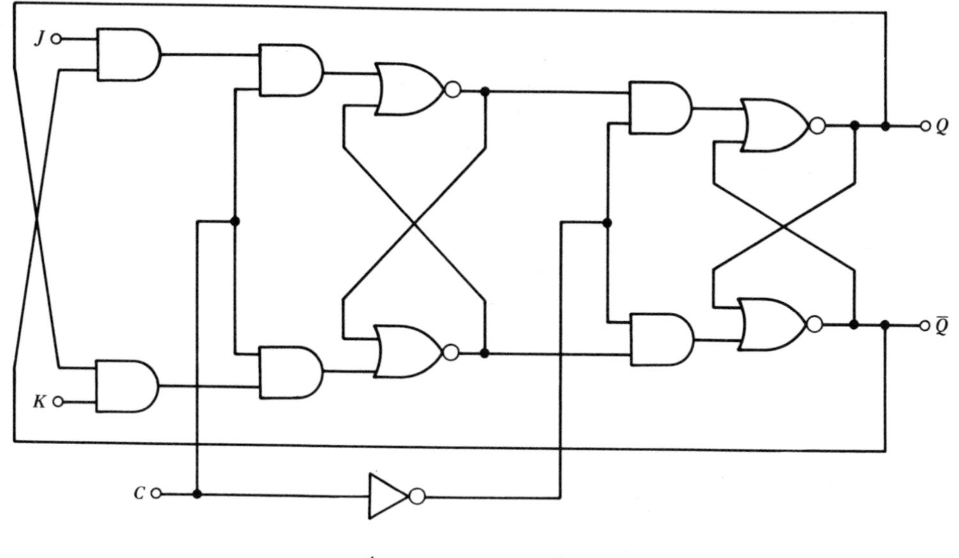

8-7. $f_{\max} = \dfrac{1}{t_{1_{\min}} + t_{0_{\min}}} = \dfrac{1}{20\ \text{ns} + 20\ \text{ns}} = 25\ \text{MHz}.$

8-9. The interval is 16 ns long and it starts 10 ns before the positive-going edge of the clock.

8-11. The transition described in the first row of Fig. P8-1 corresponds to the transition in the first row of Fig. 8-16*b*, the transition described in the second row of Fig. P8-1 to the transition in the second row of Fig. 8-16*b*, the transition described in the third row of Fig. P8-1 to the transition in the third row of Fig. 8-16*b*, and the transition described in the fourth row of Fig. P8-1 to the transition in the fourth row of Fig. 8-16*b*.

8-13. The transition described in the first row of Fig. P8-3 is given in the first *and* second rows of Fig. 8-33*c*, the transition described in the second row of Fig. P8-3 in the third *and* fourth rows of Fig. 8-33*c*, the transition described in the third row of Fig. P8-3 in the sixth *and* eighth rows of Fig. 8-33*c*, and the transition described in the fourth row of Fig. P8-3 in the fifth *and* seventh rows of Fig. 8-33*c*.

8-15. The transition described in the first row of Fig. P8-4 is given in the first *and* second rows of Fig. 8-15*a*, the transition described in the second row of Fig. P8-4 in the third row of Fig. 8-15*a*, the transition described in the third row of Fig. P8-4 in the sixth row of Fig. 8-15*a*, and the transition described in the fourth row of Fig. P8-4 in the fifth *and* seventh rows of Fig. 8-15*a*. Note that the fourth and eighth rows of Fig. 8-15*a* have question mark entries for Q_{n+1} and have no corresponding rows in Fig. P8-4.

CHAPTER 9

9-1. See Fig. 9-5.

9-3.

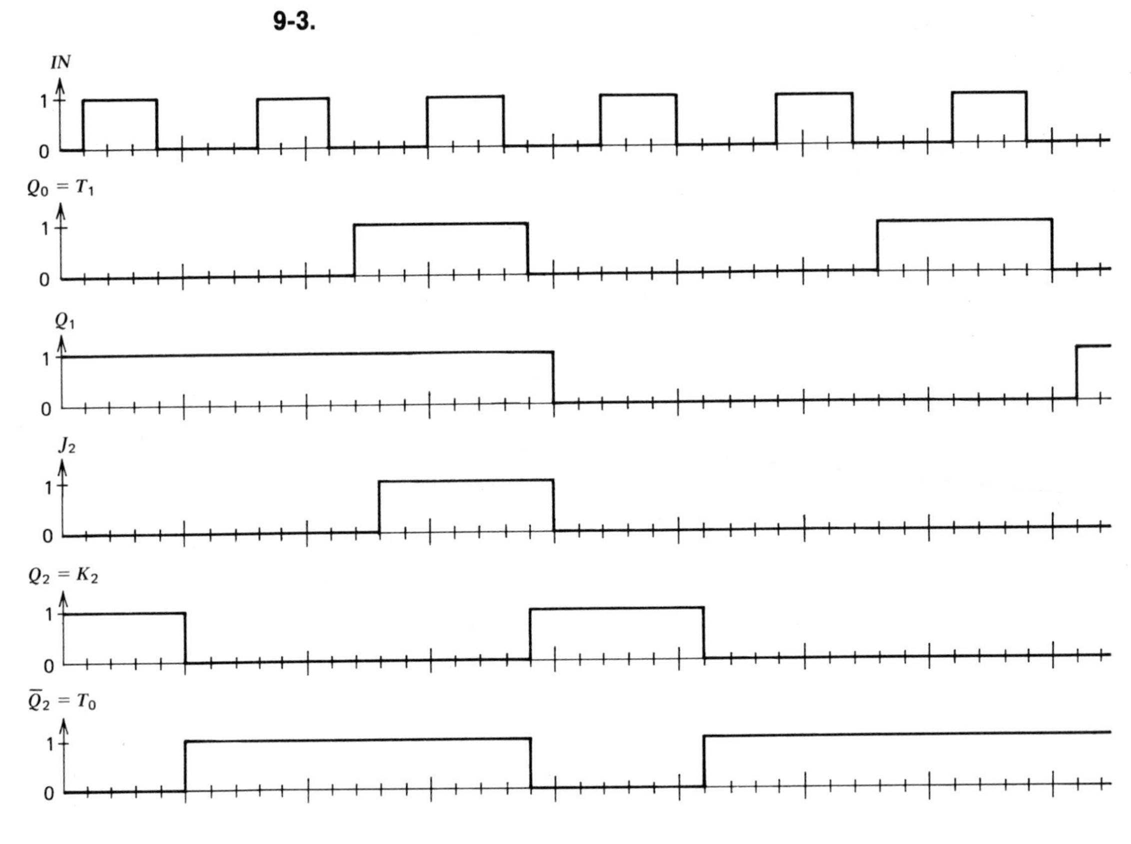

9-5.

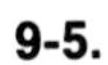

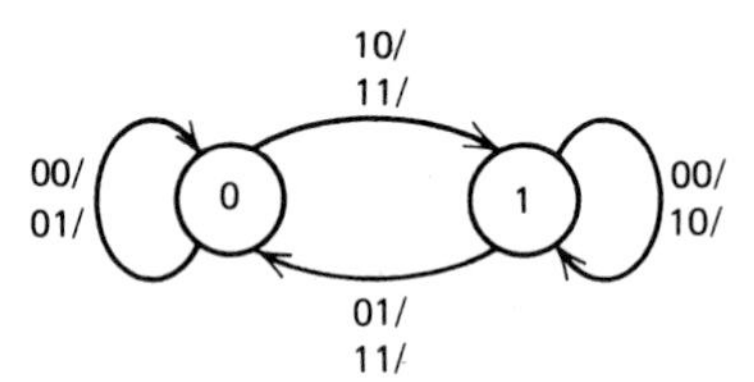

9-7.

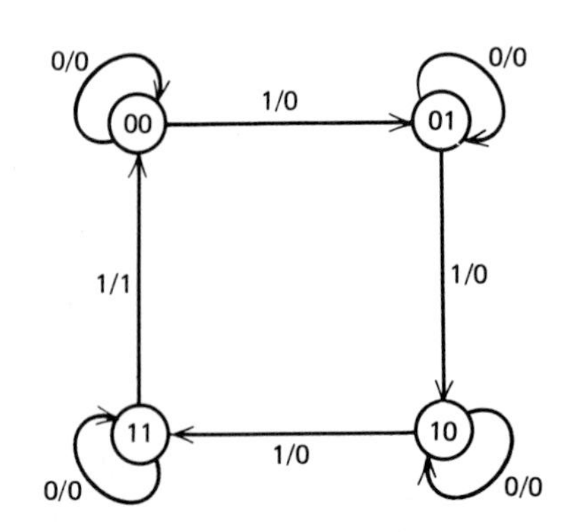

9-9.

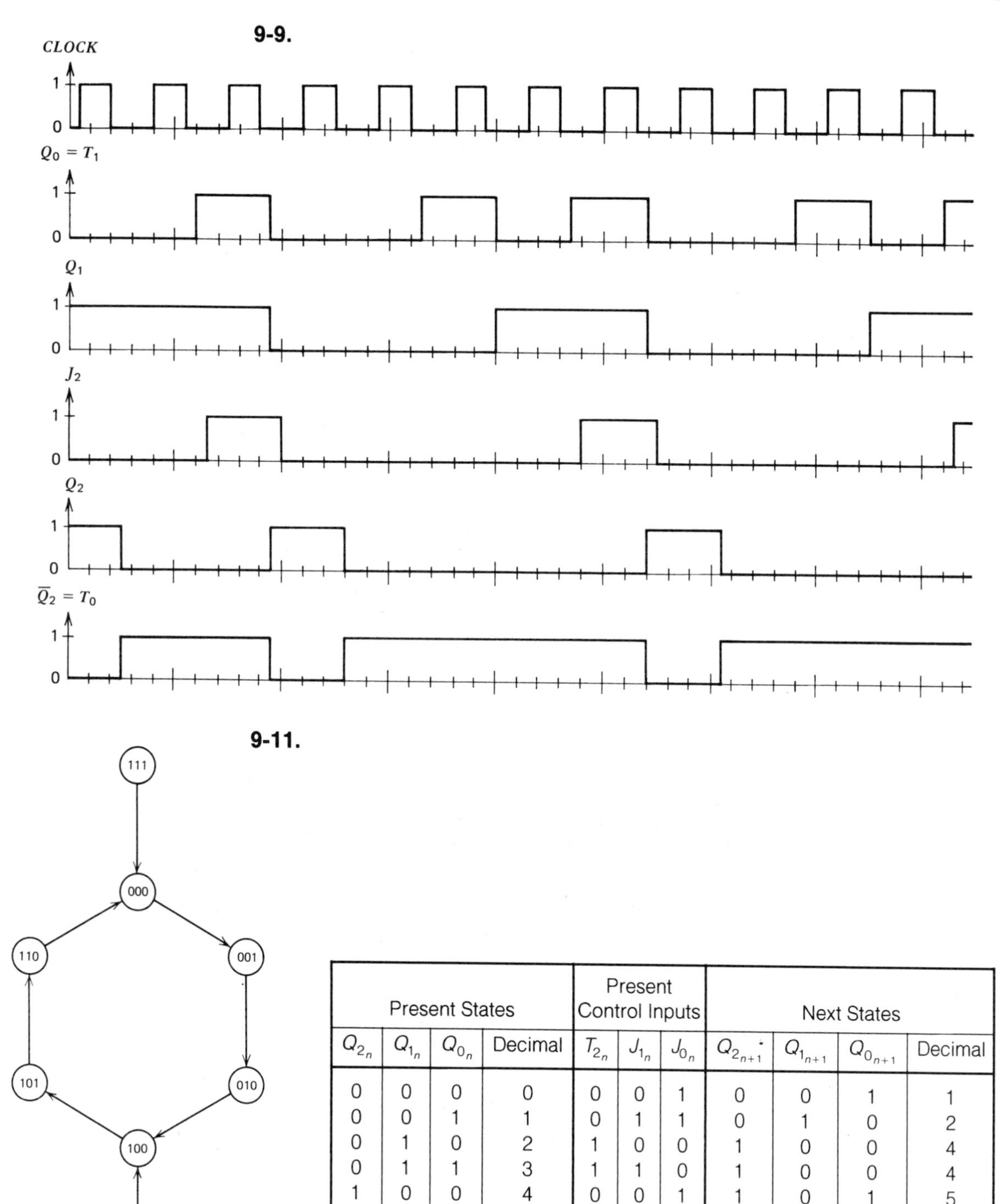

9-11.

Present States				Present Control Inputs			Next States			
Q_{2_n}	Q_{1_n}	Q_{0_n}	Decimal	T_{2_n}	J_{1_n}	J_{0_n}	$Q_{2_{n+1}}$	$Q_{1_{n+1}}$	$Q_{0_{n+1}}$	Decimal
0	0	0	0	0	0	1	0	0	1	1
0	0	1	1	0	1	1	0	1	0	2
0	1	0	2	1	0	0	1	0	0	4
0	1	1	3	1	1	0	1	0	0	4
1	0	0	4	0	0	1	1	0	1	5
1	0	1	5	0	1	1	1	1	0	6
1	1	0	6	1	0	0	0	0	0	0
1	1	1	7	1	1	0	0	0	0	0

9-13.

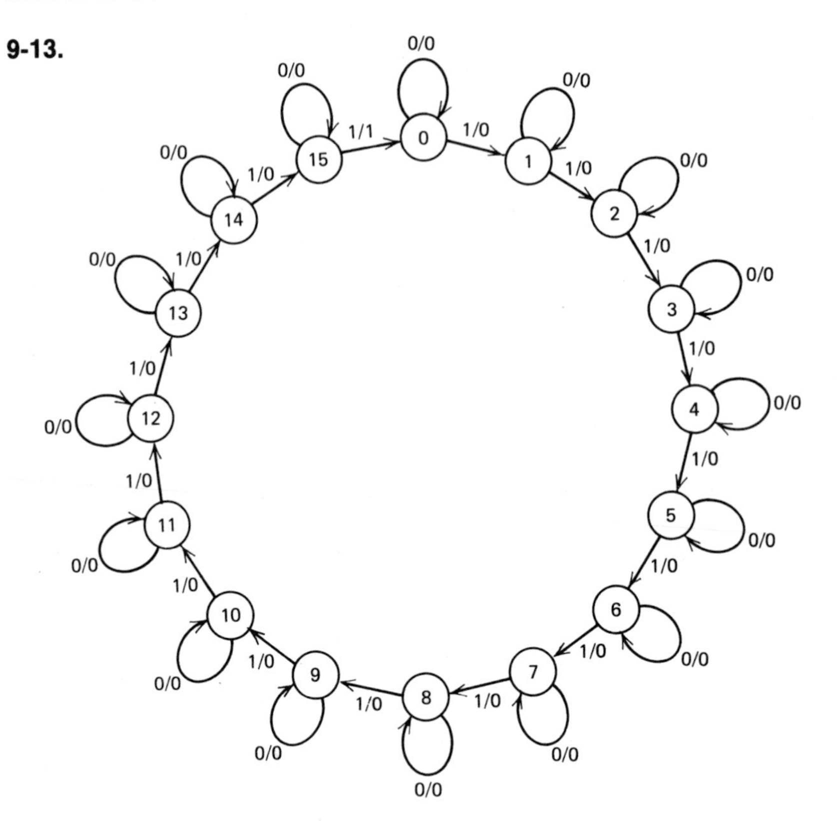
0/0
1/0
1/1
0
1
2
3
4
5
6
7
8
9
10
11
12
13
14
15

9-15.

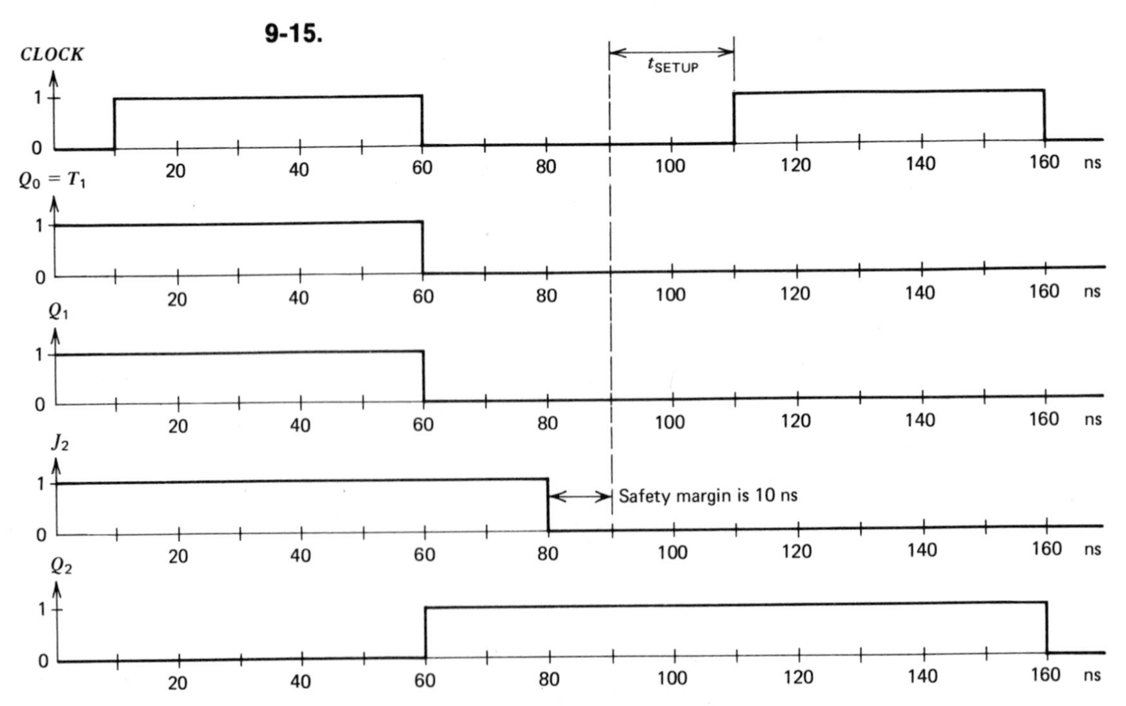
CLOCK
t_{SETUP}
$Q_0 = T_1$
Q_1
J_2
Safety margin is 10 ns
Q_2
1
0
20
40
60
80
100
120
140
160
ns

9-17. A $T = 1$ FF is characterized by $Q_{n+1} = \overline{Q}_n$. These transitions appear in the second and third rows of Fig. P8-1. In the second row $Q_n = 0$ and $D_n = 1$ are required to effect the desired transition. In the third row $Q_n = 1$ and $D_n = 0$ are required to effect the desired transition. Thus we require $D_n = 1$ when $Q_n = 0$ and $D_n = 0$ when $Q_n = 1$, that is, we require $D_n = \overline{Q}_n$—as is the case in Fig. 9-2*c* and Fig. 9-2*d*.

9-19. We add an additional column to the state table of Fig. 9-20*b* as shown below. The added column shows the required values of T_n in accordance with Fig. P8-2. The column also shows that we require $T_n = CARRY\text{-}IN_nQ_n$—as is the case in Fig. 9-20*a*. Also, the state table of Fig. 9-20*b* specifies $CARRY\text{-}OUT_n = CARRY\text{-}IN_nQ_n$—which is the case in Fig. 9-20*a*.

$CARRY\text{-}IN_n$	Q_n	$CARRY\text{-}OUT_n$	Q_{n+1}	T_n
0	0	0	0	0
0	1	0	1	0
1	0	0	1	0
1	1	1	0	1

CHAPTER 10

10-1.

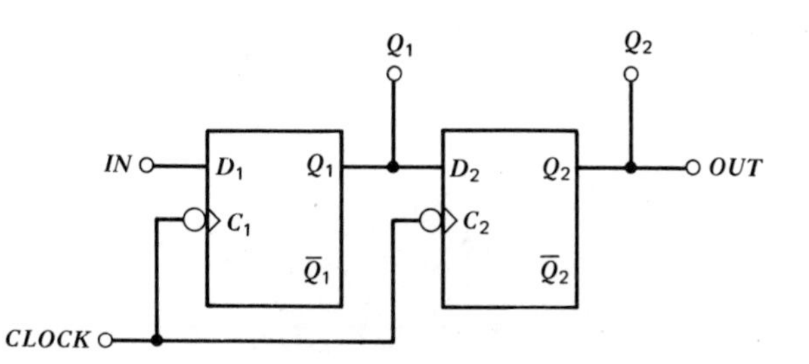

Present States		Present Inputs	Present Control Inputs		Next States	
Q_{1_n}	Q_{2_n}	IN_n	D_{1_n}	D_{2_n}	$Q_{1_{n+1}}$	$Q_{2_{n+1}}$
0	0	0	0	0	0	0
0	0	1	1	0	1	0
0	1	0	0	0	0	0
0	1	1	1	0	1	0
1	0	0	0	1	0	1
1	0	1	1	1	1	1
1	1	0	0	1	0	1
1	1	1	1	1	1	1

10-3.

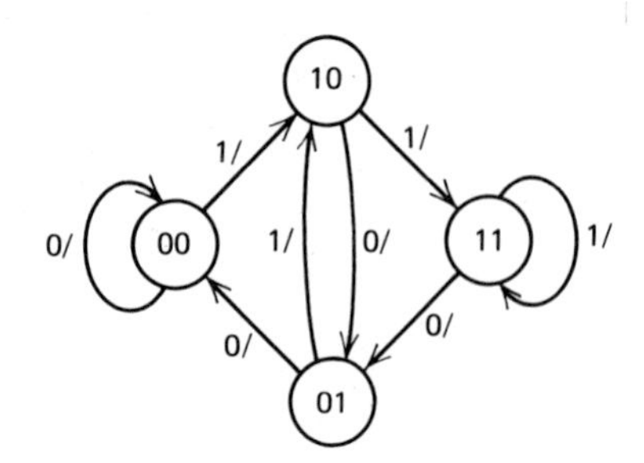

10-5. (a)

Clock Pulse	1	2	3	4	5
IN	1	0	0	0	0
D_1	1	0	0	0	0
Q_1	0	1	0	0	0
D_2	0	1	0	0	0
Q_2	0	0	1	0	0
D_3	0	0	1	0	0
Q_3	0	0	0	1	0

(b)

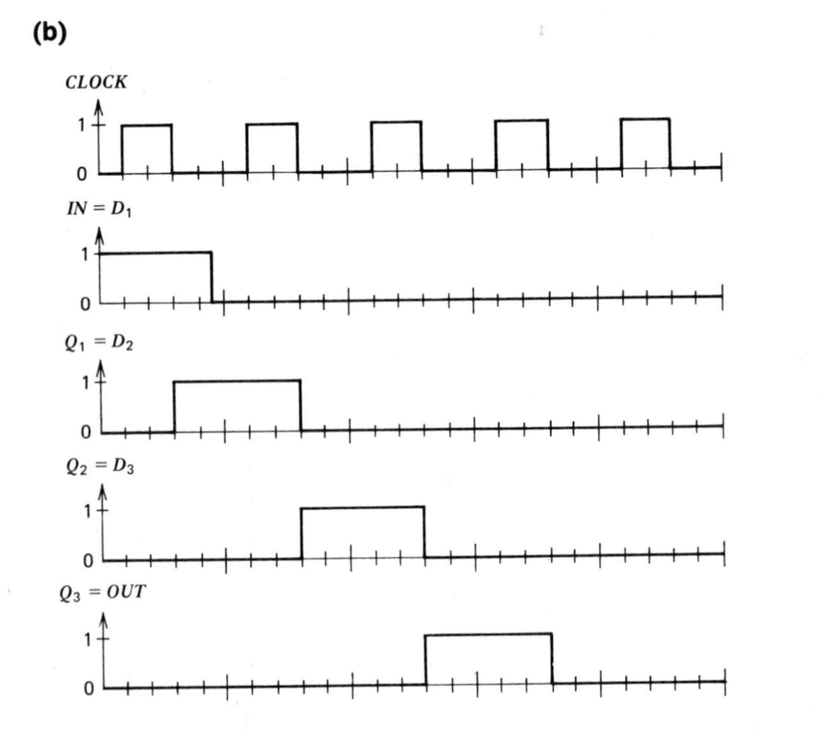

10-7.

Present States				Present Control Inputs				Next States			
Q_{1_n}	Q_{2_n}	Q_{3_n}	Q_{4_n}	D_{1_n}	D_{2_n}	D_{3_n}	D_{4_n}	$Q_{1_{n+1}}$	$Q_{2_{n+1}}$	$Q_{3_{n+1}}$	$Q_{4_{n+1}}$
0	0	0	0	1	0	0	0	1	0	0	0
0	0	0	1	1	0	0	0	1	0	0	0
0	0	1	0	0	0	0	1	0	0	0	1
0	0	1	1	0	0	0	1	0	0	0	1
0	1	0	0	0	0	1	0	0	0	1	0
0	1	0	1	0	0	1	0	0	0	1	0
0	1	1	0	0	0	1	1	0	0	1	1
0	1	1	1	0	0	1	1	0	0	1	1
1	0	0	0	0	1	0	0	0	1	0	0
1	0	0	1	0	1	0	0	0	1	0	0
1	0	1	0	0	1	0	1	0	1	0	1
1	0	1	1	0	1	0	1	0	1	0	1
1	1	0	0	0	1	1	0	0	1	1	0
1	1	0	1	0	1	1	0	0	1	1	0
1	1	1	0	0	1	1	1	0	1	1	1
1	1	1	1	0	1	1	1	0	1	1	1

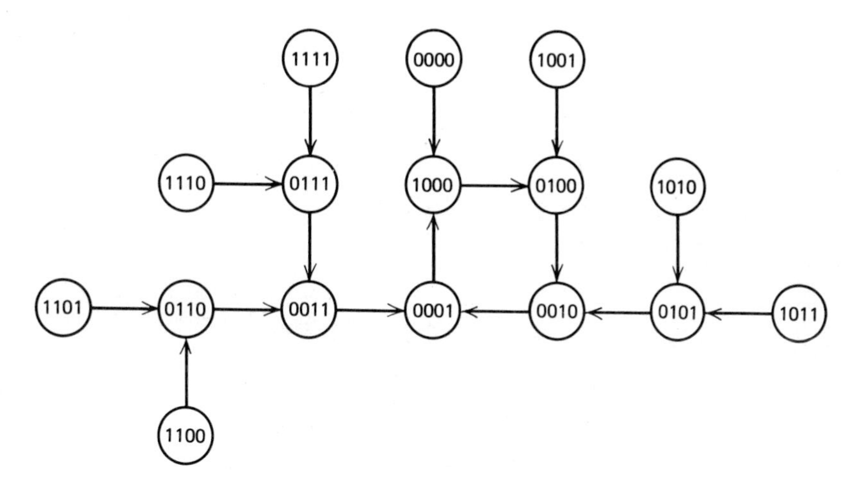

10-9.

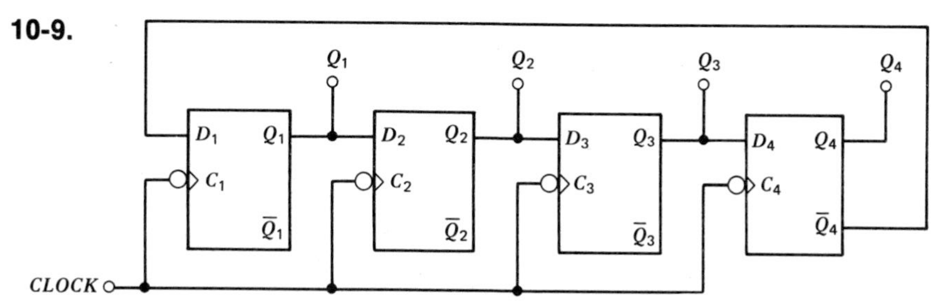

Present States				Present Control Inputs				Next States			
Q_{1_n}	Q_{2_n}	Q_{3_n}	Q_{4_n}	D_{1_n}	D_{2_n}	D_{3_n}	D_{4_n}	$Q_{1_{n+1}}$	$Q_{2_{n+1}}$	$Q_{3_{n+1}}$	$Q_{4_{n+1}}$
0	0	0	0	1	0	0	0	1	0	0	0
0	0	0	1	0	0	0	0	0	0	0	0
0	0	1	0	1	0	0	1	1	0	0	1
0	0	1	1	0	0	0	1	0	0	0	1
0	1	0	0	1	0	1	0	1	0	1	0
0	1	0	1	0	0	1	0	0	0	1	0
0	1	1	0	1	0	1	1	1	0	1	1
0	1	1	1	0	0	1	1	0	0	1	1
1	0	0	0	1	1	0	0	1	1	0	0
1	0	0	1	0	1	0	0	0	1	0	0
1	0	1	0	1	1	0	1	1	1	0	1
1	0	1	1	0	1	0	1	0	1	0	1
1	1	0	0	1	1	1	0	1	1	1	0
1	1	0	1	0	1	1	0	0	1	1	0
1	1	1	0	1	1	1	1	1	1	1	1
1	1	1	1	0	1	1	1	0	1	1	1

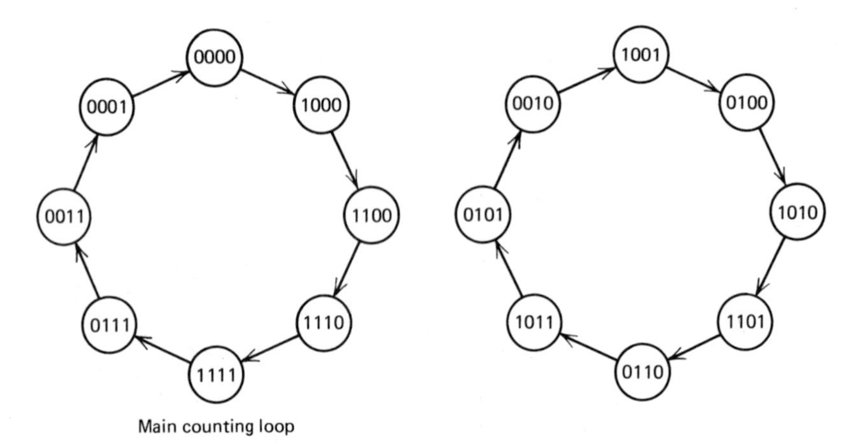

Main counting loop

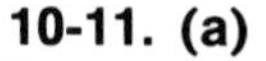

10-11. (a)

Present States					Present Control Inputs					Next States				
Q_{1_n}	Q_{2_n}	Q_{3_n}	Q_{4_n}	Q_{5_n}	D_{1_n}	D_{2_n}	D_{3_n}	D_{4_n}	D_{5_n}	$Q_{1_{n+1}}$	$Q_{2_{n+1}}$	$Q_{3_{n+1}}$	$Q_{4_{n+1}}$	$Q_{5_{n+1}}$
0	0	0	0	0	1	0	0	0	0	1	0	0	0	0
1	0	0	0	0	1	1	0	0	0	1	1	0	0	0
1	1	0	0	0	1	1	1	0	0	1	1	1	0	0
1	1	1	0	0	1	1	1	1	0	1	1	1	1	0
1	1	1	1	0	1	1	1	1	1	1	1	1	1	1
1	1	1	1	1	0	1	1	1	1	0	1	1	1	1
0	1	1	1	1	0	0	1	1	1	0	0	1	1	1
0	0	1	1	1	0	0	0	1	1	0	0	0	1	1
0	0	0	1	1	0	0	0	0	1	0	0	0	0	1
0	0	0	0	1	0	0	0	0	0	0	0	0	0	0

10-11. (b)

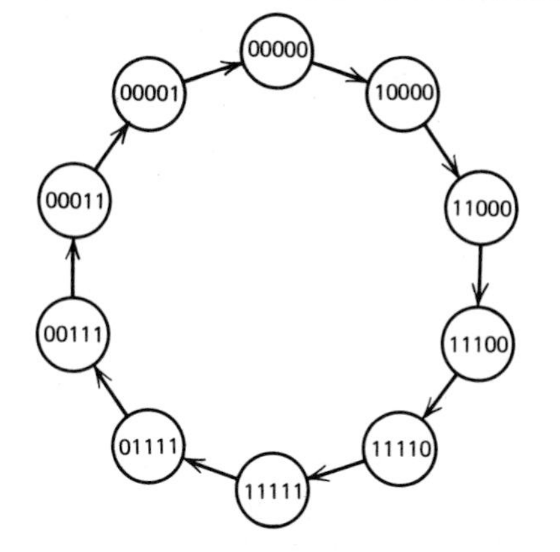

10-13. 25 MHz.

10-15. 20 MHz.

CHAPTER 11

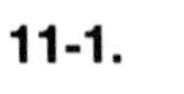

11-1.

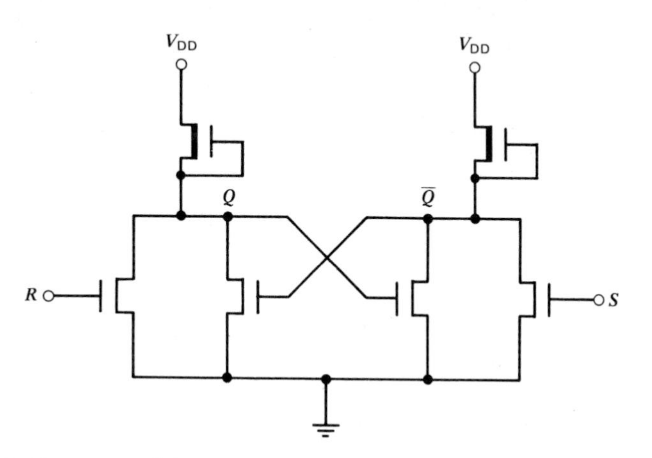

11-3. **(a)** 64 rows and 64 columns; **(b)** 6 bits to the row encoder, 6 bits to the column encoder.

11-5.

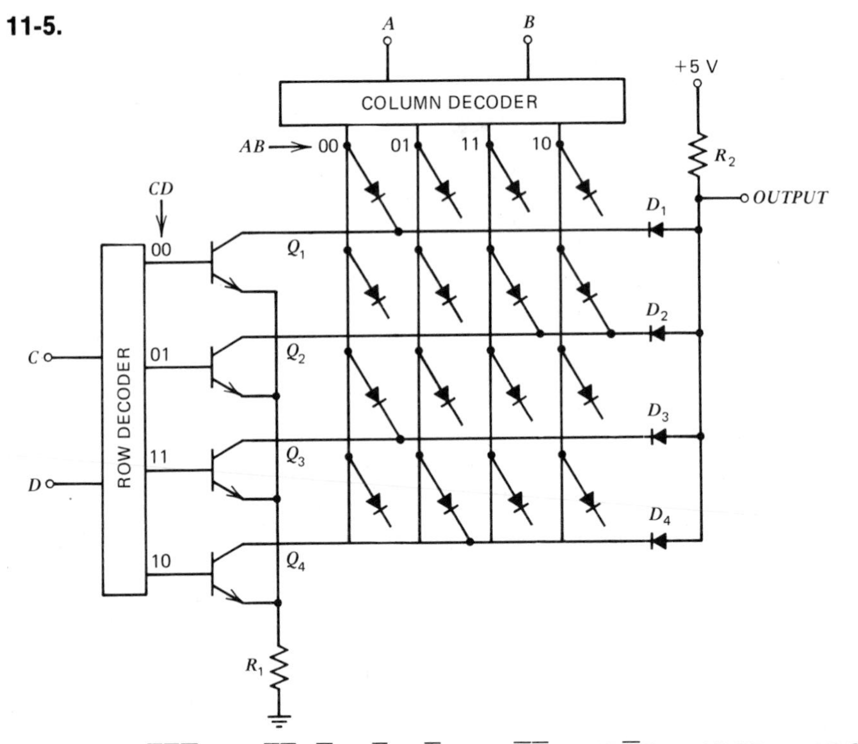

11-7. **(a)** $\bar{A}\bar{B}\bar{C}D + \bar{A}\bar{B}C\bar{D} + \bar{A}BC\bar{D} + AB\bar{C}\bar{D} + AB\bar{C}D + ABCD + ABC\bar{D} + A\bar{B}CD + A\bar{B}C\bar{D}$.

(b) The ROM for the above function is shown in Fig. 11-9*a*.

11-9. Remove from the OR plane the MOS device that has its gate connected to the horizontal P_1 line and its drain connected to the vertical line leading to the input of the logic inverter whose output is OUT_1.

***11-11.** 000101.

11-13. 6.25 mm × 6.25 mm = 39 mm^2.

11-15. 125 mW.

11-17. 1 W.

CHAPTER 12

12-1.

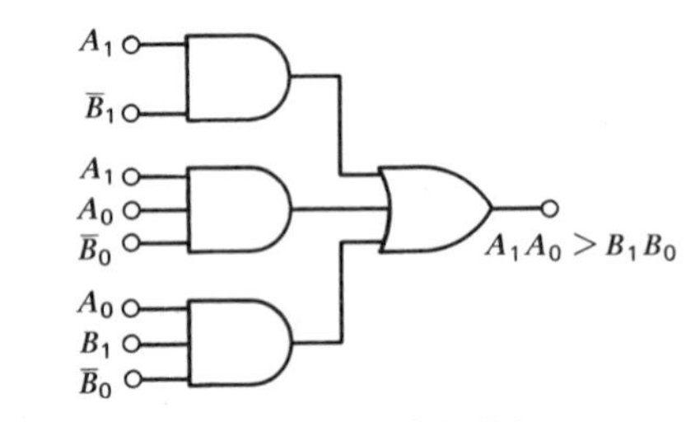

12-3. $S_0 = A_0 \oplus B_0$; $S_1 = A_1 \oplus B_1 \oplus (A_0B_0)$;
$C_{OUT} = A_1B_1 + [(A_1 \oplus B_1) \cdot A_0B_0]$.

*Optional problem.

12-5. $1010_2 + 0011_2 = 1101_2$.

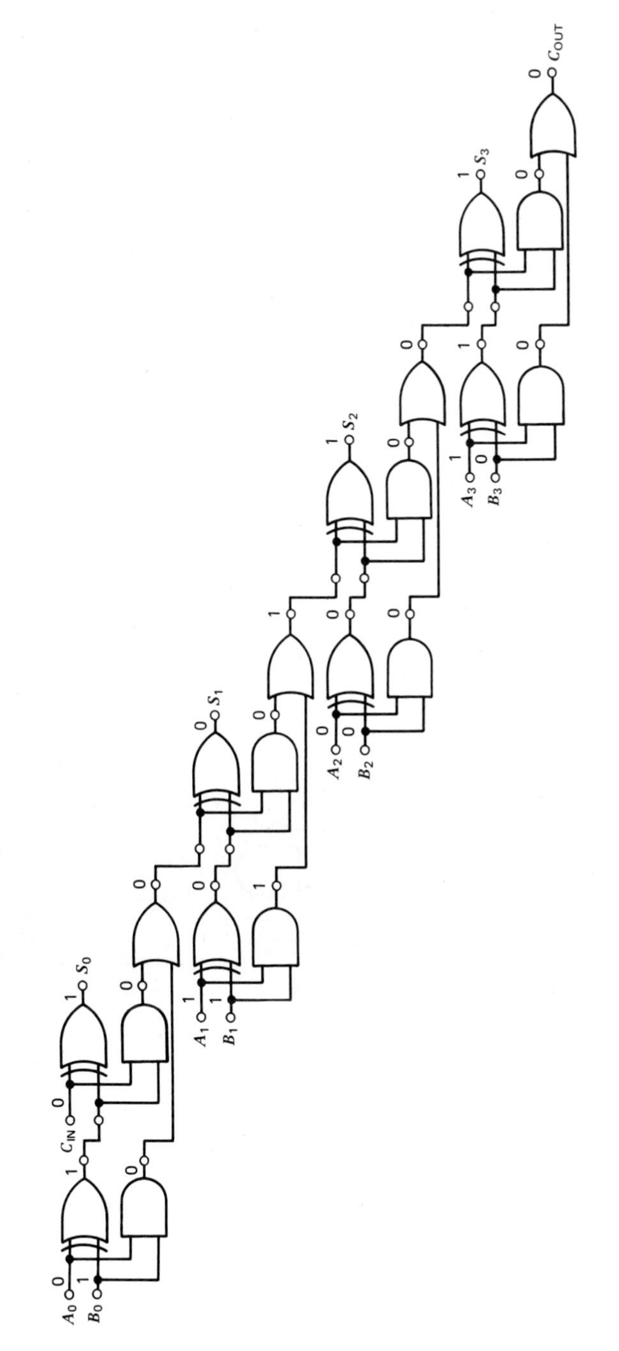

12-7. $0101_2 - 0011_2 = 5_{10} - 3_{10} = 0010 = 2_{10}$.

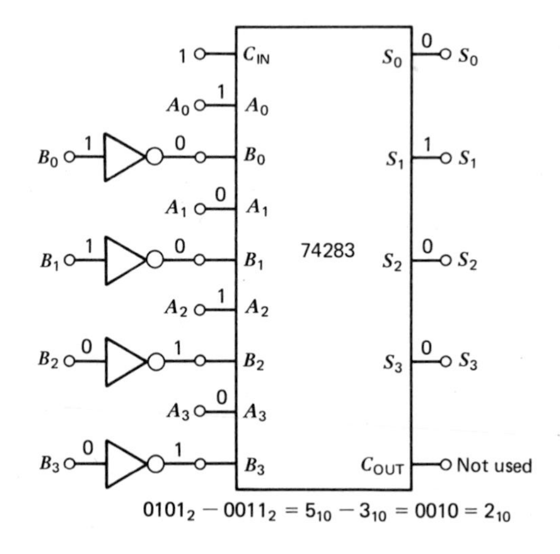

12-9. $1010_2 \times 11_2 = 10_{10} \times 3_{10} = 011110_2 = 30_{10}$.

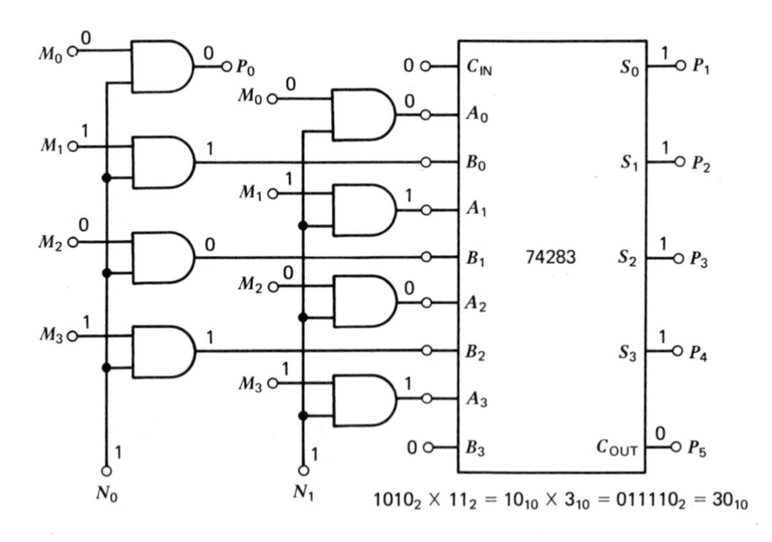

12-11. 44 ns.

CHAPTER 13

13-1. (a) $G_3 = B_3 + B_2B_0 + B_2B_1$.

$G_2 = B_1 + \overline{B}_3B_0 + B_2 + B_3\overline{B}_0$.

$G_1 = B_2B_1B_0 + \overline{B}_2\overline{B}_1 + \overline{B}_2\overline{B}_0$.

$G_0 = B_1$

(b) 1010, 1011, 1100, 1101, 1110, 1111; don't care conditions.

13-3. **(a)** $B_3 = X_3X_2 + X_3X_1X_0$.
$B_2 = X_2X_1X_0 + \overline{X}_2\overline{X}_0 + X_3\overline{X}_1X_0$.
$B_1 = X_1\overline{X}_0 + \overline{X}_1X_0$.
$B_0 = \overline{X}_0$.

(b) 0000, 0001, 0010, 1101, 1110, 1111.

***13-5.** 12-bit number = 4096 combinations (words) × (3 × 4 + 3) bits = 4096 words × 15 bits.

13-7. 1000110001.

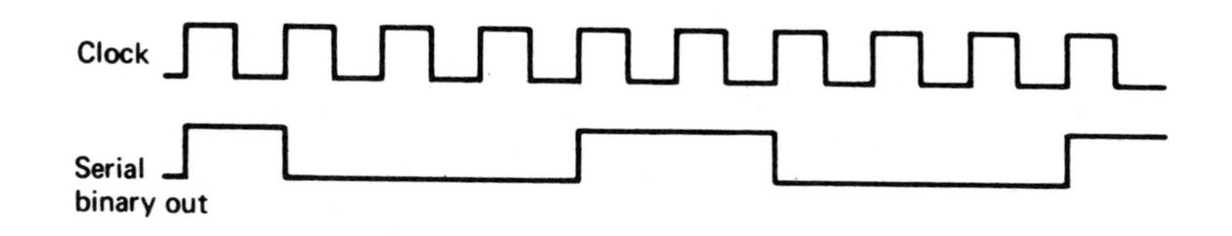

13-9. See Ex. 13-9.

***13-11.** Inputs b_0 through b_4 represent 2^5 combinations, hence 32 words. The highest numerical value is 31 = 0011 0001, requiring 6 BCD bits. (Leading zeros do not require ROM bits.)

13-13. One possible set of solutions is:

$0 = \overline{D}\overline{C}B$ $\quad$ $1 = \overline{D}\overline{B}\overline{A}$
$2 = \overline{D}\overline{B}A$ $\quad$ $3 = \overline{D}B\overline{A}$
$4 = CBA$ $\quad$ $5 = \overline{C}\overline{B}\overline{A}$
$6 = D\overline{B}A$ $\quad$ $7 = DB\overline{A}$
$8 = DBA$ $\quad$ $9 = DC$

13-15. The FF must be reset to its 0 state before parity checking commences. Data and check bits are applied serially to control input *T*. After all bits have been examined, the output of the FF should be at logic 1 for correct odd parity check and at logic 0 for correct even parity check.

CHAPTER 14

14-1. 0000 0101, 0000 1011, 1111 1111.

14-3. **(a)** No data flow; **(b)** from the left to the right.

*Optional problem.

14-5.

Q_n	D_n	E_n	Q_{n+1}
0	0	0	0
0	0	1	0
0	1	0	0
0	1	1	1
1	0	0	1
1	0	1	0
1	1	0	1
1	1	1	1

E_n	Q_{n+1}
0	Q_n
1	D_n

14-7. 16.

***14-9.** 164_{10}, $42\ 075_{10}$.

14-11. *RECEIVE 1* = 0, *SEND 1* = 1, *RECEIVE 2* = 1, *SEND 2* = 0, *RECEIVE 3* = 1, *SEND 3* = 0.

14-13. Unidirectional.

CHAPTER 15

15-1. 0.00152%, 152 μV.

15-3. −4.375 V.

15-5. 5.12 MΩ.

15-7. **(a)** +9.990 V; **(b)** +5.0 V; **(c)** +9.77 mV.

15-9. **(a)** 0.0061%; **(b)** 0.01%; **(c)** the 14-bit binary DAC has a better resolution than the 16-bit BCD DAC.

15-11. −1.875 V.

15-13. **(a)** 1 mA; **(b)** 1.96875 mA.

15-15. 0.024%.

15-17. 5.45 μV.

15-19. 79.365 μs.

15-21. **(a)** 14 μs; **(b)** no.

15-23. $1/256 \approx 0.4\% = \pm 0.2\%$.

CHAPTER 16

16-1. 40.

16-3. 20.

16-5. 0.25 V.

*Optional problem.

16-7.

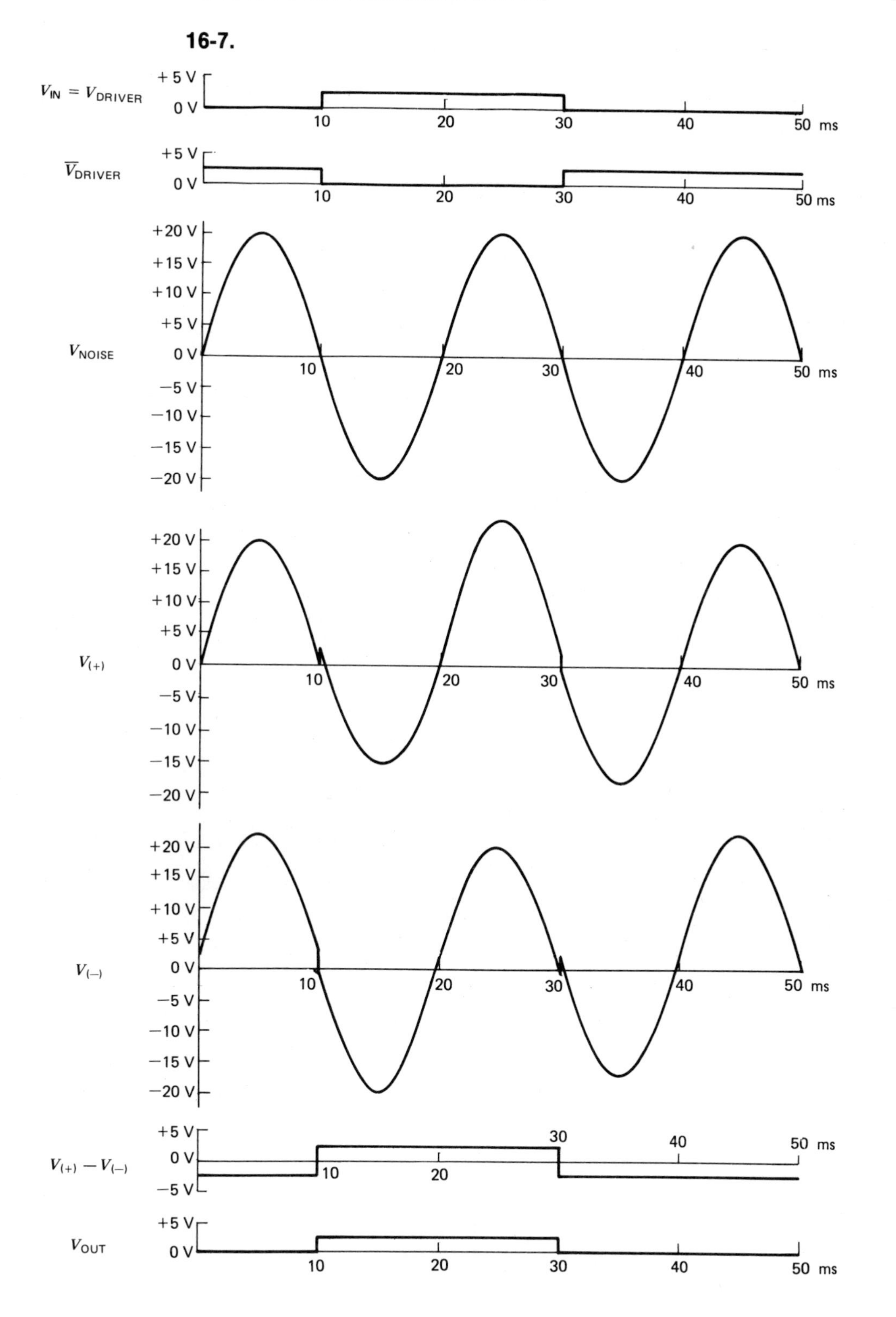
$V_{IN} = V_{DRIVER}$
+5 V
0 V
10
20
30
40
50 ms
$\overline{V}_{DRIVER}$
V_{NOISE}
+20 V
+15 V
+10 V
−5 V
−10 V
−15 V
−20 V
$V_{(+)}$
$V_{(-)}$
$V_{(+)} - V_{(-)}$
V_{OUT}

16-9.

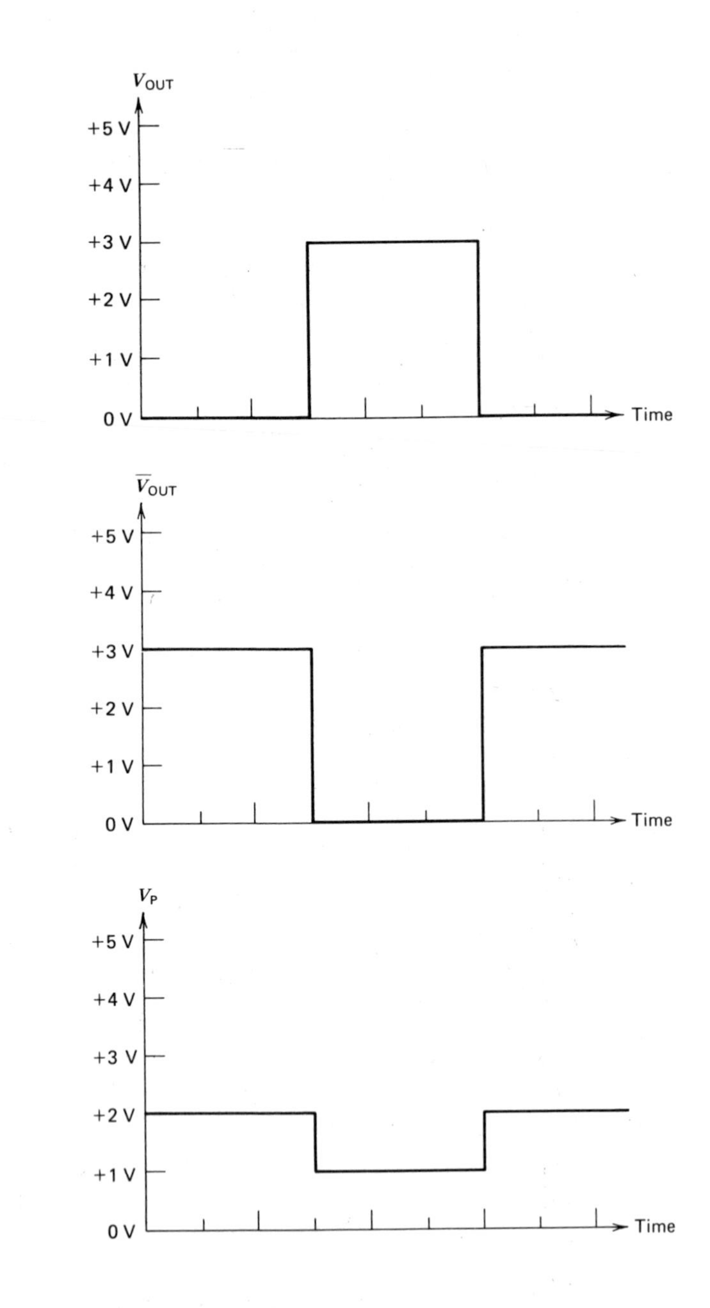

16-11. **(a)** 70 ms; **(b)** $\pm 1\%$.

16-13. 0.25 s.

16-15.

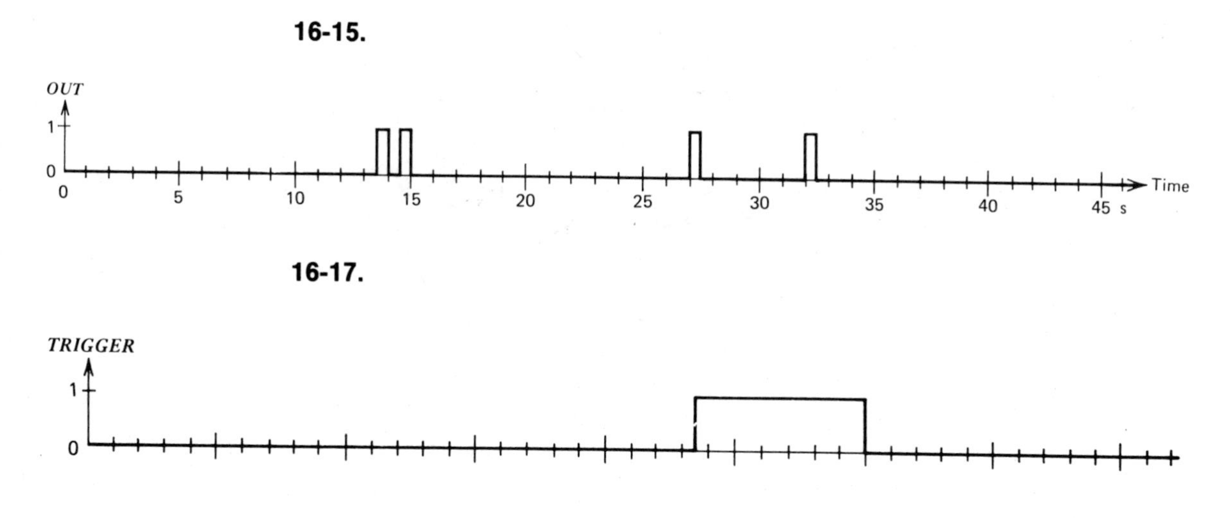

16-17.

16-19. Digital delay, display generator, word triggering, and internal clock source.

16-21. Current tracer.

APPENDIX

Table of Powers of 2

2^n	n	2^{-n}
1	0	1.0
2	1	0.5
4	2	0.25
8	3	0.125
16	4	0.0625
32	5	0.03125
64	6	0.01562 5
128	7	0.00781 25
256	8	0.00390 625
512	9	0.00195 3125
1024	10	0.00097 65625
2048	11	0.00048 82812 5
4096	12	0.00024 41406 25
8192	13	0.00012 20703 125
16,384	14	0.00006 10351 5625
32,768	15	0.00003 05175 78125
65,536	16	0.00001 52587 89062 5
131,072	17	0.00000 76293 94531 25
262,144	18	0.00000 38146 97265 625
524,288	19	0.00000 19073 48632 8125
1,048,576	20	0.00000 09536 74316 40625
2,097,152	21	0.00000 04768 37158 20312 5
4,194,304	22	0.00000 02384 18579 10156 25
8,388,608	23	0.00000 01192 09289 55078 125
16,777,216	24	0.00000 00596 04644 77539 0625
33,554,432	25	0.00000 00298 02322 38769 53125
67,108,864	26	0.00000 00149 01161 19384 76562 5
134,217,728	27	0.00000 00074 50580 59692 38281 25
268,435,456	28	0.00000 00037 25290 29846 19140 625
536,870,912	29	0.00000 00018 62645 14923 09570 3125
1,073,741,824	30	0.00000 00009 31322 57461 54785 15625
2,147,483,648	31	0.00000 00004 65661 28730 77392 57812 5
4,294,967,296	32	0.00000 00002 32830 64365 38696 28906 25

Glossary

Accumulator (register A). The principal register in a central processor unit (CPU).

AC power dissipation. The power dissipation resulting from a circuit being switched.

Acquisition time. In a sample-and-hold circuit, the time that elapses between the sample command and the instant the output voltage tracks the input voltage to the specified accuracy.

Active-high logic. A logic system in which the higher voltage level (HIGH) represents logic 1.

Active-low logic. A logic system in which the lower voltage level (LOW) represents logic 1.

ADC. See *analog-to-digital converter*.

Adder. A circuit performing addition.

Address. A number specifying a location in the memory.

Adjacency. A neighboring cell in a Karnaugh map.

Aliasing. The delivery of an output signal with a frequency that is lower than the frequency of the input signal, as a result of too low a sampling rate in an ADC.

Alphanumeric code. A code that represents alphabetic characters, numbers, punctuation marks, and often also control characters.

ALU. See *arithmetic logic unit*.

Analog-to-digital converter. A circuit that delivers digital outputs in response to an analog voltage input.

Analog voltage comparator. A high-gain dc-coupled amplifier with an analog input and a binary output: logic 0 or 1.

Analysis. For a switching circuit it results in an equation describing a given function.

AND gate. An electronic circuit whose output is at logic 1 if, and only if, all inputs are at logic 1.

AND-OR-INVERT gate. A logic circuit that is equivalent to two or more AND gates followed by a NOR gate.

Aperture time. In a sample-and-hold circuit, the time following the hold command by which time the sampling process is completed.

Arithmetic-logic unit (ALU). A digital circuit performing arithmetic and logic operations.

ASCII. *American Standard Code for Information Interchange*.

Assembler. Translates a program from an assembly language to a machine language.

Assembly language. A programming language in which each machine-language instruction code is replaced by a mnemonic abbreviation.

Associative memory. A memory that includes logic at each word. It has READ, WRITE, and MATCH operations.

Band diagram. A diagram showing noise margins.

Base. The *number of symbols* used in a given number system. The decimal number system has a base of 10, the octal a base of 8, the hexadecimal a base of 16, and the binary a base of 2.

Baudot code. A 5-bit alphanumeric code.

BCD. See *binary-coded decimal*.
BCD code, 8-4-2-1. A code in which the binary bits are weighted 8, 4, 2, and 1 according to position.
BCD counter. A counter that counts in a BCD code.
Bidirectional shift register. A shift register whose contents can be shifted right or left.
Bilateral transmission gate. A controlled floating electronic switch.
Binary. In the binary number system *two* symbols are used, 0 and 1.
Binary-coded decimal. Coding the 10 decimal numerals with various combinations of 1s and 0s.
Binary point. The point that separates the *integer* and *fractional* parts of a binary number.
Bipolar RAM. A random-access memory, which is made of bipolar transistors.
Biquinary counter. A counter that counts in the biquinary code with weights of 5-4-2-1.
Bit. Contraction of the two words: *BI*nary digi*T*.
Block parity. Horizontal and vertical parity. May be even parity or odd parity.
Boolean constant. One of the two binary values 0 and 1.
Boolean product. Two or more variables connected by the logic operator AND.
Boolean sum. Two or more variables connected by the logic operator OR.
Boolean variable. A variable that can assume either 0 or 1.
Branch. In a state diagram, it represents transitions between states.
Branching. Decision making between sequences of instructions in a program.
Buffer circuit. A circuit designed to drive heavy loads.
Bus. Interconnection between parts of a computer.
CAM (content-addressable memory). See *associative memory*.
Central processor unit (CPU). Part of a computer that includes registers, the arithmetic-logic unit (ALU), and the control unit.
Clock. A recurring signal that initiates transitions in a digital system.
CMOS. See *complementary metal-oxide semiconductor*.
Code converter. A combinational or sequential logic circuit that delivers a coded word in response to an input of a word presented in a different code.
Coding. Expressing a letter, a number, or a control character by combinations of 1s and 0s.
Collector-dot. See *wired-OR*.
Combinational logic circuit. A circuit giving a result that depends on the states of the input variables, but not on how these states were reached.
Common-mode input impedance. Measured between ground and the two parallel-connected input terminals of an operational amplifier or an analog voltage comparator.
Compiler. Translates a program from a higher-level programming language into a machine language.
Complementary metal-oxide semiconductor (CMOS) logic. A logic circuit using both n-channel and p-channel MOS transistors.
Content-addressable memory (CAM). See *associative memory*.
Control input. In a FF, an input that governs its transitions.
Conversion word rate. The number of digital words of the required resolution converted by an ADC.
Counter. A digital circuit that counts in a given sequence.
CPU. See *central processor unit*.
Current-switching pair. Consists of two transistors with their emitters connected together.

Current tracer. A digital diagnostic tool that indicates current by detecting the magnetic field in the vicinity of a current-carrying conductor.

DAC. See *digital-to-analog converter*.

Dead time. In a monostable multivibrator, the time period following an output pulse during which triggering is not possible.

Decimal. In the decimal number system 10 symbols are used: 0 through 9.

Decoder. A digital circuit that delivers a unique output for each combination of input conditions.

Decoding. Interpreting a combination of 1s and 0s as a numerical value, a letter of the alphabet, a control character, or another symbol.

Differential input impedance. The impedance measured between the two input terminals of an operational amplifier or an analog voltage comparator.

Differential nonlinearity. In an ADC or a DAC it is the measure of the size of each step in the converter.

Digit. A number symbol. For example, the decimal system uses 10 digits 0 through 9.

Digital-to-analog converter. A circuit that accepts a parallel digital input and delivers a proportional analog output.

Digital comparator. A digital circuit performing a comparison of two numbers.

Digital delay. A shift register used for delaying clocked information.

Digital demultiplexer. A combinational logic circuit that routes data from a single source to a number of destinations, the destinations being selected by control inputs.

Digital multiplexer. A combinational logic circuit that routes data from several sources to one destination, the sources being selected by control inputs.

Diode logic. A logic circuit using diodes.

Diode-transistor logic (DTL). A logic circuit using diodes and an *npn* transistor that is either cut off or saturated.

DIP. See *dual in-line package*.

Direct clear. An input that forces a FF to its 0 state irrespective of the states of the control inputs and the clock.

Direct memory access (DMA). Provides for direct data transfers between the main memory and an input or output device.

Direct preset. An input that forces a FF to its 1 state irrespective of the states of the control inputs and the clock.

DMA. See *direct memory access*.

Don't care condition. A term that represents an unallowed input combination. In a Karnaugh map it may be used either as a 0 or as a 1.

DTL. See *diode-transistor logic*.

Dual in-line package (DIP). A rectangular integrated circuit package with the pins arranged in two parallel lines.

Dynamic MOS gate. A logic gate having two or more inputs and one output. It is based on the dynamic MOS inverter.

Dynamic MOS inverter. A MOS logic inverter having two clock pulses applied to it.

Dynamic MOS shift-register. A LSI in which data are shifted serially from one element to the next in response to a 2-phase clock. The binary information is retained on stray capacitances and, therefore, it is volatile.

EAROM. See *electrically alterable read-only memory*.

EBCDIC. Extended Binary Coded Decimal Interchange Code.

ECL. See *emitter-coupled logic*.

EEROM. See electrically alterable read-only memory.

Electrically alterable read-only memory (EAROM, EEROM). A read-only memory in which the binary content can be altered by the application of an electrical pulse of well-defined shape lasting several milliseconds.

Emitter-coupled logic (ECL). A logic circuit using *npn* transistors with their emitters connected together. Also includes output emitter followers.

End-around carry. A carry that is obtained in 1's complement subtraction.

EPROM. See *erasable programmable read-only memory.*

EQUALITY COMPARATOR. See *EXCLUSIVE-NOR.*

Erasable programmable read-only memory (EPROM). A read-only memory in which information has been entered by applying electrical pulses to selected cells to obtain "avalanche induced migration." It can be "erased" by applying ultraviolet light to the EPROM via a quartz window that is provided for this purpose.

Error-correcting code. A code that has sufficient redundancy to find the location of one or more errors.

Error-detecting code. A code that has sufficient redundancy to detect the presence of an error.

Excess-3 code. A code that is obtained by adding 11_2 to the 8-4-2-1 BCD code.

EXCLUSIVE-NOR. A circuit that delivers a logic 1 when the two inputs applied to it have the same binary values.

EXCLUSIVE-OR. A circuit that delivers a logic 1 when the two inputs applied to it have complementary binary values.

Expander. Two or more AND gates that can be applied to expandable AND-OR-INVERT gates.

Expander circuit. A circuit that increases the number of inputs to a logic gate.

Fall time. The transition time of a negative-going transient between stated levels.

FF. See *flip-flop.*

Firmware. A ROM or a PLA used for data path control.

Flip-flop (FF). A circuit with two stable states.

Function table. An abbreviated state table.

Flow chart. A diagram that shows a sequence of operations and branching.

Full-adder. A digital arithmetic circuit that accepts three inputs, A, B, and a carry input C_{IN}; it delivers two outputs, one being the modulo-2 sum of A and B, the other being the carry-out.

Gate array. A fixed array of logic gates that can be interconnected by customized metalization.

Gray code. A number code in which only one binary digit changes in the representation of adjacent numerals.

Half-adder. An adder that adds 2 bits without a carry input.

Hexadecimal. In the hexadecimal number system 16 symbols are used: 0, 1, 2, 3, 4, 5, 6, 7, 8, 9, A, B, C, D, E, and F.

Hexadecimal point. The point that separates the *integer* and *fractional* parts of a hexadecimal number.

Higher level programming language. A programming language in which several instructions are combined into a statement.

High output voltage. The more positive of the two output voltage levels.

Hold time. A time interval following a specified clock pulse edge of a FF. The control inputs must remain stationary until the end of this time interval.

Hybrid counter. A counter in which the states of some, but not all, FFs change simultaneously.

Hysteresis. Backlash. In a Schmitt trigger circuit, it is the difference between the positive-going threshold and the negative-going threshold. An analog voltage comparator with hysteresis has different input voltages for turning ON and OFF.

IC. Integrated circuit.

IIL. See *integrated injection logic*.

INCLUSIVE-OR operator. A logic operator in a logic proposition of either or both. Also called OR operator.

Input bias current. Current flowing into the input terminals of an operational amplifier or an analog voltage comparator.

Input device. A device through which data can be entered into a computer.

Input offset current. The difference in bias currents flowing into the terminals of an operational amplifier or an analog voltage comparator.

Input offset voltage. The voltage required at the input of an operational amplifier to bring the output to 0 V.

Input–output (I/O) section. Interfaces the computer with the input and output devices.

Instruction. Describes a specific operation or branching in a program.

Integral nonlinearity. In an ADC or a DAC it is a measure of the linear relationship between the output and the input.

Integrated injection logic (I^2L). A logic circuit using grounded-emitter *npn* transistors with their collectors in parallel as logic, and grounded-base *pnp* transistors as current sources.

Intrinsic fall time. The fall time with no load capacitance.

Intrinsic-OR. See *wired-OR*.

Intrinsic propagation delay. The propagation delay with no load capacitance.

Intrinsic rise time. The rise time with no load capacitance.

Intrinsic transition time. The transition time with no load capacitance.

I/O section. See *input–output section*.

Johnson counter. A divide-by-*n* counter where *n* is twice the number of FFs. It is built by connecting the complementary output of the last FF in a shift register to the *D* input of the first FF.

Karnaugh map. A graphic method used for the simplification of Boolean functions.

Ladder network. A resistor network in which the resistors have values *R* and 2*R*.

Line driver. A circuit that transmits digital information onto a balanced transmission line.

Line receiver. An analog voltage comparator circuit that receives digital information from a balanced transmission line.

Load current. Current flowing *out* of the output of a circuit.

Logic analyzer. A digital diagnostic instrument that may include word triggering, digital delays, and storage registers.

Logic clip. A digital diagnostic tool that displays the logic states at all pins of an IC package.

Logic family. A collection of logic circuits using the same technology. For example: DTL, TTL, ECL, I^2L, or CMOS logic family.

Logic gate. An electronic circuit performing a logic operation. It includes inverters, AND gates, OR gates, NAND gates, and NOR gates.

Logic inverter. The simplest logic gate. The output voltage of a logic inverter is HIGH when the input voltage is LOW, and conversely it is LOW when the input voltage is HIGH.

Logic operator. A word required to form a complex logic sentence. For example, AND and OR are logic operators.

Logic probe. A digital diagnostic tool that displays the logic state applied at the tip of the probe.

Logic proposition. A sentence that can be answered by TRUE or FALSE.

Logic pulser. A pulse generator that can override the low-impedance output of a digital circuit.

Look-ahead carry adder. A 3-stage adder that is a compromise in complexity and propagation delay between a ripple adder and a 2-stage adder.

Low output voltage. The more negative of the two output voltage levels.

lsb. Least significant bit.

lsd. Least significant digit.

LSI. Large-scale integrated circuit; large-scale integration.

Machine language. A programming language that is suitable for direct operation of a computer.

Magnitude relation. One of the five relations of greater than, less than, equal, greater than or equal, less than or equal.

Majority voter. A circuit in which the output always agrees with the majority of the inputs.

Maximum counting speed. The maximum frequency of the input waveform at which the counter *counts* correctly.

Metal-oxide semiconductor (MOS). A *p*- or *n*-doped semiconductor channel, the transconductance of which is modulated by a voltage applied to an insulated metal or polysilicon gate.

Modulo-2 addition. Binary addition disregarding the carry.

Monostable multivibrator. A circuit that generates output pulses with specified duration.

MOS. See *metal-oxide semiconductor.*

MOSFET. Metal-oxide semiconductor field-effect transistor.

MOS logic. Logic circuits using *n*-channel *or* *p*-channel MOS transistors.

MOS RAM. A random-access memory that is made of metal-oxide semiconductor transistors. A MOS RAM may be either static or dynamic.

msb. Most significant bit.

msd. Most significant digit.

MSI. Medium-scale integrated circuit; medium-scale integration.

Multiemitter transistor. Equivalent to several transistors with bases and collectors connected in parallel.

Multiplier. A circuit performing multiplication.

Multiplying DAC. A digital-to-analog converter delivering an output that is proportional to the product of a digital input and an analog voltage input.

NAND. A logic operator that is equivalent to an AND operator followed by an inversion.

NAND gate. A logic gate performing the same function as an AND gate followed by a logic inverter.

Negative logic. See *active-low logic.*

Next state. The state of a FF after a clock pulse.

Nine's (9's) 8-4-2-1 BCD complement. An 8-4-2-1 BCD number obtained by subtracting from 1001_2 the number to be 9's complemented.

Noise margin. An operating margin that ensures proper operation of a digital circuit in the presence of noise and other undesired interference.

NOR. A logic operator that is equivalent to an OR operator followed by an inversion.

NOR gate. A logic gate performing the same function as an OR gate followed by a logic inverter.

Normally closed (n.c.) contact. A relay contact that is closed when the relay is not energized.

Normally open (n.o.) contact. A relay contact that is open when the relay is not energized.

Octal. In an octal number system 8 symbols are used, 0 through 7.

Octal point. The point that separates the *integer* and *fractional* parts of an octal number.

Offset-binary. A binary code similar to the 2's complement code but with the msb complemented.

One's (1's) complement. Obtained in a binary number when a 0 is substituted for each 1 and a 1 is substituted for each 0.

One-shot. See *monostable multivibrator*.

Open-collector output. The collector of a transistor that is available as output and that has no other internal connections.

Open-loop gain. The gain of an amplifier without feedback.

Operational amplifier. A dc-coupled, high-gain, differential amplifier.

Optical isolator. A device that transmits and receives digital information via light, permitting complete electrical isolation between the transmitter and the receiver.

OR gate. An electronic circuit whose output is at logic 1 if at least one input is at logic 1.

Output device. A device through which data are retrieved from a computer.

Output function. The binary value of the output that depends on the states of the input variables.

Parity bit. A bit deliberately added to a binary word to make the total number of bits either even or odd.

Parity detector. A circuit that inspects received information bits and the parity bit for even or odd parity.

Parity generator. A circuit that inspects transmitted information bits and generates an even or odd parity bit.

Parallel input. Input through which external data can be jammed into an individual FF in a counter or a shift register.

Parallel loading. Jamming data into the individual FFs in a counter or a shift register.

Parallel output. The output of a FF in a shift register.

PLA. See *programmable logic array*.

POS. See *product-of-sums*.

Positional notation. The ordering of digits (of a number in any number base) whereby the leftmost digit carries the highest weight and the rightmost digit carries the lowest weight.

Positive logic. See *active-high logic*.

Present state. The state of a FF before a clock pulse.

Priority encoder. A combinational logic circuit that delivers an output that represents the code of the highest priority input.

Product-of-sums (POS). A Boolean equation that first uses the OR operator on the variables and then combines the ORed terms via the AND operator, for example, $(A + B) \cdot (C + D)$.

Product term. A Boolean product of some but not all the variables, complemented or uncomplemented, of a function.

Programmable logic array (PLA). A logic circuit consisting of an AND matrix and an OR matrix. The particular AND and OR functions are determined (programmed) by selectively programming links at the inputs of the logic gates. The outputs can be similarly programmed for positive or negative logic.

Programmable read-only memory (PROM). A read-only memory that in the unprogrammed state has all cells either at logic 0 or at logic 1. To program the memory, links are selectively broken to obtain the desired binary patterns.

PROM. See *programmable read-only memory*.

Propagation delay. The time required for the output voltage to reach a specified level after a transition at the input.

Quad current source. Four binary weighted sources of current.

Quiescent power dissipation. The power dissipation when a circuit is not switched.
RAM. See *random-access memory*.
Random-access memory (RAM). A memory in which the read and write access times to each cell are the same as to any other cell.
Read-only memory (ROM). A memory that contains fixed information imparted during the manufacturing process. Only read operations can be executed in a read-only memory; write operations have no meaning.
Redundant state. A binary combination that is not utilized.
Redundant term. In a Boolean expression, an additional term the presence of which does not change the function.
Reflected-binary code. Same as *Gray code*.
Register. Fast-access digital storage.
Resolution. As applied to an ADC or DAC, it is the value of the least significant bit relative to the full scale.
Ring-counter. A divide-by-*n* counter where *n* equals the number of FFs. It is built by connecting the TRUE (uncomplemented) output of the last FF in a shift register to the *D* input of the first FF.
Ripple adder. A multibit adder built from 1-bit adders.
Ripple comparator. A multibit comparator built from 1-bit comparators.
Ripple counter. A counter in which the clock input of each FF is connected to an output of the preceding FF, in some cases via logic gates; the clock input of the first FF is the input of the counter.
Rise time. The transition time of a positive-going transient between stated levels.
ROM. See *read-only memory*.
Sample-and-hold. A circuit that samples an analog voltage for a short duration and holds the sample during conversion by an ADC.
Schmitt trigger. A circuit with hysteresis that is suitable for reshaping noisy signals.
Schottky-TTL. Schottky-diode clamped TTL.
Settling time. The time elapsed from the application of a step voltage input to the time when the output of an operational amplifier has entered and remained within a specified error band symmetrical about its final value.
Serial input. An input to a shift register through which data can be shifted in serially.
Serial output. An output of a shift register through which data can be shifted out serially.
Setup time. A time interval preceding a specified clock pulse edge of a FF. The control inputs must be stable from the beginning of this time interval until the end of the hold time interval.
Shift-register. A chain of FFs through which data can be shifted serially.
Single-shot. See *monostable multivibrator*.
Sink current. The current flowing *into* the output of a circuit.
Slew rate. The maximum rate of change at the output of an operational amplifier or an analog voltage comparator in response to a large step voltage input.
SOP. See *sum-of-products*.
Speed-power product. The product of the propagation delay and the power dissipation.
SSI. Small-scale integrated circuit. Small-scale integration.
Standard product term. A Boolean product of all the variables, complemented or uncomplemented, of a function.
Standard sum term. A Boolean sum of all the variables, complemented or uncomplemented, of a function.
State diagram. A diagram showing the states, transitions, and outputs of a digital circuit or system.

State table. Describes all possible states of a circuit.

Storage latch. R-S storage FF.

Subroutine. A sequence of instructions that is referenced repeatedly during the execution of a program.

Subtractor. A circuit performing subtraction.

Sum-of-products (SOP). A Boolean equation that first uses the AND operator on the variables and then combines the ANDed terms via the OR operator, for example, $A \cdot B + C \cdot D$.

Summing junction. The negative input terminal of an operational amplifier.

Sum term. A Boolean sum of some but not all of the variables, complemented or uncomplemented, of a function.

Switching function. A Boolean expression applied to a switching circuit.

Switching circuit. A circuit in which switches or logic gates are interconnected to perform a given function.

Symbolic logic. A mathematical structure used to express reasoning.

Synchronous counter. A counter in which the states of all FFs change simultaneously.

Synthesis. For a switching circuit it results in a logic diagram realizing a given function.

$T = 1$ FF. See *toggle FF*.

Ten's (10's) 8-4-2-1 BCD complement. An 8-4-2-1 BCD number obtained by subtracting from 1010_2 the number to be 10's complemented.

Theorem. A mathematical law.

Three-state TTL. TTL with three output states: LOW, HIGH, and open circuit.

Toggle FF ($T = 1$ FF). A FF with control input T permanently connected to logic 1.

Totem-pole output. An output stage consisting of two stacked transistors.

Transfer characteristic. The output voltage of a logic gate as a function of the input voltage.

Transient power dissipation. See *ac power dissipation*.

Transistor-transistor logic (TTL). A logic circuit using *npn* transistors that are either cut off or saturated.

Transition time. The time spent by a voltage between specified levels during a transition.

Truth table. A systematic listing of all possible combinations of the states of the variables, including inputs and outputs.

TTL. See *transistor-transistor logic*.

Twisted-ring counter. See *Johnson counter*.

Two's (2's) complement. Obtained by adding 1 to the 1's complement.

Two-stage adder. An adder consisting of a 2-stage logic circuit generating the sum output and of a 2-stage logic circuit generating the carry output.

Two-stage comparator. A digital comparator consisting of a 2-stage logic circuit.

Unallowed state. A state from which there is no transition back to the regular operating sequence.

VLSI. Very-large-scale integrated circuit; very-large-scale integration.

Weight. In positional notation each position has a weight attached to it. In the decimal system these weights are powers of 10, for example, 1000, 100, 10, and 1 starting from the most significant digit (msd).

Wired-AND. An AND function attained by connecting outputs of logic gates in parallel.

Wired-OR. An OR function attained by connecting outputs of logic gates in parallel.

Word triggering. A method whereby a trigger pulse is generated only for a given combination of several input bits.

Index